THE SOUNDSCAPE OF MODERNITY

ARCHITECTURAL ACOUSTICS AND THE CULTURE OF LISTENING IN AMERICA, 1900–1933

EMILY THOMPSON

The MIT Press   Cambridge, Massachusetts   London, England

First MIT Press paperback edition, 2004

© 2002 Massachusetts Institute of Technology

All rights reserved. No part of this book may be reproduced in any form by any electronic or mechanical means (including photocopying, recording, or information storage and retrieval) without permission in writing from the publisher.

This book was set in Bembo by The MIT Press.

Printed and bound in the United States of America.

V-Room is a trademark of the Wenger Corporation, Owatonna, Minnesota.

Library of Congress Cataloging-in-Publication Data

Thompson, Emily Ann.
　The soundscape of modernity : architectural acoustics and the culture of listening in America, 1900–1933 / Emily Thompson.
　　p. cm.
　Includes bibliographical references and index.
　ISBN 978-0-262-20138-4 (hc. : alk. paper)—ISBN 978-0-262-70106-8 (pb. : alk. paper)
　1. Architectural acoustics. 2. Music—Acoustics and physics. I. Title.

NA2800 T48 2002
690'.2—dc21
　　　　　　　　　　　　　　　　　　　　　　　　　　　　　　　　　　　　　2001044533

10 9 8 7 6

To family and friends,
teachers and students.

The production of this book has been generously supported by the Graham Foundation for Advanced Studies in the Fine Arts.

CONTENTS

ACKNOWLEDGMENTS vii

CHAPTER 1: INTRODUCTION: SOUND, MODERNITY, AND HISTORY 1

CHAPTER 2: THE ORIGINS OF MODERN ACOUSTICS 13
    I   INTRODUCTION: OPENING NIGHT AT SYMPHONY HALL 13
    II ACOUSTICS AND ARCHITECTURE IN THE EIGHTEENTH AND NINETEENTH CENTURIES 18
    III WALLACE SABINE AND THE REVERBERATION FORMULA 33
    IV MUSIC AND THE CULTURE OF LISTENING IN TURN-OF-THE-CENTURY AMERICA 45
    V   CONCLUSION: THE CRITICS SPEAK 51

CHAPTER 3: THE NEW ACOUSTICS, 1900–1933 59
    I   INTRODUCTION 59
    II SABINE AFTER SYMPHONY HALL 62
    III THE REVERBERATIONS OF "REVERBERATION" 81
    IV NEW TOOLS: THE ORIGINS OF MODERN ACOUSTICS 90
    V   THE NEW ACOUSTICIAN 99
    VI CONCLUSION: SABINE RESOUNDED 107

CHAPTER 4: NOISE AND MODERN CULTURE, 1900–1933 115
    I   INTRODUCTION 115
    II NOISE ABATEMENT AS ACOUSTICAL REFORM 120
    III NOISE AND MODERN MUSIC 130
    IV ENGINEERING NOISE ABATEMENT 144
    V   CONCLUSION: THE FAILURE OF NOISE ABATEMENT 157

CHAPTER 5: ACOUSTICAL MATERIALS AND MODERN ARCHITECTURE, 1900–1933 169

    I   INTRODUCTION 169

    II   ACOUSTICAL MATERIALS AT THE TURN OF THE CENTURY 173

    III   ACOUSTICAL MATERIALS AND ACOUSTICAL MODERNITY: ST. THOMAS'S CHURCH 180

    IV   ACOUSTICAL MATERIALS AND MODERN ACOUSTICS: THE NEW YORK LIFE INSURANCE COMPANY BUILDING 190

    V   MODERN ARCHITECTURE AND MODERN ACOUSTICS: THE PHILADELPHIA SAVING FUND SOCIETY BUILDING 207

    VI   CONCLUSION 227

CHAPTER 6: ELECTROACOUSTICS AND MODERN SOUND, 1900–1933 229

    I   INTRODUCTION: OPENING NIGHT AT RADIO CITY 229

    II   LISTENING TO LOUDSPEAKERS: THE ELECTROACOUSTIC SOUNDSCAPE 235

    III   THE MODERN AUDITORIUM 248

    IV   ARCHITECTURAL ELECTROACOUSTICS: THEATER AND STUDIO DESIGN 256

    V   ELECTROACOUSTIC ARCHITECTURE: SOUND ENGINEERS AND THE ELECTRICAL CONSTRUCTION OF SPACE 272

    VI   CONCLUSION: REFORMULATING REVERBERATION 285

CHAPTER 7: CONCLUSION: ROCKEFELLER CENTER AND THE END OF AN ERA 295

CODA 317

NOTES 325

BIBLIOGRAPHY 425

INDEX 471

Acknowledgments

This project could never have been accomplished without the financial, institutional, and personal support that I have received over the years. The National Science Foundation provided a Graduate Fellowship that first allowed me to begin this intellectual journey as a graduate student in history of science at Princeton University over fifteen years ago. More recently, the NSF provided a Research Fellowship in Science and Technology Studies that enabled me to draw that same journey to a close by supporting my work on the development of sound motion pictures. Along the way, the Mellon Foundation sponsored a year of postdoctoral study at Harvard University and the National Endowment for the Humanities provided a summer stipend that enabled me to experience, as well as to study, the noises of New York. A fellowship from the Department of History at Princeton University subsidized a semester's leave from teaching when I needed it most, and the Graham Foundation for Advanced Studies in the Fine Arts underwrote my procurement of the visual illustrations that, perhaps ironically, add so much to a book about sound.

The Sirens of the history of sound have beckoned from far and wide, leading me places I never would have anticipated visiting when I began this project. In each case, I was guided safely through the uncharted waters by librarians, archivists, and historians who helped me to follow the elusive strains of that always compelling song. Special thanks to Janet Parks at the Avery Architectural and Fine Arts Library of Columbia University; Sheldon Hochheiser at the AT&T Archives; Emily Novak and Karen Finkston at the New York Life Insurance Company; James Reed at the Rockefeller Center Archive Center; Charles Silver at the Celeste Bartos Film Study Center of the Museum of Modern Art; Carol Merrill-Mirsky at the Edmund D. Edelman Museum of the Hollywood Bowl; Steven Lacoste at the Los Angeles Philharmonic Archives; Jane Ward and Bridget Carr at the Boston Symphony Orchestra Archives; Nicholson Baker at the American Newspaper Repository; Alex Magoun at the

David Sarnoff Library; John Kopec and David Moyer at the Riverbank Acoustical Laboratories; and Kathleen Dorman at the Joseph Henry Papers of the Smithsonian Institution.

Thanks also to the staff at the Harvard University Archives and the Baker Library of the Harvard Business School; the Marquand Library at Princeton University; the Museum of the City of New York; The New York Municipal Reference Library and Archives; the New-York Historical Society; the Hagley Museum and Library; the Office of the Architect of the Capitol; the Neils Bohr Library of the American Institute of Physics; the American Philosophical Society; the Thomas A. Edison National Historic Site; the UCLA Film and Television Archive; the University of Illinois at Urbana-Champaign Archives; the Case Western Reserve University Special Collections; the Society of Motion Picture and Television Engineers; the Institute of Electrical and Electronic Engineers; and the Acoustical Society of America.

Because my study is centered around the material culture of sound and listening, it is particularly gratifying to thank those who contributed to the material construction of the physical artifact that you now hold in your hands. For their efforts in photography, indexing, and proofreading, I thank John Blazejewski, Dwight Primiano, Carol Thompson, Martin White, and Laraine Lach. Amanda Sobel and Jason Rifkin also provided helpful research assistance. The MIT Press makes publishing a pleasure and I particularly thank Larry Cohen, Michael Sims, and Yasuyo Iguchi for their invaluable contributions to the final product.

It is also a pleasure, as well as a privilege, to express my gratitude to those acousticians who shared with me their technical expertise and personal memories. Sincere thanks to Dr. Leo Beranek, Professor Cyril Harris, Russell Johnson, David W. Robb, Robert M. Lilkendey, Raymond Pepi, Gerald Marshall, and Thomas Horrall. Special thanks to Carl Rosenberg, who has been a good friend as well as a valuable technical consultant.

My colleagues in the Department of History and Sociology of Science at the University of Pennsylvania—students, staff, and faculty alike—have always encouraged me to do my best, and the book that follows is better for having been written in such a collegial and stimulating environment. I also thank my fellow-travelers in aural history and in the history of technology, Leigh Schmidt, Douglas Kahn, Bill Leslie, and Susan Douglas, for helping me to hear the signal of my story amid the noise of history, and for the friendship they have offered along with their support.

From my very first days as a graduate student, Charles Gillispie has educated and encouraged me in ways that I can never begin to repay. Peter Galison has generously presented me with valuable opportunities to push my work in new directions and he has always provided invaluable guidance along the way. Charles Rosenberg has a wonderful ability to identify what is most important about a story; my own story has benefitted enormously from his scrutiny and counsel, and I appreciate even more his conviction that it matters.

Finally, without the friendship of John Carson, Angela Creager, and Carolyn Goldstein, as well as the love of my family, it would all be nothing but noise.

CHAPTER 1 INTRODUCTION: SOUND, MODERNITY, AND HISTORY

The Soundscape of Modernity is a history of aural culture in early twentieth-century America. It charts dramatic transformations in what people heard, and it explores equally significant changes in the ways that people listened to those sounds. What they heard was a new kind of sound that was the product of modern technology. They listened in ways that acknowledged this fact, as critical consumers of aural commodities. By examining the technologies that produced those sounds, as well as the culture that consumed them, we can begin to recover more fully the texture of an era known as "The Machine Age," and we can comprehend more completely the experience of change, particularly technological change, that characterized this era.

By identifying a soundscape as the primary subject of the story that follows, I pursue a way of thinking about sound first developed by the musician R. Murray Schafer about twenty-five years ago. Schafer defined a soundscape as a sonic environment, a definition that reflected his engagement with the environmental movements of the 1970s and emphasized his ecologically based concern about the "polluted" nature of the soundscape of that era.[1] While Schafer's work remains socially and intellectually relevant today, the issues that influenced it are not what has motivated my own historical study, and I use the idea of a soundscape somewhat differently. Here, following the work of Alain Corbin, I define the soundscape as an auditory or aural landscape. Like a landscape, a soundscape is simultaneously a physical environment and a way of perceiving that environment; it is both a world and a culture constructed to make sense of that world.[2] The physical aspects of a soundscape consist not only of the sounds themselves, the waves of acoustical energy permeating the atmosphere in which people live, but also the material objects that create, and sometimes destroy, those sounds. A soundscape's cultural aspects incorporate scientific and aesthetic ways of listening, a listener's relationship to their environment, and the social circumstances that dictate who gets to hear what.[3] A soundscape, like a landscape, ultimately has more to do with civilization than with nature, and as such, it is constantly under construction and always undergoing change. The American soundscape underwent a particularly dramatic transformation in the years after 1900. By 1933, both the nature of sound and the culture of listening were unlike anything that had come before.

The sounds themselves were increasingly the result of technological mediation. Scientists and engineers discovered ways to manipulate traditional materials of architectural construction in order to control the behavior of sound in space. New kinds of materials specifically designed to control sound were developed, and were soon followed by new electroacoustic devices that effected even greater results by converting sounds into electrical signals. Some of the sounds that resulted from these mediations were objects of scientific scrutiny; others were the unintended consequences—the noises—of an ever-more mechanized society; others, like musical concerts, radio broadcasts, and motion picture sound tracks, were commodities consumed by an acoustically ravenous public. The contours of change were the same for all.

Accompanying these changes in the nature of sound were equally new trends in the culture of listening. A fundamental compulsion to control the behavior of sound drove technological developments in architectural acoustics, and this imperative stimulated auditors to listen more critically, to determine whether that control had been accomplished. This desire for control stemmed partly from new worries about noise, as traditionally bothersome sources of sound like animals, peddlers, and musicians were increasingly drowned out by the technological crescendo of the modern city. It was also driven by a preoccupation with efficiency that demanded the elimination of all things unnecessary, including unnecessary sounds. Finally, control was a means by which to exercise choice in a market filled with aural commodities; it allowed producers and consumers alike to identify what constituted "good sound," and to evaluate whether particular products achieved it.

Perhaps the most significant result of these physical and cultural changes was the reformulation of the relationship between sound and space. Indeed, as the new soundscape took shape, sound was gradually dissociated from space until the relationship ceased to exist. The dissociation began with the technological manipulations of sound-absorbing building materials, and the severance was made complete when electroacoustic devices claimed sound as their own. As scientists and engineers engaged increasingly with electrical representations of acoustical phenomena, sounds became indistinguishable from the circuits that produced them. When electroacoustic instruments like microphones and loudspeakers moved out of the laboratory and into the world, this new way of thinking migrated with them, and the result was that sounds were reconceived as signals.

When sounds became signals, a new criterion by which to evaluate them was established, a criterion whose origins, like the sounds themselves, were located in the new electrical technologies. Electrical systems were evaluated by measuring the strength of their signals against the inevitable encroachments of electrical noise, and this measure now became the means by which to judge all sounds. The desire for clear, controlled, signal-like sound became pervasive, and anything that interfered with this goal was now engineered out of existence.

Reverberation, the lingering over time of residual sound in a space, had always been a direct result of the architecture that created it, a function of both the size of a room and the materials that constituted its surfaces. As such, it sounded the acoustic signature of each particular place, representing the unique character (for better or worse) of the space in which it was heard. With the rise of the modern soundscape this would no longer be the case. Reverberation now became just another kind of noise, unnecessary and best eliminated.

As the new, nonreverberant criterion gained hold, and as the architectural and electroacoustic technologies designed to achieve it were more widely deployed, the sound that those technologies produced now prevailed. The result was that the many different places that made up the modern soundscape began to sound alike. From concert halls to corporate offices, from acoustical laboratories to the soundstages of motion picture studios, the new sound rang out for all to hear. Clear, direct, and nonreverberant, this modern sound was easy to understand, but it had little to say about the places in which it was produced and consumed.

This new sound was modern for a number of reasons. First, it was modern because it was efficient. It physically embodied the idea of efficiency by being stripped of all elements now deemed unnecessary, and it exemplified an aesthetic of efficiency in its resultant signal-like clarity. It additionally fostered efficient behavior in those who heard it, as the connection between minimized noise and maximized productivity was convincingly demonstrated. Second, it was modern because it was a product. It constituted a commodity in a culture increasingly defined by the act of consumption, and was evaluated by listeners who tuned their ears to the sounds of the market. Finally, it was modern because it was perceived to demonstrate man's technical mastery over his physical environment, and it did so in a way that transformed traditional relationships between sound, space, and time. Technical mastery over nature and the annihilation of time and space have long been recognized as definitive aspects of modern culture. From cubist art and Einsteinian physics to Joycean stream-of-consciousness storytelling, modern artists and thinkers were united by their desire to challenge the traditional bounds of space and time. Modern acousticians shared this desire, as well as the ability to fulfill it. By doing so, they made the soundscape modern.

Telling the story of the complicated transformations outlined above presents its own challenge to the writer who strives to control a narrative that moves through historical time and space. The story that follows begins in 1900 and ends in 1933, but it traverses this chronological trajectory several times over, returning to the start to explore new themes and phenomena, reexamining recurrent phenomena along the way, reiterating central themes, and ultimately—I hope—creating a resounding whole in which all the disparate elements combine to characterize fully and compellingly the construction of the modern soundscape.

I begin at the turn of the century with opening night at Symphony Hall in Boston, and I end with Radio City Music Hall in New York, which opened just as the Machine Age in America came to a grinding halt at the close of 1932. Symphony Hall was a secular temple in which devout listeners gathered to worship the great symphonic masterpieces of the past, particularly the music of Ludwig van Beethoven, whose name was inscribed in a place of honor at the center of the gilded proscenium. Radio City Music Hall, in contrast, was a celebration of the sound of modernity. Its gilded proscenium was crowned, not with the name of some long-dead composer, but with state-of-the-art loudspeakers that broadcast the music of the day to thousands of auditors gathered beneath it.

Yet, even as Symphony Hall was dedicated to the music of the past, it heralded a new acoustical era, an era in which science and technology would exert ever-greater degrees of control over sound. Symphony Hall was recognized as the first auditorium in the world to be constructed according to laws of modern science. Indeed, it not only embodied, but instigated, the origins of the modern science of acoustics. When a young physicist at Harvard University named Wallace Sabine was asked to consult on the acoustical design of the hall, he responded by developing a mathematical formula, an equation for predicting the acoustical quality of rooms. This formula would prove crucial for the subsequent transformation of the soundscape into something distinctly modern.

While Radio City Music Hall was intended to celebrate that soundscape, facing optimistically toward the future rather than gazing longingly back at the past, it actually signaled the end of this period of change. Radio City demonstrated an unprecedented degree of control over the behavior of sound, but this demonstration was no longer compelling in a culture now facing far greater challenges. In America in 1933, the technological enthusiasm that had fed the long drive for such mastery was fundamentally shaken. The Machine Age was over, and Radio City was immediately recognized as a relic of that bygone era.

Since Wallace Sabine's work on Symphony Hall was recognized at the time as something distinctly new, it must be examined closely in order to understand its significance for what would follow. Chapter 2 presents this examination by exploring the scientific details of Sabine's research and his application of those results to the design of Symphony Hall. The equations and formulas he developed are crucial historical artifacts for the story that follows and it would be inappropriate not to include them, but their importance will be fully explained in nonmathematical prose, for readers not accustomed to confronting scientific equations.

Just as important for understanding the nature and reception of Sabine's work is the context in which it took place, so chapter 2 also presents a brief survey of earlier efforts to control sound, and it considers why Sabine's work was perceived to be valuable by both architects and listeners. Finally, an examination of the critical reception of the acoustics of Symphony Hall demonstrates the complicated combination of social, cultural, and physical factors that go into the process of defining, as well as creating, "good sound."

Chapters 3 through 6 cover the period 1900–1933 from four different perspectives. Chapter 3 focuses on the work of the scientists who, following Sabine's lead, devoted their careers to the study of sound and its behavior in architectural spaces. Like Sabine before them, these men were initially frustrated by a lack of suitable scientific tools for measuring sound. With the development of new electrical instruments in the 1920s, not only did it become possible to measure sound, but the tools also stimulated new ways of thinking about it. Scientists drew conceptual analogies between the sounds that they studied and the circuits that measured those sounds, and the result was a new interest in the signal-like aspects of sound. By 1930, new tools, new techniques, and a new language for describing sound had fundamentally transformed the field of acoustics. "The New Acoustics" was proclaimed, and its success as a science and a profession was acknowledged with the founding of the Acoustical Society of America.

The New Acousticians of the modern era sought a larger sphere in which to apply their science, to attract public attention to that science and to earn respect for their expertise and their efforts. The problem of city noise provided a challenging and highly visible forum. Chapter 4 thus moves out into the public realm and charts changes in the problem and meaning of noise.

While noise has been a perennial problem throughout human history, the urban inhabitants of early-twentieth-century America perceived that they lived in an era unprecedentedly loud. More troubling than the level of noise was its nature, as traditional auditory irritants were increasingly drowned out by the din of modern technology: the roar of elevated trains, the rumble of internal combustion engines, the crackle and hiss of radio transmissions. As the physical nature of noise changed, so, too, did attempts to eliminate it. At the turn of the century, noise abatement was a type of progressive reform where influential citizens attempted to legislate changes in personal behavior to quiet the sounds of the city. As the sounds of modern technology swelled, it became clear that only technical experts could quell these sounds, and in the 1920s, acousticians were called upon to reengineer the harmony of the modern city.

While the majority of those who engaged with noise sought to eliminate it, some were stimulated more creatively by the sounds that surrounded them. The modern soundscape was filled with music as well as noise, and chapter 4 considers how both jazz musicians and avant-garde composers redefined the meaning of sound and the distinction between music and noise. Acousticians did much the same thing, but with scientific, rather than musical, instruments.

Noise abating engineers ultimately failed, however, to master the modern urban soundscape. Their new ability to measure noise only amplified the problem and did not translate into a solution within the public sphere of legislation and civic action. Nonetheless, a private alternative would succeed where this public approach did not, and chapter 5 retreats back indoors to consider how the technology of architectural acoustics was deployed to alleviate the problem of noise and to create a new modern sound.

Chapter 5 follows the rise of the acoustical materials industry, charting the development of a range of new building technologies dedicated to isolating and absorbing sound. Acousticians devised new materials and supervised their installation in offices, apartments, hospitals, and schools, as well as in traditional places of acoustical design like churches and auditoriums. These sound-engineered buildings offered refuge from the noise without, and transformed quiet from an unenforceable public right into a private commodity, available for purchase by anyone who could afford it.

Acoustical building materials demonstrated technical mastery over sound and embodied the values of efficiency. By minimizing reverberation and other unnecessary sounds, the materials created an acoustically efficient environment and engendered efficient behavior in those who worked within it, and began the process by which sound and space would ultimately be separated. Through a series of case studies of representative materials and the buildings in which they were installed, chapter 5 will describe the architectural construction of modern sound and will conclude by demonstrating how that sound made an integral contribution to the establishment of modern architecture in America.

With the silencing of space came a desire to fill it with a new kind of sound, the sound of the electroacoustic signal. Chapter 6 examines how electroacoustic technology moved out of the lab and into the world, and, by returning to performance spaces, emphasizes how much things had changed since 1900. Microphones, loudspeakers, radios, public address systems, and sound motion pictures now filled the soundscape with new electroacoustic products. Consumers of those products, like acoustical scientists and engineers, learned to listen in ways that distinguished the signals from the noise. This distinction became a basis for defining what constituted good sound: clear and controlled, direct and nonreverberant, denying the space in which it was produced.

This modern sound was not exclusively the product of electrical technologies, and it was constructed architecturally in auditoriums where loudspeakers were neither required nor desired. Nonetheless, most Americans heard this sound most often on the radio or at the movies, and chapter 6 focuses on the transformation of motion picture theaters and studios as both were wired for sound.

The technologies of electroacoustic control that were developed in the sound motion picture industry highlighted questions about the relationship between sound and space, forcing sound engineers and motion picture producers alike to decide just what their new sound tracks should sound like. The technology also provided new means by which to construct the sound of space, as engineers learned to create electrically a spatialized sound that we would call "virtual." The sound of space was now a quality that could be added electrically to any sound signal in any proportion; it no longer had any relationship to the physical spaces of architectural construction. This new sound bore little resemblance to that which had been heard in 1900. It was so different, Wallace Sabine's fundamental reverberation equation failed to describe it. Sabine's equation was revised to fit the modern soundscape, and with this revision, the transformation was complete.

The revision of Sabine's equation expressed the transformation of the soundscape in a cryptic mathematical language that spoke only to acousticians and sound engineers. That same transformation was more widely and unmistakably heard in the sounds and structures of Rockefeller Center, and *The Soundscape of Modernity* closes by examining the critical reception of the center in order to understand the conclusion of the era that defined the modern sound.

From the office spaces of the RCA tower to the NBC studios to the auditorium of Radio City Music Hall, the modern soundscape was epitomized and celebrated. Even before the construction of the center was complete, however, such celebration was immediately perceived to be inappropriate and outdated. New economic conditions and new attitudes regarding the previously unquestioned promise of modern technology brought the era of modern acoustics to a close. The Machine Age was now over, and the modern soundscape would begin to transform itself again into something new.

With the basic outline of the story in place, it is useful to consider briefly how this story will relate to others doubtlessly more familiar to its readers. What does *The Soundscape of Modernity* accomplish, beyond providing a sound track to a previously silent historiography? Most basically, my story builds and expands upon past histories of the science and technology of acoustics. Much of this work has been written by practitioners, and they have constructed a solid foundation upon which I have built my own understanding of the intellectual developments of the field.[4] Historians of science have only recently begun to turn their attention to the science of sound, and have so far focused on periods that precede my own.[5] These studies have offered important insights into general questions concerning the rise of modern science and the role of scientific instruments in its creation. The history of twentieth-century acoustics similarly addresses fundamental questions about the relationships between science, industry, and the military, and it elucidates the instrumental connections between the material culture of science and its intellectual accomplishments.[6] My work only begins to examine these issues, but it demonstrates the fruitfulness of the history of acoustics in a way that may encourage others to follow.

As a contribution to the history of technology, my story is situated at the intersection of two different, but equally important, strands of scholarship. While some of the best work in this field has been devoted to the history of radio, the accomplishments of Hugh Aitken and Susan Douglas have recently been complemented by the output of an emerging community of scholars focusing upon a whole range of technological topics associated with music and sound.[7] My work adds architectural acoustics to this mix, but perhaps more importantly addresses the history of listening in a way that may influence our understanding of the entire range of acoustical technologies currently being explored.[8]

The environmental trend in the history of technology is equally vibrant and particularly valuable for its consideration of the urban context.[9] My examination of the problem of noise in American cities builds upon the work of others who have explored this phenomenon, but my perspective is distinct. Instead of drawing upon late-twentieth-century concerns about pollution and the degradation of the environment, I turn instead to the cultural meaning of noise in the early decades of the century, to demonstrate how musicians and engineers created a new culture out of the noise of the modern world.[10] By doing so, I hope to argue more generally that culture is much more than an interesting context in which to place technological accomplishments; it is inseparable from technology itself.

The history of acoustics intersects with the history of the urban environment not only through the problem of city noise, but also through technologies of architectural construction, and my work addresses an aspect of construction long neglected by visually oriented architectural historians. I challenge these historians to listen to, as well as to look at, the buildings of the past, and I thereby suggest a different way to understand the advent of modern architecture in America. As an outsider to this field, I leave it to others to evaluate the usefulness of my approach and its conclusions.[11]

I am similarly an outsider to the field of film studies, but some of the most interesting and thoughtful work on the history of sound technology and the culture of listening is found here, and my own work has benefited enormously from the insights of this scholarship.[12] Still, here, as in architectural studies, many historians continue to operate with a predominantly visual orientation, understanding sound film primarily in its relation to the earlier traditions of silent film production. In contrast, I approach sound film from the perspective of the wider range of acoustical technologies that were developed and deployed alongside it. By doing so, I am able to demonstrate that, in deciding what sound film should sound like, filmmakers functioned in a larger cultural sphere. The decisions they made reflected not only the conditions of their own industry, but the larger soundscape in which that industry flourished.

Any exploration of a soundscape should ultimately inform a more general understanding of the society and culture that produced it. The reverberations of aural history within the larger intellectual framework of historical studies are just beginning to be heard, but the successes already accomplished speak well for the future of this approach. Leigh Schmidt, for example, has examined the meaning of sound in the American Enlightenment, and has thereby not only recovered the sensory experience of religion in American history, but also documented the forging of both science and popular culture out of those experiences. Mark Smith has identified a previously unacknowledged site of sectional tension in antebellum America by reconstructing the soundscapes of slaves, masters, and abolitionists.[13] And such studies of soundscapes are by no means limited to the American context. Bruce Smith has restored the lost sound of Shakespearian drama as it originally reverberated through the Globe Theatre and across Early Modern England, and in those reverberations he hears the transition from oral to literate culture. James Johnson has detected the rise of romanticism and bourgeois sensibility within the soundscape of the French concert hall, and Alain Corbin has perceived in the peals of village bells in nineteenth-century France the changing structures of religious and political authority.[14]

Clearly, these histories have much to say about the larger historical processes at work within their soundscapes, and all highlight themes and issues that historians have long considered to be constitutive of the rise of modern society and culture in the West.[15] Until recently, that long-term process of modernization was perceived as a particularly visual one, but the new aural history now demonstrates that, to paraphrase Schmidt, there is more to modernity than meets the eye.[16] This is particularly true for the period of so-called high modernism, and the long-standing absence of the aural dimension in cultural histories of the late nineteenth and early twentieth centuries is perhaps most striking of all.

"Modernism has been read and looked at in detail but rarely heard," concludes Douglas Kahn, in spite of the fact that this culture "entailed more sounds and produced a greater emphasis on listening to things," and on "listening differently" than ever before.[17] Those new sounds, and that different way of listening, were created and constructed through new acoustical technologies. James Lastra also asserts that "the experience we describe as 'modernity'—an experience of profound temporal and spatial displacements, of often accelerated and diversified shocks, of new modes of society and of experience—has been shaped decisively by the technological media."[18] To exclude acoustical technologies and sound media from scrutiny is to miss the very nature of that experience. Scholars who assume that consideration of the visual and textual is sufficient for understanding modernity, seem, well, shortsighted to say the least.

Restoring the aural dimension of modernity to our understanding of it promises not only to render that understanding "acoustically correct," it also provides a means by which to understand, more generally and significantly, the role of technology in the construction of that culture. This, after all, was the era in which the adjective *modern* achieved a new resonance through the self-conscious efforts of artists, writers, musicians, and architects, all of whose work was characterized by a pervasive engagement with technology.

Histories of modernism have long recognized the importance of technology as inspiration to the artists who are credited with creating the new culture. But these histories have too seldom engaged with technology as intensely as did those artists. Too often, the machines of the Machine Age are characterized as the uninteresting products of naive engineers that only achieved cultural significance when transmitted through the lens of art. "The impact of technology" upon these artists, not the technology itself, is what drives these accounts.[19]

It is not my intent to deny the importance of those artists and their work; indeed their music and architecture are crucial elements of the story that follows. But by juxtaposing the creations of mundane engineers with those of extraordinary artists, I implicitly argue that the works of both were equally significant and equally modern. Unremarkable objects like sound meters and acoustical tiles have as much to say about the ways that people understood their world as do the paintings of Pablo Picasso, the writings of John dos Passos, the music of Igor Stravinsky, and the architecture of Walter Gropius. All are cultural constructions that epitomized an era defined by the shocks and displacements of a society reformulating its very experience of time and space.

Karl Marx had these displacements in mind when he famously summarized the condition of modernity by proclaiming, "All that is solid melts into air."[20] Marx had very particular ideas about the material aspects of life and their role in historical change, ideas not necessarily at play in the story that follows. Nonetheless, like Marx, I believe that the essence of history is found in its material. I argue against the idea of modernity as a cultural zeitgeist, a matrix of disembodied ideas perceived and translated by great artists into material forms that then trickle down to a more popular level of consciousness. In the story that follows, modernity was built from the ground up. It was constructed by the actions and through the experiences of ordinary individuals as they struggled to make sense of their world.[21]

If modern culture is not a zeitgeist, not an immaterial cluster of ideas somehow "in the air," it must be acknowledged that sound most certainly is there, in the air. This ephemeral quality of sound has long frustrated those who have sought to control it, and the architect Rudolph Markgraf expressed the frustrations of many when he complained in 1911 that "sound has no existence, shape or form, it must be made new all the time, it slumbers until it is awaken[ed], and after it ceases its place of being it is unknown."[22] Markgraf was perplexed by "the mysteries of the acoustic," and historians of soundscapes are similarly challenged by sound's mysterious ability to melt into air, its tendency—even in a postphonographic age—to efface itself from the historical record. But if most sounds of the past are gone for good, they have nonetheless left behind a rich record of their existence in the artifacts, the people, and the cultures that once brought them forth. By starting here, with the solidity of technological objects and the material practices of those who designed, built, and used them, we can begin to recover the sounds that have long since melted into air. Along with those sounds, we can recover more fully our past.

Chapter 2
The Origins of Modern Acoustics

> Symphony Hall, the first auditorium in the world to be built in known conformity with acoustical laws, was designed in accordance with his specifications and mathematical formulae, the fruit of long and arduous research. Through self-effacing devotion to science, he nobly served the art of music. Here stands his monument.
>
> Plaque dedicated to physicist Wallace Sabine
> Located in the lobby of Symphony Hall, Boston

I Introduction: Opening Night at Symphony Hall

On 15 October 1900, the doors of Symphony Hall opened wide, welcoming Boston's music lovers to their new home for orchestral music. (See figures 2.1 and 2.2.) As people entered and took their seats, they noted with approval the tasteful appointments of the interior, but "the question of greatest permanent interest" was that of "the acoustical properties of the new hall."[1] The papers reported that "The great question concerning which not only the thousands in the hall, but tens of thousands not in it, were on the tiptoe of expectation was, 'Is the hall satisfactory acoustically?'"[2] In fact, the question of acoustics had been raised long before opening night; it originated eight years earlier, when the construction of a new auditorium had first been considered.

In 1892, the administrators of the city of Boston announced plans to lay a new road through the downtown site of the city's old Music Hall. While the venerable auditorium had housed a variety of programs over the past forty years, its most noteworthy occupant was the Boston Symphony Orchestra. Wholly owned and controlled by financier and philanthropist Henry Lee Higginson, the orchestra was one of the nation's foremost musical ensembles. Higginson welcomed this opportunity to build a new, exclusive home for his musicians, and he immediately began to raise the funds necessary to construct a new hall. The

2.1

Symphony Hall, Boston (McKim, Mead & White, 1900). Exterior, c. 1900. This new home for the Boston Symphony Orchestra embodied a romantic, even religious dedication to symphonic music that characterized elite culture in turn-of-the-century America. Courtesy Boston Symphony Orchestra Archives.

2.2

Symphony Hall, Boston (McKim, Mead & White, 1900). Interior, c. 1900. To ensure that the auditorium was acoustically worthy of the great music with which it would be filled, architect Charles McKim consulted Harvard University physicist Wallace Sabine on the design of this hall. The gilded crest at the top center of the proscenium is inscribed "Beethoven." Courtesy Boston Symphony Orchestra Archives.

commission went to McKim, Mead & White, a renowned architectural firm then in the midst of building Boston's new public library. Charles McKim took charge of the new project, and Higginson immediately underscored the importance of acoustics. He wanted a hall that would shelter its audience from the "sounds from the world" and do justice to the great music of the past, particularly that of his favorite composer, Ludwig van Beethoven. "Our present hall," he informed McKim, "gives a piano better than a forte, gives an elegant rather than a forcible return of the instruments—noble but weak—I *want* both."[3]

To obtain this effect, Higginson suggested setting the stage in an alcove whose slanted roof would direct the sound of the orchestra out toward the audience. He also identified several European halls well reputed for their sound, and he encouraged McKim to visit and study these halls. McKim contacted John Galen Howard, a former employee then enrolled at the École des Beaux Arts in Paris, and instructed him to inquire into the principles of theater design. Howard spoke with musical and architectural authorities in Europe and worked up three plans, which McKim submitted to Higginson in July of 1893.[4] One plan was rectangular (a form recommended by Charles Lamoureux, director of the Paris Opéra), one was elliptical (the form preferred by Howard's architectural professor, Victor Laloux), and a third—McKim's favorite—was semicircular.

McKim developed his favorite into a more finished design in the style of a Greek theater. (See figure 2.3.) In January 1894, a model was displayed in the newly opened public library, where the patrons "expressed themselves highly pleased with the beauty, simplicity and convenience of the design."[5] Nonetheless, this building was never built, as an economic downturn that spring developed into a severe and ultimately lengthy depression. In April, Higginson informed McKim that the city's "plans of transit" were on hold, thus removing the immediate necessity to build. It was also now difficult to raise funds for a new hall, so the project was temporarily but indefinitely set aside.[6]

By 1898, conditions had improved, the city's roadway proposal reappeared, and Higginson renewed his commitment to build a new hall—but not the one McKim had earlier designed. Higginson informed his architect that, during the hiatus, the board of directors for the new hall had decided that his semicircular design was unacceptable. "While we hanker for the Greek Theatre plan," he explained, "we think the risk too great as regards results, so we have definitely abandoned that idea."[7] The "risk" to which he referred was acoustical; no concert hall had ever been built in the form of a semicircular amphitheater before, and there was no way to know ahead of time how such a hall would sound. The

2.3

Plan for the Boston Music Hall, second floor, drawing by Charles McKim, 1892. This "Greek Theater" design was ultimately rejected by the building committee for the new music hall because its semicircular form was unprecedented in an auditorium intended for symphonic music. © Collection of The New-York Historical Society.

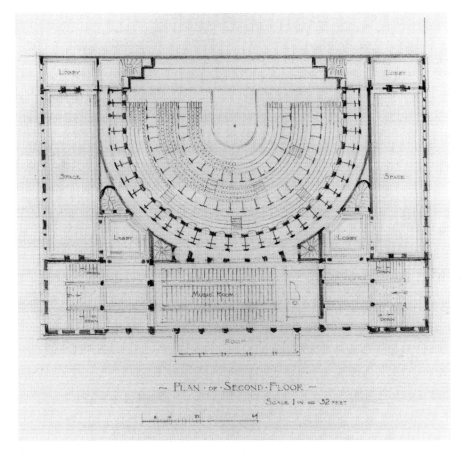

board proposed a rectangular hall, to replicate the form, and, it was hoped, the acoustical success, of the New Gewandhaus in Leipzig.⁸

McKim's own devotion to the Greek theater design had weakened over the years. While traveling in Europe during the project's hiatus, he had discussed auditorium design with a number of eminent musical directors. None could support the unusual form of his amphitheater, and one confessed, "I don't know anything about acoustics, but my first violin tells me we always get the best results in a rectangular hall."⁹

Higginson, however, required something more than a violinist's opinion to ensure that his new hall would be worthy of the great music that he so admired. After all, there were plenty of rectangular concert halls that were not considered acoustical successes. Higginson thus sought the advice of a technical expert, one who could ensure with the perceived authority of scientific laws that his hall would sound as he desired. While he had acknowledged that "musicians must decide the points eventually," Higginson confided to McKim, "I always feel like hearing their opinions most respectfully and then deciding." "Cross' opinions *seem* to me better," he admitted, citing a local scientific authority.[10] In the end, Higginson preferred the counsel of scientists to that of musicians. This preference led him to consult his friend Charles Eliot, the scientifically trained president of Harvard University. Eliot recommended that Higginson contact Wallace Sabine, a young assistant professor of physics at Harvard who had recently worked to improve the acoustics of a university lecture hall.

Wallace Sabine first met Henry Lee Higginson in January 1899. The men carefully studied McKim's plan and Sabine expressed numerous opinions regarding the length of the hall, the number of galleries, the rake of the floor, the shape of the stage, and the system of ventilation. Higginson immediately telegraphed McKim, advising: "It may be wiser to await important letter going tonight before more work on plans."[11] In that lengthy letter, he described Sabine's ideas and made clear that they were to be incorporated into the architect's design: "The room itself I think we can settle between your office, Professor Sabine's office and our office; in fact we shall have to do so." Perhaps fear of offending McKim's sense of authority led Higginson to add a short, hand-written postscript to the typed letter, reassuring the architect that "We will have a perfect hall under your guidance."[12] Any such fear must have been short-lived, however, for upon meeting Sabine, McKim was "much impressed by the force and reasonableness of his arguments, as by the modest manner in which they were presented." He also expressed his confidence that the acoustics of the hall would benefit greatly from Sabine's "counsel and advice."[13]

Sabine and McKim worked together, resolving issues raised by the design and the construction of the hall, throughout 1899 and 1900. On opening night, Higginson highlighted Sabine's contribution in his address to the hall's inaugural audience. "If it is a success," he announced, "the credit and your thanks are due to four men." He acknowledged McKim, the builder Otto Norcross, and the financial manager Charles Cotting, and he also thanked Sabine, adding, "Professor Sabine has studied thoroughly our questions of acoustics, has applied his knowledge to our problem; and I think with success."[14]

Before the nature and extent of Sabine's success can be determined, his work must be examined and contextualized in order to illuminate his accomplishments as well as his audience's expectations. To understand what Sabine accomplished, a brief survey of earlier attempts by both scientists and architects to study and to control sound will first be presented. A detailed examination of Sabine's own investigation will follow, outlining his derivation of a mathematical formula for predicting the acoustical character of rooms. A survey of musical culture in turn-of-the-century America will then consider why the audience at Symphony Hall cared so deeply about what they heard there. Finally, their evaluation of what they heard will be examined. By listening carefully to the creation and critical reception of the acoustics of Symphony Hall, we can begin to comprehend the complex conjunction of science, architecture, and music that constituted this building and this moment in America's cultural history.

II Acoustics and Architecture in the Eighteenth and Nineteenth Centuries

For as long as sound has been reflecting off the surfaces of architectural construction, auditors have reflected upon the subject of architectural acoustics. The ancient Greeks were some of the first to examine the phenomena of sound, considering how it propagated through space and questioning why it behaved differently in different kinds of spaces. In what is considered to be the oldest extant architectural treatise in the Western tradition, the Roman architect Vitruvius articulated ideas about how to control sound in theaters. Philosophers and builders alike, from ancient times through the Middle Ages and into the Renaissance, believed that the phenomena of sound and music were inherently linked to architecture through the underlying harmony of the universe. Simple numeric ratios expressed the order of the cosmos as well as the harmonies of music, and architects—whose goal was to re-create that divine order on a human scale—based their designs on those same proportions.[15]

This belief in the harmony of the universe, a belief that integrated music, architecture, astronomy, and mathematics, was gradually transformed as modern science took shape during the sixteenth and seventeenth centuries. The new science presented an understanding of the world fundamentally different from the divine ratios of the premodern cosmos. As this new way of thinking took hold, science parted ways with both music and architecture.[16]

New theories and experimental techniques enabled scientists to explore more fully the physical dimensions of sound. Mathematicians analyzed the behavior of vibrating strings via the new calculus of Isaac Newton; experimenters like Galileo Galilei and Marin Mersenne examined the motion of vibrating bodies and measured the speed of sound in different media; and countless natural historians collected anecdotes of interesting acoustical phenomena, from unusual echoes to the feats of ventriloquists and talking automata, and recorded them in the pages of new scientific journals.[17]

As modern science took shape, architecture similarly lost its cosmological significance and was recast as a set of techniques that manipulated but no longer transcended the physical world. Alberto Pérez-Gómez has shown that this new kind of architecture, which began to appear in the middle of the seventeenth century, ultimately became "thoroughly specialized, and composed of laws of an exclusively prescriptive character that purposely avoid all reference to philosophy or cosmology."[18] As science and architecture parted ways, the subject of architectural acoustics fell into the gap that opened between them.

This gap only widened over the eighteenth and nineteenth centuries, as the acoustical interests of scientists continued to diverge from the needs of architects. Mathematical elaborations of the behavior of sound reached their apotheosis in the work of Lord Rayleigh, whose *Theory of Sound* was considered the last word on the subject for many years after its publication in 1877.[19] Experimentalists continued to measure the speed of sound, and to examine vibrating bodies, contriving ingenious ways by which to render visible the minute movements of objects and air. Ernst Chladni, for example, dusted the surfaces of vibrating plates with fine sand that collected at the nodes of those plates, creating geometric patterns beautiful enough to impress an emperor. Upon viewing the phenomenon in 1808, Napoleon offered a prize to whoever could explain fully the formation of the patterns, and this prize was claimed in 1816 by the mathematician Sophie Germain.[20] Rudolph Koenig was awarded a gold medal at the 1862 Crystal Palace Exposition in London for a device that transformed vibrations of sound in air into flickering flames, and he brought this device, along with an impressive set of tuning forks and other acoustical apparatus, to America's Centennial Exposition in Philadelphia in 1876.[21] Other investigators developed means to inscribe the vibrations of sound on various media, attempting to create "sound-writing" instruments that might record sounds in a readable form, and still others continued to attempt to build talking machines.[22]

All these efforts, however, were of little use to architects. Koenig's flames failed to illuminate ideas about how best to control the behavior of sound; the talking machines remained silent on this point; and even Rayleigh's voluminous tome devoted only a few, inscrutably mathematical pages to "aerial vibrations in a rectangular chamber."[23] In 1782, the French architect Pierre Patte had searched in vain for scientific advice on the problem of acoustics, and his colleagues a

2.4

Pierre Patte's 1782 design for a theater whose elliptical shape was intended to reinforce the sound of the performers on stage. Late eighteenth-century European architects like Patte were concerned that the players would be unable to fill such a large space with sound, and they attempted to identify one best form to make the most of the sound. Reproduced here from George Saunders, *A Treatise on Theaters* (London: I. and J. Taylor, 1790), plate IV.

century later were no better off.[24] Left to their own devices, architects like Patte constructed their own creative solutions to the problem of controlling sound.

Pierre Patte's search for scientific advice at the end of the eighteenth century had been compelled by conditions that had recently rendered the need to control sound particularly acute. The commercialization of theater in Europe created new social and acoustical conditions that were perceived to demand expertise not readily available. Theaters built at this time were far larger than their royally sponsored predecessors, and their size presented unprecedented acoustical challenges. Additionally, the commercial nature of the performances taking place within them heightened the importance of delivering good sound, as this accommodation was now considered the right of a public that had paid for admission.[25]

The Margrave's Opera House at Bayreuth exemplified the older, royal tradition in theater design. Built in 1748, its 5,500 cubic meters of space were filled with an audience of just 450 courtly attendants. In contrast, Milan's La Scala, built thirty years later, filled its 11,250 cubic meters with almost 2,300 auditors who gained access not by royal invitation, but by purchasing tickets.[26] The new need for "pecuniary return,"[27] as the architect Benjamin Dean Wyatt put it, led architects to build theaters larger than ever before, but the need to build large had to be limited by the equally important requirement that every member of the audience be able to see and to hear. The goal was thus to identify "the most capacious form which can possibly be constructed, to admit of distinct VISION and SOUND."[28]

Different architects had different ideas about how to identify this form and what it might be. Some turned to analogical thinking, for example, assuming that, because a bell was a sonorous object, a bell-shaped theater would also be sonorous. Others, including the Italian Count Francesco Algarotti, considered these analogies "an absurdity," and promoted instead a more analytical approach that drew on the mathematical certainty of the principles of geometry.[29] Pierre Patte, for example, picked up his compass and rule and applied them to architectural drawings in order to determine which form was best suited to "make the most of" the power of the voice.[30]

Patte evaluated the acoustical properties of differently shaped theaters by analyzing the propagation of sound within them. He drew lines representing rays of sound emanating from a performer on stage, then, following the rule that the angle of incidence is equal to the angle of reflection, he plotted the reflections of those rays off the walls. Patte concluded that an elliptically shaped the-

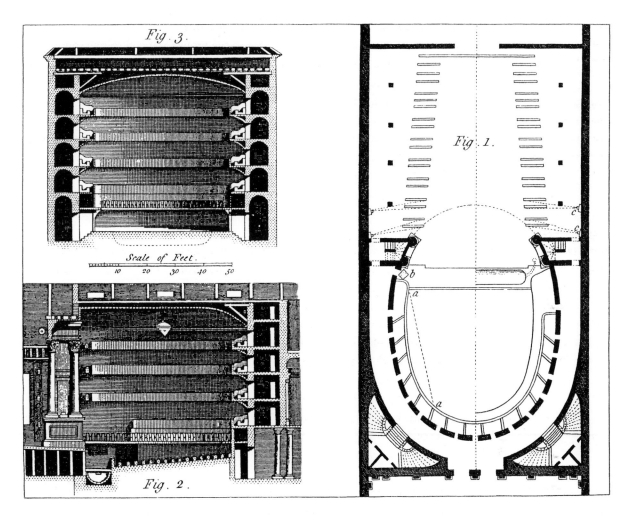

ater would generate the best acoustic effect, believing that its dual foci would actually augment the sound within. According to Patte, the rays of sound emanating from one focus (the performer on stage) would, upon reconvening at the second (in the auditorium), constitute a second source. This would effectively double the sound of the performer, which he feared would be too weak on its own to fill a large theater with sound.[31] (See figure 2.4.)

The British architect George Saunders carried out his own investigation and arrived at results different from those of Patte. Saunders was concerned with the extension, rather than reflection, of the voice. "In designing a theatre," he argued, "the first question that naturally arises is, In what form does the voice

2.5 George Saunders's analyses of the propagation of sound. His figure 6 illustrates the focusing property of ellipses that was the basis for Patte's design. Figure 4 shows the results of Saunders's own experiments on the extension of the voice, illustrating the maximum range of audibility for a listener encircling a speaker located at point "A". George Saunders, *A Treatise on Theaters* (London: I. and J. Taylor, 1790), plate I.

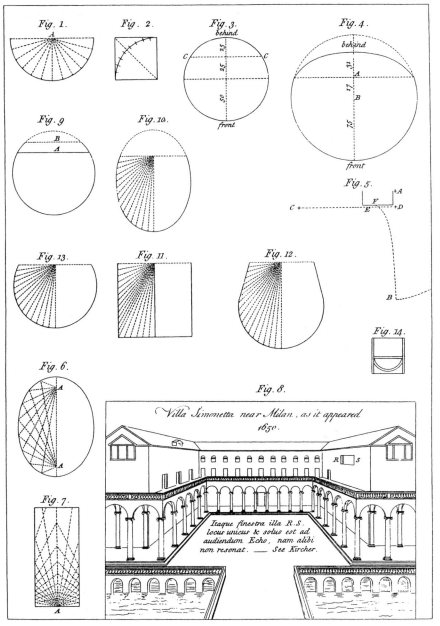

2.6
George Saunders's design for a theater, based on the results of his experiments on the extension of the voice. Both the size and the shape of his design were determined by the dimensions he had measured in his experiments. George Saunders, *A Treatise on Theaters* (London: I. and J. Taylor, 1790), plate XI.

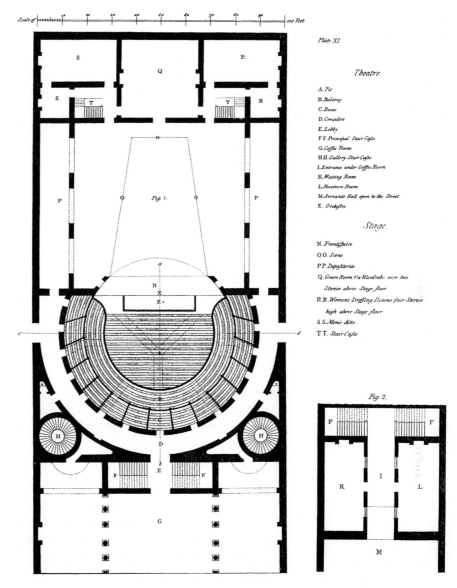

expand?"[32] To answer this question, he placed a speaker at a fixed location outdoors in open space, then had an auditor encircle the speaker, listening as he traveled in front of, around, and behind the speaker. The listener determined the most distant point from which he could hear as he encircled the speaker, thus marking out the extent of the voice in all directions. Saunders then used this figure as the basis for his design. (See figures 2.5 and 2.6.)

Algarotti promoted a semicircular theater, and Wyatt a variant of the form proposed by Saunders, but while each writer on acoustics recommended a different form, all agreed that form was the key to good sound. They shared their concern that too little sound would be generated by the performers, and they all identified as their goal the encouragement, even amplification, of the voices on stage. They also uniformly warned against the use of absorbent materials, as absorption would only impede the accomplishment of this goal.[33] Their shared geometrical approach took advantage of skills they already possessed, and was additionally reinforced by a neoclassical aesthetic that promoted the beauty of an architecture based on simple geometrical forms.[34]

The arguments of these authors, however, ultimately represent theories that thrived in books but not in buildings. Algarotti's treatise offered no specific plans for construction, while Saunders and Patte presented plans that were never built. Wyatt's ideas were realized in his Drury Lane Theatre in London; however, Drury Lane had to be completely remodeled not long after its completion, because of problems with sight and sound.[35] In fact, the acoustical realities of modern buildings were quite different from the problems that these men theorized, and the means to control those realities would ultimately prove equally different.

The American architect Benjamin Latrobe initially shared many ideas about sound with his European contemporaries, even though he was not familiar with their works. Upon engaging directly with the acoustics of an actual building, however, Latrobe reevaluated those ideas. Asked by a friend in 1803 to offer advice on the design for a Quaker meeting house, Latrobe turned to geometry to discover the best form for sound. Seeking to maximize the effect of the voice, he determined that a sphere constituted the best acoustical form, for "a ring of first echo perfectly coincident will be produced, and rings of reechoes, *ad infinitum*, many of them nearly coincident would follow." Recognizing that the sphere was not a particularly practical architectural form, Latrobe suggested, "In proportion as a room approaches this form, it approaches perfection."[36]

A few years later, as surveyor of public buildings for the United States, Latrobe supervised the construction of the Capitol Building in Washington. Shortly after its 1807 opening, the newspapers reported upon "a very material defect in the hall of the house of Representatives. The voice of the speakers is completely lost in echo, before it reaches the ear. Nothing distinctly can be heard from the chair or the members."[37] Latrobe discovered that not all echoes were beneficial, and he now sought to eliminate them. Curtains were hung, "tastefully and usefully," between the columns of the hall, and the architect reported that "though there is less sound, there is much more heard."[38] The realization that less is more came as a surprise to Latrobe, and he now emphasized that it was "the duty of the architect to suppress or exclude the echoes that would confuse the distinctness of the species of sound which it is the object of the edifice to exhibit."[39]

While Latrobe believed that his efforts to improve the acoustics of the hall had met with the "fullest success,"[40] the Congress and the press continued to complain. The troublesome echoes were eliminated temporarily in 1815 when British troops burned the Capitol to the ground during their invasion of Washington, but when the building was rebuilt in 1819, the new hall proved as unsatisfactory as its predecessor. Over the next few decades, Congress regularly solicited and received advice on how to improve the acoustics of the Hall, but to little avail.[41] One creative suggestion, actually acted upon in 1837, was to reverse the seating arrangement of the Representatives. (See figure 2.7.) The result was not considered an acoustical improvement, however, and before long Congress was back to facing forward.[42]

By mid-century the House had outgrown its still ill-sounding chamber. Plans were drawn up for the expansion of the Capitol and the construction project was assigned to the Army Corps of Engineers under the direction of Captain Montgomery Meigs. In 1853, Meigs was ordered by his commander, Secretary of War Jefferson Davis:

> You will examine the arrangements for warming, ventilation, speaking and hearing. The great object of the extension of the Capitol is to provide rooms suitable for the meeting of the two houses of Congress—rooms in which no vitiated air shall injure the health of the legislators, and in which the voice from each member's desk shall be made easily audible in all parts of the room. These problems are of difficult solution, and will require your careful study.[43]

"By direction of the President, who is desirous of obtaining the best scientific authority within reach upon this subject,"[44] Meigs invited Joseph Henry, secretary of the Smithsonian Institution, to review his ideas on sound as they applied to the new Hall of the House of Representatives. Henry, along with his scientific colleague Alexander Dallas Bache, subsequently reported to Davis that "the principles presented to them by Captain Meigs are correct, and that they are judiciously applied."[45] Nonetheless, when the new hall was finished and put to use it was found to be no better than its predecessor.

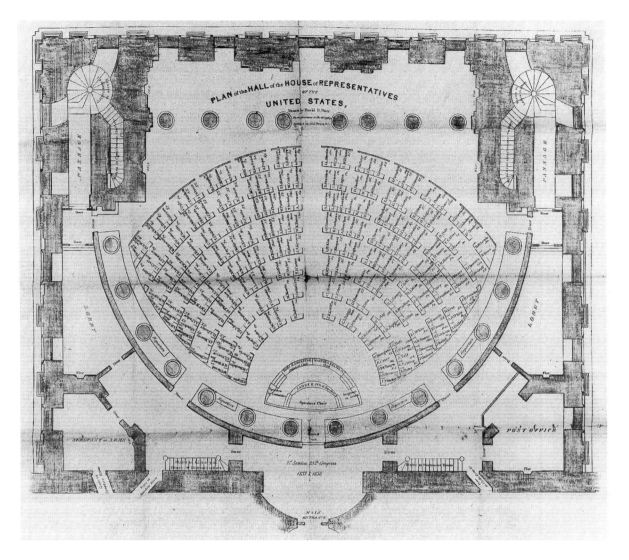

Joseph Henry's experience with the new hall may have emphasized to him that attention to form was insufficient to ensure good sound.[46] Others were certainly questioning the old approach, complaining that "form is the only point that architects seem to consider of importance."[47] While the role of materials in controlling sound had been previously acknowledged, architects seeking that control could only conclude that "the different degrees in which substances derived from the mineral, vegetable and animal kingdoms are favourable to the transmission of sound, appear to be regulated by laws not easily demonstrable."[48]

2.7
Seating Plan, United States House of Representatives, 1837–1838. This plan shows a reverse seating arrangement that was recommended by the architect Robert Mills. By having the members "speak to the curve" of the chamber's rear wall, Mills believed that the sound of the hall would be improved. The experiment was unsuccessful and the desks were returned to their normal positions in the subsequent session of Congress. Plan of the Hall of the House of Representatives 1837 & 1838, 2nd Session of the 25th Congress, drawn by David H. Burr. Architect of the Capitol.

Attempts to identify these laws were generally unconvincing,[49] but new ideas about the physical nature of sound would begin to provide a new means by which to understand the action of materials, and Henry himself would help formulate those ideas.

Shortly after his consultation on the House Chamber, Joseph Henry undertook a series of experiments to investigate the effect of materials upon sound. He sounded a tuning fork, placed the stem of the fork against the material to be tested, then measured how long the fork continued to vibrate. Believing his eyes to be more sensitive than his ears, Henry marked the cessation of vibration at the moment when he could no longer visually perceive the movement of the fork. This measure of time represented the sound-absorbing property of the different materials he tested, including cork, rubber, wood, and stone. Unlike eighteenth-century neoclassical architects, Joseph Henry had no interest in representing sound as geometric rays. As a mid-nineteenth-century physicist, he was instead committed to exploring the new idea of the conservation of energy and this energetic conception of sound was at the heart of his investigation.

According to this new way of thinking, the moving fork, the emitted sound, and the material with which the fork was in contact all contained a given amount of energy. While this energy could manifest itself in different ways, it could not be destroyed.

Henry observed that, while a vibrating fork suspended in air from a thread continued in motion for 252 seconds, the same fork vibrated for only ten seconds when placed in contact with a large thin board of pine. The board increased the volume of sound, and Henry explained that "the shortness of duration was compensated for by the greater intensity of effect produced."[50] When the fork was placed in contact with a piece of India rubber, the sound remained very feeble, yet it quickly died away. Where was the compensating effect here? Henry proved that the energy was converted to heat rather than sound, by measuring an increase in the temperature of the rubber as it absorbed the vibrations of the tuning fork.[51]

Joseph Henry's experiments constituted an innovative attempt to analyze and to quantify the sound-absorbing properties of materials, and this attempt was a direct result of a new energetic way of understanding the physical properties of sound. It is not apparent, however, that he applied his results to the design of any structure. Even though these experiments were conducted by Henry to evaluate the design of a lecture hall for his own Smithsonian Institution, Henry's practical contributions to that project focused strictly on its form. In his experiments on materials, he was ultimately more interested in tracking the conservation of energy than with generating knowledge of practical use to architects.[52]

Although Joseph Henry did not apply his new knowledge about materials directly to design of the Smithsonian lecture hall, he did use the publication of those results as an opportunity to speak out against the architecture that housed that hall. American architecture at mid-century was characterized by a historically inspired eclecticism in which virtually any style—from Gothic to Egyptian—was appropriate, as long as it was from the past. Henry disliked this approach, and he particularly disliked the crenellated castle that James Renwick had designed to house the Smithsonian Institution. As head of that organization, Henry worked and lived within its Romanesque towers, but not without complaint. "Every vestige of ancient architecture," he explained, "which now remains on the face of the earth should be preserved with religious care; but to servilely copy these, and to attempt to apply them to the uses of our day, is as preposterous as to endeavor to harmonize the refinement and civilization of the present age with the superstition and barbarity of the times of the Pharaohs." "It is only when a building expresses the dominant sentiment of an age," he continued, "when a perfect adaptation to its use is joined to harmony of proportions and an outward expression of its character, that it is entitled to our admiration."[53]

Henry's opinions about architecture were not widely shared by architects, and the historicism that he decried would become even more prevalent in the years to come.[54] Just as the geometry of neoclassicism had provided architects with a means to attempt to control sound, so, too, did the historical eclecticism of the nineteenth century offer its own approach. Practitioners of an aesthetic of imitation, not surprisingly, turned to imitation as they attempted to solve their problems of acoustical design.

At mid-century the cities of New York, Boston, and Philadelphia were all engaged in the construction of new music halls and opera houses, and in each case the architects drew on the form of an extant European theater in an attempt to re-create the acoustical qualities of that theater in their own design. The New York Academy of Music was patterned after the Berlin Opera House; the Boston Theatre after the theater at Bordeaux; and the Philadelphia Academy of Music after La Scala in Milan.[55] In no case was the attempt at imitation complete, nor were the acoustical re-creations that the architects accomplished. While these projects were more fortunate than many others in being judged acoustically successful, the method of replication was not considered a definitive approach to acoustical design. The architects of the Philadelphia Academy admitted that popular understanding of acoustics among architects was "very vague and indistinct." While they asserted that an architect who had "properly applied himself to this branch of his profession" could "certainly do a great deal toward the accomplishment of his object, especially if his study is founded upon practical experience, combined with the observations and results deducted from other buildings of a similar nature," they had to admit that "there always remains something left to chance."[56]

Almost fifty years later, Henry Higginson and Charles McKim would find few options beyond this method of replication when they sought to ensure good sound in their own music hall. This approach led Higginson to reject McKim's Greek theater plan, as it was unprecedented in housing a modern concert hall, and it drove their decision to build a rectangular hall, in imitation of the old Music Hall in Boston and the Leipzig Gewandhaus. Another precedent that Higginson surprisingly rejected was Carnegie Hall in New York. His orchestra had performed there numerous times since its opening in 1891, and he reported to McKim, "our people all think Carnegie Hall horrible." "Very noisy music produces considerable effect," he explained, "but the moment an orchestra plays the older music and relies on delicate effect, everything is gone. I have always disliked the hall very much, and I expected to like it very much before trying it."[57] Higginson's critique may have been idiosyncratic, for even if Carnegie Hall had not yet acquired the reputation it would later enjoy, the hall's acoustics were the accomplishment of an architect who, alone among his peers, was considered a master of sound.

Dankmar Adler learned his craft while rebuilding Chicago after the great fire of 1871. He established an independent practice in 1879 and received his first theater commission that same year. Adler soon promoted his talented associate Louis Sullivan to partnership, and Adler & Sullivan executed a dozen more theater and auditorium commissions over the next decade.[58] These projects were uniformly judged acoustical successes, and Adler became known as an expert on sound, serving "at various times as a consultant on acoustics."[59] One such project was William Burnet Tuthill's design for Carnegie Hall in New York.[60] His most famous accomplishment, however, was the partnership's own Auditorium Building in Chicago, which was completed in 1890.

As architects, Adler & Sullivan stood out from their colleagues by echoing Joseph Henry's earlier frustrations with the historicist tendencies of their field. Adler castigated nineteenth-century theater design for its reverence for the "his-

2.8

Auditorium Building and Theater, Chicago (Adler & Sullivan, 1889). The movable partitions that could block off the two uppermost balconies are indicated here, in both open and closed positions, with dotted lines. Dankmar Adler, "Theater-Building for American Cities," *Engineering Magazine* 7 (August 1894): 723.

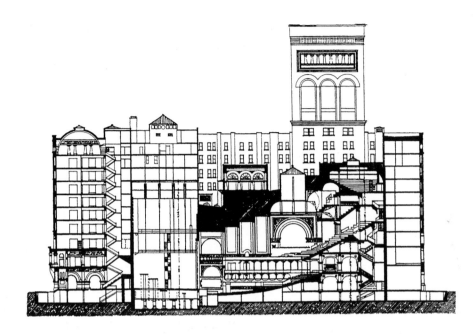

torically transmitted type," a reverence that was "the result of a mental attitude which sees in a brilliant and admirable achievement of the past, not a legitimate evolution from the conditions of its own environment, but a creation standing out for all ages to be blindly idolized and imitated."[61] The Auditorium, in sharp contrast, was a complete expression of the needs of its own environment—the excitement and energy of late nineteenth-century Chicago. It was a ballroom, a convention hall, and an auditorium for a rapidly growing city. The theater held over four thousand people, and Adler incorporated movable ceiling panels that could be pulled down to block off the two uppermost galleries and reduce the capacity when a smaller space was more appropriate. (See figure 2.8.) Adler & Sullivan surrounded the theater with a hotel and offices to render the building financially self-sustaining. Sullivan designed a simple granite facade that heightened the effect of the ornament within. The theater glittered with gilded moldings and ornate grillwork. Murals and a stained-glass skylight added color, while the whole was illuminated by a "tiara" of electric lights embedded in the ceiling.[62] (See figures 2.9 and 2.10.)

Opening ceremonies were held on 9 December 1889. President Benjamin Harrison was a special guest of honor, and a musical program was presented by Adelina Patti, opera's reigning diva. Patti pronounced, "The Auditorium is perfect. The acoustics are simply perfect," and everyone agreed.[63] Architectural

2.9

Auditorium Theater, Chicago (Adler & Sullivan, 1889). Interior, looking toward the stage. The Auditorium Theater was renowned for its excellent acoustics. Architect Dankmar Adler contested his reputation as an expert in acoustics and was ultimately unable to explain why his buildings sounded so good. *Auditorium Building* (Chicago: J.W. Taylor, c. 1890), p. 15. Courtesy Marquand Library of Art and Archaeology, Princeton University.

2.10

Auditorium Theater, Chicago (Adler & Sullivan, 1889), looking toward the rear balconies. In this photograph, the two uppermost balconies have been blocked off by movable partitions (the upper one curved, the lower one flat), thereby reducing the capacity of the hall from over 4,000 to about 2,500. *Auditorium Building* (Chicago: J.W. Taylor, c. 1890), p. 17. Courtesy Marquand Library of Art and Archaeology, Princeton University.

critic Montgomery Schuyler wrote, "It is pleasant to know that in this instance the science of acoustics, which so many architects deny to be for their purpose a science at all has been vindicated, and that the auditorium is in fact an excellent place in which to hear."[64]

Adler articulated his ideas on theater acoustics in a paper that he read to the American Institute of Architects in 1887. He offered advice on situation, construction, fireproofing, lighting, and ventilation, and concluded with the caveat that "all of these will be as naught unless the acoustic properties are such as to permit the easy and distinct transmission of articulated sound to its remotest parts."[65] In order to secure this effect, Adler proposed that the architect should avoid hard, smooth surfaces, and instead design well-broken walls and ceilings arranged to direct the sound toward the audience. The proscenium should be low, with the width and height of the hall increasing toward the rear, to promote the passage of sound.

Adler later justified these recommendations with explanations that drew upon the scientific language of the conservation of energy, but it is not apparent that the science of energy actually helped him to generate his designs. According to Sullivan, Adler's success in architectural acoustics was intuitive. "It was not a matter of mathematics, nor a matter of science," he explained. "There is a feeling, perception, instinct, and that Mr. Adler had. Mr. Adler had a grasp of the subject of acoustics which he could not have gained from study, for it was not in books. He must have gotten it by feeling."[66]

Adler himself described his technique, not as an instinctive one as Sullivan portrayed it, but as a simple program of independent thought and action. In 1894, he warned his fellow architects that he would not provide "a repository of historical information about the theaters of the past, nor a description or critical disquisition upon the theaters of the present day, nor yet a compendium of scientific formulae for solving the various problems of theater design." "With a view to stimulating original and independent thought and action," he explained, "I shall call attention to certain facts and conclusions, the recognition and formulation of which are within the reach of every intelligent observer and of every industrious student of objects and events."[67] To Adler, the theater was an "organic whole," and he took issue with those who would design a structure "in strict accordance with the tenets of any 'style,'" then leave the resolution of practical problems to "engineers and 'specialists.'"[68] He even contested his own reputation as an "alleged expert," and proposed that anyone capable of clear and incisive thought could join the ranks of such experts.[69]

But here, too, Adler's ideas were not widely shared by his colleagues. As early as 1811, Benjamin Latrobe had called for "a system by which an architect could be guided in his design,"[70] and throughout the century, architects had echoed this plea for experts to provide them with a set of "fixed rules."[71] Most shared the willingness of architect Rudolph Markgraf "to buy any books, articles, pamphlets or liter[a]ture setting force [sic] a practical method whereby to make sure of the successful properties of an Auditorium, or to employ the service of experts, if there are such experts, and if the services of such experts or specialists, can be secured at a reasonable fee and with an assurance on their part of satisfactory results."[72]

Adler's assertion that every architect could be his own acoustical expert fell on deaf ears, and Adler's success in this field remained uniquely his own. While he used the language of science to describe his approach to the problem of acoustics, he failed to provide a scientifically based system of design, and there was no means by which he could share his success with others. Adler passed away in 1900, and his acoustical expertise died with him. At the time of his death, however, architects were suddenly presented with a new means by which to achieve that success for themselves. Just a few pages away from Adler's obituary in the *American Architect and Building News*, American architects would encounter the first of a series of papers on acoustics by Wallace Sabine. Like Adler's intuitive approach, the system that Sabine outlined would consistently produce acoustically successful structures. But Sabine would additionally succeed where Adler had failed, by offering architects a compendium of scientific formulae that he, as a specialist, could simply and easily apply to their designs.

III Wallace Sabine and the Reverberation Formula

Wallace Sabine was born in 1868 in Richwood, Ohio. He was an intelligent child with an ambitious mother who apparently demonstrated an "abnormal conscientiousness in the exercise of her maternal duties."[73] Mrs. Sabine was certainly intent upon providing Wallace with every opportunity to develop his abilities. She enrolled her young son at Ohio State University, where he studied physics with Thomas Corwin Mendenhall and graduated in 1886 at the age of eighteen. Mrs. Sabine then left her less ambitious husband behind and moved with her son and daughter to Boston so that both could continue their studies, Wallace at Harvard University and his sister Ann at the Massachusetts Institute of Technology.[74]

Sabine received his M.A. from the Department of Physics at Harvard in 1888, and he subsequently collaborated with his senior colleague John Trowbridge on a series of studies exploring different aspects of electricity.[75] One investigation followed the research of Heinrich Hertz, who had recently produced the first evidence for the existence of electromagnetic waves. Hertz's work had drawn upon analogies to sound, and Trowbridge and Sabine followed suit when they concluded that Hertz's equations did not fully represent the behavior of electrical oscillations in air:

> Since the latter writer has taken the term *resonance* from the subject of acoustics, and has given it a new significance in relation to electrical waves, we are tempted to draw also an analogy from the subject of sound. Laplace showed that the discrepancy between the value for the velocity of sound in air calculated from the theoretical equation, and that obtained by experiment, was due to a transformation of energy in heating and cooling the air during the passage of the sound wave. Our experiments on the transmission of electrical waves through the air show also that the values calculated from the theoretical equation do not agree with the experimental values. The discrepancy, we believe, can be explained also by a consideration of the transformation of energy in the dielectric.[76]

Almost fifty years earlier, Joseph Henry's exploration of the acoustical properties of materials had constituted an early foray into the new energetic physics. Now, physicists like Sabine thought nothing of drawing upon the properties and principles of energy to connect phenomena as diverse as light, heat, electricity, and sound. Sabine was studying electricity, however, not sound, and this analogical thinking was about as close as he came at this time to the science of acoustics.[77] When he turned to acoustics just a few years later, however, and initiated what would become a lifelong investigation of the behavior of sound, this energetic framework would prove crucial in shaping his work.

In 1895, Sabine was asked by President Eliot to improve the faulty acoustics of a university lecture hall in Harvard's new Fogg Art Museum. The room was too reverberant, generating such a prolonged echoing of sound that a speaker's voice was unintelligible to the listeners who gathered there to hear it. (See figure 2.11.) Disappointed with this loss of valuable teaching space, Eliot asked Sabine to find a way to reduce the reverberation in the room. He suggested that Sabine develop a quantitative measure of acoustical quality, in order to compare the faulty room with Harvard's acoustically superb Sanders Theatre. Eliot hoped that the new hall could then be altered to match the acoustics of the theater.[78]

2.11
Lecture Hall, Fogg Art Museum, 1895 (since demolished). Harvard's president Charles Eliot assigned the task of improving the acoustics of this excessively reverberant room to Wallace Sabine, a young assistant professor of physics at the university. Courtesy of the Harvard University Art Museums, © President and Fellows of Harvard College.

It was not obvious to Sabine what that measure should be, as the measurement of sound was a problem that had long challenged acoustical experimenters. Throughout the past century, scientists had approached this problem primarily by attempting to render visible acoustical phenomena. Sabine initially adopted this strategy and employed a variant of Rudolph Koenig's "dancing flame" device to study the sound in the Fogg Lecture Room, but there was no useful way to interpret the results. Sabine thus abandoned all attempts to look at sound, and instead chose the seemingly obvious, but long neglected, alternative of listening to it. He discovered that "the ear itself, aided by a suitable electrical chronograph," gave "a surprisingly sensitive and accurate method of measurement."[79] What Sabine chose to measure was the time of reverberation: the duration of audibility of residual sound as it echoed through the room and slowly died away.

Sabine's technique consisted of sounding a source, an organ pipe with a pitch of 512 cycles per second (cps), until a steady volume of sound was achieved in the room. He then shut off the source of sound and listened to the residual sound, or

2.12

Experimental apparatus employed by Wallace Sabine in his investigation of reverberation. The large tank of compressed air was used to sound the organ pipe mounted on top of it. Sabine then shut off the air supply and listened to the continuation of sound, or reverberation, until it was no longer audible. The chronograph on the table recorded the interval, or reverberation time. Wallace Sabine, *Collected Papers on Acoustics* (Cambridge, Mass.: Harvard University Press, 1922), p. 15.

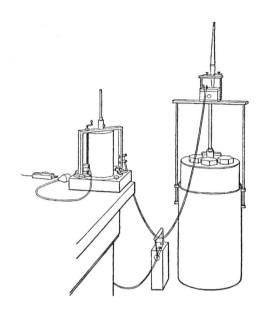

reverberation, until it was no longer audible. A torsion pendulum silently recorded the duration of audibility to hundredths of a second. (See figure 2.12.)

Sabine carefully measured the reverberation times of the Fogg Lecture Room and Sanders Theatre, and he studied numerous other rooms throughout the Harvard campus, as well as in Cambridge and Boston. In order to minimize the disturbing effects of streetcars, students, and other sources of noise, he conducted all of his research late at night.[80] Sabine emphasized to his undergraduate students the importance of experimental precision and accuracy, and he clearly practiced what he preached. He once threw out over three thousand measurements, representing several months' work, after determining that the clothing worn by the observer (himself) had a small but measurable effect upon the outcome of his experiments. Subsequently, he always wore the same outfit ("blue winter coat and vest, winter trousers, thin underwear, high shoes") when experimenting.[81]

Sabine measured the reverberation times of rooms as he found them, and he additionally manipulated those reverberation times by introducing different quantities of sound-absorbing materials. The removable seat cushions from Sanders Theatre proved conveniently portable and standardized absorbers of sound, and Sabine could be glimpsed on any given night (if one happened to be out between midnight and four o'clock in the morning) lugging heavy stacks of cushions across the dark campus in order to make his measurements.

Sabine's experimental method derived from his earlier collaborations with John Trowbridge, and was based on his fundamental assumption that sound, like virtually all other physical phenomena, was best defined as a body of energy. When Sabine studied electrical phenomena, he had focused on transformations of electrical energy in the material through which it passed. Having now turned to acoustical phenomena, Sabine retained that focus and based his examination on the transformation of sound energy in a room into heat and motion by the architectural materials of which the room was constituted. It is not evident that Sabine knew of Joseph Henry's earlier studies, but he shared Henry's emphasis on energy and materials. Sabine's work differed, however, in that the practical application of his results was always foremost in his mind.

Sabine's energetic treatment of sound was nonetheless insufficient to generate the quantitative understanding that he sought. Indeed, for a long time he was not sure what to do with his measurements, except to keep making more of them. After several years of experimentation and thousands of hours devoted to the painstaking collection of data, he was still unable to derive a fundamental mathematical relationship between the architectural properties of a room and its reverberation time. Until he had achieved that understanding, Sabine would not consider his work complete. Meanwhile, the Fogg Lecture Room remained unusable and unused.

By 1897, President Eliot had run out of patience. When he prompted the young professor for a progress report, Sabine responded, "I certainly hope to bring it to success in time, but only after a variety of experiments and a training of my hearing which will require several years, and the working of some rather remote side issues."[82] Eliot's own response was now unequivocal: "You have made sufficient progress to be able to prescribe for the Fogg Lecture Room, and you are going to make that prescription."[83] Thus forced, Sabine had panels of sound-absorbing felt hung on various wall surfaces in the lecture room, and the auditorium was finally usable, although far from the acoustical equivalent of Sanders Theatre.

The conclusion of this episode might have signaled the end of Wallace Sabine's work on acoustics.[84] It was at this time, however, that Henry Higginson approached Charles Eliot to solicit scientific advice on his new concert hall, and Eliot passed Higginson's request on to Sabine. Knowing the limitations of his understanding of sound, Sabine was initially reluctant to undertake this important new assignment. According to his biographer, he went home that evening and "devoted himself feverishly to a perusal of his notes, representing the labors of the preceding three years. Then, suddenly, at a moment when his mother was watching him anxiously, he turned to her, his face lighted with gratified satisfaction, and announced quietly, 'I have found it at last!'"[85]

What Sabine found was that when he plotted the quantity of Sanders Theatre seat cushions (x) versus the corresponding reverberation time for a room (y), the resulting graph was a rectangular hyperbola, a standard mathematical curve characterized by the equation:

$$x\,y = k,$$

where k is a constant. Sabine had graphed his data before,[86] but this time, by extrapolating beyond the points representing data that he had collected, he was able to see his experimentally derived fragment as part of a larger curve, a hyperbola. (See figure 2.13.) Sabine's earlier preoccupation with the precision and accuracy of his data points had prevented him from seeing this curve. Only after he had been forced to stop experimenting was he able to consider the data at hand without thinking about how to improve it or to collect more of it. Only then did he discover the hyperbolic relationship.

Sabine realized that his discovery was a breakthrough for his understanding of reverberation. Now eager to assume responsibility for the acoustics of Higginson's new music hall, he immediately wrote to President Eliot:

2.13
Wallace Sabine's plots of reverberation time versus the amount of sound-absorbing material in a room, 1900. The first graph shows his experimentally derived data. The second graph shows how he extrapolated this curve to discover the hyperbolic relationship between the two quantities. Wallace Sabine, *Collected Papers on Acoustics* (Cambridge, Mass.: Harvard University Press, 1922), pp. 21, 22.

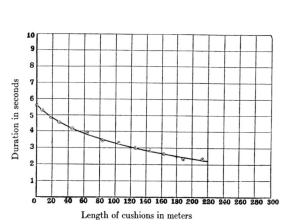

FIG. 5. Curve showing the relation of the duration of the residual sound to the added absorbing material.

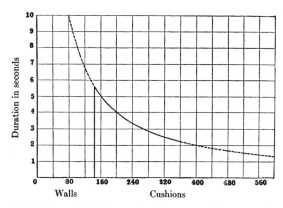

FIG. 6. Curve 5 plotted as part of its corresponding rectangular hyperbola. The solid part was determined experimentally; the displacement of this to the right measures the absorbing power of the walls of the room.

> When you spoke to me on Friday in regard to a Music Hall I met the suggestion with a hesitancy the impression of which I now desire to correct. At this time, I was floundering in a confusion of observations and results which last night resolved themselves in the clearest manner. You may be interested to know that the curve, in which the duration of the residual sound is plotted against the absorbing material, is a rectangular hyperbola with displaced origin; that the displacement of the origin is the absorbing power of the walls of the room; and that the parameter of the hyperbola is very nearly a linear function of the volume of the room. This opens up a wide field.[87]

Ever the experimenter, he added, "It is only necessary to collect further data in order to predict the character of any room that may be planned, at least as respects reverberation."[88]

Sabine's development of this wide field resulted, by 1900, in a comprehensive and quantitative analysis of reverberation.[89] He initially represented his hyperbola with the equation:

$$(a + x)\, t = k,$$

where

- a = absorbing power of room (walls, ceiling, etc.),
- x = absorbing power of materials added to the room,
- t = reverberation time, and
- k = the hyperbolic constant.

In this form, Sabine's equation differentiated the absorbing power of the room itself (a) from the absorbing power of the materials added to it (x). This distinction reflected his experimental practice, in which he first measured the reverberation time in a room, then introduced additional sound-absorbing objects to alter that reverberation time. As his focus moved away from experimentation and toward a fuller understanding of the mathematical relationship itself, the distinction between these different types of absorbing factors would become less significant.

Sabine initially expressed the total absorbing power of each room in terms of its equivalent in Sanders Theatre seat cushions. While this unit of absorption was convenient for Sabine himself, it was clearly problematic as a more general scientific standard, and Sabine replaced it with a new "open-window unit" of absorption. This unit was equivalent to the complete absorption of sound energy provided by an open window one square meter in area. Since all energy impinging on such an opening would escape to the space beyond, with no reflection back into the room, the unit represented one square meter of a perfectly absorbent material. "Hereafter," Sabine reported, "all results, though ordinarily obtained by means of cushions, will be expressed in terms of the absorbing power of open windows—a unit as permanent, universally accessible, and as nearly absolute as possible."[90]

Sabine next broke down the total absorbing power of a room into its individual components, including such items as plaster walls, wooden floors, rugs, and curtains. He expressed the absorbing power of each component with the quantity:

$$a_n s_n,$$

where

a_n = "coefficient of absorption," or absorbing power per unit area of material n, and
s_n = total surface area of material n in the room (in square meters).

Now, the total absorbing power of any room could be represented by the quantity:

$$(a_1 s_1 + a_2 s_2 + \ldots + a_n s_n).$$

For any given room, Sabine could experimentally derive the value of this sum by measuring its equivalent in Sanders Theatre seat cushions. He also knew, after making some measurements, the surface area of each different material in the room. His task was thus to determine the absorption coefficients of all those different materials. To accomplish this, Sabine set up systems of equations representing different rooms, each of which contained a different proportion of a range of materials. When he had as many equations as he had unknown coefficients, Sabine was able to solve the equations and determine the values of the different absorption coefficients. Once determined, the coefficient for a given material was available for any future calculation, and Sabine published tables of these coefficients for others to use.[91] Sample values included:

Open window 1.000
Wood-sheathing (hard pine) 061
Plaster on wood lath 034
Plaster on wire lath .033
Glass, single thickness 027
Plaster on tile 025
Brick set in Portland cement 025

These numbers may generally be interpreted as indicating the percentage of energy absorbed by each type of surface when it is exposed to sound. In other words, every time a body of sound energy encounters a surface of plaster on tile, 2.5 percent of that energy will be absorbed by the material, and 97.5 percent of the energy will be reflected off that surface back into the room. The complete absorption of an open window was represented by a coefficient of 1.00, or 100 percent.

Sabine's next task was to determine the value of the hyperbolic constant, k, for each room. By comparing hyperbolae for different rooms, he determined that the constant was directly proportional to the volume of the room. Before this proportion could be satisfactorily derived, however, Sabine had to deal with a difficult complication. His hyperbolae varied slightly from pure form in a systematic manner, and he attributed this variation to the lack of a constant initial intensity of sound in his experiments. "Each succeeding value of the duration of the residual sound was less as more and more absorbing material was brought into the room," Sabine explained, "not merely because the rate of decay was greater, but also because the initial intensity was less."[92] The lack of a suitable source, one that could generate sound of a constant intensity no matter what the condition of the room, led Sabine into a complicated side-investigation to correct for the variations that he could not eliminate or control.[93] He ultimately determined that the hyperbolic parameter k was proportional to the volume of a room according to the equation:

$$k = .164\ V.$$

Sabine's equation could now be written in the form:

$$t = \frac{.164\ V}{\Sigma\ (a_n\ s_n)},$$

where:

t = reverberation time (in seconds),
V = volume of room (in cubic meters),
a_n = absorption coefficient of material n, and
s_n = surface area of material n (in square meters).

This formula could now be used to predict the reverberatory quality of a room in advance of its construction, a privilege long sought, but never before enjoyed, by architects or their clients. The absorption coefficients of commonly employed building materials were already determined and tabulated, and values for V and s_n could be calculated off blueprints or other scaled drawings. With these known quantities in hand, the equation could generate the unknown quantity t, the reverberation time of the proposed room.

If the reverberation time that resulted from such a calculation were deemed unsatisfactory, an architect needed only to modify his design—changing the overall volume of the room, or the type or proportion of materials employed within it—until a satisfactory result was achieved. With this equation, Sabine had finally achieved the fundamental, quantitative understanding of reverberation that he had long sought, and he now welcomed the opportunity to work with Charles McKim on the design for Henry Higginson's new music hall.

When Sabine first met with Higginson in January 1899, to review McKim's design, he was unable to estimate on the spot its prospective reverberation time, as it took some time to calculate the volume of the room and the different surface areas of materials from the drawings. He nonetheless offered a number of preliminary suggestions. Most significantly, as Higginson reported to McKim, "Professor Sabine thinks the hall altogether too long. How long it should be he does not venture to say, considering that partly a matter of experiment and partly a matter of calculation, which he has not yet reached, but he is very much afraid of the long tunnel which we have laid out."[94]

While the reverberation time that Sabine later calculated from this design is not recorded, it appears not to have been in line with Higginson's acoustical criteria as embodied in the old Music Hall and the Leipzig Gewandhaus. In March, McKim informed Higginson that he would revise his design, following Sabine's suggestions. "It will be no improvement to the proportion of the large hall to cut down its length," the architect admitted, "but if, acoustically, you consider that you have reason to believe that it will be better, we shall not oppose."[95] The result was to reduce the overall volume of the hall, and thus also its reverberation. In order to maintain the original seating capacity, McKim followed Sabine's suggestion to add a second gallery to the one he had originally specified.

In his published account of the derivation of his reverberation equation and its application to the design of Boston's new music hall, Sabine outlined how he verified that this new plan would achieve the desired acoustical result.[96] He obtained scaled drawings of Boston's old Music Hall and the Leipzig Gewandhaus and he calculated their reverberation times from the data that he read off these drawings; 2.30 seconds for the former, and 2.44 seconds for the latter. (See figure 2.14.) He then turned to McKim's revised plans for the new hall, calculating its overall volume, as well as the total surface area of each of the

2.14

Architectural sections of the Leipzig Gewandhaus, the Old Boston Music Hall, and the New Boston Music Hall (Symphony Hall). The two older structures served as acoustical models for Symphony Hall. Sabine analyzed their designs and used his reverberation formula to ensure that the new hall would possess the same amount of reverberation as the models. Wallace Sabine, *Collected Papers on Acoustics* (Cambridge, Mass.: Harvard University Press, 1922), p. 66.

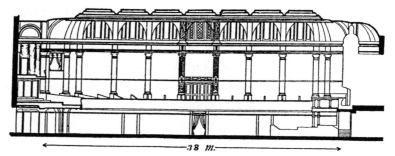

FIG. 20. The Leipzig Gewandhaus.

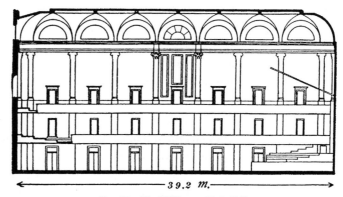

FIG. 21. The Old Boston Music Hall.

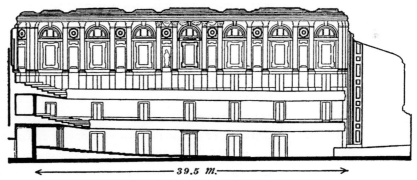

FIG. 22. The New Boston Music Hall.

THE ORIGINS OF MODERN ACOUSTICS

different materials out of which it was constructed, including plaster on lath, plaster on tile, glass, wood, and draperies. He also factored in the highly absorbent surface that the audience and orchestra members would constitute when the house was filled to capacity. Plugging all these data into his equation, he determined that McKim's hall would have a reverberation time of 2.31 seconds. The closeness of this value to those of the other halls ensured that the new hall would faithfully reproduce the amount of reverberation present in those acoustical exemplars. Sabine's technique enabled McKim to re-create the sound of past structures without having to re-create the structures themselves, and Sabine highlighted this fact when he emphasized that "neither hall served as a model architecturally."[97]

Sabine, McKim, and Higginson were in constant contact over the course of 1899 and 1900, working out the details of design and addressing new issues that arose during the construction of the hall. Sabine advised on questions ranging from where to place the organ pipes to what kind of seats should be installed.[98] Many of the questions that he addressed could not be answered simply by churning out another reverberation calculation, and he clearly drew on a more general knowledge of sound that he had gained during his years of research. Sabine even recognized the role of audience psychology in affecting judgments about the acoustical quality of the hall. When asked if a wood lining should be applied to the stage area, he informed Higginson that the small quantity of wood in question would not significantly affect the acoustics one way or another. He noted, however, that, "subjectively even this small display of wood will increase the acceptability of the hall to the public by gratifying a long established—and not wholly unreasonable—prejudice."[99]

Sabine's mathematically quantified understanding of the behavior of sound provided the basis of expertise that accredited all his suggestions, even those for which the reverberation equation itself did not provide a direct answer. It also inspired the confidence with which he rendered his advice. That advice was attractive to McKim not only because it was perceived to be scientifically authoritative, but also because it did not significantly constrain the architect's creative freedom. Sabine did not dictate one best form; his technique was applicable to any form or style of building. Although based on the manipulation of building materials, here, too, his technique laid out no strict prescriptions or proscriptions. With Sabine's technique, any desired acoustical end could be achieved through an endless variety of architectural means. If an architect were committed to one particular aspect of his design, he could simultaneously ensure the desired acoustical result by manipulating other aspects of it. This enabled Sabine to work easily with Charles McKim, as well as with many other architects who would subsequently seek his advice.

At the same time, his method clearly assigned responsibility for the final acoustical result to the consulting scientist, not the architect. Whereas Dankmar Adler had encouraged architects to take responsibility for the acoustical consequences of their designs, few shared his point of view. What architects wanted was a means by which to delegate this responsibility to an outside authority, and this was exactly what Sabine offered. Sabine's expertise was thus attractive to architects like McKim not simply because he provided an answer to long-standing questions of acoustical design, but also because his particular answer was one that architects were happy to hear.

Sabine's method not only satisfied McKim's desire to design good sound for Symphony Hall, it also served the needs of the audience who came to hear that sound. Why were the acoustics of Symphony Hall so important to those who gathered there on opening night? The development of musical culture over the past century had rendered the act of listening increasingly important, and this new culture of listening culminated in America just as Symphony Hall opened its doors to receive its audience.

IV Music and the Culture of Listening in Turn-of-the-Century America

During the eighteenth and early nineteenth centuries, music in America was performed primarily by amateurs who made music for their own enjoyment.[100] By around 1850 this local fare was regularly supplemented by the occasional performances of professional musicians—primarily visitors from Europe—who were now touring the larger cities of the United States. In 1843 the Philadelphia diarist Sidney George Fisher noted, "A love of music has grown up in this country within the last few years, and the artists of Europe find it a profitable field of operations."[101]

American-born artists as well as traveling Europeans began to profit by performing before growing audiences of eager listeners. Louis Moreau Gottschalk, perhaps the nation's first internationally recognized virtuoso, not only played in big cities like New York and Boston, but also carried his music to the hinterlands. "What singular audiences I meet with!" he proclaimed. "You can imagine what the population must be in little towns that, founded only seven or eight years ago, nevertheless give receipts of three or four hundred dollars, and sometimes more. The other evening before the concert, an honest farmer, pointing to my piano, asked me what that 'big accordion was.' He had seen square pianos and upright pianos, but the tail bothered him."[102]

While a grand piano was a novelty to the farmers of Indianapolis, in the larger cities the instruments were now commonplace. In fact, musical offerings had proliferated in American cities to the point where demand for concert space often surpassed the available supply. In Philadelphia, the 1852 charter for the new Academy of Music stated that "it cannot have escaped the observation of the merely casual observer, that the taste for and cultivation of music have rapidly increased among us within the last ten years, and we believe such an establishment as we are now laboring to obtain, would do more than anything else in guiding, fostering and sustaining a love for the most refining and humanizing of all the arts."[103] The charter also referred to the advantages "in the way of business as well as of pleasure" that the opera house would secure for the city. The population of Philadelphia then stood at half a million, and it was hoped that "all of these persons, whether possessed of a taste for music or not, would resort to a place of cheap and elegant amusement."[104] The project was as much a commercial venture as a cultural one, and openly so. The merchants who had incorporated to finance the new construction were not wealthy enough to make good any deficits that might result from poor attendance, and they were willing to accommodate any kind of performance that promised to sell tickets. At the same time, however, romantic notions of the ennobling nature of music were beginning to be heard, and these new ideas would increasingly be attached to both the performance and audition of music.

The phenomenon had already been under way for over a century in Europe. When Count Francesco Algarotti had petitioned for an acoustically controlled architecture in 1762, he pleaded as vehemently for a new attitude toward listening to accompany the sound. Algarotti longed for a rationally designed theater that would no longer constitute "a place destined for the reception of a tumultuous assembly, but as the meeting of a solemn audience."[105] His desire to control sound was paired with an equally strong desire to control the behavior of the audience. Algarotti himself already constituted such a concerted listener, and he sought an architectural means to engender this attentive way of listening in all concertgoers.

Over the course of the next century, the transformation that Algarotti longed for would indeed occur. This change was the result of complicated social and cultural forces that have been richly explored by Richard Sennett and James Johnson.[106] Urbanization, the decline of the aristocracy, the rise of the middle class, the romantic movement in arts and letters, and the development of symphonic music are just some of the factors that contributed to the gradual transformation of "the perpetual chattering of the company, in visits being made from one box to another, in supping therein, and . . . gaming"[107] into a rapt preoccupation with what was taking place on stage. In America, as Lawrence Levine has shown, these phenomena came fully into play in the nineteenth century, and resulted, by the end of that century, in a musical culture that was religious in its intensity. Listening now became a way to worship at the temple of great art.[108]

This new way of thinking about music was first and most voluably heard in Boston. At mid-century, John Sullivan Dwight began to use his *Journal of Music* "to articulate tirelessly the conception of a sacralized art: an art that makes no compromises with the 'temporal' world; an art that remains spiritually pure and never becomes secondary to the performer or to the audience; an art that is uncompromising in its devotion to cultural perfection."[109] When Boston's Music Hall opened in 1852, *Dwight's Journal* sang its praises:

> Oh fair retreat, where even now
> Art's consecrating footprints shine,
> Where Song, with her imperial brow,
> Shall hold her sway by right divine!

The commemorative poem ended several stanzas later, with "all earth's people" "kneeling near the shrine of Song."[110] But Dwight's lofty ideals for music were not yet a reality in America. Indeed, when the Music Hall was nostalgically described many years later, it was hardly remembered as a cultural shrine:

> What a versatile place was the old Music Hall,
> With its concerts and sermons and dances and all!
> Wendell Phillips has lectured there, Patti has sung,
> While the Warren Street Chapel shows captured the young.
> Crowds were drawn here by Theodore Parker, but some
> Were attracted by Mr. and Mrs. Tom Thumb.
> For a function, a fight, and a fireman's ball
> Might occur the same week in the old Music Hall.[111]

The concert halls and opera houses built in America at mid-century pointed toward a new cultural ideal but did not yet attain it. Audiences still chatted during concerts, or even whistled along (to show that they knew the tune, Gottschalk claimed), and the distinction between professional and amateur was not always clear. A rich furrier might rent out the New York Academy of Music and stage his own production of *La Traviata*, or a local shoemaker might choose to accompany a visiting virtuoso on his flute.[112] During the latter half of the century, however, musicians and music lovers like John Sullivan Dwight undertook a campaign to educate Americans to appreciate great music, and to approach it with an attitude of humility and respect.

When the French conductor Louis Antoine Jullien toured America in 1853 and 1854, he attracted large crowds by convening massive choruses and staging musical novelties like the *Fireman's Quadrille*, "which included fireworks and a simulated fire so realistic that it induced hysterical screaming and fainting spells among some in the audience."[113] When it came time to perform the music of Beethoven, however, Jullien demonstrated his reverence by donning white gloves and a special jeweled baton, and he encouraged his audiences to treat the music that his baton brought forth with equal respect.

Jullien's violinist Theodore Thomas disliked such gimmicks, and when he began touring with his own orchestra in the 1860s, he worked to develop in American audiences an appreciation for good music free of such spectacular trappings. When Thomas was appointed head of the new Chicago Symphony Orchestra in 1889, he was finally in a position to develop a relationship with a permanent ensemble of musicians as well as with a permanent audience, and he undertook to train both with equal vigor. In Boston, too, after years of pleading by John Sullivan Dwight, a permanent symphony orchestra was finally established under Higginson's sponsorship, and a series of stern German conductors similarly demanded as much of their audiences as they did of their musicians.

By 1900, these efforts had born fruit and Dwight himself, not to mention Count Algarotti, would have been pleased with the decorum and the concentrated attention to listening that now characterized the behavior of concertgoers in America. The concert hall became a solemn place, and listening became serious business. Applause was now restricted to specific places in the program, and spontaneous outbursts were discouraged. Conductors were even known to stop in the middle of a piece and reprimand audiences that talked or made other distracting noises during a performance.[114] At the 1891 opening of Carnegie Hall in New York, "a poor little girl who chanced to sneeze was regarded as a fiend incarnate."[115] A reporter for the *New York Herald* noted that the audience was "most interesting as a study of music lovers not under the pressure or mandates of fashion. The women in the boxes were in evening dress, and many were the same who nightly ornamented the *loges* of the Metropolitan Opera House, yet there was a decided change in demeanor. There was no idea of chatter or conversation."[116]

On opening night at Symphony Hall, "an inspired Harvard student" startled the audience by leaping up from his seat and calling for a volley of cheers for Henry Higginson. The audience chose not to respond, so the young man cheered alone then returned to his seat, where he sat quietly for the remainder of the program.[117] Control was the key; it was not meant to be fun. Theodore Thomas considered concertgoing "an elevating mental recreation which is not an amusement,"[118] and the *Boston Evening Transcript* editorialized proudly that "Boston does not take her music frivolously, but as a service, an education."[119]

Even in the realm of domestic music making, this sober new attitude toward music prevailed. Children were given music lessons in order to instill character and discipline, not to inspire creativity and joy; and the young women who performed in the parlors of Victorian America similarly demonstrated virtue more than virtuosity.[120] When the phonograph began to make itself heard, John Philip Sousa feared that "no one will be ready to submit himself to the ennobling discipline of learning music," and all that would be left was "the mechanical device and the professional executant."[121] But domestic music making was already on the decline, part of a larger phenomenon referred to as "The Decline of the Amateur." In 1894, the *Atlantic* magazine recalled that the adjective "amateur" had formerly signified "respect, dignity and worth." But now, "amateur has collided with professional, and the former term has gradually but steadily declined in favor; in fact, it has become almost a term of opprobrium. The work of an amateur, the touch of the amateur, a mere amateur, amateurish, amateurishness,—these are different current expressions which all mean the same thing, bad work."[122]

As amateurs gradually abandoned their own music making and listened increasingly to professional musicians, a wide chasm opened between the two groups. Amateurs who continued to make music at home found it difficult to imitate the pyrotechnic performances of turn-of-the-century virtuosi like Ignacy Jan Paderewski and Fritz Kreisler. Sheet music publishers did their best to bridge the gap, by offering "Brilliant but not Difficult"[123] versions of the most popular showpieces, but the effect of the discrepancy was gradually but effectively to silence many amateur performers of music. By the end of the century, countless parlor pianos had been replaced by automatic "reproducing" pianos or other mechanical devices that recreated the performances of great concert pianists.[124] The phonograph, too, as Sousa had feared, was now replacing self-made music with recordings by professional executants. The result of these trends was a new dissatisfaction with amateur music and, perhaps more significantly, a heightened engagement by amateurs with the experience of listening to professionals.

In 1910, for example, the social reformer Jane Addams noted a generational difference between her mother, who believed herself to have possessed musical talent but lacked opportunity to develop it, and Addams herself, who, in spite of all advantages in her youth to develop such a talent, knew herself to be lacking it. "I might believe I had unusual talent," she wistfully acknowledged, "if I did not know what good music was."[125] Concurrent with Jane Addams's youth, Edward Bellamy's best-selling novel *Looking Backward* fictionalized the same phenomenon. Bellamy told the story of Julian West, a wealthy young Bostonian who fell into a hypnotic sleep one evening in 1888 and awoke one hundred years later to find himself in the social utopia of late-twentieth-century America. West was offered music by his hostess, Miss Edith Leete:

> "Nothing would delight me so much as to listen to you," [he] said.
> "To me!" she exclaimed, laughing. "Did you think that I was going to play or sing to you? . . . Of course, we all sing nowadays as a matter of course in the training of the voice, and some learn to play instruments for their private amusement; but the professional music is so much grander and more perfect than any performance of ours, and so easily commanded when we wish to hear it, that we don't think of calling our singing or playing music at all."[126]

The music that Edith offered to Julian was a telephonic transmission of a performance that took place in one of the city's many music rooms, each "perfectly adapted acoustically to the different sorts of music."[127] Music performed by professionals in acoustically designed rooms represented the ideal for Bellamy, and for many others, in late-nineteenth-century America. The role of nonprofessionals, like Edith and Julian and the millions of Americans who read about them, was to listen intently and appreciate fully the sounds that they were privileged to hear.

Henry Higginson himself had gone to Europe as a young man hoping to become an accomplished musician. What he learned there was that he "had no talent."[128] Higginson subsequently fulfilled his love of music by sponsoring musicians more talented than himself, by listening carefully and critically to their performances, and by building a hall that would draw on scientific expertise in order to provide the best possible environment in which to listen. The two thousand others who gathered with Higginson on opening night shared his love of listening, as well as his concern over the quality of the sound that they heard.

V CONCLUSION: THE CRITICS SPEAK

Did Symphony Hall provide the acoustical environment so eagerly sought by the people who gathered there and listened so intently? The answer to this question was not immediately obvious to all who were present on opening night. William Foster Apthorp, music critic for the *Boston Evening Transcript*, dryly characterized the new building as "one of the prime fixed conditions of our hearing the larger forms of orchestral and choral music for the rest of our lives." He took very seriously his role as an arbiter of the acoustical quality of this fixed condition; so seriously, in fact, that he declined to discuss the sound of the opening night concert. Apthorp referred to McKim's and Sabine's "singleness of purpose," by which "their calculations kept but one object constantly in view: to adapting the hall to the use of the Symphony Orchestra, and to nothing else." He deferred judgment because oratorio, not symphonic music, had been performed. "I await the first symphony concert with impatience," he proclaimed, "for that will be the only real test."[129]

Apthorp's decision to withhold judgment also took into account the fact that the opening night concert had used an unusual arrangement of musicians on stage. To accommodate the large chorus required for Beethoven's *Missa Solemnis*, the first five rows of seats had been removed so that the stage floor could be extended out beyond the proscenium into the auditorium. In spite of the unusual arrangement, most critics were willing to submit their opinions of the acoustics of the hall, and their reviews were generally positive. The *Boston Herald* declared that "Symphony Hall's acoustic properties are all right, Hear, Hear!" and the out-of-town papers agreed. New York's *Evening Post* heralded the hall as "what very few concert halls are—a success acoustically," and suggested that, if an old myth that halls improved and mellowed with age proved true, it would not be surprising if "mellowing time made it a Stradivarius among halls."[130]

Henry Krehbiel, music critic for the *New-York Daily Tribune*, devoted considerable space to Sabine's work in his opening night review. "Hundreds of ears," he reported, were "alert this evening to learn whether the greatest of the problems that the construction of a music hall involves had been solved in this instance." Sabine's confidence in the result of his calculations struck Krehbiel as daring, but he concluded that it was both "justified and rewarded," for "the effects were most gratifying, and it can safely be said that for its purposes Boston has the most beautiful, appropriate and admirable hall in the United States."[131] Yet, Krehbiel suggested that until Sabine conducted a "scientific investigation after the fact," and made a precise measurement of the reverberation time in the hall, "the sceptic may not yet feel confounded." Sabine apparently never made this measurement, responding personally to Krehbiel that the only meaningful test of his work would come with the actual use of the hall.[132]

A few nights later, the first concert of the regular season was heard. The stage was restored to its normal configuration, and the orchestra was led by Wilhelm Gericke in a performance of standard works, including one of Higginson's favorites, Beethoven's Fifth Symphony. After this concert, Higginson wrote to Sabine, "Just a word to thank you for your pains and success in the Hall. Of both no doubt exists. I have never heard the music as now. You have proved here that the Science of Acoustics certainly exists in a definite form. You have done a great part of the Hall, and every one thanks you."[133]

The papers generally shared Higginson's sentiments. Philip Hale, of the *Boston Sunday Journal,* concluded that "doubt as to the acoustic properties of the hall were dispelled. Solo instruments were heard with delightful distinctness; the bite of the strings was more decided than in the old hall, and the ensemble was effective without muddiness or echo."[134] The *Sunday Herald* declared the hall "A Complete Success," noting that "The wholly favorable impression made by the acoustic qualities of the hall on the opening night was re-enforced last evening. Everything is heard with the most perfect distinctness, the contrasting timbres of the different instruments stand out clearly, and at no time, even in the heaviest fortissimos, is there any cloudiness of tone."[135] The *Herald* celebrated Sabine's work as "A Feat in Acoustics," and quoted extensively from his published article on reverberation in order to describe his work to its readers.

A new note of uncertainty was introduced, however, by other papers in response to this concert. The *Boston Post* reported that, while there was no difficulty in hearing throughout the hall, there seemed to be "less body" to the sound than had been the case in the old Music Hall. The reviewer suggested, however, that this might be due to the selections performed rather than to the hall itself.[136] William Apthorp, now finally prepared to pass judgment on the new hall, also measured its acoustical merits with ambivalence. Apthorp first noted the familiarity of the pieces on the program, "so one could give almost undivided attention to the effect of the music in the hall." As he listened to the opening number, Weber's overture to *Euryanthe*, he found the effect of the music disappointing: "Everything was clean-cut and distinct, the tone was beautifully smooth, and, so to speak, highly polished; but it had no life, there was nothing commanding and compelling about it." In contrast, the Handel Concerto for Organ that followed almost convinced him that the acoustics of the hall were "superb." But Beethoven's Fifth Symphony confirmed his initial reaction, and he reported that, while there was a "great distinctness of definition," the tone had "no body, no fulness; it is not searching; it is thin and ineffectual. Moreover, the hall itself seems perfectly dead to it, it does not awake to the orchestra's call and vibrate with it. Things that should sound heroic and awakening, seem merely polite and irreproachable."[137]

Apthorp suggested that Beethoven sounded as if he had appeared in "impeccable evening dress," freshly coiffed by the court hairdresser, the very picture of a "Brumellianly elegant" dandy, and it was obvious that the critic preferred his romantics unkempt and unruly. Still, Apthorp took pains to discount these early impressions. He emphasized that they were, above all, a reaction to the newness and unfamiliarity of the sound of the orchestra in the new hall. He confessed that he felt disoriented, seemingly in "some new musical country, never visited before, where old habits of listening needed reforming."[138] Apthorp noted that his tentative and preliminary judgments would be subject to future revision, and in his review of the next evening's concert of the Handel and Haydn Society, he did in fact revise those opinions. Now, he concluded that the effect of the music "left nothing to be desired."[139] But over time, Apthorp's fluctuating opinion of the acoustics of the hall stabilized into a decidedly critical viewpoint, and that criticism began to echo in the columns of other papers.

The *Musical Courier*, a national paper published in New York, came out strongly against the sound of the new building. Citing praise by the Boston press of Sabine's work, the *Courier* begged to differ: "We do not accept all that is said . . . as the acoustics on Saturday night were by no means satisfactory." The *Courier*'s criticism, however, was leveled not so much against the sound itself, but more philosophically against the idea that "science" could ever master anything as beautiful and ephemeral as great music:

> Sound is not music, but is merely one of music's utilizations. A voice or tone may sound scientifically correct at a given time in a given hall and may be measured and its formula fixed and established chronometrically or chronographically or in any

chronoform, but that sound or combination of sounds is not music. Music does not repeat itself; music is the moment, because music is art and art cannot be measured beforehand. . . . From the days of Pythagoras all kinds of experiments in acoustics have been facing the physicists and agitated the laboratories, but no clew has been discovered for such a science as can foretell with usual and necessary scientific accuracy how music will sound, and why not? Because if music could always sound as we before its issue could predict by formula $X + N = Y$, why then it would no longer be music.[140]

Apthorp had resisted the controlled character of the sound of Symphony Hall most strongly when it was applied to the impassioned strains of Beethoven. The reviewer for the *Courier* similarly, if more fundamentally, resisted the very idea of a scientifically controlled sound, as it contradicted his own romantic conception of the unpredictable nature of all music.

The criticism of the *Courier* represented an extreme, if revealing, reaction to the sound of Symphony Hall. Nonetheless, as time passed, a rising chorus of criticism could be heard. In March 1901, Apthorp noted, "there was much in the solo part that I could not hear well. Maybe the hall was again at fault; it is certainly not a brilliant hall,"[141] and papers that had initially approved of the sound of the hall now reported negatively. The same *Herald* that had pronounced the hall "A Complete Success" now referred to "the unfortunate acoustics of Symphony Hall,"[142] and the *Journal*, too, changed its opinion: "The acoustical properties, in spite of Mr. Sabine's brave pamphlet illustrated with diagrams and figures, are by no means satisfactory to either musicians or hearers."[143]

In May 1902, Henry Higginson received an unsolicited letter from a man named Edmund Spear, who offered his services "as an acoustician in aiding you with the remodeling of Symphony Hall which I understand has been undertaken."[144] Later that year, the writer Frank Waldo published a glowing account of Sabine's work on Symphony Hall. The *Boston Evening Transcript* excerpted Waldo's piece, and Apthorp amended a scathing postscript, condemning Sabine with perhaps the ultimate insult. He deemed "Mr. Sabine" incompetent "to express a musical opinion of any weight whatsoever," as Sabine came musically from "the amateur class." Apthorp continued, "We have not yet met the musician who did not call Symphony Hall a bad hall for music. Expert condemnations of the hall differ, as far as we have been able to discover, only in degree of violence."[145]

What did Sabine make of this expanding wave of criticism? Little evidence exists, but in a letter to Charles McKim written in May 1901, Sabine indicated that his first intimation of criticism had come just two weeks earlier, and he expressed surprise at the fact that initially positive reviews had now given way to criticism. He also took issue more specifically with what some listeners had identified as the cause of the worsening acoustics. Apparently, people were blaming the now-bad sound on the installation of statues into niches in the walls high above the second balcony. (See again figure 2.2.) The statues, which were cast plaster replicas of famous artifacts from antiquity, had been called for in McKim's original plans, but a lack of funds had prevented their procurement and installation in time for the hall's opening. They were gradually obtained and installed in the months after opening night, until this acoustical controversy brought the installations to a halt.[146] Sabine explained to McKim that the statues were part of the original plan "not only artistically in your scheme but acoustically," and he adamantly asserted, "The statues will not in the least affect the reverberation in the hall."[147]

Sabine also emphasized that he had not been the source of any musical judgment associated with the acoustical design of the hall. Reverberation, he acknowledged, was "a matter of taste." "Recognizing this," he explained to McKim,

> I sought the opinion of Mr. Gericke, and the Committee in regard to what halls were satisfactory in this respect and accepted this as the best available definition of the desired result. Then I made a special study that this above all things might be quantitative, investigated these halls, was struck by the nice agreement of the opinions expressed, and reproduced the condition in the present hall. On the certainty of my work in this respect I shall not yield."[148]

Wallace Sabine ultimately dealt with the highly subjective opinions of the critics and the public in the only way he could; he attempted to objectify them. In 1902, he embarked upon a study of "The Accuracy of Musical Taste in Regard to Architectural Acoustics," declaring this problem fundamental to any future work, "for unless musical taste is precise, the problem, at least as far as it concerns the design of the auditorium for musical purposes, is indeterminate."[149] Sabine divided the subject of architectural acoustics into two distinct lines of investigation. The first was based on the physical phenomena, and the second on their musical effect. "One is a purely physical investigation," Sabine elaborated, "and its conclusions should be based and should be disputed only on scientific grounds; the other is a matter of judgement and taste, and its conclusions are weighty in proportion to the weight and unanimity of the authority in which they find their source."[150]

To investigate the latter, Sabine had a committee of faculty members from the New England Conservatory of Music listen to piano music in five different rooms in the conservatory. He altered the reverberation time of each room by introducing varying amounts of sound-absorbing materials (the ever-useful Sanders Theatre seat cushions), and each committee member indicated when they felt each room sounded best. Sabine then evaluated the consistency of opinion expressed: the average optimal reverberation time for the five rooms was 1.08 seconds, and the average departure from this value was just 0.05 seconds. Sabine indicated that he found this high degree of accuracy in musical taste "surprising."[151]

By the time of this investigation, however, it appears that the general sentiment regarding the acoustics of Symphony Hall, if not that of William Foster Apthorp, had begun to return to a more favorable consensus. In February 1902, the chair of the statuary committee, Mary Elliot, wrote to McKim expressing her desire to resume installation of the statues in the hall. "A freind [sic] of ours," she informed the architect, "who is a Musician told me the other day that Gericke & the Musicians generally, are feeling very differently about the Acoustics of the Symphony Hall this winter, the Music sounds beautifully & they think that the general drying out of the Materials has made a great difference in the resonance."[152] It is unlikely that the drying or aging of the walls of the hall had any significant effect upon the sound. More likely, the musicians simply required time to become used to playing in the new hall. As they grew familiar with the sound of the space, they learned to adjust their technique in order to fill the space with the sound that they desired.[153]

In 1903, Theodore Thomas moved his Chicago Symphony Orchestra from Adler & Sullivan's Auditorium into the new Orchestra Hall designed by architect Daniel Burnham.[154] Thomas made clear that he would require a period of experimentation with his musicians in the new hall before he would be able to produce the sound he desired.[155] In Chicago, where the dominant personality was the conductor, the building was treated like a new instrument that Thomas had to learn to play. In Boston, in contrast, it was the owner Higginson, not any particular conductor, who defined the orchestra in the public mind.[156] Wilhelm Gericke's contribution was little acknowledged in early discussions of the acoustics of Symphony Hall, and the music that he created there was considered separately from the sound of the building itself.[157] Perhaps this distinction was a result of the fanfare over Sabine's work that had preceded the opening of the hall. It was a novelty for a scientist to be so involved in the creation of a new auditorium. How to distinguish the contribution of that scientist from all the other factors and players was an interesting new problem that appears to have been ignored.

It is also possible that the initial rejection and gradual acceptance of the sound of Symphony Hall was due to the fact that the audience required time to become used to that new sound. Apthorp had certainly acknowledged the discomfort of unfamiliarity in his early reflections upon the experience of listening in the new hall, and others may have shared his distress, perhaps without being fully aware of the reason for it. For whatever reason, as the sound of Symphony Hall grew familiar, listeners' displeasure did indeed dissipate.

While it is difficult to determine exactly when the criticism of Symphony Hall's sound was silenced once and for all, indirect evidence suggests that the hall's reputation was restored within just a few years of its opening. Sabine, for example, was soon in great demand as an acoustical consultant for architects from all over the country, and this would hardly have been the case if his work on Symphony Hall were considered a failure. McKim, Mead & White apparently never lost faith in his contribution to their work, and they were reenlisting his services as early as 1901.

At the time of his death in 1919, Sabine's eulogist could claim that the acoustics of Symphony Hall "have now been approved by the audiences of many years,"[158] and the reputation of the hall has only improved over the subsequent decades. In the 1950s, a plaque commemorating Sabine was installed in the foyer of the hall. The memorial calls attention to the building's historic status as "the first auditorium in the world to be built in known conformity with acoustical laws," but the hall itself offers its own testimony whenever music is performed within it, for Symphony Hall is considered today to be one of the best places in the world for listening to music.

The acoustical reputation of Symphony Hall is only one measure of Wallace Sabine's success, however, and for the story that follows, it is not necessarily the most important. Sabine's work succeeded in many different ways, for many different groups of people. For architects, he provided the "fixed rule" and the scientific expertise that they had long sought to guide and inform their acoustical designs. For audiences, his work endowed the spaces in which they gathered to listen with what most listeners considered to be a satisfying sense of control. And, for scientists like himself, Sabine opened up a wide new field of opportunity. His method established a research agenda and it identified new problems that now required solution. A new community of acoustical researchers would confront these problems, and would soon provide an even greater and more powerful range of solutions.

CHAPTER 3   THE NEW ACOUSTICS, 1900–1933

> Acoustics is a science of the last thirty years.[1]
>
> Dayton Miller, physicist, 1931

I INTRODUCTION

In 1901, James Loudon's presidential address to the American Association for the Advancement of Science outlined "A Century of Progress in Acoustics." Loudon opened by apologizing to the audience for his unusual choice of topic, confessing, "I am fully alive to the fact that this branch of science has been comparatively neglected by physicists for many years, and that consequently I cannot hope to arouse the interest which the choice of a more popular subject might command." "It is, however," he explained, "just because of this neglect of an important field of science that I conceive it to be my duty to direct some attention thereto."[2]

Less than thirty years later, William Eccles presented a similar address before the Physical Society in London. "The New Acoustics," according to Eccles, had "increased its bulk and scope enormously" since the turn of the century, having been invigorated by new techniques, new ideas, and "a new jargon for expressing these new things."[3] Whereas Loudon had hoped to stimulate interest in a field of study that he himself recognized as moribund, Eccles sought instead to enlist his colleagues in an ongoing and exciting new endeavor, to encourage British scientists to catch up with and join in on the vital and interesting work in acoustics that was primarily taking place in the United States.

The new vitality associated with acoustics circa 1930 was perceived not only by scientists. The public, too, had become "sound conscious,"[4] recognizing the important role that acoustical technologies and commodities now played in modern life. In 1931, children were encouraged to consider acoustical engineering as an exciting new answer to "Youth's Inevitable Question: 'What Shall I

Be?'" *Careers*, a series of publications outlining different occupations to schoolchildren, now included a pamphlet dedicated to this new field. The pamphlet described "innumerable opportunities" in this "pioneering profession," and predicted that, in the years to come, "the acoustical engineer will become more and more indispensable to civilization."[5]

Careers noted that architects "have been thoroughly won over to the science of acoustical engineering as an indispensable element in the design of a building,"[6] and the pamphlet offered advice on the college curriculum to be undertaken by an aspiring young acoustician. Courses in architectural acoustics were being taught at Harvard; the Massachusetts Institute of Technology; the Universities of Illinois, Iowa, and Indiana; and the University of California at Los Angeles. A graduate of any of these schools could then apply for employment to the many companies that manufactured acoustical materials; to architectural partnerships; to firms of contracting engineers; or to the American Telephone & Telegraph Company, "the greatest corporation in the world using the services of the acoustical engineer."[7] The student of acoustics was also encouraged to join the Acoustical Society of America, in order to make "valuable contacts among the outstanding men in his chosen profession."[8]

The Acoustical Society of America was organized in 1928, institutionally acknowledging the tremendous expansion of the field of acoustics that had occurred since the turn of the century. At its November 1932 meeting, the society's president Dayton Miller presented a special lecture on the history of acoustics, charting developments in the science of sound from the ancient ideas of Pythagoras and Aristotle to the work of Wallace Sabine. Sabine "laid the foundation" for the modern science of architectural acoustics, Miller explained, with his "epoch-making paper on 'Reverberation.'" Sabine had passed away in 1919, but Miller was certain that, had he survived another decade, he would "surely have been president of the Acoustical Society of America." "Probably not half of the members of the Society ever met him," Miller noted. "What a loss! He must not be thought of as an old man; had he lived to this day, he would be two years younger than your present president."[9]

To the members of the Acoustical Society of America, Wallace Sabine was a heroic figure from an already distant past. The transformations of the past three decades were so dramatic, acousticians hardly recognized the foundation upon which their field had been built. In order to understand how Sabine's work came so quickly to be perceived as a faint echo from a long-distant past, the development of the science of acoustics between 1900 and 1930 must be examined. By following Sabine's career after Symphony Hall, and by charting the careers of the men who followed him, it will be possible to understand just what was so new about "The New Acoustics."

During the first two decades of the century, Sabine continued his investigation of reverberation and he began to explore other aspects of the behavior of sound in rooms, including the transmission of sound through walls. He experimented with new kinds of tools for studying sound, he consulted with architects on a range of projects from large churches to private homes, and he collaborated with manufacturers of building materials on the design of new sound-absorbing materials. While he worked alongside architects and builders on the practical application of his science, as a scientist, Sabine always worked alone. Parallel to his solitary endeavors, however, a small community of acoustical researchers was beginning to take shape. The direction of their work gradually shifted away from the direction that Sabine had pursued, and after Sabine's death in 1919 this transformation would accelerate.

In the decade known as the Roaring Twenties, concern over the problem of city noise grew and the demand for sound control in buildings increased. The market for new acoustical building materials expanded, as did the need for consultants to oversee the installation of those materials. New industries dedicated to a range of acoustical products and services, especially the telephone and radio, became important sectors of the American economy and offered new opportunities and resources for the study of sound. The electroacoustic basis of these industries and their products impelled acousticians to work with, and think about, sound in new ways.

New tools for producing, modifying, and measuring sound transformed the scientific study of it. As acousticians became adept at manipulating microphones, amplifiers, loudspeakers, and the electrical signals that these devices employed, they began to reconceptualize acoustical phenomena as electrical phenomena. Electrical analogies now provided fruitful new ways to understand the behavior of sound. They provided a powerful sense of control, and they stimulated new ideas about what constituted "good sound." These analogies, along with the tools that had elicited them, constituted the innovative ideas and techniques that heralded Eccles's New Acoustics.

Wallace Sabine had been aware of these material and intellectual transformations, but during his lifetime these changes were just beginning to occur. At the time of his death, he stood tentatively poised between two worlds, uncertain about what the future of acoustics would hold. Acousticians who came of age during and after the First World War, in contrast, enthusiastically embraced that future. Young men like the physicist Vern Knudsen constituted a new generation of acoustical scientists whose careers were built upon the innovations and opportunities that came out of the electroacoustic industries. Knudsen's career, when compared to Sabine's, highlights the remarkable changes that occurred in the science and practice of acoustics in the 1920s.

Knudsen and his colleagues acknowledged and celebrated these changes by founding the Acoustical Society of America. But while the New Acoustics was exciting, it was not unproblematic. Most notably, the founders of the society struggled to gain the respect of their scientific colleagues in physics, colleagues who disparaged the applied and commercial nature of their expertise. In the early histories that these acousticians wrote of their new discipline, the tension between the ideals of pure science and the realities of their own commercially oriented careers was palpable. To resolve this tension, those same histories reconstructed Wallace Sabine's life and work in ways that rendered him heroic, but also archaic.

II Sabine After Symphony Hall

Wallace Sabine's initial investigation of reverberation raised as many questions as it answered, and after the opening of Symphony Hall in October 1900, Sabine turned to those questions seeking answers. He first convinced himself of the accuracy and consistency of musical taste through his experiments with the faculty at the New England Conservatory of Music. He then returned to the more physical aspects of architectural acoustics.

In 1904, Sabine began to expand on his earlier study of reverberation by examining the frequency dependence of the sound-absorbing powers of materials. Sabine's earlier work had focused exclusively on the effect of materials upon a sound of frequency C_4, or 512 cps, and he now set out to discover whether or not a given material absorbed sounds of different frequencies to differing degrees. This study followed the same method as his earlier work, supplementing the data collected at 512 cps with data for six other frequencies ranging from 64 to 4,096 cps. Sabine discovered that the absorbing properties of materials varied considerably over this range, and since the variations were not simple functions of frequency, he plotted the result for each material as a curve. (See figure 3.1.)

In the course of this investigation, Sabine utilized the equations that he had derived while working with his original source of 512 cps, although he

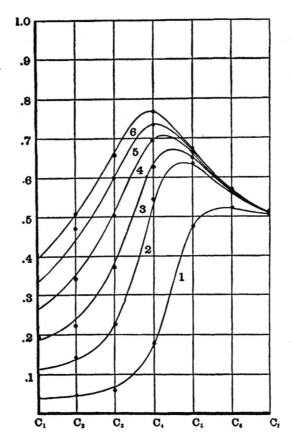

3.1
Curves showing the frequency-dependence of the sound-absorbing power of felt, as determined by Wallace Sabine, c. 1906. Curve 1 is for a single layer of felt, 1.1 cm thick. Each successive curve is for additional layers. The frequency ranges from C_1 = 64 cps to C_7 = 4,096 cps, and the absorbing powers vary considerably over this range. Wallace Sabine, *Collected Papers on Acoustics* (Cambridge, Mass.: Harvard University Press, 1922), p. 99.

acknowledged that these equations might, in fact, not be valid for other frequencies. Referring to the reverberation formula:

$$t = \frac{.164\ V}{a},$$

Sabine admitted, "It is debatable whether or not this definition should be extended without alteration to reverberation for other notes than C_4 512. There is a good deal to be said both for and against its retention. The whole, however, hinges on the outcome of a physiological or psychological inquiry not yet in such shape as to lead to a final decision. The question is therefore held in abeyance, and for the time the definition is retained."[10]

The psychological inquiry to which Sabine referred was a determination of the frequency-dependence of the human sense of loudness. Sabine's method of measuring reverberation—and the experimental technique embedded in his reverberation equation—required that an auditor determine the time at which a sound in a room became inaudible. At this time, it was assumed that the sound had dropped to one-millionth of its original intensity. If sounds of different pitches were perceived as inaudible at different intensity levels, this difference would somehow have to be taken into account. Only then would the equation be valid for all frequencies within the range of human hearing.

It was apparent to Sabine that human hearing was indeed variably sensitive to sounds of different frequencies, and he needed to understand this variability if he were to continue to depend on the ear as his instrument of detection. In 1910, Sabine published a brief memorandum on the results of a preliminary investigation into the perception of loudness. He tested a number of auditors to determine the relative energy required, at each of seven frequencies, to produce a sensation of equal loudness for each sound.[11] Even as he attempted to objectify the subjectivities of the human ear, however, Sabine encountered new obstacles. In this experiment, as in virtually all of his work, Sabine could only express the intensity of a sound relative to the minimum audible intensity for each pitch. There was no way to measure the absolute intensity of a sound, nor even to produce consistently a sound of constant intensity from a single source. "It is very unfortunate indeed," Sabine lamented, "that there are no standard sources of sound."[12] The limitations of the available sources and detectors impelled Sabine to reconsider the utility of techniques for visually representing sound, and he returned to the tradition of looking at sound in order to explore local effects in rooms such as echoes and interference patterns.

In order to understand the propagation of sound and the creation of distinct echoes, Sabine built scaled models of rooms and employed the "Toeppler-Boys-Foley method" to photograph the movement of sound waves through these models. (See figure 3.2.) As Sabine himself described it, "the method consists essentially of taking off the sides of the model, and, as the sound is passing through it, illuminating it instantaneously by the light from a very fine and somewhat distant electric spark. After passing through the model the light falls on a photographic plate placed at a little distance on the other side. The light is refracted by the sound-waves, which thus act practically as their own lens in producing the photograph."[13]

3.2
Photographic series showing the propagation of sound through a scaled model of the New Theater (Carrère & Hastings, New York, 1909), taken by Wallace Sabine, c. 1913. The New Theater (later known as the Century Theater) was plagued by numerous problems, some of them acoustical, including the echoes depicted here. It was demolished in 1930. Wallace Sabine, *Collected Papers on Acoustics* (Cambridge, Mass.: Harvard University Press, 1922), p. 185.

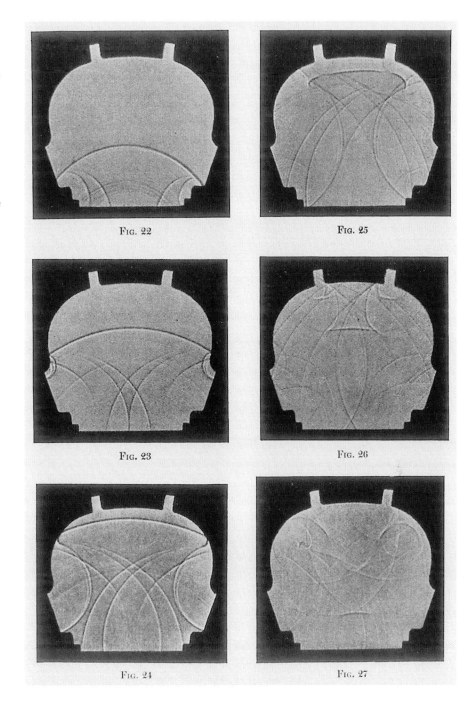

Sabine's interest in local acoustical effects also led him to devise a means by which to visualize the spatial variations of sound intensity that resulted from the interference of direct and reflected waves of sound in a room. In 1910, he constructed a map of the Constant Temperature Room of the Jefferson Physical Laboratory, "in which the intensity of the sound has been indicated by contour lines in the manner employed in the drawing of the Geodetic Survey maps."[14] (See figure 3.3.) Although Sabine's goal was to understand the variation of sound intensity, the means by which he generated this map are perhaps more interesting than the map itself, for this investigation appears to constitute Sabine's first significant engagement with electroacoustical tools.

3.3
Wallace Sabine's map representing the distribution of sound intensity in the Constant Temperature Room of the Jefferson Physical Laboratory, Harvard University, c. 1910. This horizontal cut shows the intensity at head-level for a sound of 248 cps. The units, from 0 to 12, are relative measures, not calibrated to any absolute physical standard. Wallace Sabine, *Collected Papers on Acoustics* (Cambridge, Mass.: Harvard University Press, 1922), p. 152.

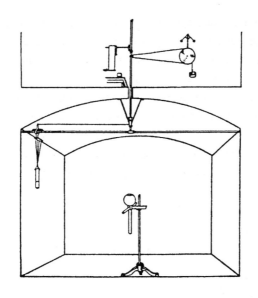

3.4
Wallace Sabine's experimental setup for mapping the distribution of sound intensity in the Constant Temperature Room, c. 1910. The source of sound, an electrically driven tuning fork of 248 cps, was mounted on the stand in the center of the room. The apparatus suspended from the ceiling simultaneously rotated and drew inward the telephonic detector suspended from the left side of the pole. Paul Sabine, *Acoustics and Architecture* (New York: McGraw-Hill, 1932), p. 41.

3.5
Fragment of Wallace Sabine's motion picture film record of the sound intensity registered by the electroacoustic detector as it moved through the Constant Temperature Room. The image shows the magnitude of vibration of the silvered string of a galvanometer connected to the detector. The vertical lines allowed Sabine to map this image to specific points in the spiral path of the detector. Paul Sabine, *Acoustics and Architecture* (New York: McGraw-Hill, 1932), p. 42.

In this study, Sabine did not employ an air-driven organ pipe as his source of sound; he instead used an electrically driven tuning fork. The detector—usually his own two ears—was, in this case, a telephone receiver or earpiece.[15] The tuning fork was placed at the center of the room and covered with an amplifying resonator. The receiver was rigged to a complicated mechanism that was just two waltzing mice short of a Rube Goldberg machine. A falling weight caused the long pole on which the receiver was mounted to rotate; at the same time, the rotary motion caused the receiver to be gradually pulled from the end to the center of the pole. The result was that the receiver traveled a continuous spiral path through the room at a constant height. (See figure 3.4.) The telephonic receiver generated an electrical current that represented the variations in sound intensity it encountered as it spiraled through space. That current was then fed to a sensitive "Einthoven string dynamometer," where it set up vibrations of varying amplitude in a silvered string. Sabine rigged a motion picture camera to photograph the image of the vibrating string onto a strip of film (see figure 3.5), and the constantly changing intensity of vibration could then be read off the

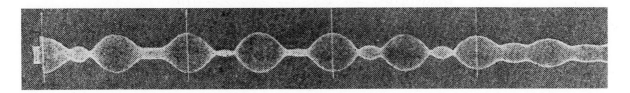

3.6
Wallace Sabine's plot of relative sound intensities in the Constant Temperature Room, with values read off the motion picture film and mapped to their corresponding locations along the spiral path of the detector. By drawing smooth lines connecting points of equal amplitude, Sabine created the map shown in figure 3.3. Paul Sabine, *Acoustics and Architecture* (New York: McGraw-Hill, 1932), p. 44.

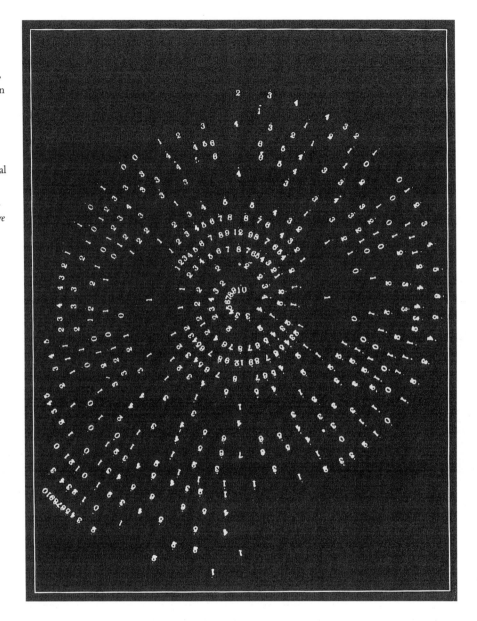

developed image on that film. Sabine mapped those intensities back onto the spiral path traversed by the receiver, to create a point-by-point plot of the relative sound intensity in the room. (See figure 3.6.) Finally, by connecting locations of equal intensity, Sabine created the contour map illustrated in figure 3.3.

While Sabine published his map in 1912, he chose not to include an account of how he obtained it.[16] This suggests that he was not fully comfortable with his new electroacoustic technique, and he apparently remained uncomfortable with it throughout his career. Sabine did not know if his telephone receiver responded uniformly to sounds of different frequencies, or if it—like the human ear—was particularly sensitive to sounds of particular pitches. Nor was his electrical source obviously an improvement over the organ pipe he generally depended on. For Sabine, these devices were undependable, and he used them only to generate images of sound. These images enabled Sabine to begin to understand qualitatively the local behavior of sound in rooms, but the electrical signals he used to obtain them were otherwise of little use or interest to him.[17]

The complicated spatial effects registered in Sabine's contour map were primarily an artifact of the laboratory. Under normal circumstances, the sounds that an auditor encounters are not pure, steady-state tones generating stationary interference patterns, but complex and constantly varying combinations of sound waves of different frequencies and intensities. In a typical room filled with music or speech, interference patterns continually shift and change, and most local effects are fleeting or they average out over time. Thus, while Sabine labored in his laboratory to understand the full complexity of the behavior of sound, he simultaneously was able to work in the world outside his laboratory with a far more generalized model of that behavior. Sabine's reverberation equation remained an extremely powerful tool, and he applied it to an increasing number of architectural projects.

Sabine kept a list of the architects with whom he corresponded, and by 1916, this list contained eighty-four names.[18] He worked with many of the most eminent architectural firms of the day, and he treated with equal care and attention the inquiries of less renowned individuals. McKim, Mead & White, for example, continued to turn to Sabine for acoustical advice after the completion of Symphony Hall. In 1901, Sabine advised Charles McKim how best to reduce the reverberation in the Rhode Island Hall of the House of Representatives at Providence.[19] (See figure 3.7.) In 1903, Stanford White sought ideas about how to remove a prominent echo from the indoor tennis courts he had built for John Jacob Astor in Rhinebeck-on-Hudson, New York. "Although it has an earth floor," White wrote, "the echo and reverberation are very unpleasant. The only reason I am anxious about this is that high-born gentlemen 'holler,' and very beautiful ladies 'scream,' and get their remarks back in their faces from the vaulted wall! What shall we do about this?"[20] That same year, William Mead

consulted Sabine on how best to create a soundproof room for Joseph Pulitzer in his New York townhouse. "We have been building a house for Joseph Pulitzer, who is a nervous wreck and most susceptible to noises," Mead explained, "and he has discovered many real and imaginary noises in his house. Some of them are real and can be obviated, and we have great confidence that you can discover the cause and a remedy for them."[21]

Sabine worked even more extensively with Cram, Goodhue & Ferguson, architects who specialized in building neo-Gothic churches and university buildings. Sabine advised them on the acoustics of St. Thomas's Church and the Cathedral of St. John the Divine (both in New York), as well as numerous other projects. In 1916, Bertram Goodhue asked Sabine for advice on his plans for St. Bartholomew's Church in New York and for a music hall at the Throop College

3.7

Wallace Sabine's acoustical correction of the Hall of the House of Representatives in the Rhode Island State Capitol, c. 1901 (McKim, Mead & White, Providence, 1895–1903). The hall was too reverberant, so Sabine recommended the installation of sound-absorbing felt on the wall between the pilasters seen here on the right. The felt was covered with a tapestry to create the trompe-l'oeil effect of a garden. Wallace Sabine, *Collected Papers on Acoustics* (Cambridge, Mass.: Harvard University Press, 1922), p. 135.

of Technology (later the California Institute of Technology) in Pasadena, California. "As you know," Goodhue wrote, "I am one of those who do not move a step in such matters without your approval. We have sketch plans completed and if you could examine and approve them, my mind would be very much relieved."[22]

Sabine developed close personal relationships with regular clients like Goodhue, Cram, and McKim. He acknowledged to Goodhue that "It has been one of my keen pleasures during recent years to enjoy the acquaintance and to see the work of a few of the most eminent architects of this country." "Of all these," he confided, "your work as well as your personal friendship has given me the greatest satisfaction."[23] More typically, Sabine worked in a less personal vein, consulting with clients on a one-time basis, and by written correspondence only.

Upon receipt of blueprints, Sabine would evaluate the design by calculating the overall volume of the room and determining the square footage of each of the various materials that constituted its surfaces. Those data, along with the absorption coefficients for the different materials, would then be plugged into his reverberation formula to calculate the reverberation time of the room. In cases where he analyzed designs in advance of construction, Sabine would determine whether the expected value was satisfactory. If not, he would recommend architectural changes to bring the calculation in line with the desired result. For cases where he was asked to improve the faulty acoustics of an extant structure, Sabine used his equation to inform recommendations on how best to modify the room to transform its sound. In most cases, these rooms were overly reverberant, and Sabine recommended the installation of a specific amount of sound-absorbing materials in particular locations to reduce the reverberation in the room, as in figure 3.7. He also addressed problems of echo and other matters resulting from the form of the room, but analysis of reverberation virtually always constituted the core of his evaluation.

The 1912 inquiry of architects Stevens & Nelson, of New Orleans, was representative of many received by Sabine. "We have recently been reading of experimental tests that have been conducted by you on acoustical effects in auditoriums," they wrote, "and as we have had the sad experience of probably a great many others in experiencing unsatisfactory results in some of our auditorium work, are writing to ask if we may not procure some assistance from you."[24] In 1911, H. Osgood Holland of Buffalo wrote, "I have lately erected an auditorium which is giving trouble acoustically. The Building Committee originally rejected my proposal to employ an acoustical expert, but are now considering the matter."[25] Sabine responded to Holland that his fee was two hundred dollars "for the study of the problem and for recommendation for the correction so drawn that contractors, local or otherwise, could bid on and execute the work."[26] This fee, which appears to have remained unchanged from 1909 onward, was to be paid by the owner of the building. When Alfred Altschuler attempted to bargain Sabine down in price, explaining that he received no compensation for this expense, Sabine replied to the architect that he was unable to lower his fee. He continued, "Had I known that the expense would be borne by you personally I think I should have been unwilling to undertake it under any circumstances. In all previous experience this additional expense has been borne, as I think it should be borne, by the owner."[27]

Sabine was, in fact, acutely uncomfortable with the commercial value of his expertise, and he struggled throughout his career to balance the economic aspects of his research and consulting with a rather idealistic vision of a scientific practice that transcended the material world of dollars and cents. From his earliest days as an acoustical investigator, Sabine had to be forced by President Eliot to submit receipts for the reimbursement of research expenses.[28] As a consultant, Sabine was equally particular. "Let me repeat," he wrote to architect R. Clipston Sturgis in 1910, "that the only condition on which I am willing to make any charges whatever is that they neither are paid by the architect nor are embarrassing in their transmission."[29]

The possibility of such embarrassment was particularly problematic with those regular clients whom Sabine considered as friends. He declined numerous times to accept payment or even reimbursement of expenses from Cram, Goodhue & Ferguson. It was his pleasure, for example, to inspect the Cathedral of St. John the Divine free of any charge, out of his "warm regard and admiration for Mr. Cram."[30] Architect Winthrop Ames recalled that, while Sabine "went to great trouble and pains to help us solve our problems, he always gave us the impression that our problems were so interesting that it was we who were conferring a favor upon him by giving him an opportunity of helping us solve them."[31]

Sabine was indeed eager to take on new consulting projects, and he used his publications in architectural journals to solicit work that might shed light on problems of particular interest. In a 1914 article in the *Brickbuilder*, for example, Sabine characterized his paper as "a report of progress as well as an appeal for further opportunities, and it is hoped that it will not be out of place at the end of the paper to point out some of the problems which remain and ask that interested architects call attention to any rooms in which it may be possible to complete the work."³² Sabine was particularly interested in determining the absorption characteristics of glass and "old plaster," and he specified that it was "necessary for such experiments that rooms practically free from furniture should be available and that the walls and ceiling of the room should be composed in a large measure of the material to be tested."³³

The analysis of differently constructed walls would have been a costly and time-consuming investigation to carry out in the laboratory, and Sabine solicited opportunities for field work in order to make such an investigation economically practical. An alternative to this approach was presented in 1912, when a manufacturer of hydrated lime offered to subsidize Sabine's research on the acoustical properties of lime plaster walls, but the offer proved problematic. As Sabine recalled, he was "asked to take up this investigation by the National Lime Manufacturers Association with the proposal that they should bear the cost of the research, which I had placed at seven hundred dollars, although I have since found that that was an under estimate. When I stated, however, that the only condition on which I would undertake the work was that the results, whether favorable or unfavorable to them, should be published, they did not wish to carry on the investigation."³⁴

Just as he was reluctant to profit from his scientific consulting, Sabine was acutely sensitive to the possibility of undue influence—or even the appearance of such influence—on his scientific research. He guarded with jealousy his reputation as a pure and disinterested investigator, refusing even offers of material assistance that came with no strings attached. Sabine articulated his preference for independence from sponsorship in 1916 to a representative of the Gypsum Industries Association. "I am conducting these tests entirely on my own initiative and responsibility and at my own expense," he explained. "Though a number of firms have offered to bear the expense of the tests of their own materials, I have thought it best that comparative tests of commercial products should be free from any possible bias."³⁵

Although Sabine questioned the propriety of accepting corporate subsidies to underwrite the evaluation of commercial materials, his attitude toward working with manufacturers to develop new materials was, in contrast, one of eager willingness. In 1911, Sabine began to work with the ceramic tile manufacturer Raphael Guastavino to develop sound-absorbing tiles for use in the vaults of large churches and other spaces in which reduced reverberation was desired. Guastavino perceived a strong potential market for such a material and Sabine was generously compensated for this work, receiving an initial payment of three thousand dollars as well as a royalty based on the square footage of tile actually installed in buildings.[36] In sharp contrast to his dealings with the lime manufacturers, Sabine reported here that "the attitude of the Guastavino Company has been exceedingly satisfactory. Not once have they shown the slightest inclination to exploit my experiments" with embarrassing propaganda or advertising.[37] A similar collaboration with the H.W. Johns-Manville Co. led Sabine to express his indebtedness to the company, "not merely for having placed at my disposal their materials and technical experience, but also for having borne the expense of some recent investigations looking toward the development of improved materials, with entire privilege of my making free publication of scientific results."[38]

The commercial aspects of Sabine's scientific expertise were thus a benefit to be enjoyed, but also a potential problem that required close monitoring lest he undercut the intellectual value of his science or undermine his own scientific reputation. While Sabine was clearly a modest man, that reputation was extremely important to him. It constituted in his mind the true reward for his hard work and commitment to the ideals of science. "For financial return I am not eager," Sabine explained to architect Albert Kahn in 1911. "On the other hand, I do earnestly desire recognition for such scientific services as I may have rendered the architectural profession."[39]

Sabine's avowal to Kahn was provoked by a correspondence in which the architect had initially inquired about the acoustical feasibility of a 4,000-seat auditorium for the University of Michigan. Kahn ultimately associated himself with another acoustical consultant, a man named Hugh Tallant, and Sabine recalled to Kahn an uncomfortable episode in which he had been compelled to ask Tallant for adequate recognition in a survey of acoustics that Tallant was about to publish in the *Brickbuilder*.[40] Sabine's concern ultimately proved misplaced, however, and he was equally chagrined to learn that, on the completion of the auditorium at Michigan a few years later, it was he, not Tallant, who received credit for it. "I can quite understand why your name has been brought up in connection with the Hall," Kahn explained, "although I have always been very careful to mention that Mr. Tallant has been our Consulting Engineer in connection with the work. No doubt the general knowledge of the fact that the remarkable development[s] in the science of acoustics are due in the largest measure to your own research work, is responsible for this impression."[41]

Nor was this encounter with Tallant the only occasion on which Sabine was forced to respond to the actions of others who sought to take credit for his work. In 1911, Sabine learned that a man named Jacob Mazer had recently applied for a patent on the general technique of acoustical control made possible by the application of Sabine's reverberation equation. Only the hasty intervention of Charles Eliot and Henry Higginson prevented the granting of this patent and ensured that Sabine's formula remained free for all to use at will.[42]

Both Hugh Tallant and Jacob Mazer had initially sought advice directly from Sabine, who had responded by freely sharing his knowledge with both men. Perhaps the resulting incidents left Sabine wary of collaborating with others. Perhaps he was simply a solitary soul. Whatever the reason, although a small community of acoustical consultants and researchers in sound was clearly taking shape in the years after 1900, Sabine never perceived himself as a part of that community. Although articles on acoustics by other authors were beginning to appear in architectural and scientific journals alongside his own, Sabine never referred to this literature in his publications.[43] He expressed his sense of isolation in 1912, in a letter to the organ builder Robert Hope-Jones. "I wish," Sabine wrote, "that we might get together oftener, there are so few that are scientifically engaged on the subject of acoustics in any form, either here or abroad, that we rarely meet."[44] No school of acousticians developed around Sabine at Harvard, and just one man, Clifford Swan, could claim to have studied sound with him. Swan spent several years as his "graduate student and associate," and Sabine admitted, "He is the only student I have had, and, as matters now stand, my sole hope of making the subject of Architectural Acoustics an engineering science."[45]

That Sabine failed, or at least neglected, to train a succeeding generation of acoustical researchers at Harvard is not as surprising as it might initially seem. In fact, Sabine would have had little time for such an undertaking even if he had desired it. In addition to teaching undergraduate physics courses and pursuing his research and consulting, Sabine took on extensive administrative duties. From 1897 to 1899, he had served as an unofficial assistant to President Eliot, and in the years after the turn of the century he continued to advise Eliot, particularly on how best to integrate Harvard's Lawrence Scientific School more fully into the university. When a Graduate School of Arts and Sciences was founded in 1906, Sabine was appointed its first dean. He served in this position for almost ten years, working to establish the new school on a firm basis by attracting new faculty and purchasing state-of-the-art laboratory equipment. His efforts came to naught, however, when the school was dissolved in a short-lived merger between Harvard and the Massachusetts Institute of Technology.[46]

In 1915, with his administrative obligations behind him, Sabine accepted an invitation to lecture on acoustics at the Sorbonne in Paris. He arrived in war-torn Europe with his family in the summer of 1916. The lectures did not commence until November, so Sabine and his physician wife spent the summer working for the Rockefeller War Relief Commission. Wallace investigated French facilities for the treatment of tubercular patients, while Jane Kelly Sabine headed a committee that supervised the care of Belgian refugee children. The Sabines, along with their two young daughters, moved to Paris in the fall of 1916, but a sudden and serious kidney infection prevented Wallace from presenting his lectures at that time. He spent the winter in a Swiss sanitarium, recovering his health and improving his French. He returned to Paris the following spring and delivered his lectures.[47] Sabine remained in Europe through the summer of 1917, traveling extensively to advise the French, British, and Italian governments on a number of different war projects, from sound-ranging techniques for the location of enemy artillery, to submarine detection, to aviation and aerial photography.[48]

Back in the United States in the fall of 1917, Sabine commuted between Boston and Washington as he served on both the Aviation Section of the Signal Corps and the Department of Technical Information of the Bureau of Aircraft Production. In 1918, he was appointed a member of the National Advisory Committee for Aeronautics by President Woodrow Wilson. The health problems that had plagued him in Europe returned, but Sabine refused to relinquish his growing responsibilities. Surgery was recommended, but he stubbornly, if patriotically declined: "Not while the War is on and other lives are in danger."[49] By December 1918, with the war finally over, Sabine was willing to schedule a hospital stay for the upcoming university holiday. His health had weakened, however, to the point that recovery was no longer possible, and Wallace Sabine passed away on 10 January 1919.[50]

At the time of his death, Sabine had been looking forward to returning to his research in architectural acoustics. After years of preoccupation with administrative responsibilities and the exigencies of war, he was eager to resume his scientific studies. The war had exposed Sabine, along with many other acoustical investigators, to new acoustical technologies and particularly to the growing potential of electroacoustic devices as tools for studying and controlling sound. But while others would eagerly embrace these new tools, Sabine apparently planned to return to his old tools and techniques. And, while the development of war-related acoustical technologies additionally fostered a nascent sense of community among acoustical investigators in America, Sabine remained an outsider to that community. In fact, after the war he was planning to isolate himself even further, by relocating his investigations to a new facility that had been constructed specially for him. This laboratory, built by an eccentric but generous patron, was located far from the hubbub of Harvard, in the quiet countryside of Illinois.

George Fabyan was a scion of an old Massachusetts textile family who, exercising a streak of adolescent independence, had moved west in 1883 at the age of sixteen. He worked in a variety of industries for a decade or so before returning to the family fold to run the Chicago office of the Bliss Fabyan Company. Around this time, he purchased 600 acres of land along the Fox River, west of Chicago in the town of Geneva, and established a country estate called Riverbank. Fabyan ruled Riverbank like the lord of a medieval manor, and his fiefdom eventually included a working Dutch windmill, a lighthouse, a Japanese garden, a colonnaded Roman pool, and a menagerie of exotic animals.[51]

In addition to collecting alligators and bears, Fabyan had a hobby of deciphering secret codes and he invited an elderly woman, Mrs. Elizabeth Wells Gallup, to come live at Riverbank to help him pursue this hobby.[52] Mrs. Gallup had made a name for herself by decoding secret, "bilaterally coded" messages that she (and many others) believed that Francis Bacon had placed in the first printed edition of the plays of William Shakespeare. The messages were allegedly reports of scientific experiments carried out by a secret society, the Rosicrucians. One such message decoded by Mrs. Gallup described a cylindrical device surrounded by stretched and tuned wires. When the wires were sounded, according to the message, the cylinder would levitate. Fabyan was intrigued by this report. In 1913, he had the device built, and when it didn't levitate he sought to discover why. Fabyan's brother Marshall was affiliated with Harvard University through the family's philanthropy, so when George contacted Marshall hoping to identify an expert to solve his acoustical mystery, Marshall referred him to Wallace Sabine.

Sabine's response to George Fabyan's inquiry is unfortunately unrecorded, but as a result of their correspondence, Fabyan became interested in Sabine's acoustical research.[53] When he learned how Sabine had to carry out his experiments late at night in order to minimize interference from city noise, Fabyan generously offered to build the physicist an acoustical laboratory at Riverbank, far from the disturbances of traffic, trains, and nightlife.

In fact, noise was becoming an increasingly significant factor in Sabine's research and consulting. William Mead's 1903 request for advice on how to soundproof Joseph Pulitzer's New York townhouse was just the first of a growing number of inquiries concerning the isolation, rather than the reverberation, of sound.[54] In 1914, Sabine identified noise as a "modern acoustical difficulty," and he noted that,

> Coincident with the increased use of reënforced concrete construction and some other building forms there has come increased complaint of the transmission of sound from room to room, either through the walls or through the floors. Whether the present general complaint is due to new materials and new methods of construction, or to a greater sensitiveness to unnecessary noise, or whether it is due to greater sources of disturbance, heavier traffic, heavier cars and wagons, elevators, and elevator doors, where elevators were not used before,—whatever the cause of the annoyance there is urgent need of its abatement.[55]

Stimulated by this new problem, Sabine planned to study systematically all sorts of wall constructions, to examine the transmission of sound through them as well as the absorption and reflection of sound off their surfaces. Even his work on the control of reverberation was affected by this growing concern over the problem of noise. In 1914, a reporter for *System: The Magazine of Business* interviewed Sabine on the problem of office noise and reported that "several large industries and banks," as well as the general offices of a Chicago meat packer, had already benefitted from Sabine's expertise by utilizing sound-absorbing materials to quiet the noise.[56] Sabine contributed further to the elimination of office noise when he advised the Remington Typewriter Company on how best to reduce the noise produced by their typewriters.[57]

Fabyan's proposal to build a quiet retreat from which to study sound and noise was an offer that Sabine could not refuse. Plans were drawn up in 1916 and the building was completed in 1918, just as Sabine's war work was coming to a close. (See figure 3.8.) Sabine apparently intended to work at the laboratory himself during university holidays, and to supervise indirectly the work of others there during the school year, while he was resident in Cambridge.[58] He had just begun to calibrate the organ pipes that were to be installed in the new laboratory when his final illness took hold.

After Sabine's death, George Fabyan once again turned to his brother Marshall for advice, this time on how to find a replacement for the seemingly irreplaceable Sabine. Marshall Fabyan put him in touch with a distant cousin of

3.8
Plan and section of the Riverbank Acoustical Laboratory, Geneva, Ill. The laboratory was built for Wallace Sabine and to his specifications by George Fabyan, a wealthy patron whose Riverbank Estate was located on the rural outskirts of Chicago. Alan E. Munby, "American Research in Acoustics," *Nature* 110 (28 October 1922): 576.

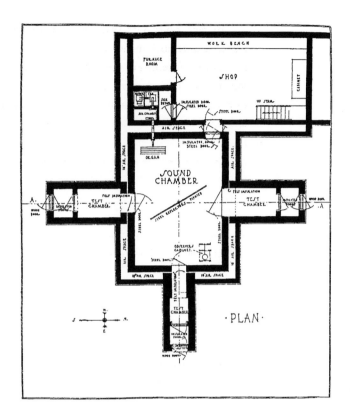

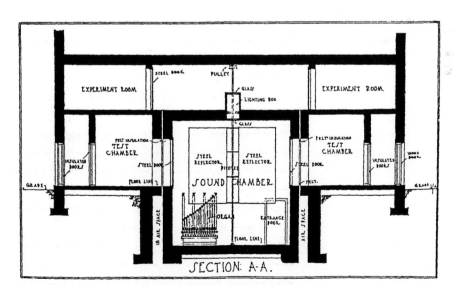

3.9

Riverbank Acoustical Laboratory, Geneva, Ill., c. 1918. The lab became an important facility for the testing of acoustical materials and products, and it continues to operate today as a part of the Illinois Institute of Technology Research Institute. Courtesy Riverbank Acoustical Laboratories, IIT Research Institute.

Sabine who had recently received his Ph.D. in physics from Harvard. Paul Sabine had studied spectroscopy, not acoustics; it is not even clear that he knew his cousin Wallace very well. Nonetheless, he accepted Fabyan's offer to come to Riverbank and supervise the new facility.[59] In 1919, Paul Sabine introduced the new laboratory to readers of the *American Architect*. He invited architects to direct their queries and problems in acoustics to its staff, and the Riverbank Laboratory soon became a major facility for acoustical research and for the independent testing of building materials and other commercial products.[60] (See figure 3.9.)

The solitary trajectory of Wallace Sabine's career from 1900 to 1919 creates the impression that he alone was working to forward the study and application of the science of architectural acoustics. In fact, this was not at all the case. Almost immediately after the publication of Sabine's first paper on reverberation, a small but growing community of acoustical researchers began to develop. At the time of his death this group was just reaching a critical mass, and it would flourish during the 1920s. While Sabine himself appeared to be largely unaware of this nascent community of scholars, its members, in contrast, all recognized Sabine's work as the origin of, and stimulus to, their own interest in acoustics.

III THE REVERBERATIONS OF "REVERBERATION"

Not long after its publication in 1900, Wallace Sabine's work on reverberation was being cited in physics textbooks, in architectural journals, and in a small but growing number of scientific articles dedicated to the topic of architectural acoustics.[61] In 1902, a theoretical derivation of Sabine's experimentally determined reverberation equation was presented by William S. Franklin, a physicist at Lehigh University. Franklin verified the form of Sabine's equation, as well as the value for the constant k that Sabine had obtained experimentally.[62] George Stewart, a physics instructor at Cornell University, was the first to repeat Sabine's experimental method for determining the acoustical properties of materials. In 1903, Stewart confirmed Sabine's reverberation equation in the new Sibley Auditorium at Cornell, and he measured the absorptive power of cocoa-matting, adding it to Sabine's table of absorption coefficients.[63]

Stewart, like Sabine, struggled with the inadequacies of acoustical instrumentation. A wooden organ pipe, blown by mouth, served as his source of sound and elicited the complaint that "the initial intensity produced is not known." Stewart could only relate his results to Sabine's by comparing his own source directly to that which Sabine had employed, and Sabine generously lent his apparatus to Stewart to allow him to make this comparison. "I was thus enabled to compare the two organ pipes," Stewart explained, "and, since the rate of production of his was known, the initial intensity produced by the wooden pipe could be computed."[64]

Another physicist who modeled his own acoustical researches after Sabine's was Floyd Watson. Watson's interest in sound originated around the turn of the century, when he was a graduate student at Cornell.[65] While his curiosity may have been piqued by observing George Stewart's work in the new Sibley

Auditorium, Watson only began to study sound himself in 1908. Now an assistant professor of physics at the University of Illinois, Watson—like Sabine before him—was asked by his president to examine and improve the poor acoustics of a new auditorium. He spent over six years investigating the University Auditorium, and its many faults provided an opportunity to study not just reverberation, but also echo formation, the focusing of sound by curved surfaces, the effect of ventilation systems on sound, and the use of sounding-boards to improve the intelligibility of a speaker.[66]

While Watson's introduction to the study of architectural acoustics was strikingly similar to Sabine's, the ways in which the two men carried out their studies were just as strikingly different. Sabine was reluctant to publish or report on any preliminary results of his research, preferring to wait until each investigation was fully complete. Watson, in contrast, preferred to present his work to colleagues as it progressed. He regularly delivered papers at the meetings of the American Physical Society and published numerous articles along the way.[67] While Sabine had almost always worked alone, Watson was eager to enlist the help of students, and the auditorium project yielded one graduate and two undergraduate theses.[68] Finally, while Sabine had felt the distinct lack of an intellectual community with which to exchange ideas, Watson, in contrast, quickly identified just such a community. As early as 1911, he was referring to "the field of Architectural Acoustics" in a way that suggests a growing awareness of other researchers, and in 1914, with the University Auditorium work complete, Watson published a summary article whose bibliography listed over thirty twentieth-century sources on architectural acoustics.[69] Of course this bibliography included Sabine's articles, and Watson additionally had the opportunity to meet Sabine in person, sometime over the winter holiday of 1909–1910. The two discussed their researches in acoustics, and Watson reported, "Professor Sabine finds as I do, that many obstacles beset the path of the experimenter in acoustics."[70]

Perhaps the greatest obstacle besetting Watson and Sabine was the difficulty of measuring sound. Although Sabine did tentatively explore new tools to accomplish this task, he continued to depend on his ears as detectors in spite of his awareness of the frequency-dependence of their perception of the loudness of different sounds. Others, including Watson, sought to avoid the subjectivity of the human ear and turned instead to instrumental detectors that measured the physical intensity, rather than the perceived loudness, of a sound. But here too—as Sabine had recognized—numerous obstacles still beset those who chose to use such ostensibly objective devices.

In 1901, James Loudon had called attention to the "great lacuna in our acoustical knowledge" that resulted from the lack of tools to measure the intensity of sound.[71] Sabine's colleague in the Department of Physics at Harvard, George Pierce, sought to redress this deficiency in 1908. Pierce borrowed from Sabine's own laboratory an electromagnetic telephone receiver, and he put it to use as a detector in a new instrument he designed to measure the intensity of sound.[72] The receiver consisted of a light metal diaphragm mounted within a magnetic field created by a permanent magnet and an electrical circuit. When the diaphragm vibrated under the influence of impinging waves of sound, it altered the strength of the magnetic field and generated a fluctuating electrical current in the circuit. This signal was fed to a galvanometer, which indicated the varying voltage of the electrical signal, and this measurement corresponded to the intensity of the original sound. Electromagnetic telephone receivers were not very sensitive detectors of sound, however, and they created very weak electrical signals. To compensate for this insensitivity, Pierce tuned his electrical circuit so that it would resonate at the frequency of his source of sound (an organ pipe of 705 cps). By doing so, he increased the sensitivity of his detector, but he also narrowed its applicability to the measurement of sounds of just this one frequency.

Pierce used his apparatus to sample the spatial variations in sound intensity in the Constant Temperature Room of the Jefferson Physical Laboratory. He did not construct an intensity map, as Sabine would do several years later, because Pierce, unlike Sabine, was not particularly interested in the patterns of sound in the room. George Pierce was primarily an electrical researcher, not an acoustician, and as such, he was far more interested in his apparatus than in the phenomena that it was measuring. In 1910, Pierce would publish *Principles of Wireless Telegraphy*, one of the first scientific treatises dedicated to the new subject of radio, and his subsequent career would be equally divided between the theoretical elaboration of technologies of electrical communication and the invention of numerous devices that made such communication possible. Pierce's new sound measuring instrument was a variation of a device that he had previously designed to detect electromagnetic waves, and the idea to tune his circuits to resonate with his source of sound certainly came from his background in radio, where the practice of *syntony*, the tuning of circuits, was well established.[73]

Pierce acknowledged, however, that there were limitations to his new device. Not only was it was designed to measure sounds of just one frequency, but even at that frequency, the instrument indicated only qualitatively the variation in sound intensity; the galvanometer readings could not be converted into absolute physical measurements. In spite of its limitations, the device suggested to Pierce the potential of "phono-electric"[74] instruments for measuring sound, and he was not alone in recognizing this potential.

At the 1909 meeting of the American Physical Society, Floyd Watson described his own design for an electrical sound detector. Watson's instrument, like Pierce's, consisted of a telephone receiver connected to a galvanometer, with the circuitry tuned to resonate at the frequency of the source of sound. "By means of this apparatus," Watson reported, "maxima and minima of sound were easily detected in a small laboratory, and a series of standing waves near a wall measured."[75] W. M. Boehm, working at the University of Pennsylvania, devised another instrument, similar to those of Watson and Pierce. Boehm experienced problems working with his device however, as noise from outdoors—the "incidental disturbances which occur several times a second in a large city"—intruded on his experiments. Physical noises were transformed into electrical noises that interfered with the sound signal he sought to measure. Boehm solved his problem by modifying his apparatus into a hybrid of electrical and optical elements. His circuits were redesigned to vibrate a small mirror, and the reflection of a bright beam of light off the vibrating mirror was then observed. With this setup, Boehm was able to distinguish visually the signal from the noise in a way that he could not accomplish when he scrutinized the readout of an electrical meter. He explained that "Accidental vibrations are easily distinguished from steady ones. Generators or motors in the building or the blast of a locomotive may interfere sufficiently to make observations impossible but traffic along the street produced less annoyance than a person walking over the floor."[76] Like Sabine, Boehm struggled against the encroachment of noise on his investigations of sound and he turned to techniques of visualization to redress the shortcomings of the new electrical instruments.

While Boehm, Watson, Pierce, and Sabine were exploring new tools in order to measure the intensity of sound in space, others were devising new means to measure the sound-absorbing properties of materials, but here, too, problems arose. Sabine's method for measuring the absorption coefficients of materials required a full-sized room possessing a significant amount of the material to be tested. The absorbing power of the material was calculated from the reverberation time of the room. Other investigators sought more convenient ways to evaluate much smaller samples of materials. Instead of measuring the reverberation times of rooms, they sought to measure directly the intensity of sound passing through or reflecting off their samples, and they, too, were therefore faced with the challenge of finding a way to measure the intensity of sound.

In 1902, F. L. Tufts of Columbia University published the results of experiments he had carried out on the transmission of sound through solid materials, and he described his frustration over the lack of a technique for the absolute measurement of sound intensity. While his investigation was stimulated by the problem of constructing a soundproof telephone booth for use in noisy cities, Tufts never considered using the telephone itself as a tool in his investigation. Instead, he listened, through a stethoscopic device, to sounds transmitted through small samples of different materials. His setup allowed him to compare directly the loudness of sounds transmitted through two different samples. By making a series of comparisons, Tufts was able to rank qualitatively a range of materials for their ability to transmit sound.[77]

In 1911, C. S. McGinnis and M. R. Harkins turned to the telephone itself as a measuring tool, using a detector based on Pierce's design in their experiments on the transmission of sound through materials.[78] Two years later, however, Hawley Taylor of Cornell rejected electrical tools when he devised his own method of determining the sound-absorbing power of small samples of different materials. "In the search for means for measuring the intensity of sound," Taylor explained, "tests were made of everything of any promise, and telephone receivers and transmitters, strong and weak field galvanometers, molybdenite and silicon rectifiers, barretters and microradiometers all figured. The Rayleigh disc was finally adopted as the most reliable and sensitive sound measuring instrument."[79]

Taylor comprehensively surveyed the many different means of electrically measuring sound, but he ultimately returned to an older technique of visual representation to make his measurements. The problem of measuring sound was, circa 1910, at the "forefront"[80] of acoustical research. But while many had begun to explore the new realm of electrical instrumentation, the limitations of these new instruments were both apparent and significant, and many investigators—like Taylor and Sabine—ultimately remained committed to the older tradition of rendering visible the vibrations of sound in air.

One of the most useful optical devices in acoustical research, as Taylor recognized, was the Rayleigh disc. Introduced by its inventor, Lord Rayleigh, in 1882, the instrument consisted of a horizontal tube in which was suspended, at an angle of 45 degrees to the axis of the tube, a lightweight mirror. When placed near a source of sound, the longitudinal vibrations of sound within the tube caused the mirror to pivot, with the force of rotation dependent on the amplitude of the sound wave. When a beam of light was reflected off the mirror and projected onto a distant scale, the degree of rotation, and thus the amplitude of the sound wave, could easily be measured.[81]

Arthur Webster's phonometer, like the Rayleigh disc, optically registered the disturbance of an object set in motion by sound waves as a means to measure the intensity of sound,[82] and Dayton Miller's phonodeik also followed this tradition. (See figure 3.10.) Invented in 1908, the phonodeik consisted of a sound-collecting horn with a thin glass diaphragm at its apex. One end of a silk string was attached to the center of the diaphragm; the other was wound around a tiny

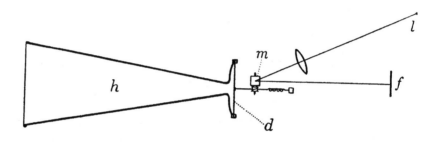

3.10
Schematic of Dayton Miller's Phonodeik, invented in 1908 for creating visual images of sound vibrations. Sound entered the horn "h" and vibrated the diaphragm "d," pushing and pulling on the tense string attached to it. The string, wound around a jewel-mounted spindle onto which was attached a tiny (about 1 mm square) mirror "m," caused the mirror to rotate. Light from a source "l" was reflected off the mirror and onto a distant scale "f," which amplified the movement and thus rendered visible the vibrations of sound. Dayton C. Miller, *The Science of Musical Sounds* (New York: MacMillan, 1916), p. 79.

spindle resting on jeweled bearings, and the string was held in tension by a small spring. The vibration of the diaphragm under the action of sound waves thus caused the spindle to rotate. As in the Rayleigh disc, light reflecting off a mirror attached to the spindle amplified this motion and registered it on a distant scale.

Miller adapted his apparatus to create photographic images of sound vibrations, and he traced these photographs with a mechanical harmonic analyzer to determine the frequency content of the sounds that he captured on film. In this painstaking and time-consuming way, he was able to analyze the sounds produced by different musical instruments, including the human voice. Perhaps because the procedure was so laborious, Miller's technique was not widely used by other acousticians in their studies of sound. The phonodeik did, however, have an impact beyond the scientific sphere when Miller devised a means to project its optical output in real time before an audience. These moving images of sound soon captured the attention of the general public, and Miller became known as "The Wizard of Visible Sound" as he traveled across the country demonstrating his device. In 1914, Wallace Sabine invited Miller to give a series of public lectures at the Lowell Institute, and the *Boston Evening Transcript* reported that the audience was "fascinated" by his graphic representations of music and noise. Miller's sound images were seen by millions more when they appeared in newspaper advertisements for the Aeolian-Vocalion phonograph.[83] (See figure 3.11.)

By 1915, the study of sound was clearly a growing field of scientific inquiry. While Sabine himself failed to recognize the emerging community of acoustical researchers, he crossed paths with many of its members. Like Sabine, these men struggled with the fundamental problem of how best to measure sound. While some began tentatively to explore new electrical tools, most—like Sabine—remained committed to the more traditional means of listening directly or generating optical representations of sound. Also like Sabine, many of these men would spend the next several years applying their expertise in sound to the problems of war. Unlike Sabine, however, most of these men would survive their war work. The Great War served as a catalyst to their sense of community as well as to that community's output, and these men would construct an entirely new world of acoustical tools, concepts, and problems to pursue in the years immediately following the Armistice.

Scientists of all sorts contributed their expertise to the prosecution of the First World War, but the impact of the war on the field of acoustics—and vice versa—was particularly strong. It would be difficult to prove that this war was

3.11
Phonodeik image of a sound wave, as reproduced in an advertisement for the Aeolian-Vocalion Phonograph, *New York Times* (21 February 1915), sect. I, p. 5. This image depicts the sound of an Aeolian-Vocalion recording of Tchaikovsky's *March Slav*.

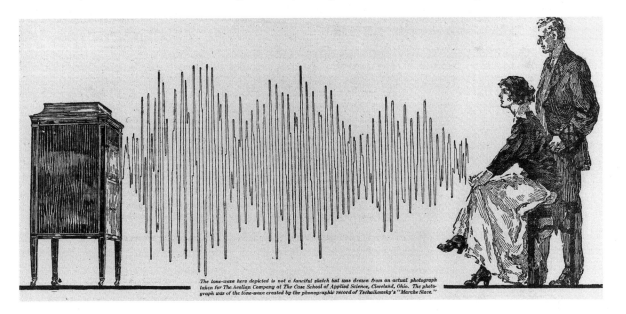

actually louder than any previous conflict, but it is clearly the case that this was a war in which people listened more intently than ever before. Soldiers on the ground pointed large arrays of acoustical horns toward the sky and listened for the faint but telltale drone of distant engines in order to defend themselves against encroaching enemy aircraft.[84] "Sound ranging" systems were devised in which microphones were strung out across European battlefields to register the reports of enemy guns. The different time of arrival at each microphone of the sound of a firing gun provided data that could be triangulated to locate, and then target and destroy, the enemy artillery.[85] In the trenches, Allied and German soldiers alike learned to distinguish the myriad sounds of different kinds of incoming shells. Some who survived the shells themselves were psychologically felled by the constant barrage of noise and were sent home as victims of "shell shock."[86]

"Modern trench-warfare demands knowledge and experience," explained Paul Bäumer, the fictional soldier created by the novelist and war veteran Erich Maria Remarque.

> A man must have a feeling for the contours of the ground, an ear for the sound and character of the shells, must be able to decide beforehand where they will drop, how they will burst, and how to shelter from them.
> The young recruits of course know none of these things. They get killed simply because they hardly can tell shrapnel from high-explosive, they are mown down because they are listening anxiously to the roar of the big coal-boxes falling in the rear, and miss the light, piping whistle of the low spreading daisy-cutters.[87]

Bäumer's own skill at listening ultimately failed to save him, however, and he was killed by a lone sniper's bullet on a day when all was quiet on the Western Front.

Perhaps the war's most deadly silence, and its most intensive listening, occurred at sea. In order to locate the submerged German U-boats, the Allies dedicated tremendous resources to the development of sensitive underwater sound detectors. Patrol boats were equipped with listening devices that enabled their crews to hunt down the invisible enemy craft and destroy or disperse them with depth-charges. Distinguishing the harmless noises of the patrol boat itself, the turbulence of the sea, and even the sounds of passing schools of fish from the quiet but deadly throb of a U-boat's propeller required extremely sensitive detectors, as well as specially trained operators. In the United States, a combination of navy officers, industrial researchers, and academic physicists at Nahant,

Massachusetts, and New London, Connecticut, produced this equipment and trained the personnel to operate it. At the close of the war, the New London Experimental Station was staffed by thirty-two physicists and over 700 enlisted men, and the listening devices that these men deployed were credited with helping the Allies to win the war.[88]

If acoustical research helped the Allies achieve victory in the war, it was also the case that the war, in turn, served as an equally valuable catalyst for the fledgling new field. Annual meetings of the American Physical Society prior to 1919, for example, never included more than four papers on acoustical topics. In 1919, fourteen such papers were delivered; in 1920, there were nineteen, and these numbers were generally sustained over the next decade.[89] The National Research Council, an organization of scientists formed in 1916 to address questions of national security, also proved crucial for fostering the community of acoustical researchers in America.[90]

After the war, the council rededicated itself to the peacetime application of scientific expertise, and in 1922 its new Committee on Acoustics met at George Fabyan's Riverbank Laboratory to evaluate and summarize the state of "Certain Problems in Acoustics." Members of the committee included Floyd Watson, Dayton Miller, and George Stewart (now a professor at the University of Iowa), all of whom had been involved in acoustical research projects during the war, as well as Paul Sabine, Arthur Gordon Webster of Clark University, Arthur Foley of Indiana University, and Louis King of McGill University.[91]

The committee identified thirteen different subfields of acoustical research, then summarized the salient problems in each. Their bibliographic research made fully evident the increasing attention to the study of sound that had occurred in recent years. Nonetheless, their report also made clear that most of the obstacles faced by prewar investigators remained in place. While the committee surveyed a wide range of topics—from audition to acoustics in navigation to the study of musical sounds—many of the problems identified in each area ultimately came back to the fundamental difficulty posed by the lack of suitable instrumentation.

Arthur Webster and Dayton Miller reported on the "Detection and Measurement of Sound," and concluded that "probably the instruments available to the physicists for the detection and measurement of sound are less satisfactory than those for any other field of research."[92] Webster and Paul Sabine focused on "The Measurement of Sound Intensity in Absolute Units," and were equally discouraged. They described the phonometer, the Rayleigh disc, and other prewar instruments, then lamented the shortcomings of each, as well as the incommensurability of results obtained from different instruments. "The need," they concluded, "is for a single carefully organized research in which results with different instruments and by different methods are secured under conditions so nearly identical as to make these results comparable." "The problem," they continued, "is peculiarly fundamental for real progress in experimental acoustics."[93]

Alongside their descriptions of the various unsatisfactory devices, the committee also noted the more recent appearance of a new kind of measuring tool, the condenser transmitter. Although the committee hardly realized it at the time, this device was about to usher in "a new and thrilling era for the quantitative measurement of acoustical phenomena."[94] The impact of this new tool would extend far beyond the scientific community, too, for it not only set a new standard for the scientific measurement of sound, but also helped to stimulate the development of a whole range of innovative new sound technologies that would ultimately transform the American soundscape.

IV NEW TOOLS: THE ORIGINS OF MODERN ACOUSTICS

The condenser transmitter was not the product of university research; its inventor, Edward Wente, was a researcher in the engineering department of the Western Electric Company, the manufacturing division of the American Telephone & Telegraph Company. While the National Research Council played a valuable role during the war by integrating academic scientists more fully into the governmental war effort, equally valuable was its role in breaking down the barriers between academic and corporate research programs. Corporate research had, in fact, grown up alongside the field of acoustics during the early years of the century, and some of the earliest and most innovative industrial research laboratories were established by companies committed to the design and delivery of acoustical products, including the telephone services of AT&T and the radio divisions of General Electric and Westinghouse.[95]

The telephone industry had, of course, long been interested in electroacoustic transducers to convert sound energy into electrical energy and vice versa. Alexander Graham Bell developed a variety of transducers when he undertook his first telephonic experiments in the mid–1870s. He soon settled on a design using an electromagnetic transducer to serve as both the transmitter (mouthpiece) and receiver (earpiece) of his telephone. At the transmitter end, sound waves were transformed by a vibrating diaphragm and an electromagnet into a varying electrical current that represented the sound. When this current arrived at the receiver, the process was reversed: A varying electromagnetic field generated by the current pushed and pulled on a magnetized diaphragm, whose movement in air subsequently re-created the vibrations of the original sound.

Bell's electromagnetic transmitter was not very sensitive, and the voice signal it generated was subsequently weak and difficult to transmit over telephone lines of any considerable length. In 1877, Thomas Edison devised a far more sensitive mouthpiece, rendering the telephone a much more practical device. Edison replaced Bell's rigid diaphragm with a button of compressed carbon granules. The carbon button constituted an electrically resistive element of the telephone circuit, and its resistance varied depending on the pressure to which it was exposed. When the carbon button was exposed to the pressure of impinging sound waves, its changing resistance modified the current in the circuit, creating a signal that represented the sound. The sensitive carbon transmitter generated a voice signal significantly stronger than that generated by Bell's original design, and this signal was far more successfully transmitted over commercial telephone lines.[96]

Bell's transducer remained useful as a receiver, however, and these two devices—the carbon transmitter and the electromagnetic receiver—constituted the technological core of the telephone system that grew out of Bell's experiments and Edison's improvements. Numerous other improvements to the telephone system were introduced in the 1880s and 1890s, and these improvements were accompanied by just as many lawsuits, as inventors like Elisha Gray, Emile Berliner, and countless others challenged the increasing power of the Bell Telephone System. The Bell System defended its claims in court, absorbed its competitors, purchased the equipment manufacturer Western Electric, and eventually became the monopoly known as the American Telephone and Telegraph Company.[97]

In 1907, AT&T President Theodore Vail consolidated the engineering departments of Western Electric and Bell, moving them from Chicago and Boston to corporate headquarters in New York. John J. Carty was placed in charge of the new department, whose mission was to improve the quality and range of telephone service. By encouraging the in-house development of telephonic technologies, Vail and Carty hoped to free the company of its long-standing dependence on outside inventors like Thomas Edison. At the same time, a new threat to AT&T's monopolistic network of telephone wires was presented by the wireless technology of radio, and Carty's staff was additionally expected to find a way to gain control over this new technology by inventing and patenting crucial new components for wireless systems of communication.[98]

Carty identified the problem of amplification as crucial to both telephony and radio. As the telephone company extended its long-distance lines over greater distances, the electrical resistance of mile after mile of wire gradually attenuated even the strongest voice signals and an amplifying "repeater" was required to boost the signal strength at intervals during its journey. The signals generated by radio receivers were also often unacceptably weak, and Carty realized that a high-quality amplifier could solve the critical problem of weakened signal strength in both wired and wireless applications. In 1909, he dedicated his department to the challenge of developing an amplifying repeater, and, to spur them on, he announced that AT&T would have transcontinental telephone service in place at the upcoming Panama-Pacific Exposition. The exposition was scheduled for 1914, so his staff had less than five years to develop the amplifier that would be necessary to accomplish coast-to-coast service.

By 1910, the engineering department had little to show for its labors. Carty's assistant Frank Jewett suggested that they hire some academic physicists and establish a research department dedicated to fundamental investigations of physical processes in order to meet the challenge of devising the device. In 1911, the University of Chicago–trained physicist Harold Arnold was hired, but he, too, was unable to discover a means by which to amplify weak electrical signals without distorting them beyond recognition. A year later, however, Jewett and Arnold were shown a device that had been developed by the independent inventor Lee de Forest, and they immediately recognized its potential to solve their problem of amplification.

The origins of Lee de Forest's device date back to the incandescent lightbulb first invented by Thomas Edison in the 1870s. Around 1880, Edison had observed dark streaks on the inner surfaces of his lightbulbs. He added a second electrode to a bulb, and found that he could use it to control and measure the flow of whatever it was that was creating those streaks. John Ambrose Fleming, an employee of the British Edison Electric Light Company, investigated this "Edison effect" and patented a modification of Edison's dual-electrode lightbulb to function as a device for rectifying current in wireless applications. In 1906, Lee de Forest modified this "Fleming valve" by adding a third electrode, a small wire grid that enabled the tube to act as a nondistorting amplifier of electrical signals. Historian Hugh Aitken has called the audion, as de Forest named his device, "without hyperbole, one of the pivotal inventions of the twentieth century."[99]

Harold Arnold later recalled his reaction to de Forest's audion. "I was amazed," he admitted. "I had made a study of repeaters and I thought that I had pretty well sized up all the repeater possibilities in the world at that time . . . and when I went into the room and saw this thing and saw how it worked I was much astonished and somewhat chagrined because I had overlooked the wonderful possibilities of that third electrode operation, the grid operation of the audion."[100] He quickly overcame his chagrin. As Frank Jewett put AT&T's lawyers to work arranging the purchase of the rights to de Forest's audion, Arnold began to think about how to improve and modify the device to meet the needs of the telephone company. By increasing the level of vacuum in the tube and by redesigning its electrodes and filaments, Arnold created a remarkably distortion-free signal amplifier that enabled the expansion of the long-distance network. In January 1915, Alexander Graham Bell, who was in New York, called his former assistant Thomas Watson in San Francisco at the Panama Pacific Exposition, and the two men re-created the historic phone call that had initiated a new era in communication back in 1876. With this call, Carty's goal of coast-to-coast telephone service became a reality.

Even as he was transforming de Forest's audion into a high-quality signal amplifier for use in long-distance telephony, Harold Arnold convinced his employers to establish a new research program to investigate the fundamental phenomena of speech and hearing, in order to have a sound basis from which to determine how best to improve the overall quality of the telephone system.[101] The physicist Irving Crandall was hired in 1913 to oversee this effort, and he quickly discovered what other academic acousticians already knew: Fundamental research was hampered by a lack of suitable tools. The first task facing Crandall's new group was thus to develop such tools for themselves.

Crandall and Arnold collaborated on the design of one of the first new tools to come out of their lab. The thermophone consisted of a wide but thin ribbon of platinum through which was passed an oscillating electrical current of known frequency. The current induced a rapid heating and cooling of the ribbon, and the temperature variation expanded and contracted the air proximate to the ribbon's surface, creating a sound wave of the same frequency as the electrical signal. The thermophone thus constituted a highly precise and controllable source of sound for use in the acoustical laboratory, and the simplicity of its physical design further enabled the physicists to calculate the absolute intensity of the sound it produced.[102]

One of Irving Crandall's first hires was Edward Wente, who was assigned the task of developing a laboratory-quality detector of sound. Wente rejected the electromagnetic instruments that George Pierce, Floyd Watson, and others had attempted to use. The insensitivity of these devices and the subsequent need to tune them to detect sounds of just one frequency, as well as their inability to measure in absolute physical units, constituted unacceptable limitations for a device of the quality and general utility that Wente sought. Carbon transmitters, while far more sensitive, were infamous for the inconsistency of their behavior. They worked well enough within the telephone system, transmitting the human voice with sufficient strength and quality to be audible and intelligible at the receiving end, but their behavior was far too unpredictable for use in laboratory investigations, as the constant movement of carbon granules under the influence of sound made a device respond differently every time it was used. Harvey Fletcher, another physicist who had joined Crandall's group in 1916, recalled studying sound with carbon microphones during his first year at Western Electric. When asked whether any of this research was ever published, Fletcher responded, "There was nothing to publish! No repeatable data!"[103]

Wente wanted a device that would combine the sensitivity of a carbon transmitter with the consistent and repeatable behavior of an electromagnetic receiver. Thanks to the efforts of de Forest and Arnold, as well as AT&T's legion of patent lawyers, Wente had at his disposal the means to create such a device. He realized that AT&T's new nondistorting vacuum-tube amplifier could provide the signal strength he required. He could thus focus on designing a highly accurate transducer—something superior to both the carbon transmitter and the electromagnetic receiver—without having to worry about the magnitude of its output.

Wente subsequently designed a microphone that used the property of electrical capacitance to register the effect of sound. A capacitor, or condenser, consists of two plates of electrically conductive material separated by air or another nonconductive material. When the two conductive elements are connected to an externally powered electrical circuit, a layer of charge builds up on each: positive charge on one plate, negative charge on the other. If the physical parameters of the condenser are changed, for example, if the distance separating the plates changes, the amount of charge stored in the device changes accordingly, and a current is created in the circuit as the device gains or loses charge. Wente designed a condenser in which a stationary steel plate was separated from a thin, flexible steel diaphragm by an air gap of several thousandths of an inch.

Impinging waves of sound vibrated the diaphragm, altering the width of the air gap and thus changing the charge-carrying properties of the condenser. In this way it created an electrical signal representing the sound. While the signal generated by the condenser was extremely weak, the new vacuum-tube amplifier was available to amplify it without distortion, no matter what the frequency of sound and signal. Wente combined his condenser transducer and vacuum-tube amplifier into a single unit and published a description of the new device in 1917.[104] (See figure 3.12.)

3.12
Cross-section of Edward Wente's condenser transmitter or microphone. The diaphragm and the plate "B," separated by a thin layer of air, created a capacitor in the electrical circuit to which the device was connected. When sound caused the diaphragm to vibrate, increasing and decreasing the width of the air gap, the capacitance of the device changed and thus changed the current in the circuit. The vacuum-tube amplifier is not shown. Edward Wente, "A Condenser Transmitter as a Uniformly Sensitive Instrument for the Absolute Measurement of Sound Intensity," *Physical Review,* 2d ser., 10 (July 1917): 43.

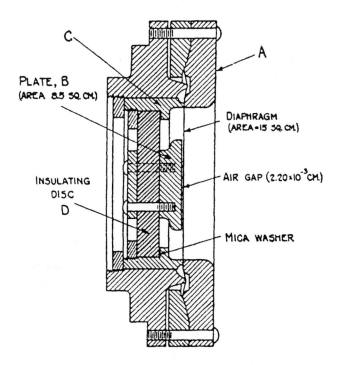

Wente's transmitter, or microphone, constituted a perfectly reproducible instrument whose measurements were equally reproducible, just what the fledgling field of acoustical research required. Furthermore, it could be calibrated with Arnold and Crandall's thermophone so that its output could be registered in absolute physical units. Wente thus provided the long-sought answer to the long-standing question of how best to measure sound. "The condenser transmitter," one enthusiast waxed, "is the most nearly perfect electro-acoustical instrument in existence." "Modern acoustics," another asserted, "have begun with his invention of the condenser microphone."[105]

The condenser transmitter formed the basis for a whole range of powerful new tools that were developed over the course of the 1920s. It served as the detector for a new electrical sound meter that measured the intensity of sound in absolute units. When attached to an oscillograph, a device that generated visual displays of electrical signals, it provided phonodeiklike records of complex sounds. When attached to tunable circuits it became a frequency analyzer, detecting and measuring the individual frequency components of complex sounds. All of these devices, which originated as laboratory prototypes, were soon being manufactured and sold as commercial products that acousticians could purchase and put to use, and by 1930 their laboratories were filled with such electroacoustic instruments. In 1900, Wallace Sabine had depended on organ pipes and human ears for his studies of reverberation. His cousin Paul indicated that, in the 1930s, the "commonplace equipment of every acoustical laboratory" consisted of "linear response microphones, vacuum tube amplifier[s] and oscillators, sensitive alternating current meters, and telephonic loud speakers."[106]

These tools not only provided acoustical researchers new means by which to study sound, they also provided new models for thinking about it. As electroacoustic transducers transformed acoustical energy into electrical signals and vice versa, the scientists who used these tools began to effect similar transformations between sounds and signals in their minds, developing new ideas about the behavior of sound and the physical objects that produced it. In the 1920s, conceptual analogies between acoustical systems and electrical circuits "sprang up spontaneously in so many places at about the same time that it seems as useless as it would be difficult to establish who did it first."[107]

Sound waves in a medium can be mathematically represented by systems of linear differential equations. These equations, and the variables within them, are analogous to the differential equations that are used to represent certain kinds of electrical circuits. As electroacoustic tools increasingly blurred the distinction between sounds and circuits, scientists began to use this analogy to transfer expertise in circuit theory to the frontiers of acoustical research. William Eccles characterized this analogy as "a language for thinking and talking" that helped "to clear the mind and assist reasoning."[108]

The key to the analogy between electrical circuits and sound was the concept of impedance. Introduced by Oliver Heaviside in the 1890s, electrical impedance was defined to be a measure of a circuit's resistance to the flow of current. In 1912, George Pierce and his Harvard colleague Arthur Kennelly studied the behavior of telephone receivers and linked the electrical impedance of the instruments to their mechanical properties.[109] In a 1914 study of the behavior of acoustical horns, Arthur Webster introduced the concept of acoustical impedance and thereby provided a further means to connect conceptually the behavior of sounds and signals.[110] Just as Heinrich Hertz had earlier drawn on his knowledge of sound to understand the new phenomena of electromagnetic waves, acousticians could now apply the mathematical equations that represented electrical circuits to problems of mechanical acoustical systems, and a body of well-established expertise in electrical theory could be drawn on to explain the behavior of those systems.

In 1925, two researchers at the newly named Bell Telephone Laboratories (formerly the research department of Western Electric) took the analogy one step further and used their understanding of circuit behavior to design a phonograph that reproduced sounds with far less distortion than any model currently available. Joseph Maxfield and Henry Harrison explained:

> The economic need for the solution of many of the problems connected with electric wave transmission over long distances coupled with the consequent development of accurate electric measuring apparatus has led to a rather complete theoretical and practical knowledge of electrical wave transmission. The advance has been so great that the knowledge of electric systems has surpassed our previous engineering knowledge of mechanical wave transmission systems. The result is, therefore, that mechanical transmission systems can be designed more successfully if they are viewed as analogs of electric circuits.[111]

By establishing the electrical analog of the mechanical phonograph, Maxfield and Harrison transformed the challenging problem of how to build a better phonograph into the straightforward task of optimizing the frequency response of the equivalent circuit. They translated their circuit back into a mechanical system, and the result was a (non-electric) phonograph that reproduced sound with much less distortion than had previous designs. The design technique that the two men employed was just as significant as the product that resulted, and, to those who studied sound, suddenly, "the whole body of electric communication network theory . . . came within the domain of acoustical engineering."[112] (See figure 3.13.)

By the mid–1920s, acoustical research was fundamentally different from what it had been circa 1900. The changes—both material and conceptual—were so dramatic, some members of the old guard were in danger of being left

3.13
Electromechanical analogies of Joseph Maxfield and Henry Harrison, 1926. Realizing that similar types of mathematical equations represent the behavior of both mechanical acoustical systems (like the phonograph sound box shown here) and certain kinds of electrical circuits, Maxfield and Harrison constructed an electrical analog of the phonograph as a means to understand and improve its performance. J. P. Maxfield and H. C. Harrison, "Methods of High Quality Recording and Reproducing of Music and Speech Based on Telephone Research," *Transactions of the American Institute of Electrical Engineers* 45 (February 1926): 343, 344. © 1926 AIEE, now IEEE.

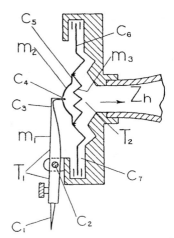

Fig. 15—Diagrammatic sketch of the mechanical system of the phonograph

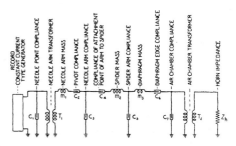

Fig. 16—Electric equivalent of the system shown in Fig. 15

behind. Arthur Webster, for example, continued to promote his phonometer and to discourage the use of electrical instruments even as his colleagues were rapidly abandoning the former for the latter. In 1919, Webster presented an account of the latest version of his phonometer to the American Institute of Electrical Engineers. In the discussion that followed, an engineer pointed out to Webster that his colleagues, "from the prejudice of their training, very much prefer to read their results on electric instruments when it is possible, rather than to observe them through a microscope as mechanical displacements." He called Webster's attention to the new electrical tools of Arnold, Crandall, and Wente, and the physicist responded, "I believe I can give more satisfactory answers to all of these telephone engineer's queries than can be got by the instruments he gets up himself. They are handy, no doubt, and all that. I remember Lord Kelvin seeing one of my instruments several years ago, and he said 'It was important that sound could be measured by electrical reading apparatus.' I do not do it this way."[113] As historian and acoustician Harry Miller put it, "The handwriting was now on the wall, but Webster would not look."[114]

If some members of the older generation of acousticians were struggling to comprehend the transformation of their field, a new generation would engage with the new instruments and concepts almost effortlessly. These men, like the young physicist Vern Knudsen, would apply the new tools to the solution of equally new problems, and would construct careers in acoustics very different from those of their predecessors.

V THE NEW ACOUSTICIAN

Vern Knudsen was born in 1893; thus a full generation separated him from Wallace Sabine, Dayton Miller, Arthur Webster, and other acoustical pioneers who had been born in the years immediately after the Civil War. As an undergraduate at Brigham Young University, Knudsen was introduced to physics by Harvey Fletcher, a young professor who had recently received his Ph.D. under the supervision of Robert Millikan at the University of Chicago. Fletcher, like Knudsen, was a Utah-born Mormon, but in 1916 he chose to leave behind his home state and his position at Brigham Young to join Harold Arnold and Irving Crandall at the Western Electric Research Laboratories in New York. A few years later, the now-graduated Knudsen followed him there.[115] Knudsen found the telephone company in the midst of mobilization for war, and he was initially assigned to a project in which Western Electric's newly invented public address systems were installed on airplanes, so that high-flying commanders could announce orders to troops on the ground. This was pioneering work in the early days of the electrical amplification and reproduction of sound, and Knudsen devoted himself to understanding the vacuum-tube amplifiers and electro-acoustic transducers that constituted this system.[116]

After a year in industry, Knudsen decided to return to school, and, as he recalled, "Chicago was the place to work for that coveted Ph.D. degree in Physics."[117] Like his mentor Fletcher, Knudsen worked under Robert Millikan's supervision, and Millikan assigned his new student the task of determining the contribution of electrons to the specific heats of metals. Knudsen was unenthusiastic about taking on a problem whose solution had eluded some of the best scientific minds of the day, so he took advantage of an unsupervised interval when Millikan was away in Europe to undertake a very different project, a study of the ability of the ear to distinguish very small differences in the intensity and frequency of sounds. He drew on his experience with electroacoustic tools to devise a means of analysis superior to the "crude" (as he put it) studies of hearing that had preceded his own, studies that had depended on old-fashioned instruments like tuning forks and bowed strings.[118] "I worked furiously," Knudsen later remembered, "using the vacuum tube technique I had acquired at Western Electric Research Laboratories. . . . Three months later Millikan returned to Chicago. My impudence seemed to startle him, but he acquiesced."[119]

Knudsen received his Ph.D. in 1922 and immediately turned down an offer to return to the research department at Western Electric. Preferring to raise his young family in southern California rather than New York City, he accepted an instructorship at the University of California at Los Angeles. University president William Campbell had written to Knudsen of the Southern Branch (as the Los Angeles campus was known): "It's only a junior college now and it ought not to be anything more, but the Chamber of Commerce of Los Angeles and other boosters down there are determined to make it a real university."[120] Before heading west, however, Knudsen joined some of his Chicago professors in a visit to the Riverbank Estate of George Fabyan, and this visit would have a profound impact on the young physicist's subsequent career.

Fabyan was most interested in showing his guests the secret messages encoded in his Shakespeare folios, but for Knudsen, the visit to the acoustical laboratory was the high point of the day. After observing the experiments and hearing of Wallace Sabine's work, Knudsen declared, "'Well, I'm going to get hold of the *Collected Papers* of Wallace Clement Sabine.' . . . This book I practically memorized. I read it and reread it. This book really influenced my career, I think, as much as anything else. I could to this day tell you almost verbatim much of what's in the Wallace C. Sabine book."[121] While Knudsen had been studying sound and hearing for the past several years, this chance encounter at the Riverbank Laboratory introduced him to architectural acoustics. Sabine's papers had originally appeared in a variety of different scientific, engineering, and architectural journals and were increasingly hard to access by the early 1920s. Fortunately for Knudsen, and for other young scientists interested in the subject, those papers had just been compiled and published in a collected edition by the Harvard University Press.[122] Knudsen was thus easily able to learn more about Sabine's work, and the more he learned, the more he was intrigued.

Upon arriving at the junior college in Los Angeles, Sabine's *Collected Papers* in hand, Knudsen soon learned that there was neither the space nor the funds necessary to carry out research in architectural acoustics. He resourcefully approached the Los Angeles Board of Education and soon had all the (apparently bad-sounding) high school auditoriums of the city at his disposal for experimentation. He also took a step toward bridging the gap between his academic salary of $2,400, and the $4,000 offer from Western Electric that he had declined, as he was paid between $100 and $200 for each auditorium whose acoustics were improved through his intervention. Twenty years earlier, Wallace Sabine had been embarrassed by the financial aspects of his acoustical research. Knudsen, in contrast, saw consulting as a viable and legitimate source of income, indeed he had counted on this when he turned down the better-paying position in industry.[123] Later recalling those early consultations, Knudsen emphasized that "the acoustical requirement of highest priority" was to obtain the right amount of reverberation. "The Sabine Formula was used for making these calculations and making the adjustments. And that plus a little attention to avoiding shapes that we knew would give rise to echoes was about the extent of it."[124]

Like Sabine, Knudsen treated these consultations as opportunities to carry out original research on the behavior of sound in rooms, and he focused his research on measuring the effect of reverberation on the intelligibility of speech. Once again drawing on his experience in the telephone industry, he borrowed a technique called articulation testing that had been developed at Western Electric for quantitatively analyzing the sound quality of telephone systems. Knudsen used the technique to analyze the acoustical quality of auditoriums.[125]

Knudsen also initiated a program to measure the acoustical properties of different architectural materials. The building materials industry in the 1920s was offering increasing numbers of sound-absorbing materials to architects for use in acoustical design. The staff of the Riverbank Laboratory, under Paul Sabine's supervision, was kept busy measuring and evaluating the performance of these new products, and the National Bureau of Standards in Washington also established an acoustical division in 1922 to test these new sound products.[126] Knudsen's first acoustical lab was, however, far less impressive than either of these facilities. It was, in fact, a converted men's washroom on the university campus. His circumstances began to improve in 1925, when the Simpson Brothers Cal-Acoustic Plastering Company provided him with a special testing room to measure the absorption coefficients of building materials, including their own. This laboratory was built on Central Avenue, in the heart of the manufacturing district of the city.[127]

Here, Knudsen employed the electroacoustic tools he had learned to use at Western Electric in order to measure the sound-absorbing properties of different materials. Knudsen's industrial experience not only influenced the way he studied sound, but it also shaped his attitude concerning the role that industry could play in scientific research. Wallace Sabine had been uncomfortable with any offer of financial assistance from commercial manufacturers, striving to protect his scientific reputation by avoiding even the appearance of undue influence. Knudsen, in contrast, was perfectly comfortable working in a facility constructed for him by one of the very manufacturers whose products he was to evaluate. The laboratories of Western Electric had constituted for Knudsen a context in which legitimate science and corporate concerns not only coexisted, but mutually prospered. Perhaps for this reason, Knudsen was never concerned with the conflicts of interest, real or imagined, that had so consumed Wallace Sabine.

Indeed, Knudsen maintained his working relationship with his friends back east at Western Electric, and they provided him with state-of-the-art equipment for use in his research. In 1925, for example, the physics department at UCLA received a gift of new loudspeakers from the telephone company. These speakers were used, not just for acoustical research, but also to broadcast the inaugural address of President Calvin Coolidge to students and faculty gathered in the university auditorium.[128] The acoustical products of Western Electric would play an even greater role in Knudsen's career a few years later, when the new sound motion picture system recently developed by the telephone company arrived in Hollywood.

When Western Electric first presented its new sound system to the motion picture industry in 1925, the response of the studios was a deafening silence. Countless past efforts to make the movies talk had resoundingly failed, and there was no reason to think that this latest attempt would be any different. Only one decidedly second-rank studio, Warner Brothers, was willing to experiment with the new technology. By 1928, however, the industry's initial reluctance had fallen away, and virtually every studio was now trying to catch up with Warner Brothers and create for themselves the phenomenal success of that studio's new talking films. These sound movies had been produced in New York, in studios proximate to the technical expertise of the telephone engineers, but within a few years, all of the major studios were building new soundstages in Hollywood, and all were in need of acoustical expertise. Knudsen, fortuitously located in Los Angeles, was willing and able to provide it.

The first studio to seek his advice was Metro-Goldwyn-Mayer, and Knudsen's success there led to similar consultations with Paramount, Fox, Universal, and Warner Brothers.[129] The income from these projects was substantial; MGM alone paid over $2,500 for his services. Knudsen recalled his financial dealings with the studio's business manager, Eddie Mannix:

> I had charged Metro-Goldwyn-Mayer $12.50 an hour, which figured out at $100 a day. It was considered that this was a fair amount for a top consultant in those days. I considered it very good pay for a young assistant professor. But after I had completed my work on stages A and B, Mr. Mannix called me in just for a personal conversation, and he said, "Knudsen, I want to give you some personal advice." He said, "Your services on the stages A and B were worth so much more than you charged us that I think hereafter you should make your charge on the basis of a fixed fee for the services you are going to perform." And he further said, "We want you to help us with the design of some more sound stages. I think $2500 would be a more reasonable fee."[130]

Nor was soundstage design the only opportunity for Knudsen to capitalize on his expertise. The silence of the new stages revealed noise from the air-conditioning system that was required to counteract the heat of the studio lights. Eddie Mannix informed the Carrier Corporation that they would have to find a way to design the noise out of their equipment, and he recommended Knudsen as the man to do it for them. Knudsen was retained by Carrier for a fee of $3,000 per year plus $100 per working day. Knudsen recalled, "For a young assistant professor who was getting probably $2,700 or at most $3,000 a year at this time [in salary], this, my first retainer, was a real windfall. The retainer may sound very high for a professor, but the retainer and per diem were suggested by Carrier themselves. I worked for them three years (probably five to ten days a year) and they survived it very well."[131]

While Sabine had been well paid for his collaboration with the Guastavino Company in 1911, Knudsen now had far more opportunities to garner even greater pay circa 1930. The world was now filled with sound products that had not existed when Sabine was alive. These products not only shaped the contours of his scientific research, they also provided a market for acoustical expertise that was filled with new opportunities for entrepreneurial consultants.

Knudsen thrived in this world and was energized by the commercial application of his expertise. While he acknowledged that Wallace Sabine's work had originally inspired him to make a career in acoustics, he also recognized that he lived in a different age. "This modern era of acoustics," he argued, "began in 1915, when the thermionic vacuum tube and the high-quality microphone became practical devices for research as well as for telephony and radio communication."[132] His career in acoustics was as different from Sabine's as Los Angeles in the 1920s was different from Boston circa 1900, and that difference was evident even in how the two men spent their money. Sabine had joined St. Botolph's Club, a stately Back Bay institution whose membership had included generations of Cabots and Adamses.[133] Knudsen, in contrast, indulgently spent some of his own consulting windfall on a lifetime membership to the Gables Beach Club in Santa Monica, just south of the magnificent oceanfront mansion that William Randolph Hearst had built for his movie star mistress, Marion Davies.[134]

In 1928, Floyd Watson and his former student Wallace Waterfall happened to be visiting the West Coast, and Knudsen invited the men to join him for dinner at the Gables Beach Club. Waterfall, who worked for a manufacturer of sound-absorbing building materials, had proposed the idea of forming "some sort of organization that would foster both research and the exploitation of acoustical materials in the treatment of rooms," and Knudsen was eager to discuss this idea with his colleagues.[135] All three men agreed that the field of acoustics had flourished to the point where it now seemed both useful and possible to form a new scientific society. In addition, there was another, somewhat more troubling reason for considering the formation of an acoustical society.

Knudsen recalled that, in the 1920s, acoustics was considered by many fellow scientists to be a "has-been branch of physics." His colleagues at the University of Chicago had thought that he was "off the beam" when he chose to pursue acoustical research for his doctoral dissertation, and many continued to believe that Rayleigh's monumental *Theory of Sound* had pronounced "the last word in acoustics" at the end of the previous century. By 1925, new fields like relativity and quantum mechanics constituted the cutting edge of modern physical research, and acoustical physicists like Knudsen began to feel a "second-rate citizenship" within the American Physical Society.[136] Harvey Fletcher recalled similar feelings of frustration, noting that, when he gave papers at meetings of the American Physical Society, "nobody seemed to be interested; nobody would listen to them."[137]

If relatively young men like Knudsen and Fletcher were feeling cut off from the cutting edge of physics, it was even more difficult for the older generation of acousticians to adjust. Dayton Miller, who had been dubbed the "Wizard of Visible Sound" in the teens, became equally well known in the 1920s when he claimed that he had disproved Einstein's theories of modern physics.[138] Like Miller, Arthur Webster was also fundamentally unable or unwilling to grasp the new physics, in this case with tragic results. Apparently afflicted with severe bouts of depression, Webster committed suicide in 1923, leaving behind a note explaining that "physics had gotten beyond me, and I can't catch up."[139]

Thus, when Knudsen, Watson, and Waterfall met at the Gables Beach Club in 1928 to discuss forming a new society, they not only sought to organize the growing community of acoustical researchers. They also wanted to create a place where they could be judged on their own merits, free from the criticism of others who might look down on the inherently applied nature of their work or look askance at the distance that separated it from the exciting new theoretical developments in relativity and quantum mechanics. While Knudsen, Waterfall, and Watson all specialized in architectural acoustics, the new organization would be open to all scientists and engineers generally interested in sound.[140]

Harvey Fletcher was an early enthusiast for the project, and he offered to sponsor an organizational meeting at the Bell Telephone Laboratories in New York. Forty academic and industrial scientists and engineers came together there in December 1928 and formally established the Acoustical Society of America. (See figure 3.14.) A membership drive resulted in a charter membership of 457 in 1929. By 1932, there were almost 800 members who constituted "a mingling of many disciplines besides acoustical engineers and acoustical physicists; there were psychologists; there were musicians, otologists, phonoticians, and you name almost anything associated with acoustics, and there was representation there."[141]

The new organization was as fiscally secure as it was diverse, with corporate support coming from musical instrument manufacturers (American Piano Co., Baldwin Co., C.G. Conn Ltd.); manufacturers of architectural materials and products (American Seating Co., Celotex Co., Johns-Manville); several corporate divisions of AT&T; and the United Research Corporation, an industrial laboratory devoted to sound reproduction.[142] The presence of industry in the new organization was evident not only in the list of sponsors, but also within the ranks of the members themselves. At least 80 percent of the charter members were affiliated with corporations offering different acoustically based products and services.[143]

The first official meeting of the Acoustical Society of America was held, again at Bell Laboratories in New York, in May 1929. After an introductory

3.14
Organizational meeting of the Acoustical Society of America, December 1928. The meeting took place at Bell Laboratories in New York City. The first row includes Dayton Miller (3d from left), Wallace Waterfall (4th), Vern Knudsen (5th), Harvey Fletcher (6th), and Floyd Watson (3d from right). Edward Wente is the second man from the right in the top row. Reprinted with permission from the *Journal of the Acoustical Society of America* 26 (1954): 882. © 1954, Acoustical Society of America.

presentation by Harold Arnold, in which an early form of stereophonic recording was demonstrated in a joint session with the Society of Motion Picture Engineers, the new society got to work. The first regular session was a symposium on the various methods for measuring the absorption coefficients of materials. Paul Sabine, Vern Knudsen, and Edward Wente each presented different techniques for determining the sound-absorbing properties of materials. This concern over tools and techniques and the establishment of standard practices was evident throughout the meeting, as over half of the twenty-two papers dealt in some way with the measurement of acoustical phenomena.[144] The society's new journal, which was largely composed of published versions of the papers presented at the society's meetings, disseminated this same concern with tools and techniques to its readers. With the common forum of a professional society and journal now in place, however, it would not be long before these issues of standardization would be resolved, and by 1934, acoustical standards—of nomenclature, instrumentation, and methodology—were fully codified by the Acoustical Division of the American Standards Association.[145]

When William Eccles heralded "The New Acoustics" to the Physical Society of London in 1929, he was describing the work of the people who came together to form the Acoustical Society of America. When he highlighted the new techniques, ideas, and jargon that now characterized studies of sound, he identified the elements that had helped to forge that community. When Dayton Miller presented a historical address to the Acoustical Society in 1932, he provided one final but equally crucial element, the construction of a common heritage. The early histories that acousticians chose to tell about themselves say as much about their situation circa 1930 as they do about their past, thus these stories deserve a careful hearing. When one listens closely, it becomes apparent that the dominant theme of optimism so harmoniously expressed in these accounts was accompanied by the occasional dissonance, and a subtle counter melody in a decidedly minor key. Even the newest of New Acousticians recognized the unfamiliarity of the place in which they found themselves, and their histories provided a strategy for establishing a sense of permanence in a rapidly changing world.

VI Conclusion: Sabine Resounded

In 1933, a biography of Wallace Sabine appeared. Its introductory chords set the celebratory tone for the 350 pages that followed:

> The life of Wallace Sabine embraces the fundamental history of a new science and the romantic story of its discovery. What Morse did for the Telegraph, what Edison did for the Electric Light, what Alexander Bell did for the Telephone, what Marconi did for the Wireless—Sabine did for the Science of Acoustics, by solving the mystery of the intricacies of Sound which had baffled investigators from the time of ancient Greece.[146]

It would be easy to dismiss William Dana Orcutt's hagiographic volume, since Sabine's widow apparently "approved every line in it."[147] Yet Orcutt clearly identified the kind of story that his audience wanted to hear, and his account resonated strongly with acousticians in search of a history for their discipline. While Orcutt's biography thus explicitly tells the story of Wallace Sabine, it also speaks—more implicitly but more interestingly—of the lives of those who followed. Perhaps most telling is the tension, evident throughout the account, between Orcutt's desire to emphasize the practical nature of Sabine's accomplishments and an equally strong desire to portray the physicist as a "pure" scientist, isolated from and innocent of the larger world that lay beyond his experiments and ideas.[148]

Orcutt opened his account by placing Sabine in a pantheon of great inventors. Modern acousticians would have recognized those figures—Morse, Bell, Edison, Marconi—as the very men whose ingenuity had initiated the new era of electroacoustic technology that increasingly shaped their own lives. Commercial products like telephones, phonographs, and radios, as well as the new scientific tools based on those same technologies, defined the contours of their careers. But what did Sabine really have in common with these inventors, and how were his scientific accomplishments related to their technological innovations?

The awkward way in which Orcutt rhetorically implied that Sabine had invented the "Science of Acoustics" just as Bell had invented the telephone only hinted at the difficulties that lay ahead, as the author attempted to construct an account that offered the best of both worlds. Sabine's accomplishments, according to Orcutt, were just as practical as these revolutionary inventions, but they simultaneously constituted "Science" in a way that technological devices and commercial products clearly did not. Modern acousticians accepted this equation because it enabled them to connect Sabine's story directly to their own technologically and commercially based careers, while still allowing them to claim a scientific pedigree. With such a pedigree, they could alleviate that sense of second-rate citizenship in the community of physicists that Knudsen and Fletcher had articulated. Vern Knudsen recalled that, when the Acoustical Society of America was being organized, there were lengthy discussions over what would constitute an appropriate balance between physics and engineering, and he struggled to achieve a similar balance in his own career.[149]

In the early 1930s, while investigating the effect of humidity on reverberation, Knudsen decided to expand the scope of his study to explore more fundamentally the absorption of sound by gases, thereby moving his work into the realm of molecular physics. He did this partly to achieve a basic understanding of the acoustical phenomena that he studied, but also because he wanted to prove to himself—and to others—that he was capable of carrying out pure scientific research that would have an impact in physics beyond his immediate and practically oriented community of architectural acousticians. Knudsen recalled this period as the intellectual high point of his life, and he relished the prize that he received when he presented this work to the American Association for the Advancement of Science.[150]

Having proved his mettle as a pure scientist, Knudsen was subsequently more comfortable with the practical orientation of his career. Orcutt's characterization of Sabine similarly, if awkwardly, combined the perceived virtues of pure science with those of utility, and later histories of Sabine's career would echo this refrain. "There was never, I suppose, a more thoroughly scientific mind than his, in point of the eagerness with which he pursued the truth," Paul Sabine eulogized.

> But that eagerness was excited only by a problem whose solution was of more than purely academic interest. Knowledge which could be translated into terms of practical utility and human betterment was what he sought.
>
> This rare combination of the completely scientific and the intensely practical in Sabine's mental equipment characterized all of his scientific work.[151]

For those unable to achieve this rare combination for themselves, accounts like Sabine's and Orcutt's allowed them to experience it vicariously.

Still, the utilitarian nature of Sabine's work had to be treated delicately in an era in which the boundaries between acoustical science and commerce were hard to distinguish. Sabine himself had struggled to discern how best to enjoy the commercial benefits of his expertise while maintaining his scientific reputation, and his biographers similarly struggled to strike an appropriate balance in their accounts of his life. Sabine's obituary noted that, while the main purpose of his work was, "of course, utilitarian," it was so only "in a highly refined sense," whatever that might mean.[152] More typically, Sabine's biographers solved this problem by simply denying him any real monetary reward for his commercial endeavors.

Orcutt portrayed Sabine as uninterested in and incapable of profiting from his expertise. As a friend of the physicist put it, "Sabine changes his personality when he takes off his laboratory coat and puts on his business suit." According to Orcutt, this business suit didn't fit well at all, for Sabine "could not bring himself to charge proper fees for his own services." Winthrop Ames spoke for many architects who were required to solicit bills from a reluctant Sabine when Orcutt quoted him as saying, "I was very much impressed with the complete absence of any commercial instinct in Professor Sabine's make-up," and Orcutt further claimed that the income generated from Sabine's patents with Guastavino was devoted solely to "furthering his experiments."[153]

The young readers of *Careers* considering a career in the field of acoustics in 1931 were likewise informed that Sabine had singlehandedly brought architectural acoustics from the stage of "rule-of-thumb practice" to the "status of a reasoned science and a precise art" with so little financial assistance that "he was probably a poorer man by thousands of dollars than he would have been had he never attempted it. Moreover, he had published his formulae and procedures freely to the world, for anyone to use who could."[154] Yet, *Careers* also made clear that the careers of future acousticians would certainly differ from Sabine's. What awaited them was not the intellectual and altruistic adventure of new scientific discoveries, but rather "innumerable opportunities for the application of acoustical engineering," particularly in service to the large corporations dedicated to the commercial value of the control of sound that Sabine had ostensibly refused to acknowledge.[155]

The careers of modern acousticians were defined by new markets for sound control. These men dedicated themselves to the manufacture and application of sound-absorbing building materials; the reduction of noise on city streets and in offices and apartments; the reception and reproduction of sound signals in radios, phonographs, and telephones; and the installation of new systems for talking motion pictures. This was a world that Sabine had glimpsed but never inhabited. After the war, he was precariously poised to enter it, but his death prevented him from taking a decisive step forward into the realm of modern acoustics. Those who followed would necessarily enter and engage with this new world, but, while they would generally thrive there, they were nonetheless impelled to look back with longing to an earlier era.

Sabine's biographers might have pulled him forward, emphasizing what he had in common with the members of the Acoustical Society of America. Instead, they chose to push him backward into the past, to emphasize the differences until "he seemed almost of another age and civilization."[156] By doing so, they created a deeper history for their new profession, establishing an anchorage in a sea of change. By dissociating Sabine from his commercial associations, they projected their scientific origins back to a mythic time in which the lines between science and business were easily drawn and seldom crossed. They also increased the historical distance between Sabine and themselves by focusing, not on his later career, but instead on his early work in the Constant Temperature Room of the Jefferson Physical Laboratory, his discovery of the hyperbolic relationship behind the reverberation equation, and its application in Symphony Hall. In this way, they not only told and retold the moment of discovery from which all their careers had sprung, but they also described a young and innocent nineteenth-century professor of physics. Over the next twenty years of his life,

Sabine would mature and change; he would live amid and contribute to the creation of twentieth-century culture. But this modern aspect of Sabine was always overshadowed by the youthful investigator of the Fogg lecture room.

Even in his obituary, Sabine—a fifty-year-old man at the time of his death—was characterized as an "elfin being," in possession of a "still youthful face."[157] When that face was reproduced in articles and books, it was virtually always depicted with a photograph that had been taken back in 1906; a portrait of an earnest, old-fashionedly attired young man. (See figure 3.15.) Orcutt used this image of Sabine as his frontispiece, but his biography also included a photograph of Sabine that had been taken in 1918. This image, buried deep in the back of his text, portrayed a much older-looking man—now balding, no longer slender, in a far more contemporary style of dress—but such a modern image of

3.15
Wallace Sabine in 1906. In this portrait, the most frequently reproduced image of Sabine, the sober young scientist in old-fashioned attire appears as a figure from a long-distant past. William Dana Orcutt, *Wallace Clement Sabine* (Norwood, Mass.: Plimpton Press, 1933), frontispiece.

3.16

Wallace Sabine in 1918. The mature Sabine depicted in this portrait, with his receding hairline and modern attire, is seldom encountered in early historical accounts of the field of acoustics. Reprinted with permission from the *Journal of the Acoustical Society of America* 26 (1954): 887. © 1954, Acoustical Society of America. Photograph courtesy Riverbank Acoustical Laboratories, IIT Research Institute.

Sabine would not appear elsewhere for many years.[158] The modern Sabine was more problematic, and the problems that he had encountered were yet to be resolved by his followers. (See figure 3.16.)

Perhaps the most astute chronicler of the dilemmas facing modern acousticians was not a scientist at all, but a perceptive outsider. In his Pulitzer Prize-winning novel of 1925, Sinclair Lewis portrayed the heroic struggles of a medical scientist named Martin Arrowsmith. The odyssey of *Arrowsmith* is not only a quest for scientific truth, but also a search for the proper place to pursue that truth, a place free of all influence except the drive to know.[159]

Over the course of his career, Arrowsmith moves from the inauspicious beginnings of small-town doctoring, to the political limelight of the public health department of a midwestern city, to the penthouse-suite laboratory of a private research institute in New York City. The intellectual ideals toward which Arrowsmith strives are represented by his mentor Max Gottlieb, who began his own scientific career by studying acoustics with Hermann Helmholtz. Arrowsmith struggles to achieve Gottlieb's ideal of pure science amid the materialism that pervades American culture, and he ultimately concludes that the modern world offers no haven to the scientist. The corrupting influences of profit-seeking corporations, of politics, and of publicity-seeking philanthropy are everywhere. Arrowsmith must ultimately leave this modern world behind and escape to a pastoral past—a rustic cabin deep in the woods of Vermont—to create the pure science that fills his dreams.

Like Arrowsmith, Sabine struggled to find a way to create pure science in the midst of an impure world, and, like Arrowsmith, he ultimately planned to retreat from that world in order to accomplish his goals. Not to the Vermont woods, but instead to an isolated acoustical laboratory on the banks of the Fox River in rural Illinois. Sabine died before he could make that retreat, but his historians effected the isolation nonetheless. Unable to move him to Riverbank, they instead returned Sabine to the site where their science had originated, the Constant Temperature Room of the Jefferson Physical Laboratory at Harvard. By constantly retelling the story of the origins of architectural acoustics, they preserved the image of a youthful investigator cut off from noise, corruption, and worldliness in an isolated subterranean chamber. The birthplace of their science became Sabine's tomb, the "shrine of all acoustical engineers."[160] Even as they buried Sabine, however, and left his youthful ghost to haunt that silent chamber, the New Acousticians moved out into the noise and complexity of the modern world. The transformations that had occurred within their scientific community were only instantiations of much larger changes at work in that world, and the sounds of modern acoustics echoed far beyond the walls of their laboratories, constituting a pervasive new soundscape that the modern acousticians eagerly claimed as their own.

CHAPTER 4   NOISE AND MODERN CULTURE, 1900–1933

> "What news from New York?"
> "Stocks go up. A baby murdered a gangster."
> "Nothing more?"
> "Nothing. Radios blare in the street."[1]
>
> F. Scott Fitzgerald, "My Lost City," 1932

I INTRODUCTION

Writing from the depths of the Great Depression in 1932, F. Scott Fitzgerald looked back on the decade that had roared. He recalled that roar as so characteristic, so ubiquitous as to be remarkably unremarkable. Fitzgerald's contemporaries may have been less blasé, but many shared his belief that New York was defined by its din. In 1920, a Japanese governor visiting the city for the first time noted, "My first impression of New York was its noise." While initially appalled by the clamor that surrounded him, he soon became enamored of the task of listening to the noise and identifying individual sounds within the cacophony. "[W]hen I know what they mean," he explained to a reporter, "I will understand civilization."[2] The pervasive din of New York was, for Fitzgerald, foreign visitors, and countless others, the keynote of modern civilization. Some chose to celebrate this noise, others sought to eliminate it. All perceived that they lived in an era uniquely and unprecedentedly loud.

Yet it seems that people have always been bothered by noise. Buddhist scriptures dating from 500 BCE list "the ten noises in a great city," which included elephants, horses, chariots, drums, tabors, lutes, song, cymbals, gongs, and people crying "Eat ye, and drink!"[3] And complaints of noises similar to those compiled by the Buddha (excepting perhaps the elephants) have been voiced continually over the course of the centuries. The ruins of ancient Pompeii include a wall marked by graffiti that pleads for quiet.⁴ An anonymous fourteenth-century European poet complained that "Swart smutted smiths, smattered with smoke, Drive me to death with the din of their dints."⁵ The din of eighteenth-century London was well captured by William Hogarth (see figure 4.1), and the acoustical distress experienced by his "Enraged Musician" was suffered by countless other urban inhabitants as cities' populations increased more rapidly than their geographies expanded.⁶ As the congestion resulting from urbanization further concentrated the noises of everyday life in the nineteenth century, the frequency (as well as the urgency) of complaint rose. Goethe hated barking dogs; Schopenhauer despised the noise of drivers cracking their horsewhips.⁷ Thomas Carlyle was compelled to build a soundproof room at the top of his London townhouse to escape from the sounds of the city streets.⁸

The sounds that so bothered Carlyle and Goethe were almost identical to those that had been identified by the Buddha centuries earlier: organic sounds created by humans and animals at work and at play. These sounds constitute the

4.1
"The Enraged Musician," William Hogarth (1741), as engraved by W. H. Watt. Hogarth's print vividly evokes the noise of an eighteenth-century city street. It further indicates the almost exclusively organic nature of that noise by casting people and animals as its primary source. William Hogarth, *Hogarth Moralized* (London: J. Major, 1831), facing p. 138. Graphic Arts Collection. Department of Rare Books and Special Collections. Princeton University Library.

constant sonic background that has always accompanied human civilization. With urbanization they were certainly concentrated; with industrialization, however, new kinds of noises began to offend. The sound of the railroad, for example, became a new source of complaint. The noise of its steam whistle was disturbing not only for its loudness but also for its unfamiliarity. Carlyle could only express his distress at its mechanical scream in terms of his old, familiar enemies, comparing it to the screech of ten thousand cats, each as big as a cathedral. Over the course of the nineteenth century, the clanking din of the factory, the squeal of the streetcar, and other new sounds were increasingly incorporated into the soundscape.[9] In spite of the presence of these new sounds, however, lists of complaint continued to emphasize the traditional noises of people and animals. In America at the turn of the twentieth century, this emphasis remained.

When Dr. J. H. Girdner cataloged "The Plague of City Noises" in 1896, almost all the noises he listed were traditional sounds: horse-drawn vehicles, peddlers, musicians, animals, and bells. "Nearly every kind of city noise," he reported, "will find its proper place under one of the above headings."[10] Less than thirty years later, however, this plague had mutated into a very different organism; indeed, by 1925 it was no longer organic at all:

> The air belongs to the steady burr of the motor, to the regular clank clank of the elevated, and to the chitter of the steel drill. Underneath is the rhythmic roll over clattering ties of the subway; above, the drone of the airplane. The recurrent explosions of the internal combustion engine, and the rhythmic jar of bodies in rapid motion determine the tempo of the sound world in which we have to live.[11]

Not long thereafter, the amplified output of electric loudspeakers was added to the score, and the transformation was complete. When New Yorkers were polled in 1929 about the noises that bothered them, only 7 percent of their complaints corresponded to the traditional sounds that Girdner had emphasized in 1896. The ten most troubling noises were all identified as the products of "machine-age inventions," and only with number eleven, noisy parties, did "the sounds of human activity" enter the picture.[12] Clearly, the sound world circa 1930 had little in common with that of 1900. (See figure 4.2.) To those who lived through that transformation, the change was dramatic and deeply felt. Some were energized, others enervated; all felt challenged to respond to the modern soundscape in which they now lived.

That challenge was stimulated not simply by the noise itself, but also by social and cultural forces at work in urban America. To those who perceived it as

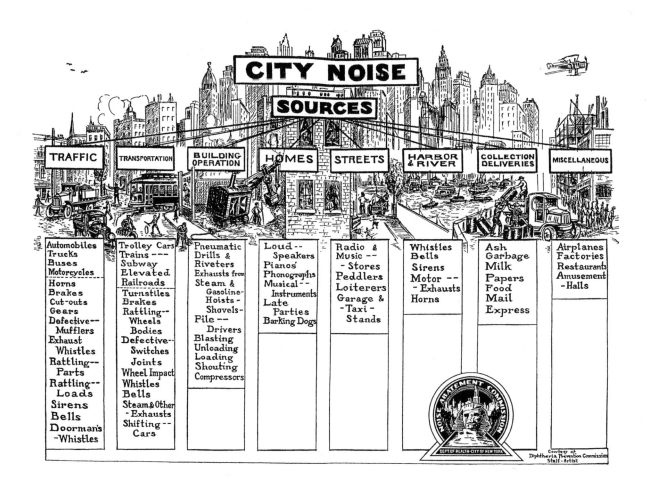

4.2
"City Noise." The frontispiece for the 1930 report of the Noise Abatement Commission of New York City made clear that the soundscape of the modern city was no longer dominated by the sounds of humans and animals, but instead by the noises of modern technology. Edward Brown et al., eds., *City Noise* (New York: Department of Health, 1930).

a problem, noise was just one of the many perils of the modern American city, including overcrowded tenements, epidemic disease, and industrial pollution. "Noise reform" was part of a larger program of reform that included urban planning, public health programs, and other progressive efforts to apply expert knowledge to the problems of the modern city.[13] Doctors warned of the danger that noise posed to physical and mental health, while efficiency experts proclaimed the deleterious effect of noise upon the nation's productivity. Concerned citizens pushed for antinoise legislation in an effort to impel what they considered appropriate behavior, to guarantee the public its right to an environment free of unnecessary noise. The efforts and actions of these noise-abaters, well covered in newspapers and magazines, drew increased attention to the problem of noise.

Not everyone recoiled from the soundscape of the modern city, however, and some were more constructively stimulated by the new sounds that surrounded them. Jazz musicians and avant-garde composers created new kinds of music directly inspired by the noises of the modern world. By doing so they tested long-standing definitions of musical sound, and they challenged listeners to reevaluate their own distinctions between music and noise. Some of these listeners met the challenge and embraced the new music, while others refused to listen.

The problem of noise was further amplified in the 1920s by the actions of acoustical experts. Like the musicians, these men constructed new means for defining and dealing with noise in the modern world. For the first time, scientists and engineers were able to measure noise with electroacoustical instruments, and with this ability to measure came a powerful sense of mastery and control. Acousticians were eager to step into the public realm, to display their tools, and to demonstrate their expertise as they battled the wayward sounds. Their unprecedented ability to quantify the noise of the modern city further heightened public awareness of the problem as well as expectation of its solution.

That solution would prove elusive, however, as even the most technically proficient campaigns for noise abatement struggled to effect change within the public soundscape. By the end of the decade, urban dwellers were forced to retreat into private solutions to the problem of noise. Acoustical expertise was brought back indoors, and acousticians devoted themselves to the construction of soundproof buildings that offered refuge from the noise without.

Thus, while noise has always been a companion to human activity, and while it has always been a source of complaint, the particular problem of noise in early-twentieth-century America was historically unique. The physical transformation of the soundscape, as well as the social and cultural transformations taking place within it, combined to create a culture in which noise became a defining element. Noise was now an essential aspect of the modern American experience. It generated an "intense American excitability;" it was an "American symptom."[14] "There is nothing fanciful," the *Saturday Review of Literature* editorialized, "in the assertion that the pitch of modern life is raised by the rhythmic noise that constantly beats upon us. No one strolls in city streets, there is no repose in automobiles or subways, nor relaxation anywhere within the range of a throbbing that is swifter than nature. Our nervous hearts react from noise to more noise, speeding the car, hastening the rattling train, crowding in cities that rise higher and higher into an air that, far above the grosser accidents of sound, pulses with pure rhythm."[15] Simply put, the Roaring Twenties really did roar. By listening to that roar as acutely as did that Japanese visitor many years ago, we can understand more fully the civilization that produced it, as well as the culture that civilization constructed to comprehend it.

II Noise Abatement as Acoustical Reform

In 1853, Henry David Thoreau was awakened from his agrarian reverie at Walden Pond by the screaming whistle of a passing train. Yet, as he listened, Thoreau realized that it was not just the train that was passing, but also the old ways of life he was attempting to perpetuate. As Leo Marx has shown, Thoreau, Nathaniel Hawthorne, Ralph Waldo Emerson, and many other nineteenth-century American writers struggled with mixed emotions about the coming of industry. The steam whistle, which announced the arrival of both railroad and factory, constituted the acoustic signal of industrialization. Writers used it to punctuate their stories of the American pastoral experience, and to delineate what they perceived to be "the opposing forces of civilization and nature."[16]

But generally speaking, most nineteenth-century Americans celebrated the hum of industry as an unambivalent symbol of material progress.[17] Complaint might be voiced, but few were willing to slow the machines of progress to appease the complainants. In 1878, the noise of the new elevated trains in New York was dismissed with the simple statement that it "has to be."[18] Noise nuisance lawsuits were easily defended, as it was only necessary "for lawyers in such cases to establish as a defense against a plaintiff that the noise was a part of the very necessary industrial processes and that the industry was a very necessary part of the community and therefore the noise had to be tolerated as a necessary evil."[19] This association of noise with progress and prosperity echoed well into the twentieth century, and in 1920 noises were still being celebrated as "the outward indications of the qualities of civilization." "Civilization," it was argued, "the greatest of all achievements, is by that token, of all, the most audible. It is, in fact, the Big Noise."[20]

Well before 1920, however, many Americans had begun to argue the opposite, that noise was the enemy of progress, the sign of a distinct lack of civilization. The *Nation* asserted in 1893 that "the progress of a race in civilization may be marked by a steady reduction in the volume of sound which it produces."[21] There are some, argued Mrs. Isaac Rice in 1907, "who claim that racket and prosperity are synonymous." But, she continued, "the 'hum' of industry has now made way for the shriek of industry, and it is perhaps well to call attention here to the fact that noise is not an essential part of progress."[22] In fact, Mrs. Rice took on this task herself, becoming the leader of an influential group of citizens who called attention to the problem of noise and attempted to regulate the soundscape of the modern city.

Julia Barnett Rice was the medically trained wife of the wealthy businessman and publisher Isaac Rice, whose magazine, the *Forum*, would become a resounding voice for noise reform once Mrs. Rice made this her mission.[23] The Rices resided in an Italianate mansion at Riverside Drive and 89th Street in New York, but the tranquil life at Villa Julia was increasingly disrupted after 1905 by the piercing steam whistles of tugboats on the Hudson River. Mrs. Rice hired students from Columbia University to monitor the situation, and they counted almost 3,000 whistles in just one night. While obviously motivated by her own family's discomfort, Mrs. Rice was more concerned about the effect of this noise upon the many patients in hospitals that were within earshot of the city's rivers. Interviews with riverboat captains convinced her that the majority of whistles were social calls not relating to navigation or safety, and she thus began a campaign to eliminate them.

Over the next year, Mrs. Rice was directed from one bureaucratic office to another, as each official—city dock commissioner, warden of the port, police commissioner, steamboat inspector, U.S. secretary of commerce and labor—assured her that someone else was responsible for the problem. She succeeded in attracting attention to her cause, if not in eliminating the noise. Numerous doctors attested to the harm that the whistles caused their patients, and 3,000 assumedly healthy neighbors of Mrs. Rice signed a petition against the noise that was delivered to the Board of Health. By the end of 1906, New York congressman William Bennet had joined the campaign, and he introduced federal legislation that forbade the unnecessary blowing of whistles in ports and harbors. The Bennet Act became law early in 1907, and Mrs. Rice experienced her first taste of victory.[24]

In December 1906, seeking to expand the field of engagement, Mrs. Rice organized the Society for the Suppression of Unnecessary Noise. By enlisting the support of "scores of prominent men and women," she hoped particularly to improve circumstances for the sick and mentally ill by focusing on the prevention of noise in and around the city's hospitals.[25] Dr. George Hope Ryder, of the Sloane Maternity Hospital at 59th St. and Amsterdam Ave., described the noises that plagued his patients in prose that brings to mind the modern poetry that fellow physician William Carlos Williams was then just beginning to write:

> Electric cars crash by with whining motors and the pounding of flattened wheels. Wagons rattle past over cobblestones. Automobiles flash by, blowing horns or siren whistles. Drunken people argue and fight on the sidewalks. Children shout, pound on tin cans, and even set off firecrackers under the windows. Hucksters stand and cry their wares in front of the buildings.[26]

"These noises," Ryder explained, "are not merely an annoyance; they are a serious menace to the health of sick patients."[27] To tackle this menace, the society enlisted the support of Health Commissioner Thomas Darlington, as well as doctors from sixteen of the city's hospitals, Congressman Bennet, several university presidents, and the novelist William Dean Howells, who declared, "You can hardly voice my protest against unnecessary noises too strongly. The volume of sound seems to be increasing year by year."[28]

Although the papers described the organization as an "anti-noise" society, Mrs. Rice emphasized that its efforts would be dedicated to eliminating only unnecessary noises. The society recognized the fact that much noise was simply unavoidable, and its members had no desire to interfere with the vital commerce and business of the city. This emphasis enabled them to enlist the support of business organizations that might otherwise have resisted their efforts. It also tapped into a larger cultural trend that was increasingly valorizing the principle of efficiency and its corollary, the elimination of all things unnecessary.

As early as 1888, noise had been recognized as unnecessary to the performance of most useful work. "It means waste, wear and tear in the majority of cases," Dr. Walter Platt reported. "The most perfect are the most noiseless machines, and this applies to the social organism as well."[29] William Dean Howells argued that it was "the needlessness of most noises that renders them unsufferable," and the *New York Times* agreed that needless noises "should be dealt with on the plain ground that they are needless, and by that fact objectionable."[30] Noise was compared to smoke, and campaigns for noise abatement were clearly inspired by earlier efforts toward the abatement of smoke. In these campaigns, the popular perception of smoke had been transformed from an indicator of industrial prosperity to a sign of industrial waste, untapped resources, and poorly designed processes. The same rhetorical strategies were employed in the fight against noise.[31]

Historians of noise abatement, particularly those who wrote in the 1970s, have emphasized the connection between noise and smoke in a way that may say more about their own historical context than that of their subjects. Raymond Smilor, for example, identifies the problem as "noise pollution" in a way that connects noise abatement directly to the antipollution movements of his own era, and R. Murray Schafer's writings are similarly imbued with an environmental strain.[32] While noise reformers did compare noise to smoke, it is not evident that they did so with late-twentieth-century ideas of pollution in mind.[33] It thus seems more appropriate to situate noise abatement in the cultural context of efficiency, in order to convey best how the noise abaters understood themselves and their actions.

Historian Samuel Haber has asserted that "an efficiency craze—a secular Great Awakening" occurred at the turn of the century, as "efficient and good came closer to meaning the same thing in these years than in any other period."[34] Waste, whether of natural resources, human labor, or time, was the enemy and efficiency the means by which to conquer it. "With the spreading of the movement toward greater efficiency," *Harper's Weekly* proclaimed in 1912, "a new and highly improved era in national life has begun."[35] If efficiency was a religion, its high priest was the mechanical engineer Frederick Winslow Taylor, who dedicated his life to rooting out the inefficiencies of industrial America. By applying what he believed were scientific analyses to the tools and techniques of industrial labor and management, he promised to end the waste and to usher in a new era of productivity and prosperity for all.[36]

But if this culture of efficiency drew strength from its origins in the ostensibly objective realm of engineering and scientific management, words like "needless" and "unnecessary" were clearly subjective. They not only highlighted the difficulty of defining noise objectively, but also invited the selective identification of targets upon which noise reformers could focus their efforts. While Mrs. Rice and her colleagues sincerely believed that they represented those who were not powerful enough to speak out against noise—the sick, the poor, the city's children—this kind of noise reform, like many other such progressive efforts, would affect different classes of people in very different ways.[37]

Laws newly passed or newly enforced at the urging of noise abaters typically identified relatively powerless targets, noisemakers who impeded, in ways not just acoustical, the middle-class vision of a well-ordered city. In June 1907, for example, the commissioner of police placed a ban on the use of megaphones by the barkers at Coney Island. "Cut out the megaphones? Impossible!" cried Pop

Hooligan, the oldest barker on the island. "What would Coney Island be without megaphones? How are you going to get a crowd to come in and see the boy with the tomato head and the rest of the wonders if you don't talk to them? I will see this Czar and make him abrogate his order."[38] In spite of Pop's defiant stance, however, the police order was at least temporarily effective, and it foretold of far more ambitious efforts to regulate and harmonize the sonic disorder of urban life.

John Kasson has described how amusement parks like Coney Island exerted a "special fascination" upon progressive reformers interested in transforming the social environment. To the working class, Coney Island was an urban oasis of food, music, spectacle, and especially the mechanized rides in which riders were challenged to maintain their balance as they were whirled, spun, and tossed about. To middle-class progressives, in contrast, the park was "a vast laboratory of human behavior" where they sought to achieve an equally precarious balance amid the much larger forces pulling at modern urban society.[39] The muzzling of barkers was just one of numerous efforts to "clean up" the park, and the barkers clearly understood this context. Their organized response to the police order was intentionally enacted in front of a freak show on the Bowery, one of the few places "where evidences of the old Coney Island which have escaped the regenerating whitewash brush" still remained. The men donned placards, not to advertise the spectacle of a tomato-headed boy, but instead to decry the censorship to which they were now subject. "Talking is a crime" read one sign; "They have taken our calling away" proclaimed another.[40] The long-term effectiveness of this police act is not evident, but within a year, this kind of noise reform would move out of the laboratory setting of Coney Island into the streets of the city itself.

In 1908, the health commissioner of New York joined forces with Police Commissioner Thomas Bingham to combat the problem of city noise. Bingham issued General Order 47, which called for the enforcement of the numerous and typically unenforced ordinances against particular kinds of noises already written into the city's legal codes. Noises so targeted included the shouts and bells of street vendors, the cries of newsboys, whistles on peanut roasters' carts, and the assorted sounds of roller skaters, kickers of tin cans, automobile horns, automobiles operated without mufflers, and flat-wheeled streetcars. Yet, reports of arrests made subsequent to the order indicated that vendors, musicians, and shouters, not motorists or streetcar companies, were the only targets actually pursued by the police.[41] In 1909, a new ordinance went after the vendors specifically, stipulating:

> No peddler, vender, or huckster who plies a trade or calling of whatsoever nature on the streets and thoroughfares of the City of New York shall blow or use, or suffer or permit to be blown upon or used, any horn or other instrument, nor make, or suffer or permit to be made, any improper noise tending to disturb the peace and quiet of a neighborhood for the purpose of directing attention to his ware, trade or calling, under penalty of not more than $5 for each offense.[42]

The *Times* labeled the new ordinance "the iron law of silence," and the peddlers and hucksters were distinctly displeased. "The whole thing is a move to add power to the janitor," declared the scissors grinder Isaac Leschatzsky.

> They will say that all the tenants who want butcher knives or scissors ground must tell the janitors in advance, and then we may go in and ask the janitor about it.
> Won't the janitor come in on the graft? We will have to make ourselves solid with the janitor or we won't get anything. I see it all. It's a plot of theirs. And think of the time lost asking each one of them. Is this a free country? I ask it. It is not.[43]

These noise bans were ultimately a means to accomplish the more general goal of clearing the streets of vendors altogether, and—as at Coney Island—the vendors were well aware of what was really at stake. The "Ole Clo'" men who bought and sold old clothing presented an organized, if unsuccessful, challenge to Bingham's order, and peddlers in Chicago responded to a similar ordinance three years later by rioting.[44] Daniel Bluestone has described how these pushcart bans served gradually to remove many "vital social and economic activities" from city streets. "In short," he concludes, "the bans sought to accommodate a vision of streets as exclusive traffic arteries that simply would not have been conceivable in earlier cities."[45] Ironically, by silencing peddlers and then removing them from the streets altogether, city officials only cleared the way for the more powerful noises of motorized traffic.

Another strategy in the war against noise was to create special zones of quiet in particular areas of the city. Zoning in general was an attempt to legislate the landscape of urban life, to control not only its physical appearance but also the behavior of those who inhabited it. By geographically separating the different social functions that unplanned cities naturally superimposed—residential, commercial, industrial—city planners sought to rationalize the urban environment in a way that would improve the performance of each sector. The numerous "City Beautiful" movements of the late nineteenth and early twentieth centuries additionally sought to enhance the aesthetic appeal of the urban environment. By combining the morally improving qualities of art with the rationalizing order of science, proponents of these movements presented their work as a powerful "antidote to urban moral decay and social disorder."[46]

The first item on the agenda of the Society for the Suppression of Unnecessary Noises was to designate special quiet zones around New York's hospitals, legally defining spaces in which a range of noises would be rendered illegal as a result of their proximity to the ill. In June 1907, "Little Tim" Sullivan introduced such legislation to the city's governing board, and the Aldermen quickly passed the bill.[47] A similar bill passed in Philadelphia in 1908 at the urging of Imogen Oakley. Like Mrs. Rice, Oakley was a prosperous and experienced organizer who established a Committee on Unnecessary Noise within the Civic Club of Philadelphia. Also like Rice, she enlisted the support of the city's medical authorities when petitioning for the protection of the ill.[48] The sick, however, were not the only members of society who required protection from noise. When Mrs. Rice undertook a tour of New York's schools in order to teach children about the importance of respecting the hospital quiet zones, she was dismayed to discover that the schools themselves suffered as much as did the city's hospitals from the noises that surrounded them. A campaign to establish quiet zones around schools was soon under way, not just in New York but across the nation.[49]

Schools suffered from their proximity to noisy work sites like garages and factories, as well as from the noises of vendors and traffic. Teachers grew hoarse struggling to be heard over the din, and when they closed their classroom windows to shut out the noise, the children's health and intellectual vigor were compromised by the lack of fresh air. "It is no exaggeration," Mrs. Rice argued, "to say that noise robs class and teachers of 25 per cent. of their time. The work of both pupils and teachers would be increased in efficiency and made easy by anything that would tend to reduce the din."[50]

By 1914, numerous American cities had established quiet zones around both hospitals and schools, and Baltimore even designated the nation's first exclusive "Anti-Noise Policeman" to patrol and enforce the hospital zones of that city. Over the course of one week, Officer Maurice Pease confronted and eliminated the noises of streetcar bell-ringers and squeaky-wheeled trolleys, a baker noisily unloading bread from his wagon, a shouting fishmonger, raucous school children, three roosters, six cats, another noisy baker, twenty-four more cats, newsboys, a scissors grinder, and several rag-and-bone collectors.[51]

Quiet zones like that policed by Officer Pease designated special places in the city where noise was considered particularly noisome. But the problem of noise was also recognized in the more general zoning policies that established distinct districts for residence, commerce, and industry in American cities. The most famous piece of such zoning legislation was enacted in New York in 1916. This law is perhaps best known for requiring the city's ever-taller skyscrapers to set back, or recede, as they reached higher into the air, in order to ensure the availability of daylight and fresh air on the ground. The Commission on Building Districts and Restrictions that wrote this law also acknowledged, however, that the soundscape of the city had to be similarly controlled. The unregulated development of tall buildings exacerbated the problem of noise and congestion on the streets, and the juxtaposition of noisy businesses and factories with residential structures also required regulation to prevent its continuation. "Quiet," the commission concluded, "is a prime requisite. The zone plan, by keeping business and industrial buildings out of the residential streets, will decrease the street traffic," and thereby protect "the quiet and peace of the residential street."[52]

While zoning laws like that in New York recognized the problem of noise and sought to map its solution, and while these laws doubtless had a long-term effect on the soundscape of American cities, there was no law requiring extant workshops to vacate the newly designated residential districts; thus they did not provide an immediate remedy to the problem at hand. Nor did the new legislation present any means to solve the problem of noise within exclusively residential neighborhoods. Many annoying sounds simply came from other residents, and these noises were not often covered under specific antinoise ordinances. In such cases, the acoustically aggrieved had no choice but to appeal to general nuisance laws. While noise reformers had hoped to regulate the soundscape in a way that recognized the larger social benefits of a city free of unnecessary noises, citizens were ultimately left to their own devices and forced to act as individuals responding to the particular noises that intruded upon their lives.

When a person filed a noise complaint, a lengthy procedure was set in motion that seldom concluded satisfactorily. Complaints were directed to the Department of Health, which dispatched to the scene a sanitary inspector or a member of the health squad, a special unit of the police force dedicated to enforcing the sanitary codes of the city. In 1912, inspections regarding noise complaints were tallied and divided into two categories, "machinery, motor boats and pumps" and "dogs, horses and animals." There were 668 registered complaints against the former, and 491 against the latter. Official responses to these complaints ranged from "No Cause for Complaint" to "Not Complied

With," "Abated by Personal Effort," and—for half of the total—the indeterminate "Held for Observation."⁵³ A year later, the health department recorded almost 5,000 inspections of noise complaints, but just one arrest, and no issuance of fines whatsoever.⁵⁴

Persistent persons might have chosen to take the offending party to court. While certain kinds of noises were specifically outlawed in the sanitary code of the city, most were not, and the situation encountered by Emanuel Gogel in 1930 was typical of that which had prevailed for the past few decades. When Gogel complained to the health department about the noise of construction by a private contractor near his home in Brooklyn, he was informed that:

> The jurisdiction of the Department of Health over noises is conferred by the Sanitary Code and the provisions of same do not comprehend the character of the noises of which you complained.
>
> There is a remedy and it is by resort to a summons for a violation of Sections 1530 and 1532 of the Penal Law which is known as the Public Nuisance Act.
>
> The persons discommoded must apply to the nearest Magistrate's Court for a summons for the contractor making the noise and requiring him to appear and answer before the Magistrate. The statute requires that a "considerable number of persons" must be shown to be discommoded and deprived of comfort, health and repose. The courts have held that more than three constitute a considerable number and you can doubtless get more than three persons who will appear to testify in reference to the noises.⁵⁵

It is not evident whether Dr. Gogel ever followed through with this procedure and presented his complaint to the courts. For those who did, it is apparent that satisfaction was by no means assured.

In the spring of 1921, for example, Mrs. Richard T. Wilson was taken to court by her downstairs neighbor Francis Newton, who had filed a complaint against her frequent late-night musical soirées. Most recently, a party on February 20th had included music that continued well past midnight. At that time, Newton, along with the Wilsons' upstairs neighbor, the painter Childe Hassam, complained to the police. When the officers arrived at the party, Mrs. Wilson asked her guests to lower the volume of their music and conversation, and she thought this was the end of the matter. But it was not, as the subsequent summons to court made clear.

At the trial Mr. Hassam declared, "I am kept awake by an absolute riot." He confessed a desire to "rig up a pounding machine" over the Wilsons' bedroom ceiling, to prevent their sleep just as their loud parties prevented his. Mr.

4.3
Cartoonist John Held, Jr.'s interpretation of the conflict between Mrs. Richard Wilson, high-society sponsor of late-night musicales, and her acoustically aggrieved neighbor, the painter Childe Hassam. *New York Times* (20 March 1921): sect. 3, p. 8. Courtesy American Newspaper Repository, photo by Russell French.

Newton more judiciously pointed out that the co-op building in which they all lived had rules forbidding music after 11:00 P.M. In her defense, Mrs. Wilson (who was the sister-in-law of a Vanderbilt) brought forth a parade of witnesses, the socially prominent friends who regularly attended her parties. They all testified that the music performed was of the best "artistic character," and therefore could not constitute noise at any time of day or night. The judge agreed, and the case was dismissed.[56] (See figure 4.3.)

The Wilson case was not unique for placing the nature of the sound at the heart of the matter. In 1925, Mrs. Martha Sanders, superintendent of an apartment house in Queens, took her tenant Arthur Loesserman to court, complaining that the music student constantly "pounded on the piano and scratched the fiddle." Mrs. Sanders produced two witnesses to corroborate her complaint. In his defense, Mr. Loesserman brought only his violin. Upon hearing his rendition of "Ave Maria," the audience in the courtroom burst into applause. The court attendant, a musician himself for sixty of his eighty-two years, declared the boy a genius and the judge dismissed the complaint.[57] In another case, Miss Veronica Ray defended the late-night sounds of the Russian Music Lovers' Association by arguing: "Why, we number among our members Feodor Chaliapin and other

singers of fame. Their music is music at any time and at any place." This time the judge disagreed, and he stipulated that the music must stop at 11:00 P.M.[58]

While each court case constitutes only anecdotal evidence, their cumulative coverage in the newspapers suggests that these conflicts exemplified frustrations common to many city dwellers. Indeed, the *Times* noted that "practically everybody in the city, rich, poor and those in between, must have felt what was or amounted to a personal interest in the case of Mrs. Richard T. Wilson.... The same quarrel has arisen innumerable times before."[59] The problem was not simply the disturbance of noise, but the failure of the legal system to provide a consistent and satisfactory means by which to adjudicate such situations. Not only was it inconvenient and expensive to take a noisy neighbor to court, but there was no objective basis for anticipating the outcome of these cases. Just as the subjective definition of what constituted an "unnecessary" noise had led to the selective targeting of noisemakers during crusades for public noise reform, defining what constituted a noise in the more private dealings of the courts was equally subjective. Judges were free to decide for themselves, and the decisions they rendered varied greatly from case to case. Clearly, the problem of defining what constituted a noise had to be resolved before the problem of noise itself could be solved.

While most people interested in defining noise were motivated by their desire to eliminate it, some were more constructively stimulated by the sounds of the modern city. In his testimony against the Wilsons, Francis Newton had specified that "a great deal of the music was of a jazz character," and when Childe Hassam was asked to describe the music that so bothered him, he responded emphatically, "Ragtime. I should say cacophony."[60] While ragtime and jazz were perceived as noise by listeners like Newton and Hassam, to many others they constituted a musical rendition of the soundscape of the modern city. Classically trained composers, too, were similarly inspired by their new aural environment to redefine the very meaning of music. Thus, not only in courts of law, but also in nightclubs and concert halls, the distinction between music and noise was tested and transformed.

III NOISE AND MODERN MUSIC

The connection between jazz and the sounds of the city was evident to virtually all who listened in. Joel Rogers located the roots of jazz in African music, but he also acknowledged the influence of "the American environment," and that environment was filled with noise. "With its cowbells, auto horns, calliopes, rattles, dinner gongs, kitchen utensils, cymbals, screams, crashes, clankings and monotonous rhythm," Rogers remarked in 1925, jazz "bears all the marks of a nerve-strung, strident, mechanized civilization."[61] The result of that influence can be heard in the music itself, from the police siren that closes Fats Waller's "The Joint Is Jumpin'" to the symphonic evocations of subways, nightclubs, and other urban sounds that constitute James P. Johnson's *Harlem Symphony* and Duke Ellington's *Harlem Air Shaft*. "So much goes on in a Harlem air shaft," Ellington explained. "You get the full essence of Harlem in an air shaft. You hear fights, you smell dinner, you hear people making love. You hear intimate gossip floating down. You hear the radio. An airshaft is one great big loudspeaker."[62] The connection between jazz and urban noise that Ellington celebrated was, however, far more frequently invoked by those who condemned it.

Critics of jazz articulated their disdain for the new music in a curious conjunction of racism and antimechanism. Jazz was attacked "not only for returning civilized people to the jungles of barbarism but also for expressing the mechanistic sterility of modern urban life."[63] It was perceived to reflect "an impulse for wildness" even as it was "perfectly adapted to robots."[64] It stimulated "the half-crazed barbarian to the vilest deeds" while simultaneously constituting "the exact musical reflection of modern capitalistic industrialism."[65] This curious conjunction of things seemingly primitive with those technologically advanced drove not only critics, but also the most fervent enthusiasts of a culture self-consciously defining itself as "modern."[66] Alain Locke recognized jazz as a "symptom of a profound cultural unrest and change," and historian Kathy Ogren has concluded that, "to argue about jazz was to argue about the nature of change itself."[67] The change that such arguments focused upon was both racial and technological.

The racist aspect of the criticism of jazz reflected the distress that many Americans felt with the rapidly changing demography of the cities in which they lived. The widespread migration of African Americans from the rural south to the industrial cities of the north in the early decades of the century heightened racial tensions between blacks and whites in those cities.[68] It also engendered discomfort in some black intellectuals whose hard-won claims to cultural legitimacy were perceived to be threatened by these newcomers. Of all the writers whose work came to constitute the Harlem Renaissance, poet Langston Hughes was virtually alone in the respect he accorded jazz musicians, and he took his colleagues to task for their neglect of the Renaissance in music: "Let the blare of Negro jazz bands and the bellowing voice of Bessie Smith singing Blues penetrate the closed ears of the colored near-intellectuals until they listen and perhaps understand."[69]

As Hughes acknowledged, closing one's ears was a futile attempt to shut out the sound of jazz, as futile as attempting to elude the din of the modern city. The technological changes driving that crescendo were as disconcerting as was the new racial geography, and the technological aspects of the criticism of jazz only echoed larger concerns of people who were struggling to make sense of the new industrial soundscape of their cities. Both types of changes were dramatic and unsettling to all parties involved. Indeed, the African Americans who migrated from rural southern counties to large industrial cities would have experienced an aural transformation far more dramatic than that experienced by virtually any other group of Americans at this time. The city itself was an engine of changes both social and technological, and the agents of change that operated within it, from jazz musicians to internal combustion engines, were what made the decade roar. The Machine Age was simultaneously the Jazz Age; the machinery and the music together defined the new era and filled it with new kinds of sounds.

At the foundation of debates over the musical and cultural value of jazz was an assumption of a fundamental dichotomy between music and noise. Music was legitimate sound and noise was not. Music was harmonious, regular, and orderly; noise was discordant, irregular, and disorderly. This definition of noise had long been asserted by classically trained musicians and was backed by the authority of science. As Hermann Helmholtz had explained in 1877:

> The first and principal difference between various sounds experienced by our ear, is that between *noises* and *musical tones*. The soughing, howling, and whistling of the wind, the splashing of water, the rolling and rumbling of carriages, are examples of the first kind, and the tones of all musical instruments of the second. . . . [A] musical tone strikes the ear as a perfectly undisturbed, uniform sound which remains unaltered as long as it exists, and it presents no alteration of various kinds of constituents. To this then corresponds a simple, regular kind of sensation, whereas in a noise many various sensations of musical tone are irregularly mixed up and as it were tumbled about in confusion.[70]

Helmholtz's elaboration drew exclusively upon a naturalistic, preindustrial repertoire of noises that would soon be overwhelmed by the sounds of industry and technology. More significant, the unquestioned authority of long-standing scientific definitions such as this would also soon become a relic of the past. In the early twentieth century, it was not unusual for such definitions to be questioned, challenged, even overturned. Newtonian conceptions of inflexible space and immutable time were replaced by the supple space-time continuum of Einstein's relativistic physics. Cartesian certainty was replaced by the Uncertainty Principle of Werner Heisenberg. And the physical distinction between noise and music was similarly challenged, not only amid the gin and smoke of jazzy nightclubs but also from within the realm of elite musical culture itself, as a new generation of classically trained composers self-consciously turned to noise for inspiration and brought it directly into the concert hall.

"The Joys of Noise" were what inspired composer Henry Cowell to explore a "little-considered, but natural, element of music." "Music and noise," he wrote in 1929, "according to a time-honored axiom, are opposites."

> If a reviewer writes "It is not music, but noise," he feels that all necessary comment has been made.
>
> Within recent times it has been discovered that the geometrical axioms of Euclid could not be taken for granted, and the explorations outside them have given us non-Euclidean geometry and Einstein's physically demonstrable theories.
>
> Might not a closer scrutiny of musical axioms break down some of the hard-and-fast notions still current in musical theory?[71]

By 1929, those axioms had, in fact, already been considerably weakened. Some composers used traditional musical instruments to represent the noises of the modern world. Others incorporated noisemaking machines into their orchestrations. Still others sought entirely new instruments to create totally new sounds. In all cases, their intent was to redefine the very meaning of music and to transform the ways that people listened to both music and noise.

As early as 1906, Charles Ives had incorporated representations of city noises into his composition *Central Park in the Dark*. In this piece, Ives employed an orchestra of traditional instruments to evoke the cacophony of sounds experienced by a nocturnal visitor to the heart of New York. Street singers, late night whistlers, shouting newsboys, the elevated train, a streetcar, a fire engine, and dueling player pianos pumping out popular songs of the day all compete with, then gradually overpower, the gentle, natural, insectlike drone of the night. The noises accumulate and build to a loud climax, but, when they finally and abruptly fall away, the drone of the night is once again audible, and the transcendental peace of nature ultimately triumphs over the acoustical distractions of man.[72]

By 1912, however, Ives appears to have felt differently. He now declared New York a "Hell Hole," and spent as much time as possible at his estate in Connecticut, to escape the din of the city. Even this rural retreat could not offer sanctuary, however, and when his idyll was interrupted by the noise of a low-flying airplane, he would shake his cane in the air with disgust.[73] Ives would later recall affectionately the discordant array of sounds he had captured in *Central Park in the Dark,* and he parenthetically suggested his discontent with the modern soundscape when he described the piece as "a picture-in-sounds of the sounds of nature and of happenings that men would hear some thirty or so years ago (before the combustion engine and radio monopolized the earth and air)."[74] Ives's music existed only on the margins of American musical culture during the composer's lifetime, but it is now recognized as constituting "the beginnings of a trend increasingly evident in the early twentieth century," a trend in which "the metropolitan experience" impelled composers toward "a more radical musical language."[75] The development of this new musical language, like noise itself, was not an exclusively American phenomenon, and some of its earliest articulations occurred in Europe. Nonetheless, many of the most challenging examples of modern music, even works composed by Europeans, explicitly drew on the excitement of American technology and the new modern soundscape epitomized in American cities.[76]

In 1907, Ferruccio Busoni articulated a dissatisfaction that many composers were beginning to share when he wrote of "the narrow confines of our musical art." "The gradation of the octave is *infinite*," he proclaimed, so "let us strive to draw a little nearer to infinitude."[77] To do this, new instruments were required. Busoni had experimented with voice and violin to create partial tones, notes located in the interstices of the tempered system ("between" the keys of a piano, so to speak), but without much success. More promising was a report from America of a new invention by Dr. Thaddeus Cahill, "a comprehensive apparatus which makes it possible to transform an electric current into a fixed and mathematically exact number of vibrations." With Cahill's machine, Busoni hoped, "the infinite gradation of the octave may be accomplished by merely moving a lever."[78]

While Busoni theorized a new music, his own compositions never really fulfilled these ideas, and others were able to break more fully with the traditions of the past. The Italian Futurists, for example, eagerly embraced an art that would "mock everything consecrated by time."[79] An enthusiasm for all things new, and particularly for new technologies, infused their efforts to revolutionize poetry, painting, and music. The movement was heralded in 1909 by the poet Fillipo

Tomasso Marinetti. While art historians have emphasized the importance of dynamism—speed and motion—in the Futurist aesthetic, it is also clear that noise was a paramount source of inspiration. In any medium of Futurist art—literary, visual, or musical—the noise of the modern world could always be heard.[80]

Marinetti, for example, sought to free poetry from the strictures of tradition and convention, to "enrich lyricism with brute reality," including the reality of noise.[81] In *Zang-Tumb-Tumb*, written while he was a correspondent covering the siege of Adrianople during the Balkan Wars of 1912–1913, Marinetti vividly imparted the auditory chaos of modern warfare:

> every 5 seconds siege cannons gutting space with a chord ZANG-TUMB-TUUUMB mutiny of 500 echos smashing scattering it to infinity. In the center of this hateful ZANG-TUMB-TUUUMB area 50 square kilometers leaping bursts lacerations fists rapid fire batteries. Violence ferocity regularity this deep bass scanning the strange shrill frantic crowds of the battle Fury breathless ears eyes nostrils open![82]

Futurist words became physical sounds when these poems were performed live in theaters, read out loud—loudly—and accompanied by sound effects and music. Wyndham Lewis attended a performance of *Zang-Tumb-Tumb* in London and later recalled that "even at the front, when bullets whistled around him, he had never encountered such a terrifying volume of noise as Marinetti produced."[83] Unappreciative audiences frequently responded to these performances with noises of their own, only adding to the aural chaos.

Futurist visual art similarly strove to represent the sounds of the modern world. In his 1913 manifesto "The Painting of Sounds, Noises and Smells," Carlo Carrá proclaimed that Futurist painting must express "the plastic equivalent of the sounds, noises and smells found in theatres, music-halls, cinemas, brothels, railway stations, ports, garages, hospitals, workshops."[84] Carrá's plea was taken to heart in such works as Luigi Russolo's *La Música* (1911–1912); Fortunato Depero's *Plastic Motor-Noise Construction* (1915); and Umberto Boccioni's *The Noise of the Street Penetrates the House* (1911).

In such an acoustically conscious environment, a Futurist music was bound to appear. In 1911, the composer Balilla Pratella published a "Technical Manifesto of Futurist Music," in which he proclaimed:

> All forces of nature, tamed by man through his continued scientific discoveries, must find their reflection in composition—the musical soul of the crowds, of great

industrial plants, of trains, of transatlantic liners, of armored warships, of automobiles, or airplanes. This will unite the great central motives of a musical poem with the power of the machine and the victorious reign of electricity.[85]

It was not Pratella but his colleague Luigi Russolo who would turn these ideas into sounds, creating music out of the noise of the modern world.

While Russolo began his Futurist career as a painter, the modern soundscape in which he worked soon impelled him away from the visual arts and into music, in spite of (or perhaps because of) the meagerness of his formal training in that arena.[86] Disappointed by Pratella's dependence on traditional musical instruments to create untraditional music, Russolo began immediately to theorize, and then to build, new kinds of instruments that he called "noise-intoners" (*intonarumori*).

Russolo's inevitable manifesto "The Art of Noises" appeared in 1913.[87] "Noise is triumphant," he proclaimed, "and reigns sovereign over the sensibility of men." Russolo argued that the musical tones that had been employed by musicians for hundreds of years were now so familiar as to have lost all power to stimulate the listener. "Today," he explained, "the machine has created such a variety and contention of noises that pure sound in its slightness and monotony no longer provokes emotion."[88] "Away!" he exclaimed, abandoning those sterile tones for the vital sounds of life itself, the noises of the modern city:

> Let us cross a large modern capital with our ears more sensitive than our eyes. We will delight in distinguishing the eddying of water, of air or gas in metal pipes, the muttering of motors that breathe and pulse with an indisputable animality, the throbbing of valves, the bustle of pistons, the shrieks of mechanical saws, the starting of the tram on the tracks. . . .[89]

Machines, having sapped all vitality from the old music, would now become the basis for a vital new music.

Even as he composed his manifesto, Russolo was hard at work building his new instruments.[90] Housed in wooden boxes with protruding acoustical horns, the noise-intoners looked like strange mutations of the ordinary phonograph. Russolo named the different instruments according to the sound that each produced: howler, roarer, crackler, rubber, hummer, gurgler, hisser, whistler, burster, croaker, and rustler. All employed a drumheadlike diaphragm to produce the sound vibrations. Via a hand crank or a battery-powered motor, a different kind of mechanism set the diaphragm in motion in each device, creating the different types of sounds. Each instrument also possessed an adjustable lever that varied the tension of the diaphragm, allowing it to produce noises over a range of frequencies.[91]

After several months' work, Russolo had constructed an orchestra of sixteen different instruments. He presented a private concert at Marinetti's home in Milan, featuring two of his own compositions written for the occasion, *Awakening of a City* and *Meeting of Automobiles and Airplanes*. A reporter for a London newspaper described the experience of *Awakening*:

> At first a quiet even murmur was heard. The great city was asleep. Now and again some giant hidden in one of those queer boxes snored portentously; and a newborn child cried. Then, the murmur was heard again, a faint noise like breakers on the shore. Presently, a far-away noise rapidly grew into a mighty roar. I fancied it must have been the roar of the huge printing machines of the newspapers.
>
> I was right, as a few seconds later hundreds of vans and motor lorries seemed to be hurrying towards the station, summoned by the shrill whistling of the locomotive. Later, the trains were heard, speeding boisterously away; then, a flood of water seemed to wash the town, children crying and girls laughing under the refreshing shower.
>
> A multitude of doors was next heard to open and shut with a bang, and a procession of receding footsteps intimated that the great army of bread-winners was going to work. Finally, all the noises of the street and factory merged into a gigantic roar, and the music ceased.
>
> I awoke as though from a dream and applauded.[92]

Although Russolo had emphasized the abstract over the imitative quality of his music, listeners were apparently compelled to understand this new music in terms of its direct resemblance to the actual noises of the modern world. While the reporter for the *Pall Mall Gazette* seemed to enjoy this resemblance, others felt differently.

The first public performance of the noise orchestra took place on 21 April 1914 at the Teatro dal Verme in Milan. According to Russolo, the audience of conservative critics and musicians came only "so that they could refuse to listen."[93] As soon as the orchestra began to play, the crowd broke into a violent uproar. The musicians continued undaunted while fellow Futurists hurled themselves into the audience and defended the Art of Noises with their fists. In the end, eleven people were sent to the hospital, none of them Futurists, as belligerence was a central component of the Futurist approach to art and life, and many were talented boxers.[94] A subsequent concert in Genoa was more politely received and was followed in June by a series of twelve concerts in London, where Russolo claimed he was besieged by the press as well as by enthusiastic auditors. The London *Times* suggested it was the audience that had been besieged, however, and reported that, after just one piece, the "noisicians" were greeted with "pathetic cries of 'no more.'"[95] For better or worse, the noise orchestra had certainly captured the public's attention.

Russolo argued that "the constant and attentive study of noises can reveal new pleasures and profound emotions," and he described how his own musicians had "developed" their ears by playing and listening to his instruments. After four or five rehearsals, "they took great pleasure in following the noises of trams, automobiles, and so on, in the traffic outside. And they verified with amazement the variety of pitch they encountered in these noises." As Russolo explained, "It was the noise instruments that deserved the credit for revealing these phenomena to them."[96]

Russolo hoped to impart this aural education to his audience as well as his musicians, to teach all to perceive music within the noise of the modern world. He planned a grand tour, but the fall of 1914 turned out not to be a good time for a concert tour of Europe. As Russolo put it, "The war caused it all to be postponed. . . . I left for the front. . . . And I was lucky enough to fight in the midst of the marvelous and grand and tragic symphony of modern war."[97] Wounded in battle at the end of 1917, Russolo returned home to his music hoping to pick up where he had left off three years earlier. But the loud noises of war had apparently deafened the European audience that had previously been so intrigued by his work, and he never recaptured the fame and infamy that he had enjoyed in 1913.[98]

Another musician whose life was fundamentally changed by the war was the French composer Edgard Varèse. Like Russolo, Varèse had been searching for a music in which all sounds were possible. Varèse was no belligerent Futurist, however, and when the war came he did not enlist but instead withdrew to America, arriving in New York in December 1915. The soundscape of New York stimulated the composer to create the new music that he had only been able to hypothesize in Europe, and Varèse's first major composition, *Amériques,* was a tribute to his new home. "I was still under the spell of my first impressions of New York," Varèse later recalled. "Not only New York seen, but more especially heard. For the first time with my physical ears I heard a sound that had kept recurring in my dreams as a boy—a high whistling C-sharp. It came to me as I worked in my Westside apartment where I could hear all the river sounds—

the lonely foghorns, the shrill peremptory whistles—the whole wonderful river symphony which moved me more than anything ever had before."[99]

Completed in 1921, *Amériques* was scored for a full orchestra of 142 instruments, including two sirens.[100] The size and complexity of this score rendered its production prohibitively expensive, however, and it would not be premiered until 1926. In the meantime, Varèse composed several smaller works that were more economically performed. *Hyperprism*, a short piece scored for a small orchestra of brass and winds accompanied by a siren and a prominent array of percussion instruments, premiered in March 1923, causing "the first great scandal in New York's musical life."[101] At its conclusion, the audience broke out into a raucous medley of laughter, hisses, and catcalls. As music critic Paul Rosenfeld later recalled, one sound in particular, a piercing note emitted by the siren, had evoked nervous laughter from the auditors.[102] It was the same C-sharp that Varèse had dreamt of as a boy and now heard rising above the cacophony of New York. While Varèse had been able to transform that noise into music, his audience—who lived amid that same din—apparently could not. Their nervous laughter suggests that, consciously or unconsciously, they recognized this particular sound and were uncomfortable with its new context in the concert hall.

Hyperprism was performed again in November by Leopold Stokowski and the Philadelphia Orchestra, with a siren borrowed from a local fire company. The Philadelphia premiere went "splendidly," according to the conductor; "practically all the audience remained to hear it." Olin Downes, music critic for the *New York Times*, could only describe it as a medley of "election night, a menagerie or two, and a catastrophe in a boiler factory," but others were more willing to accept the piece on its own terms. The *Herald-Tribune*'s Lawrence Gilman thought the work "a riotous and zestful playing with timbres, rhythms, sonorities." While the audience "tittered a bit" during the performance, Gilman noted, after its conclusion they "burst into the heartiest, most spontaneous applause we have ever heard given to an ultra-modern work."[103] Paul Rosenfeld argued that Varèse never simply imitated the sounds of the city. "He has come into relationship with the elements of American life, and found corresponding rhythms within himself set free. Because of this spark of creativeness, it has been given him to hear the symphony of New York."[104]

When Varèse's true symphony of New York was finally undertaken by Stokowski and the Philadelphia Orchestra in 1926, the ensemble required an unprecedented sixteen rehearsals to prepare the demanding score.[105] The premiere of *Amériques* was presented at the Academy of Music in Philadelphia, to a

Friday-afternoon audience famous for being more elderly, female, and conservative than that which came out on other nights. Varèse's music provoked these "sedate-looking ladies" to indecorous catcalls, whistles, and hisses. Indeed, "jeers and cheers, hisses and hurrahs, made the audience's reception of this radical work almost as deliriously dissonant as was the 'music' itself."[106]

While an almost circuslike atmosphere apparently accompanied many performances of Varèse's works, the composer himself was serious, sincere, and even scientific in his approach to his music. He prefaced his score to *Arcana* with an epigram from the sixteenth-century alchemist Paracelsus, and the alchemical idea of transmutation was at the heart of this piece. A simple eleven-note passage is introduced at the outset; it then travels throughout the orchestra and undergoes "melodic, rhythmic and instrumental transmutation."[107] Music critic W.J. Henderson confessed, "The present writer does not know how to describe such music." "There is portent and mystery in this music," Lawrence Gilman concluded. "It is good to hear it and thus to be perturbed."[108]

Paul Rosenfeld heard, amid *Arcana*'s alchemical evocation of past centuries, a distinctly contemporary resonance, "a passion for discovery." He noted that, for Varèse, "the exciting scientific perspectives of the day related to his new emotional and auditory experiences."[109] Indeed, ever since his arrival in America, Varèse (whose father was an engineer and who had been encouraged to become one himself) had been looking for scientists and engineers with whom to collaborate. "Our musical alphabet must be enriched," Varèse had pronounced to a New York reporter back in 1916. "We also need new instruments very badly.... Musicians should take up this question in deep earnest with the help of machinery specialists." "What I am looking for," he explained, "are new technical mediums which can lend themselves to every expression of thought and can keep up with thought."[110]

At that time, Varèse had sought out Cahill's Dynamophone, the electrical instrument that had excited Ferruccio Busoni. Upon hearing it, however, Varèse did not detect in its tones the music he sought to create, and he did not pursue composing music for the device. In 1922, he reiterated his desire for a new instrument, and he acknowledged that "the composer and the electrician will have to labor together to get it."[111] Varèse's dependence on the siren, in *Amériques* and other works, was not intended to re-create the sounds of fire engines or ambulances, but rather to bring into his music those sounds he could not achieve with traditional instruments. It was a necessary compromise, a *trompe l'oreille*, that would increasingly frustrate the composer as time passed.

In 1927, Varèse began corresponding with Harvey Fletcher at Bell Laboratories, hoping to enlist the physicist and the financial resources of AT&T in his mission to develop "an instrument for the producing of new sounds."[112] He also contacted motion picture producers, hoping to gain access to the technological tools of their new sound studios. These attempts to engage in technological collaboration ultimately came to naught, however, and only after the Second World War would Varèse finally realize his dream to work with skilled technicians and new technologies to create modern music.[113]

In 1927, however, the composer was still full of hope and at the height of his renown. When *Arcana* was premiered by Stokowski in April, it received the enthusiastic praise of a small but growing group of advocates, and it also provoked the begrudging acceptance of at least some of his ever-present critics. Perhaps the critics and concertgoers were developing "new ears," gradually learning—like Luigi Russolo's noise musicians—to listen in new ways.[114] The cultural legitimacy of Varèse's music was also highlighted by its juxtaposition to the most infamous example of noise-music of the 1920s, the *Ballet Mécanique* of George Antheil.

George Antheil was in many ways a mirror image of Varèse. Whereas Varèse had been born in France and moved to America to further his musical career, Antheil was a product of the industrial town of Trenton who moved to Europe in 1920 to make his name as a concert pianist. Antheil spent several years touring the continent, after which he settled in Paris. He rented an apartment above a bookstore that was renowned as a gathering-place for expatriate artists, literary moderns and their friends, including James Joyce, Gertrude Stein, and Pablo Picasso. Antheil's work, like that of Varèse, was shaped by the same combination of the American soundscape and the ideas of the European avant garde. For him, the sequence of experiences was simply reversed.[115]

Antheil's compositions featured the piano, but he treated it more like a percussion instrument than a keyboard, demanding player-piano-like precision and speed of the performer. His early works drew the attention of Ezra Pound, who began vigorously to promote the young composer. Pound declared that Antheil had "invented new mechanisms of this particular age." He used machines, "actual modern machines" to create musically "a world of steel bars, not of old stone and ivy."[116] When Antheil's *Symphony for Five Instruments* was presented at a private salon in 1924, another enthusiast proclaimed: "America's sky-scrapers found their musical expression in Paris." His music represented "the rhythm of modern America with a strange combination of esthetic beauty and sheer cacophony."[117]

The public premiere of Antheil's piece *Mechanisms* occurred at the Théâtre des Champs-Elysées in 1923. Antheil himself performed his piece as a musical prelude to a performance by the Ballet Suédois, an innovative dance troupe that had attracted to the theater the most avant of the Parisian garde. As soon as he began to play, bedlam ensued. Man Ray started throwing punches; Marcel Duchamp argued loudly while Erik Satie applauded and shouted "Quel precision!"[118] The police arrived, arrests were made, and, as Antheil's friend Aaron Copeland later exclaimed, "George had Paris by the *ear!*"[119]

With his fame—or infamy—secured, Antheil was invited to expand *Mechanisms* into a larger work, and the result, *Ballet pour Instruments Mécanique et Percussion,* was brought to America in 1927. His European escapades had been well covered by the American press, thus Antheil's reputation preceded his return and his homecoming concert was advertised in the *New Yorker* as "an event no New Yorker can afford to miss."[120] Tickets for the April 10th performance at Carnegie Hall quickly sold out, but an atmosphere of musical scepticism permeated the hall that night.

Eugene Goossens led an orchestra that included Antheil, as well as Aaron Copeland, among the musicians. Among the audience was the poet William Carlos Williams, who reflected on the traditional role of the great hall, and the music with which it was typically filled, in the midst of the modern city:

> Here is Carnegie Hall. You have heard something of the great Beethoven and it has been charming, masterful in its power over the mind. We have been alleviated, strengthened against life—the enemy—by it. We go out of Carnegie into the subway and we can for a moment withstand the assault of that noise, failingly! as the strength of the music dies. . . .
>
> But as we came from Antheil's "Ballet Mechanique," a women of our party, herself a musician, made this remark: "The subway seems sweet after that."[121]

Scored for six pianos, one Pianola or mechanical piano-player, bass drums, xylophones, whistles, rattles, electric bells, sewing machine motors, an airplane propeller, and two large pieces of tin, Antheil's *Ballet* was a far—and loud—cry from the charming strains of Beethoven.[122]

The next day's *Herald Tribune* headlined "Boos Greet Antheil Ballet of Machines," and the boos were supplemented with meows, whistles, hisses, and a deluge of paper airplanes. The woman seated behind William Carlos Williams kept repeating "It's all wrong, it's all wrong," and a "lantern jawed young gentleman" stumbled out of the auditorium, shaking his head and bellowing "like a tormented young bull." Another waved a white handkerchief tied to a cane, signaling his surrender to the enemy sounds issuing from the stage.[123]

Lawrence Gilman required four columns to dismiss Antheil's work, finding the *Ballet* "a brainless and stupid nullity." The ruckus (by the audience) within the hall seemed "suspiciously manufactured in character," and Gilman reported that, at the program's conclusion, "an infinitely wearied audience" exited into the "hideousness and wonder and incomparable fascination" of the real New York having "rebuffed the mechanistic wooing of this troubadour from Trenton."[124] One wonders, however, whether Gilman or others in the audience had found such wonder and fascination in the sounds of New York prior to Antheil's aural assault. William Carlos Williams was convinced that "many a one went away from Carnegie Hall thinking hard of what had been performed before him." When his companion remarked "The subway seems sweet after that," Williams replied "Good." He explained:

> I felt that the noise, the unrelated noise of life such as this in the subway had not been battened out as would have been the case with Beethoven still warm in the mind but it had actually been mastered, subjugated. Antheil had taken this hated thing life and rigged himself into power over it by his music. The offence had not been held, cooled, varnished over but annihilated and life itself made thereby triumphant. This is an important difference. By hearing Antheil's music, seemingly so much noise, when I actually came upon noise in reality, I found that I had gone up over it.[125]

Like Russolo's musicians, who had learned to hear noise in new ways by performing on and listening to the noise-intoning instruments, Williams was able to conquer noise, to transcend its offensive character, by hearing it in a new way, a hearing that Antheil's music had enabled.

Paul Rosenfeld later echoed Williams's ideas, as he, too, found that the new music enabled him to hear noise in new ways. For Rosenfeld, it was the music of Varèse, not Antheil, that had transformed his perception of the urban soundscape in which he lived:

> Following a first hearing of these pieces, the streets are full of jangly echoes. The taxi squeaking to a halt at the crossroad recalls a theme. Timbres and motives are sounded by police-whistles, bark and moan of motor-horns and fire sirens, mooing of great sea-cows steering through harbor and river, chatter of drills in the garishly lit fifty-foot excavations. You walk, ride, fly through a world of steel and glass and concrete, by rasping, blasting, threatening machinery become strangely humanized

and fraternal; yourself freshly receptive and good-humoured. A thousand insignificant sensations have suddenly become interesting, full of character and meaning; gathered in out of isolation and disharmony and remoteness; revealed integral parts of some homogeneous organism breathing, roaring and flowing about.

For the concert-hall just quit, overtones and timbres and rhythms corresponding to the blasts and calls of the monster town had formed part of a clear, hard musical composition; a strange symphony of new sounds, new stridencies, new abrupt accents, new acrid opulencies of harmony. Varèse has done with the auditory sensations of the giant cities and the industrial phantasmagoria, their distillation of strange tones and timbres much what Picasso has done with the corresponding visual ones. He has formed his style on them. Or, rather, they have transformed musical style in him by their effect on his ears and his imagination.[126]

To composers like Antheil and Varèse, the noises of the modern city inspired the creation of a new kind of music. When this music was performed in places like Carnegie Hall, audiences were challenged to test their ideas about the distinction between music and noise. Some—including critics like Gilman and Rosenfeld, as well as other perceptive listeners like Williams—clearly developed a new way of listening, learning not only to celebrate the noise in music, but also to appreciate the music in noise. This was not, however, the only way to test the definition of noise. Acousticians and engineers were also redefining the meaning of sound, with new instruments of their own. When they took those tools out of the laboratory and put them to work in a world filled with sound, they, too, challenged listeners to listen in new ways.

IV ENGINEERING NOISE ABATEMENT

On 27 April 1932, a sound engineer from General Electric entered the radio broadcast booth at the Metropolitan Opera House in New York to set up some new equipment. The "electric ear" that he installed had originally been developed by GE for use in the "location, measurement and control of insidious noises that affect the nervous system," and later that night he would point it at Lily Pons.[127] The next day's paper reported that the famed diva was "noisier than a street car," having hit a peak of 75 decibels during her aria, "Caro nome," in Giuseppe Verdi's *Rigoletto*. Miss Pons was bested by her leading man, however, for Beniamino Gigli topped out at 77 dB, "midway between the streetcar, rating 65 decibels, and the subway, rating 95."[128] The engineer, M. S. Mead, candidly admitted that the experiment had no immediate practical value, but this did not prevent the editors of the *Times* from editorializing. "For real decibels," they suggested, "bring on Stravinsky, or, better still, Antheil, with that battery of pneumatic riveters which made his 'Ballet Mécanique' so ear-splitting."[129]

While this event was strictly a musical and technological curiosity, it nonetheless highlights the role that new tools and terminology—and the technicians who wielded them—played in transforming the meaning of noise. Just as musicians were devising new instruments and developing a new vocabulary, so, too, were scientists and engineers. Indeed, it sometimes became difficult to distinguish between musical instruments and their scientific counterparts. Like Stravinsky and Antheil, sound engineer Mead and his acoustical colleagues were at the forefront of cultural change, actively constructing the physical sounds of the modern soundscape along with new ways to understand them.

From the late nineteenth century on, efforts to control urban noise had been accompanied by attempts to measure that noise. In 1878, when a group of doctors complained before a grand jury of the noise created by the trains of the Metropolitan Elevated Railway Company in New York, the company asked Thomas Edison to study the problem and to recommend a remedy. Edison made inscriptions of the noise with a phonautograph, a device that rendered visual but nonreproducible records of sound. His tools were described as a "sorcerer's kit" with which he cast "metrophonic spells," but in fact, Edison's spells were powerless to characterize the noise in a meaningful way, let alone to eliminate it.[130] Mrs. Rice later turned to Edison's phonograph to spread the word about the problem of noise. When she organized the Society for the Suppression of Unnecessary Noise in 1906, she enlisted the Columbia Phonograph Company to make recordings of the noise around New York's hospitals, in order to convince city authorities of the severity of the problem.[131] The problem of measuring sound that plagued professors of physics like Wallace Sabine and Floyd Watson was clearly not just academic, and a 1917 report on the "Progress of the Anti-Noise Movement" could only conclude that "as to measurement of noise disturbance and the establishment of standards to show what degrees of noise are and are not endurable, the anti-noise movement can show no advance." "Noise," the report continued, "not only has no instrument of measurement but it is even without a satisfactory definition."[132]

Not long after this complaint was registered, however, the predicament would be resolved. With the development of high-quality microphones, vacuum-tube amplifiers, and other electroacoustical devices in the 1920s, powerful new weapons were enlisted in the campaign against noise. The technicians who wielded them were similarly perceived as formidable allies. By 1930, the

Saturday Evening Post could highlight the fact that "the fight against wasteful racket is out of the hands of cranks and theorists and is being directed by trained technical minds." "These hard-headed experts," the report continued, "clearly recognize that reform movements never amount to much until they get away from hasty assumptions founded on guessing and are established upon the conclusive results of extended tests."[133] These tests, the equipment with which they were executed, and the technicians who executed them were primarily the progeny of the radio and telephone industries.

As the American Telephone and Telegraph Company undertook to improve the quality of its aural products and services in the teens and twenties, acoustical researchers at Western Electric and Bell Laboratories investigated the phenomena of noise and hearing in order to determine how best to improve the performance of the telephone system. Telephone engineers devised tools for measuring the electrical noise that hampered the intelligibility of speech on telephone lines, and researchers like Irving Crandall and Harvey Fletcher also designed instruments to measure the character of speech and hearing. These tools were subsequently adapted to measure the sounds and subjects of the non-telephonic world.

The 1-A Noise Measuring Set of 1924, for example, measured electrical noise in a telephone circuit. A technician listened, through a telephone earpiece, alternately to the circuit under investigation and to a source of electrically generated noise. The latter was gradually attenuated in volume by means of a potentiometer until the two sounds were perceived to be equally loud, and the setting of the potentiometer (scaled in arbitrary "noise units") indicated the level of noise in the circuit. According to its instruction manual, the device required a skilled operator since "noise in telephone circuits varies greatly in quality under different conditions." "For this reason," the manual explained, "a comparison is frequently one which depends a great deal on individual judgement, and whenever possible should be made by those accustomed to the use of this apparatus."[134] The telephone engineers who used these devices developed a skilled way of listening to noise, a skill that the instruments themselves engendered.

Researchers at Western Electric also developed new tools for testing the sensitivity of the human ear. At the request of psychologists and otologists, Harvey Fletcher designed an audiometer to measure hearing loss at different frequencies.[135] Fletcher's work resulted, by 1923, in a range of commercial products, from the armoire-sized professional model 1-A to the simplified and portable 3-A. The 1-A generated pure tones at variable intensities, and the subject listened to these tones, one after another, through a headset. The investigator gradually increased the intensity of each tone until it was just audible by the subject, and the amplitude of the signal at this point indicated the sensitivity of the subject to sounds of that frequency.[136] The devices were calibrated to give intensity readings as sound pressure measurements in dynes per square centimeter, but users typically referred to a new scale inscribed on the instrument by which the range of audible intensity was broken down into "sensation units," each of which constituted a just-perceptible increase or decrease in sound intensity. About 120 such units covered the range of normal human hearing, from the threshold of audibility to the threshold of pain.

By testing the hearing of thousands of listeners, from school children to industrial workers (see figure 4.4), a typical response curve for normal human hearing was determined. This curve indicated that human hearing, which generally ranged between 16 and 16,000 cps, was most sensitive to sounds of around 2,000 cps. Sensitivity fell off gradually below this pitch, and more rapidly above it. While the basic parameters of the limits of human hearing had been known before, the large-scale precision testing made possible by the new audiometer

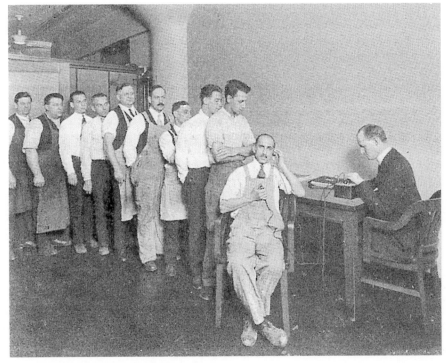

4.4

Industrial workers undergoing hearing exams with the Western Electric 3-A Audiometer, c. 1923. The examiner varied the frequency and intensity of a sound signal that was transmitted to the earpiece that the subject is holding up to his ear. The subject pressed a button to indicate when the signal became audible, and the signal strength at this moment indicated the sensitivity of the subject's hearing at that frequency. "The No. 3-A Audiometer," n.d., p. 1. Photo #00-0684. Property of AT&T Archives. Reprinted with permission of AT&T.

endowed this curve with a statistical relevance that it had not previously possessed. The experience of being tested additionally became a new element of aural culture for increasing numbers of people over the course of the decade.[137]

In 1926, Edward Elway Free, the science editor of *Forum* magazine, used a Western Electric audiometer to undertake a "scientific investigation" of noise in New York, "the first investigation of its kind anywhere."[138] While Isaac Rice no longer controlled the magazine (he had died in 1915), it is likely that Free was familiar with the past efforts of Mrs. Rice and other noise reformers, but he had little use for those efforts. "When we set out to accumulate information on this subject," Free informed his readers, "we discovered that practically none was in existence. No one had determined, by unquestionable physical tests, just how much noise there is on a city street." "People had impressions on these points," he continued. "We had some ourselves. But these were rough ear-impressions only; they had not been checked and corrected by data which exact physical science could respect. Accordingly we set out to get this data."[139]

In order to measure city noise with the audiometer, Free used the device in much the same way that telephone engineers measured electrical noise on transmission lines. He listened to the audiometer tone by applying the earpiece of the instrument to one ear, and his other ear was left open to the noise of the city. He then increased the intensity of the audiometer tone until it was just loud enough to mask the city noise, and the audiometer reading thereby indicated in sensation units the loudness of the city noise.

With this new technique—"the most modern of physical methods"—Free measured noise levels at hundreds of sites all over Manhattan, and he concluded that the main source of city noise was its street traffic. "Most New Yorkers," he asserted, "would probably say, as we did before we knew, that the elevated trains make more noise than anything else from which the city suffers."[140] But Free's measurements proved that this was not the case; at street level the noise of automobiles and especially of chain-driven trucks exceeded that produced by the elevated trains. Even more surprising was the realization that horse-drawn traffic was actually louder than automobiles or trucks. The apparent increase in the city's noise—which seemed obvious to all even if it had not been measured before—was thus not the result of the replacement of horse-powered traffic by cars and trucks, but was instead due simply to the tremendous increase in the amount of traffic. The noisiest spot measured by Free was one of the city's busiest traffic intersections, at 34th Street and Sixth Avenue, with a noise level of 55 sensation units.[141]

While the quantification of noise in Free's report was novel, his conclusion was not. For the past several decades, the sounds of traffic had been moving steadily up toward the top of lists of noise nuisances. Earlier lists had cited the rattle of horse-drawn wagons, but this noise was soon drowned out by the scraping screech of the flattened metal wheels of streetcars. Complaints of unmuffled, or "cut-out," automobiles began to appear as early as 1911, and both the frequency and despair of these complaints increased dramatically in the 1920s.[142] Motorcycles, automobile horns, and chain-driven trucks were added to the litany, and the noises of motorized traffic dominated listings by 1925. At this time, the *Saturday Review of Literature* observed that "the air belongs to the steady burr of the motor" and "the recurrent explosions of the internal combustion engine."[143]

The *New York Times*, perceptively responding to Free's conclusion that horse-drawn traffic was actually louder than automobile traffic, suggested that perhaps it was not the level of noise that was the crux of the problem, but rather the nature of the sounds. The problem was that "the machine age has brought so many new noises into existence, the ear has not learned how to handle them. It is still bewildered by them."[144] Whereas in 1905 the paper had illustrated the problem of noise with a variety of harmless—if irritating—human agents, by 1930 New York's papers depicted the enemy as a machine-age beast that threatened to overpower any human foolish enough to stand in its path. (See figures 4.5 and 4.6.) This changing character of the soundscape, as much as any actual or perceived increase in overall loudness, was fundamental to the growing concern over the problem of noise. Like Edgard Varèse, the *Times* challenged its readers/listeners to retrain their ears in order "to handle" the new soundscape of their city.

While the noise of traffic had gradually crept up on listeners over the course of a decade or more, a new noise that announced its presence far more abruptly was the amplified output of electroacoustic loudspeakers. Ironically, or perhaps fortuitously, the same electroacoustic industry that was responsible for developing new noise-measuring instruments was also guilty of providing one of the worst producers of noise to measure. While everyone enjoyed listening to his or her own favorite music or radio programs, hearing a neighbor's favorites through the wall or an open window was entirely different, especially late at night. Radio retailers who installed loudspeakers above their shop doors, to broadcast their wares out into the streets, were even worse offenders to those who lived or worked nearby.[145] Worst of all were the advertising airplanes that

4.5
Comic illustration of the sources of noise in New York, 1905. City noises in early twentieth-century America were identified as the products of individual, if annoying, people. "New York the Noisiest City on Earth," *New York Times* (2 July 1905): part 3, p. 3.

4.6
By 1930, noise was depicted exclusively as the product of modern technology. This cartoon by Robert Day, which originally appeared in the *New York Herald Tribune,* was reproduced in the report of the Noise Abatement Commission of New York. Edward Brown et al., eds., *City Noise* (New York: Department of Health, 1930), p. 255.

flew low over the city for hours at a time, broadcasting slogans, jingles, and ditties down on the acoustically helpless multitudes below.[146]

When New Yorkers were polled about the noises that bothered them in 1929, over thirteen hundred complaints (12 percent of the total received) cited the noise of loudspeakers.[147] Acoustically aggrieved citizens had begun writing letters of complaint about "the *enfant terrible* of the present electrical age" as early as 1922,[148] and in 1930, it was noted that the "annoyance has increased since the powerful electro-dynamic loud-speakers became the vogue."[149] One creative complainant devised a "violet ray device" that emitted electromagnetic interference, rendering his neighbors' radios useless and forcing them to find other (presumably quieter) means of nocturnal entertainment. In Chicago, angry neighbors bombed a woman's apartment when their complaints about her noisy radio brought no relief.[150] Fortunately, few were willing to undertake such extreme measures to abate the noise, and the law-abiding citizens of New York received at least some respite from their plight in 1930 when Alderman Murray Stand introduced a bill to regulate the use of outdoor loudspeakers.

"In the last few years," Stand explained, "a particular noise nuisance has sprung up, causing great disturbance to large numbers of people. They cannot escape from this tremendous din—the like of which was impossible until modern ingenuity produced the electrical magnification of sound."[151] Stand's bill required anyone desiring to operate a loudspeaker out of doors to obtain a permit from the city. Although the public hearing on the bill had to be postponed—the noise of an impromptu concert by the Sanitation Department Brass Band outside City Hall made it impossible to hear testimony in the committee room—it eventually passed and on 5 June 1930, Joseph Krauss, the owner of a radio and phonograph store on 2d Avenue at 86th Street, had the dubious honor of being the first person taken to court for violating the new law.[152]

Even before Alderman Stand's bill had become law, the Department of Health amended its Sanitary Code with Section 215a, which stated more generally that:

> No person owning, occupying or having charge of any building or premises or any part thereof in the city of New York shall cause, suffer or allow any loud, excessive or unusual noise in the operation or use of any radio, phonograph or other mechanical or electrical sound making or reproducing device, instrument or machine, which loud, excessive or unusual noise shall disturb the comfort, quiet or repose of persons therein or in the vicinity.[153]

This new amendment was successfully tested in May 1930 when the neighbors of Thomas Hill, proprietor of a music store in the Bronx, took him to court for the disturbance that his loudspeaker caused them. Mr. Hill pleaded guilty and agreed to pay a $50 fine, and the magistrate warned that a second offense would carry a fine of $250 along with three months in jail.[154] The new, amplified sounds of loudspeakers were clearly distinctive enough to mobilize into action a legal system that had been almost uniformly unsuccessful in addressing the problem of more traditional sources of sound.

Radio loudspeakers also changed the way that people defined noise within the confines of their own homes, as the unwanted sound of a neighbor's loudspeaker was not the only kind of noise that radio produced. For those who tuned in, a whole new vocabulary was required to differentiate between the noises of electromagnetic static and other distortions that stood between a listener and the program that they sought to enjoy. Even neighborhoods free from violet-ray vigilantes suffered "The Demon in Radio," as listeners struggled to separate the signal from the noise and educated their ears to listen like skilled telephone engineers. In 1924, the *Literary Digest* classified the new pandemonium into "'grinders' or 'rollers' (a more or less rattling or grinding noise), 'clicks' (sharp isolated knocks), and 'sizzles' (a buzzing or frying noise more or less continuous)." *Century Magazine* described the noises of radio as ranging between "the hiss of frying bacon and the wail of a cat in purgatory."[155] One of the worst noises was elicited when a listener's hand approached the tuning dials of the receiver to make an adjustment. Since every radio receiver also emitted a small amount of radio-frequency energy, the introduction of a person's hand into the locally generated electromagnetic field surrounding the receiver sometimes created feedback that resulted in a hair-raising squeal. Manufacturers found a way to silence this squeal, but not before one inventive listener detected in it the means to create a new kind of music.

Just as Luigi Russolo and Edgard Varèse heard music in the mechanical din of the modern city, the engineer Leon Theremin (Lev Termen in his native Soviet Union) heard music in the feedback squeal of radio. In 1920, Theremin used the principle of this feedback as the basis for a new musical instrument. The Etherophone (later known as the Theremin Vox or Theremin) consisted of a combined radio transmitter-receiver. It was housed in a wooden box raised on legs that might have been mistaken for a lectern except for two protruding antennas. (See figure 4.7.) To play the instrument, the musician moved her hands through space, altering the electromagnetic field surrounding the device; the

4.7
Alexandra Stepanoff performing on an RCA Theremin, c. 1930. Performers created music by moving their hands in the vicinity of the Theremin's antennas, manipulating the electromagnetic field surrounding the device in ways that altered the frequency and amplitude of an electrical signal. This staged photo omits the loudspeaker that would have been required to translate that signal into audible sound. The microphone shown here had little function except to advertise NBC. George H. Clark Collection, Archives Center, National Museum of American History, Smithsonian Institution, SI negative #2000–11232.

proximity of the right hand to the vertical antenna controlled the frequency of sound, and the left hand controlled its volume via the horizontal antenna. The melodic signal generated within the circuitry was amplified by vacuum tubes and transmitted to a loudspeaker, and the unique sound that resulted captured the imagination of all who heard it. When Theremin demonstrated his device to Vladimir Lenin at the Kremlin in March 1922, the press bestowed the ultimate Soviet compliment, proclaiming, "Termen's invention is a musical tractor."[156]

Theremin emigrated to the United States in 1927 and demonstrated his musical instrument to much acclaim in high-society salons, engineering society meetings, and public concerts.[157] He received a U.S. patent in 1929, and soon thereafter representatives of the Radio Corporation of America, "chagrined that none of its engineers hit upon the idea,"[158] negotiated an agreement to manufacture and market the new instrument. "That terrible demon of the early days of the radio," the *New Yorker* reported, "still a restless and yowling house cat at times, has become an invisible piano."[159]

The heyday of the Theremin coincided with the peak of interest in the music of composers like Edgard Varèse and George Antheil. Modern composers—including Varèse—wrote for the new instrument, and Leopold Stokowski championed the Theremin as he championed all things modern. "Thus will begin a new era in music," the conductor proclaimed in 1928, "just as modern materials and methods of construction have produced a new era in architecture."[160] Other listeners, however, were more troubled by this new addition to the musical soundscape.

When the electrically generated and amplified sounds of Joseph Schillinger's *First Airphonic Suite* for RCA Theremin and orchestra were presented at Carnegie Hall in 1929, Olin Downes objected more to the fact of amplification than to the actual tone of the instrument or to the musical nature of the composition. "We do not like to think of a populace at the mercy of this fearfully magnified and potent tone that Professor Theremin has brought into the world." "The radio machines are bad enough," he complained, "but what will happen to the auditory nerves in a land where super-Theremin machines can hurl a jazz ditty through the atmosphere with such horribly magnified sonorities that they could deaden the sound of an automobile exhaust from twenty miles away?"[161]

The introduction of loudspeakers and the amplified sounds they emitted into the sacrosanct setting of Carnegie Hall was as troubling as had been George Antheil's airplane propellers and sirens two years earlier. These critical reactions to such technological breaches of that last bastion of aural refuge, the concert

4.8
Western Electric Sound Meter, 1931. The microphone on the left transformed sound into an electrical signal, which was modified in a circuit designed to imitate the frequency response of the human ear. The loudness of the sound was then indicated in decibels on the meter at the right. Not shown is the unwieldy power supply. T. G. Castner et al., "Indicating Meter for Measurement and Analysis of Noise," *Transactions of the American Institute of Electrical Engineers* 50 (September 1931): 1042. © 1931 AIEE, now IEEE.

hall, only amplified more general concerns about the noise of the city itself. As the soundscape was transformed by modern technology, it became increasingly evident that only modern technologists would be able to control that environment.

Edward Free's 1926 report on city noise in New York was soon followed by a similar survey in Chicago, where the Board of Health sponsored an investigation carried out by engineers of the Burgess Laboratories using "a newly perfected acoustimeter" of their own design.[162] Representatives of the Graybar Electric Company surveyed Washington, D.C., and numerous other noise surveys were carried out by engineers in cities across the nation, using new tools specially designed for this purpose.[163] (See figure 4.8.)

In 1928, Edward Free followed up on his "now famous" report of 1926. According to Free, knowledge of the "physical side" of the problem of city noise had made more progress in the past two years "than in all the previous history of acoustic science."[164] What remained, he argued, was the psychological side of the question: Which noises were most annoying and harmful, and what was their effect? "Nobody knows what noise costs," Free concluded—implying costs both human and economic—"and nobody is going to discover except by some more hard scientific work."[165]

One researcher who sought to answer this question was Donald Laird, an industrial psychologist at Colgate University. "Noise *Does* Impair Production," Laird announced after determining experimentally in 1927 that it could reduce manual or mental output by as much as thirty percent.[166] Laird studied the effect of noise on the physiology and working efficiency of typists by scientifically analyzing their performance under both quiet and noisy conditions. Typing and error rates were compared, and the exhalations of the typists were chemically

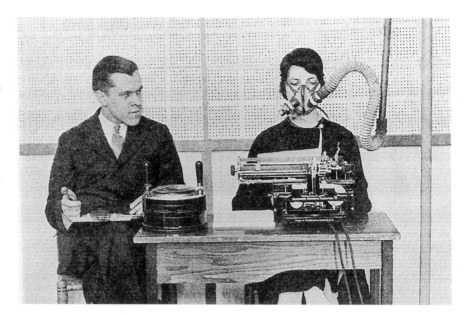

4.9
Industrial psychologist Donald Laird's study of the effect of noise on clerical workers. A typist worked under both quiet and noisy conditions, and her rate of caloric consumption was determined by chemically analyzing her exhalations—collected via the face mask—as she maintained a typing rate of 150 words per minute. Donald Laird, "Experiments on the Physiological Cost of Noise," *Journal of the National Institute of Industrial Psychology* 4 (1929): 253, figure 1. Princeton University Library.

analyzed to determine their rates of caloric consumption. Laird concluded that energy consumption increased by 19 percent when typists worked under noisy conditions, and he also demonstrated that the best typists worked about 7 percent faster in a quieter environment.[167] (See figure 4.9.)

The energy lost to production seemed to be used up in an involuntary tightening of muscle tissue, and this observation led Laird to examine more fully the physiological effect of noise. In a study of the effect of noise on stomach contractions, Laird confirmed that very loud noises had a "profound effect on involuntary muscle activities of the stomach," an effect equivalent to the primal "fear reaction."[168] New Yorkers were soon being told that their bodies responded to noise in the same way that their prehistoric ancestors had responded to the roar of a saber-toothed tiger.[169] As startling as this news may have been, Laird's measurement of the noise-induced loss of workers' productive output was equally newsworthy, for he had now documented scientifically what had long been suspected; the economic cost of noise was enormous.[170]

The inefficiency of noise had been a compelling problem earlier in the century, but the numbers now associated with it—errors per hour, percent decrease in productivity, dollars lost per day—increased the gravity of the problem. Further, the concept of efficiency itself was transformed in the 1920s in ways that invested it with an even greater cultural significance. Efficiency not only stood for the economical and moral values of productivity and prosperity, but now further constituted an aesthetic style that represented everything modern. This stylistic turn allowed the concept of efficiency to migrate into fields far removed from its technical origins in the management of industrial labor.

In 1920, for example, William Strunk's *Elements of Style* signaled the death of flowery Victorian prose with the concise dictum, "Omit needless words."[171] Library systematizer Melville Dewey became Melvil Dui in 1924, when he undertook a campaign for simplified spelling. Dui claimed that "one seventh of all English writing is made up of unnecessary letters," and he proposed to eliminate such waste from the language.[172] Women's fashions, too, were pared down to essentials. Flappers cut off their long hair and shed yards of clothing to emphasize their now-slim figures.[173]

The same reductive imperative located behind these diverse cultural phenomena also drove the desire to eliminate noise. Indeed, the justification for noise abatement was now expressed in prose that might have been written by Strunk himself: "Noise costs money. It lowers efficiency. It causes waste. It shortens life."[174]

As efficiency became a style that was celebrated throughout modern American culture, engineers became secular saviors as the bringers of that efficiency. They were cast as heroes in popular novels and movies, and the objects they designed were celebrated simply for being "engineered."[175] An engineered soundscape promised not only to recover lost dollars and to reinvigorate tired workers, but also to constitute a thing of modern beauty in and of itself. Overlooked was the fact that the engineers who would design this new soundscape were the same technicians who had created the machines that were making all the noise. More important was the belief that no one but those engineers could ever hope to regain control over those machines, to engineer an efficient soundscape in which the inhabitants of the modern city could thrive.

V Conclusion: The Failure of Noise Abatement

"The increasing number of complaints of noise and the intimate relation between noise and health" were what led New York City Health Commissioner Shirley Wynne to appoint a Noise Abatement Commission in 1929, "the first of its kind in this country."[176] The purpose of the commission was to classify, measure, and map the noises of the city, then to study extant laws and recommend new ones, along with any other measures, that promised to control or eliminate those noises. "We have been fortunate," Wynne proclaimed, "in securing for the membership of this Commission leading scientists and business men. The cost of this research work which would easily run into hundreds of thousands of dollars if engaged by the city, has been contributed to this Commission's task without cost to the city by the Bell Telephone Laboratories, the Johns-Manville Corporation and other important organizations with their facilities and scientific personnel."[177] These scientific personnel would soon turn the entire city into "a veritable laboratory for the study of sound,"[178] as they began identifying, measuring, and attempting to abate the noise of New York.

To gather public impressions of the problem of noise, the commission published a questionnaire in the major metropolitan newspapers. Responses submitted by readers confirmed that the vast majority of the noises that plagued New Yorkers were the product of modern technological inventions. (See figures 4.10 and 4.11.) Many additional complaints "poured into" the office of the commission, or were sent directly to Mayor Walker, and these letters similarly identified the machines of modern technology as the principal objects of complaint.[179] (See table 4.1.)

The commission now set out to map and measure the city's noise, and they did so in a specially equipped truck, a "roving noise laboratory," filled with state-of-the-art sound equipment and staffed with men from Bell Labs, Johns-Manville, and the Department of Health. The truck logged over 500 miles as it traveled throughout the city. Technicians, looking more like G-men than sound engineers, collected 10,000 measurements at 138 locations.[180] (See figure 4.12.)

The engineers employed two distinct kinds of measuring tools. The first was an audiometer like that used earlier by E. E. Free to measure the "deafening effect" of noise. The second was a sound meter that "listened" through a microphone and gave a direct reading of the intensity of the noise. The truck was also equipped with frequency analyzers to explore the physical makeup of specific kinds of noises. Sound meters, microphones, vacuum-tube amplifiers, and analyzers constituted "the armoury of the acoustical investigator,"[181] and these new weapons were proudly displayed by the engineers who wielded them to slay city noise. (See figure 4.13.)

Not only the tools, but even the units with which the sound was measured were new. The ambiguous "noise units," "sensation units," or "transmission units" that sound-measuring instruments had previously registered were now replaced by a new standard, the decibel, which was named in honor of the father of electroacoustics, Alexander Graham Bell.[182] In 1928, Edward Free had indicated that

NOISE ABATEMENT QUESTIONNAIRE

Use a soft pencil in filling out questionnaire. Under "Location" give the address of the source of the noises most annoying to you, and under "Hour of Day" state the time at which these noises are noticed by you.

SOURCE OF NOISE	LOCATION	HOUR OF DAY
Loud Speakers in Home		
Automobile Horns		
Trucks——Horse-Drawn		
Trucks——Motor		
Buses——Noisy Mechanism or Tires		
Automobile Cut-Outs		
Noisy Brakes on Automobiles		
Riveting		
Pneumatic Drills on Streets		
Pneumatic Drills on Excavations		
Loud Speakers Outside of Stores		
Airplanes		
Noisy Parties		
Locomotive Whistles and Bells		
Tug and Steamship Whistles		
Elevated Trains		
Subway Trains		
Subway Turnstiles		
Street Cars		
Ash and Garbage Collections		
Newsboys' Cries		
Unmuffled Motorboats		
Traffic Whistles		
Fire Department Sirens and Trucks		
Milkmen		
Factories		
What ONE noise is MOST annoying?		

If you have suggestions to offer, write a letter and attach it to your questionnaire.

Signed _____

Address _____

NOTE: Your name and address will not be used publicly in any way or at any time.

Mail this questionnaire to: NOISE ABATEMENT COMMISSION
505 Pearl Street, New York City

TABULATION OF NOISE COMPLAINTS—March 1, 1930

SOURCE	NUMBER	PERCENT
Trucks——Motor	1,125	10.16
Automobile Horns	1,087	9.81
Radios——Homes	774	7.00
Elevated Trains	731	6.62
Radios——Street & Stores	593	5.36
Automobile Brakes	583	5.27
Ash & Garbage Collections	572	5.17
Street Cars	570	5.16
Automobile Cut-Outs	504	4.55
Fire Department Sirens and Trucks	455	4.12
Noisy Parties and Entertainments	453	4.10
Milk and Ice Deliveries	451	4.07
Riveting	373	3.37
Subway Turnstiles	317	2.86
Buses	271	2.45
Trucks——Horse Drawn	268	2.41
Locomotive Whistles and Bells	238	2.15
Pneumatic Drills——Excavations	233	2.11
Tug and Steamship Whistles	223	2.01
Pneumatic Drills——Streets	213	1.93
Newsboys and Peddlers	212	1.91
Subway Trains	183	1.65
Dogs and Cats	140	1.26
Traffic Whistles	137	1.24
Factories	117	1.06
Airplanes	113	1.02
Motor Boats	66	0.59
Motorcycles	41	0.37
Restaurant Dishwashing	25	0.22
	11,068	100.00

CLASSIFICATION

SOURCE	NUMBER	PERCENT
TRAFFIC (Trucks, Automobile Horns, Cut-Outs, Brakes, Buses, Traffic Whistles, Motorcycles)	4,016	36.28
TRANSPORTATION (Elevated, Street Cars, Subway)	1,801	16.29
RADIOS (Homes, Streets & Stores)	1,367	12.34
COLLECTIONS & DELIVERIES (Ash, Garbage, Milk, Ice)	1,023	9.25
WHISTLES & BELLS (Fire Dept., Locomotives & Tugs & Steamships)	916	8.28
CONSTRUCTION (Riveting, Pneumatic Drills)	819	7.40
VOCAL, ETC. (Newsboys, Peddlers, Dogs, Cats, Noisy Parties)	805	7.27
OTHERS	321	2.89
	11,068	100.00

4.10
Questionnaire distributed in 1930, via metropolitan newspapers, by the Noise Abatement Commission of New York. Edward Brown et al., eds., *City Noise* (New York: Department of Health, 1930), p. 25.

4.11
Tabulated results of the Noise Abatement Questionnaire of 1930. Responses to the survey by New Yorkers emphasized the prevalence of technology in the modern urban soundscape. Edward Brown et al., eds., *City Noise* (New York: Department of Health, 1930), p. 27.

TABLE 4.1: NOISE COMPLAINT SUMMARY, NEW YORK, 1926-1934

	TOTAL	TOTAL %	1926	1927	1928	1929	1930	1930 %	1931	1932	1933	1934
Construction	91	15.7	2	2	6	9	57	16.9	13	2	0	0
Loudspeakers	88	15.2	0	1	2	5	64	18.9	4	9	2	1
Transportation	82	14.1	1	0	0	6	48	14.2	8	4	14	1
Commercial	78	13.4	1	3	1	3	35	10.4	13	12	10	0
Generic	67	11.6	1	1	2	8	40	11.8	5	6	4	0
Industrial	53	9.1	2	1	0	2	24	7.1	8	8	8	0
Services	46	7.9	2	1	0	1	24	7.1	10	4	4	0
People	27	4.7	1	0	0	0	15	4.4	3	1	6	1
Animals	26	4.5	0	3	1	1	11	3.3	1	5	4	0
Music	15	2.6	0	0	1	0	13	3.8	1	0	0	0
Miscellaneous	7	1.2	0	0	0	0	7	2.1	0	0	0	0
TOTAL	580		10	12	13	35	338		66	51	52	3

TABLE 4.1 KEY:

Description of Categories Listed:

Construction:	building and subway construction, riveting, steam shovels, blasting, drilling, etc.
Loudspeakers:	any electrically-amplified sound source
Transportation:	operation of cars, trucks, horns, railroads, boats, subways, garages, taxi stands
Commercial:	noises from shops, stores, restaurants, laundries, bakeries, etc.
Generic:	all unspecified noise complaints
Industrial:	noises from factories or heavy industrial machinery
Services:	milk and ice delivery, removal of ashes and garbage, fire engines, ambulances
People:	noises of human activities not falling in any other category
Animals:	noises of animal origin (dogs, cats, poultry, horses, pet hospitals)
Music:	playing of instruments and other nonamplified sources of music, bells
Miscellaneous:	whistles and sirens other than fire engines or ambulances

Sources: New York City Municipal Archives: Mayoral Papers, James Walker, Departmental Correspondence Received and Sent: "Health Department" (1926–1932); Department of Health, Administration/Subject Files: "Noise" (1929–1934).

4.12

Official Noise Measuring Truck of the Noise Abatement Commission of New York, 1930. The truck logged hundreds of miles as it measured noise levels at hundreds of sites all over the city. It was manned by sound engineers from AT&T and the Johns-Manville Company. Photo #HM46839. Property of AT&T Archives. Reprinted with permission of AT&T.

4.13

Inside the Noise Measuring Truck. The man in the white hat is listening to a standard noise signal. He varied the strength of this signal with the control in his left hand until it was just loud enough to mask the city noise that he heard in his unobstructed left ear. The signal strength at this point indicated the loudness of the city's noise. Photo #HM46753. Property of AT&T Archives. Reprinted with permission of AT&T.

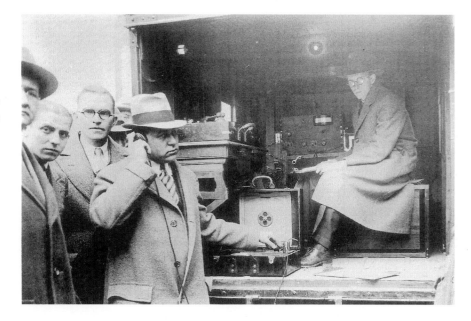

the old "noise units" meant little, "except to the acoustic expert,"[183] but now newspapers and magazines covering the activities of the Noise Abatement Commission were eager, not only to make the new units understandable to the general public, but also to provide themselves with a technically precise language for reporting on noise. In describing the commission's noise survey, for example, the *New York Times* explained the decibel in detail:

> The unit of loudness used was the decibel, described by the experts as "approximately the smallest change that the ear can detect in the level of sound."
>
> Decibels do not measure ascending steps, all of equal intensity, . . . but rather express a ratio that increases rapidly in moving up the scale. . . .
>
> According to this system of measuring, the loudness of an average conversation measured at a distance of three feet is about 60 decibels. The roar of explosives at a subway excavation in the Bronx measured 98 decibels, while riveters produced the terrific sound intensity of 99 decibels. . . . These sounds, it was pointed out, are all more than 1,000,000,000 times as loud as the faintest sound which man can hear.[184]

The Noise Abatement Commission published charts depicting the decibellic ascension of city noises both indoors and out, and such charts also appeared in popular magazines, educating readers about the new measure of sound as well as the noises that surrounded them.[185] (See figures 4.14 and 4.15.) In December 1929, Harvey Fletcher presented a radio address over WEAF in New York in which he not only explained the scientific survey of noise being carried out by the commission, but also demonstrated sounds of different decibel levels to his listening audience.[186]

When acoustical engineers from AT&T measured the noise of the subway system, the city learned that the noise sometimes reached 120 decibels, the threshold of pain for normal human beings.[187] (See figure 4.16.) When the Noise Abatement Commission measured the noise of randomly stopped trucks at York Avenue and 77th Street, the average level of 81 decibels was similarly announced to the public.[188] In June 1931, the commission investigated a new model of "semi-noiseless" ash can, and a crowd of 200 turned out to watch Nunzio Parrino—one of the sanitation department's finest—roll, toss, and manhandle the new can as the engineers measured his acoustical output. The rubber-bottomed can proved too bouncy to be practical, but the experiment determined that a rubber lining on the side of the truck would reduce the noise of collection by 11 decibels.[189] As other cities followed New York's lead and undertook their own noise surveys, a perverse kind of competition even developed, as

4.14

"Noise in Buildings," chart listing the noise level in decibels of different types of interior spaces, c. 1930. Through charts like this, New Yorkers and other Americans were taught to quantify the noises that surrounded them. Edward Brown et al., eds., *City Noise* (New York: Department of Health, 1930), p. 158.

4.15

"Noise Levels out of Doors," chart listing the noise level in decibels of different sounds typically encountered outdoors in the city. Edward Brown et al., eds., *City Noise* (New York: Department of Health, 1930), p. 131.

4.16

Sound engineers from Electrical Research Products, Inc., a division of AT&T, measuring the noise of the New York City subway system in 1931. G. T. Stanton (left) indicates the noise level reading to G. M. Purver, of the Board of Transporation. J. E. Tweeddale (rear) holds a condenser microphone in his hand. Photo #W4195. Property of AT&T Archives. Reprinted with permission of AT&T.

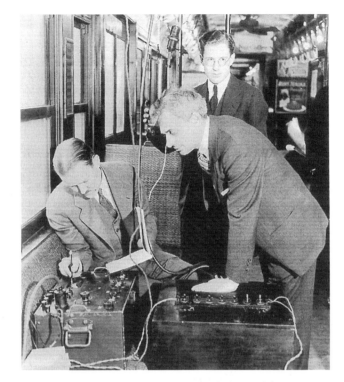

the decibellic levels of Chicago, New York, and other cities were compared and commented on in the press.[190]

The editors of the *Times* suggested that, if the commission kept up its work, New Yorkers' "trained ears will become as sensitive as a noise meter to the sound of a dropping pin" and citizens might begin to "count decibels" themselves. Letters to the editor indicate that this was indeed the case, as writers began to cite decibels when describing the noises that plagued them.[191] But the power of the language of decibels ultimately proved delusory, as this language was not easily translated into actual abatement of those sounds.

The Noise Abatement Commission of New York was active for two years, during which it transformed public perceptions of the problem of noise by scientifically demonstrating the power and pervasiveness of that problem. It heightened New Yorkers' awareness of noise and it educated them to listen in new ways. The impact of the commission went beyond the local, as its ambitions and activities were well covered, not only in New York newspapers, but in national magazines and through the widespread distribution of its first official report.[192] (See figure 4.17.) But, while the press reported energetically and

4.17
"Abating the Noise Evil," cartoon by Otto Soglow for the *New Yorker* (5 July 1930). Soglow's whimsical solutions to the problem of noise suggest that the less fanciful efforts of the Noise Abatement Commission were well known to readers of the magazine. © The New Yorker Collection, 1930, Otto Soglow, from cartoonbank.com. All Rights Reserved.

enthusiastically on the various activities of the commission, when faced with the hard question of whether those activities were actually abating the noise of New York, everyone agreed that there was little to show for all the hard work. After the commission was dissolved in 1932, people asked themselves and each other, "Of what value was our much-touted Noise Abatement Commission? In what way have conditions improved?"[193] The answer was anything but clear.

In its final report, the Noise Abatement Commission outlined the numerous concrete measures that it had initiated, executed, or supported over the past two years. Whistle-blowing traffic police had been replaced by silent traffic lights. New quiet turnstiles had replaced the perniciously loud older models at numerous subway stations. The commission had sponsored the amendment to the Sanitary Code regarding the regulation of loudspeaker noise, and had also supported Alderman Stand's efforts against this same foe. It supported the adoption of a new clause in the city's building codes that would allow the new technology of welding to silence the noise of riveting. Most ambitious of all, it sponsored a significant amendment to the Sanitary Code that would identify a wide range of noises and render them all illegal. Under the new law, perpetrators would be subject to a system of tickets and fines that would eliminate the costly and inconvenient necessity of hauling offenders into court. But this law, which would have constituted the crowning achievement of the commission, was not passed.[194]

As municipal priorities changed in the early 1930s, and as the city government itself changed hands after a corruption scandal led to the resignation of the mayor, the antinoise amendment was lost in the shuffle and sacrificed to more imperative agendas. Some blamed the depression itself for the failure of the campaign against noise. It was not evident that "The Crash" of the stock market had simply brought the noisy machines of the Machine Age to a halt, but observers did note that "opposition to unnecessary noise has been somewhat drowned out in the Big Noises of politics, repeal and national recovery. Commissions have ceased to function or to make themselves heard, and only distracted individuals complain of the continuing din."[195] "With the fading away of the Noise Abatement Committee," one such individual confessed, "I have no one to whom to tell my decibellic troubles except the *New York Times*."[196]

It is clear that the political upheavals that accompanied the rapidly changing economy stalled at least some of the efforts of the commission, and its members were particularly frustrated by the failure of the city aldermen to enact their antinoise amendment.[197] But, as the commission itself made clear, its primary role had been advisory. Its members had no direct legal power to abate the noises they studied; their job had been to recommend such action, and if those recommendations were not followed, the blame was not theirs to assume. By 1932, they were impelled to point the finger at others, as it was clear that the city's larger mission of abatement had failed. When the commission filed its final report, it was, according to historian Raymond Smilor, "the product of disappointed authors."[198]

"Law enforcing agencies are not doing their duty," the commission complained, and "governmental bodies have failed to stop the din." The report detailed the morass of bureaucracy that prevented the passage of the antinoise amendment, but government alone was not to blame. The people themselves were at least partly responsible, for, as the report bitterly reported, "most of the thousands who complained . . . were unwilling to lift a finger themselves to stop the noise nuisances they faced." "They expected the Commission to come to their rescue like a magic prince, solving their difficulties with a wave of a fairy wand and emphatically without any effort to themselves. They expected a body made up of private citizens and one part time paid executive to put a calming finger on anyone, anywhere in the many square miles of New York City."[199]

Perhaps, in the end, the experts were ill-served by their expertise. By demonstrating the power of modern technology to measure and map city noise, the commission misled the city into thinking that its engineers could just as easily eliminate it. A perceptive observer had recognized this danger in 1931, noting that "Important as the measurement of noise is for so many purposes, there is a real danger that too much attention may be focussed on it, and the suppression of unnecessary, devastating, harmful din neglected."[200] But, if modern acoustical science ultimately failed to provide a public solution to the problem of noise, it succeeded in presenting a private alternative.

Early twentieth-century efforts at noise reform, as well as the later efforts of the Noise Abatement Commission, had attempted to eliminate noise by regulating the actions of noise-making people and machinery. The goal was to control the public soundscape of the city, to enforce and ensure the civic right of all to enjoy a noise-free environment. The commission's final report blamed the public, or rather "public apathy," for the failure of this approach.[201] Simultaneous with these failed efforts to control the public soundscape, however, modern acousticians were far more successfully exerting control over the soundscape of private life. Indeed, the success of the latter may be partially responsible for the failure of the former. Even as the commission measured and charted the noise in the streets of New York, sound-absorbing building materials were being deployed to transform homes, offices, hospitals, and hotels into shelters from that noise. By manipulating and controlling private space, by turning inward and creating acoustically efficient refuges from the noises of public life, acousticians offered a compelling alternative solution to the problem of noise.

CHAPTER 5   ACOUSTICAL MATERIALS AND MODERN ARCHITECTURE, 1900–1933

> Today, architects and engineers the world over have come to recognize proper acoustics as necessary in the modern types of building construction. The importance of sound control and noise reduction, together with the great benefits in health and happiness which it gives to the human race, is something now generally accepted.[1]
>
> "Absorbex" Sales Pamphlet, 1932

I INTRODUCTION

Because the din of building construction was one of the worst of city noises, American cities simultaneously enjoyed and suffered from the building "booms" of the early twentieth century. Steam shovels chugged and scraped, and pneumatic riveters relentlessly pounded metal on metal as construction flourished across the nation. In earlier times, a church spire had typically constituted the high point of a city's skyline, but by the twentieth century commercial architecture towered over all. Cass Gilbert's sixty-story Woolworth Building was not just the tallest building in New York, but the tallest building in the world when it was completed in 1913. Its lofty height, the rich, Gothic-styled ornamentation that covered it from ground to pinnacle, and a tongue-in-cheek acknowledgment of America's true religion inspired its nickname, the "Cathedral of Commerce."[2] As new construction continued apace in the 1920s, a rising tide of stone and steel gradually encroached on the Woolworth Building's eminence until it was at last overshadowed, most notably by the Empire State Building. For a rapidly growing city located on an island, up was the only direction to go. While streetcars, subways, and automobiles now transported many formerly urban residents to new homes in the surrounding suburbs, the commercial heart of the city remained centered upon a few acres of prime real estate in Manhattan. The corporate leaders, builders, and real estate speculators who held the deeds to those valuable acres sought to extract maximum value from their holdings, and the tall building was the means to do so.[3]

The result of this upward growth, however, was to exacerbate problems on the ground. Row upon row of monolithic towers turned streets into increasingly crowded and darkened canyons. New York's 1916 zoning law required tall buildings to recede, or step back, from their ground level footprint as they rose, and thus began to restore a degree of sunlight and fresh air to the streets. But architects and builders responded with a complex calculus of design that enabled them to operate within the restrictions laid down by the law, yet still maximize the profitability of a structure by pushing the envelope, or volume, of the building to its legal maximum. The zoning law helped create the distinctively angular New York skyline, and it alleviated some of the problems that had resulted from unregulated building. Nonetheless, congestion at street level continued to worsen, and the increase in traffic—vehicular and pedestrian—contributed to an increase in noise that persisted long after the machines of construction had ground to a halt. Organizations like New York's Noise Abatement Commission attempted to eliminate that noise, but without much success. A more promising approach to the problem was to employ the science and technology of architectural acoustics to transform the buildings themselves from problem to solution.

Benjamin Betts, editor of the *American Architect*, was just one of many who pointed to "the business of sound control," the manufacture and installation of sound-absorbing and insulating building materials, to solve the problem of noise. "Through its power," he wrote in 1931, "outside noises can be shut out of offices and apartments." "The day is not far distant," Betts predicted, "when prospective buyers and tenants of buildings will ask, 'Is it soundproof?'"[4] In fact, that day was already at hand.

By 1930, dozens of different corporations were manufacturing and selling vast quantities of acoustical building materials. Akoustolith, Acousti-Celotex, Acoustone, Sanacoustic Tile, Sabinite, and Sprayo-Flake represent only a sampling of what was available. These materials were made seemingly of anything and everything: gypsum, mineral wool, volcanic silica, flax, wood pulp, sugarcane fibers, disinfected cattle hair, and asbestos. There were insulating papers, rigid wallboards, stonelike tiles, plasters, and all sorts of mechanical devices for structurally isolating floors, walls, and ceilings. By the time that Betts wrote, thousands of American buildings were already filled with these different types of acoustical products.[5]

These materials were found, not just in auditoriums and sanctuaries, but in offices, apartments, schools, and the various spaces of everyday life. As Betts acknowledged, the goal of sound control was no longer limited to the problem of creating good sound in rooms where listening was the primary activity. Now, the techniques of architectural acoustics were deployed far more widely, to minimize noise wherever it occurred and to insulate people from noises beyond their control. Through the widespread use of these architectural technologies, a new sense of mastery over the soundscape—a mastery that had ultimately eluded the noise abaters—was finally achieved.

But the story of the development of the acoustical materials industry, the rise of the business of sound control, is not simply a tale of technological triumph over noise. Just as modern technologies like pneumatic riveters, automobiles, and loudspeakers transformed the soundscape of city streets, so, too, did acoustical materials fundamentally transform the aural dimensions of interior space. These materials didn't simply eliminate the noises of the modern era, they additionally created a new, modern sound of their own.

This sound was characterized first and foremost by its lack of reverberation; unprecedentedly absorptive materials created a sound that was clear and direct. In a culture preoccupied with noise and efficiency, reverberation became just another form of noise, an unnecessary sound that was inefficient and best eliminated. Reverberation was inefficient because it interfered with the transmission of speech, like electrical noise in a telephone circuit. It also impeded the performance of work by amplifying and sustaining the cacophony of sounds that sapped workers' energy and productivity. The modern sound that resulted from the use of new acoustical materials was thus stripped not only of reverberation but also of these inefficiencies. It both constituted and signified the efficiency of the spaces in which it was heard.

The efficiently nonreverberant quality of this sound was not all that made it modern, however. It also signaled the power of human ingenuity over the physical environment. If science had failed to silence the city, acoustical technology could nonetheless create quiet places of refuge within it. The private character of many of these spaces—apartments and offices, for example—highlights the commodified nature of the new sound and this, too, made it modern. As Betts recognized, sound control was a business, and its products were not only the physical materials themselves but also the sound that those materials produced. The modern sound was achieved through private commerce, not public policy; it was experienced by individualized consumers, not citizens. The quiet, controlled efficiency of the new sound was thus modern in its economic, as well as its physical, nature.

Finally, the new sound was modern because it instantiated a distinctive cultural characteristic that has long been recognized as definitive of the era. When reverberation was reconceived as noise, it lost its traditional meaning as the acoustic signature of a space, and the age-old connection between sound and space—a connection as old as architecture itself—was severed. Reverberation connected sound and space through the element of time, and its loss was just one element in a larger cultural matrix of modernity dedicated to the destruction of traditional space–time relationships.[6] Cubist art, non-Euclidian geometry, and cinematic montage are just a few of the phenomena and artifacts that have been heralded as definitive of the modern, and modern sound should similarly be recognized as a cultural artifact at the cutting edge of change.

Yet, while American acousticians, architects, architectural critics, and the public alike applauded this modern sound, they were far less eager to embrace a similar transformation in the visual aspect of the architecture that produced it. In the midst of aural transformation, the culture of construction clung conservatively to visual vestiges of the past. As a result, throughout the teens and twenties, the modern sound was encountered in spaces that visually defied the changes taking place within and around them. In the neo-Gothic churches of Cram, Goodhue & Ferguson, the medievalesque skyscrapers of Cass Gilbert, and countless other architectural evocations of the past, acoustical materials were disguised, concealed, or simply ignored as irrelevant to the ideals of architectural beauty.

In the early 1930s, however, this disjuncture between sight and sound would finally be resolved. When the radically new look of the modern architecture that had developed in Europe finally arrived in America, its visual celebration of efficiency and technological mastery fit perfectly with the acoustical modernity already in place. Acoustical materials became an integral component of the new style, and the clean, efficient sound that they produced now resonated with an equally efficient look. Modern architecture was founded upon an ideology of environmental control, and acoustical materials transformed this ideology into architectural reality.

The development of acoustical materials and modern sound can best be charted by examining representative products and structures. In St. Thomas's Church (1913), the technological possibilities of new materials were first made evident through the application of sound-absorbing ceramic tiles manufactured by the Guastavino Company. The headquarters of the New York Life Insurance Company (1928) indicate the full incorporation of acoustical materials into the modern corporation, as the building was cloaked top-to-bottom with the felted products of the Johns-Manville Corporation. Finally, the Philadelphia Saving Fund Society Building (1932) exemplifies the perfect fit between modern sound and modern architecture, and demonstrates the perhaps surprising cultural significance of the suspended acoustical-tile ceilings of the Acoustical Corporation of America.

II ACOUSTICAL MATERIALS AT THE TURN OF THE CENTURY

Cass Gilbert's interest in controlling sound dates back to 1895. Just as Wallace Sabine was beginning his investigation of architectural acoustics, Gilbert wrote to the editors of the *American Architect and Building News*. "Can you inform me of any definite set of rules or laws of acoustics, or any treatise on the subject?" he inquired. "If you can supply such a work, I would be very glad to have you do so."[7] Gilbert—then a striving young midwestern architect—was working on his entry in a design competition for the Minnesota State Capitol, and concern about the acoustics of its legislative chamber may have prompted his inquiry.[8] The editors of the journal recommended a few well-known, if not well reputed, books on the topic, then cited a half dozen articles on acoustics that had appeared in their journal over the previous fifteen years. It is not evident how helpful Gilbert found these suggestions. He probably concluded, as did the author of one of the recommended articles, that "one cannot but feel much regret and some degree of astonishment, that this branch of applied, or perhaps I should say unapplied, science should still be in the unsatisfactory condition of uncertainty in which it is."[9]

Wallace Sabine's work soon stimulated the transformation of this uncertain, unapplied science into one increasingly certain and recurrently applied. Additionally, the nature of his work—with its emphasis upon the manipulation of the materials of architectural construction—helped stimulate the development of a new industry based on the manufacture and installation of special-purpose acoustical building materials. The very first building material to be widely advertised and sold for acoustical purposes in America, however, was developed independently of Sabine's researches. Like the reverberation formula, "Cabot's Quilt" originated in Boston at the close of the nineteenth century. Also like Sabine's equation, Quilt was soon being applied to buildings across the nation and around the world.

Samuel Cabot (1850–1906) was descended from a long line of successful New England merchants and manufacturers. A predilection for science rather than commerce led him to the Massachusetts Institute of Technology, and he continued his education abroad, studying chemistry at the Zurich Polytechnicum and visiting chemical laboratories and manufacturing plants throughout Europe. Upon returning to America in 1874, Cabot attempted unsuccessfully to introduce European chemical manufacturing techniques at a bleachery for the Lowell textile mills. He also wrote a few unremarkable scientific papers. In 1877, Cabot entered into a business partnership with the intent of transforming a coal tar distillery into a manufactory of "fine organic chemicals." This plan also proved overambitious, however, and the company instead concentrated on the production of more mundane products such as pitch, tar paper, lampblack, and creosote.[10]

As his business prospered, Cabot continued to pursue his scientific interests and in 1885 he patented sulpho-naphthol, a disinfectant derived from coal tar.[11] In 1892, he developed a new type of padded building paper with excellent properties of both heat and sound insulation. Cabot's Quilt consisted of a thick layer of cured eel grass (*Zostera marina*, a long-stranded seaweed) sandwiched between sheets of heavy building paper or asbestos sheathing. It was advertised as impervious to decay, vermin, and fire, and was presented as "the first thing that was ever scientifically made for deadening sound."[12] Quilt was typically installed within a building's walls and floors, where it provided an elastic cushion with which to isolate the structural members, preventing the transmission of sound from one room to another. Its success was attested to by architects, builders, and Rudyard Kipling, who wrote in 1895 that he "found the Quilt invaluable as a deadener of noise."[13]

Another powerful endorsement was provided by Professor Charles Norton of the Massachusetts Institute of Technology. In 1902, the trustees of the New England Conservatory of Music planned to build a new dormitory, and soundproof construction was considered essential to ensure that each student's practicing would not disturb others'. The trustees commissioned Norton to evaluate the insulating properties of various types of wall construction so they could identify the best method by which to soundproof their dormitory.

Norton's experiments were carried out in test rooms constructed in a Boston warehouse by the manufacturers of the products being evaluated.[14] Like others who were studying sound at this time, he struggled with the lack of appropriate equipment. A "microphonic apparatus" was first employed to measure the diminution of the intensity of sound as it passed through the various partitions, but its indicator fluctuated too rapidly to be useful. To carry out his evaluation, Norton instead listened, with unaided ears and with a felt-mouthed stethoscope, to the sounds of a piano, a violin, and an Italian tenor "drawn from the ranks of the laborers on the building" as they were transmitted through the various types of walls. While he ranked the different constructions on a scale from 1 to 100, he cautioned that this rating merely indicated the "order of magnitude" of the sound-isolating properties of the partitions. Norton concluded that the wall constructed with Cabot's Quilt was the most impervious to the transmission of sound, and the Cabot company touted this claim for over twenty years.[15]

While Cabot advertisements proudly cited the results of Norton's test, they neglected to mention that, when the construction of the dormitory was complete, the soundproofing proved less effective than had been anticipated. Wallace Sabine was asked in 1904 to explain why this was so. Sabine observed that the problem of sound transmission was little understood. His own research had focused upon absorption (the decay of sound energy within a room), not transmission (sound travel between rooms), and few others had pursued the latter. Sabine suggested that, while there was little direct passage of sound between the walls of the various rooms, the vibratory "responsiveness" of the light, flexible walls rendered that small amount of sound energy distractingly audible.[16]

In 1899, Sabine had mentioned the possibility of using Cabot's Quilt in the construction of Symphony Hall, and he later began to conduct experiments on its properties of sound transmission, but he generally relied upon a different type of acoustical material in his work.[17] When Sabine was asked to improve the acoustics of a poor-sounding room, he was usually called upon to lessen the reverberation or to eliminate a distinct echo. He required a sound-absorbing material that could be applied directly to an exposed wall surface, a material like the hair felt he had employed in the Fogg Lecture Room. In 1901, for example, when Charles McKim asked Sabine to prescribe for the overly reverberant Hall of Representatives in the Rhode Island State Capitol at Providence, Sabine recommended a felt of jute, cotton, and wool manufactured by C.N. Bacon of Boston.[18] By 1906, at least five other such "sound-deadening" felts were available to architects, including Florian Sound-Deadening Felt, No-Noise Deafening Felt, Keystone Hair Insulator, Kelly's Linofelt, and "Tomb" Brand Deadening Felt.[19]

In 1911, Sabine was asked, by William Mead of McKim, Mead & White, to correct the acoustics of the excessively reverberant lecture hall at the Metropolitan Museum of Art in New York. For this assignment Sabine recommended Keystone Hair Insulator, a thick felt of "thoroughly cleansed" cattle hair sandwiched between layers of fireproof asbestos sheathing. Keystone was a product of the Johns-Manville Company, the world's largest manufacturer of asbestos materials and products.[20] The choice of Keystone for the acoustical correction of the museum lecture hall, seemingly inconsequential at the time it was made, would, in fact, initiate a sequence of events that eventually attracted the attention of the president of the United States.

While the architects at McKim, Mead & White were executing Sabine's recommendation to install panels of Keystone felt on the walls of the Metropolitan Museum's lecture hall, they were suddenly served with a restraining order that prevented them from completing the work. As Sabine soon learned, a man named Jacob Mazer had apparently just recently patented the technique of applying sound-absorbing materials to wall surfaces to control the acoustics of rooms. Mazer had negotiated an agreement with Johns-Manville to exploit his patent, and Sabine's acoustical correction of the Metropolitan lecture room violated Mazer's alleged patent rights as well as his agreement with Johns-Manville.[21]

According to Sabine, Jacob Mazer had solicited advice concerning the acoustical correction of a synagogue in Pittsburgh two years earlier. Sabine sent along copies of his published articles and offered more specific suggestions, without charge, even though Mazer himself was well paid for the work. Mazer later visited Sabine "and spent two days asking all sorts of questions." Upon hearing of the legal action against McKim, Mead & White, Sabine caught the next train to their New York office while William Mead obtained a copy of Mazer's patent, which, it turned out, had not yet been officially granted by the Patent Office. Sabine discovered that the technique he had formulized, developed, and given freely to the world was now being claimed by Mazer as his own. Mazer apparently even plagiarized text and tables from Sabine's published papers in his patent application.[22]

The patent was about to receive final approval, so Sabine's influential friends had to work quickly if they hoped to redress this injustice and a flurry of letters and telegrams flew between Boston and Washington. Henry Higginson wrote to Senator Henry Cabot Lodge ("My Dear Cabot") and to James Curtis of the Treasury Department ("Dear Jim"). Charles Eliot wrote directly to President Taft. On 27 February 1911, Senator Lodge telegraphed Higginson: "I have just received the following message from the White House: The President has asked the Secretary of the Interior to order the Commissioner of Patents to withhold the issuance of a patent to Mazer until he can see him at the Cabinet Meeting." Needless to say, the patent was not granted and in that week's issue of the *Patent Gazette,* which had already gone to the printer, Mazer's entry was stamped "Withdrawn."[23] (See figure 5.1.)

5.1

Abstract of Jacob Mazer's patent for "Acoustic-Controlling Material." Mazer attempted to patent a method of sound control based on Wallace Sabine's reverberation formula, but Sabine's influential friends stepped in at the last minute to prevent this from occurring. The Patent Gazette had already gone to press, so Mazer's entry was stamped "Withdrawn" in red ink. *Weekly Gazette of the U.S. Patent Office* (7 March 1911): 147.

MARCH 7, 1911.   U. S. PATENT OFFICE.   147

tending between and held together by the plates, one of said members including a flat tubular case, and a longitu-

dinally extending flat flexible reinforcing strip formed in one piece and concealed within the sheath.

986,191. ACOUSTIC-CONTROLLING MATERIAL. JACOB MAZER, Pittsburg, Pa., assignor to The Acoustics Improvement Company, New York, N. Y., a Corporation of New York. Filed Dec. 16, 1909. Serial No. 533,441.

1. An acoustic controlling covering for auditorium interiors comprising a layer of highly sound absorbing material, in combination with an exposed layer of porous material attached thereto, said layers having varying sound absorbing efficiencies, but each at least equal to commercial cheese-cloth.
2. An acoustic controlling covering for auditorium interiors comprising an inner layer of highly sound absorbing material, in combination with an exposed layer of porous material attached thereto, each layer having a sound absorbing efficiency at least equal to cheese-cloth, but the exposed layer in less degree than the inner layer.
3. An acoustic controlling covering for auditorium interiors comprising a plurality of layers of sound absorbing material each having a sound absorbing efficiency greater than commercial cheese-cloth, but in different degrees.
4. An acoustic controlling covering for auditorium interiors comprising a plurality of layers of sound absorbing material each having a sound absorbing efficiency greater than commercial cheese-cloth, but in different degrees, the exposed layer in the least degree.
5. An acoustic controlling covering for auditorium interiors comprising a layer of highly sound absorbing material supported upon the interior surface and an exposed layer of porous material attached thereto and adapted to admit sound waves in a degree at least equal to commercial cheese-cloth.

[Claims 6 to 14 not printed in the Gazette.]

986,192. WRENCH. JOHN C. MCLEAN, Cleveland, Ohio. Filed Apr. 15, 1910. Serial No. 555,587.
1. In a wrench the combination with a stationary and movable member, of an adjusting nut connected to the movable member and having a roughened surface, a slide arranged on the stationary member and provided with a roughened surface, and a handle rotatably mounted on the stationary member and connected to the slide and adapted when turned in one direction to move the roughened surface of the slide into binding engagement with the roughened surface of the adjusting nut, for the purposes described.

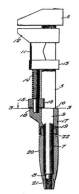

2. In a wrench the combination with a stationary and movable member; of an adjusting nut connected to the movable member and having one face thereof serrated, a slide arranged on the stationary member below the nut, a handle rotatably mounted on the stationary member having one end portion threaded into the slide and adapted when turned to move the slide into and out of engagement with the serrated surface of the nut.

986,193. INSECT-COLLECTING MACHINE. ADA MEEK, Burleson, Tex. Filed Mar. 28, 1910. Serial No. 551,951.

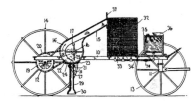

In a machine of the character described, the combination with a vehicle, a receptacle, a casing provided with an aperture in the wall thereof, a suction tube support detachably secured to said casing and provided with a plurality of apertures disposed within the limitations of the aperture in the casing when the support is secured thereto, a flange surrounding each aperture of the support, a suction tube having its upper end telescoping with and disposed exteriorly of each flange, a compression ring for detachably securing each suction tube to its respective flange, and a suction device for drawing insects and infected vegetation through said suction tubes and casing and depositing same in the receptacle.

986,194. WATER - STERILIZING APPARATUS. CLIFFORD D. MEEKER, East Orange, N. J., and CHARLES FRED WALLACE, New York, N. Y., assignors to Gerard Ozone Process Company, New York, N. Y., a Corporation of New Jersey. Filed Aug. 4, 1910. Serial No. 575,492.
1. A water sterilizer having a tank containing oil, an ozonizer element submerged in said oil, a transformer also

Regretting the role their company had played in this unfortunate affair, the executives at Johns-Manville asked Sabine what they could do to make amends and to promote his work in architectural acoustics. Sabine suggested that they establish a special department devoted to acoustical correction, and that they place his student, Clifford Swan, in charge. Johns-Manville did just that. The department was in place by the end of 1911, and in 1914 the company advertised that it was "prepared to execute contracts for the correction of defective acoustical conditions in all types of public and municipal buildings: churches, theaters, court houses, schools, colleges, hotels, offices, etc." "Our Acoustical Department," the advertisement continued, "is in charge of experts who have made a scientific study of architectural acoustics, and their knowledge is supplemented by the practical experience gained in the technique of applying the necessary corrective materials."[24]

The departmental procedure for analyzing acoustically faulty structures and recommending their correction was a straightforward application of Sabine's reverberation formula; indeed, it was all distilled onto a standardized form. (See figure 5.2.) The form provided space to record the area of each of the materials that constituted the room's different surfaces, as well as the overall volume of the room and other relevant factors. A few simple calculations, using Sabine's formula and the absorption coefficients for the different materials (which were already printed on the form), indicated how much sound-absorbing material was required to achieve an acceptable amount of reverberation. By 1919, the acoustical department at Johns-Manville had supervised over 800 such acoustical corrections.[25]

Sabine was grateful to Johns-Manville for providing this new resource for architects with acoustical problems, as his own predilection was to take on projects in advance of construction rather than to remedy the faulty acoustics of extant structures.[26] He was also interested in developing new, more structural kinds of acoustical materials. "I do not feel that we can look on the use of felt in a building which is being planned as anything but an abomination," Sabine wrote to architect Albert Kahn in 1911. "It is corrective in character and temporary in quality." He then described to Kahn a new project with which he was engaged, a project that would, he predicted, result in "materials which will be structural in character and which will enormously increase the possibilities of architectural acoustics."[27] This project, undertaken in collaboration with the builder and tile manufacturer Raphael Guastavino, did indeed fulfill Sabine's predictions. It not only led to the development of a more structural acoustical material, but also opened up entirely new possibilities for the control of sound.

5.2

Data sheet from the Acoustical Department of the Johns-Manville Co., showing data collected by Clifford Swan from the First Church of Christ, Scientist, Boston, 1918. This form simplified the analysis and corrective prescription of acoustically faulty buildings. Courtesy Riverbank Acoustical Laboratories, IIT Research Institute.

Copy

H. W. JOHNS-MANVILLE CO.
ACOUSTICAL DEPARTMENT
DATA SHEET

Name of Auditorium: *First Church of Christ, Scientist (Mother Church)*
Address: *Boston*
Architect:
Address:
Data collected by: *sketches* of _____ Branch
Date: *July 8, 1918* Computed by: *C.M.S.*

MATERIAL	LOCATION	Exposed Area in Sq. Ft.	Absorptive Power	Units Absorption	REVERBERATION		
					Number in Audience	Duration before Correction	Duration after Correction
Soft Plaster on Wood Lath			0.034				
Hard Plaster, *Stone + Glass*		*60,000*	0.025	*1500*	0	4.0	2.8
Wood Sheathing			0.06				
Other Wood			0.03		*2500*	2.7	2.1
Brick			0.025				
Concrete and Stone		*24,000*	0.015	*360*	*5000*	2.1	1.7
Glazed Tile			0.018		RECOMMENDATION OF DEPARTMENT		
Glass			0.027				
Marble			0.01				
Cast Metal			0.01				
Carpet, unlined		*1000*	0.15	*150*			
Heavy Draperies		*1000*	0.25	*250*			
Cork Tile			0.03				
Wood Seats		sittings	0.1				
Upholstered Seats		sittings	3.0				
Cushions	*5000*	sittings	2.2	*11000*			

Average Audience: *2500* persons — 2.5 / 4.6 — *1360* / *13300* Total absorption when empty *19300*
Maximum Audience: *5000* persons — 2.5 / 4.6 — *6250* *19500* Total average absorption *25500*
Average dimensions of room: — *12500* *25800* Total absorption when filled *31800*
Volume of room: *1,060,000* cubic ft.

SURFACES AVAILABLE FOR TREATMENT: (Give location and areas)

15000 sq. ft. × 0.40 = 6000 abs. units

III Acoustical Materials and Acoustical Modernity: St. Thomas's Church

In 1911, a man named Raphael Guastavino presented himself to Wallace Sabine with a letter of introduction from the architectural firm of Cram, Goodhue & Ferguson:

> My dear Mr. Sabine,
> This is to introduce you to Mr. R. Guastavino who, we are glad to say, is extremely interested in your suggestion that tile may be made a far better material, acoustically considered, than at present, and has asked for this letter of introduction to take with him when he goes on to talk over the matter with you.[28]

The architects introduced Guastavino, a tile manufacturer, to Sabine with hopes that the two men would work together to develop a stonelike but sound-absorbing material that could be employed in the neo-Gothic ecclesiastical architecture that was the specialty of their firm. With such a material, they realized, it would be possible to build a Gothic-looking church with a distinctly modern sound.

Ralph Adams Cram wanted a Gothic look because, from his earliest days as an architectural apprentice in Boston, he had subscribed to an aesthetic that looked back to the Middle Ages for spiritual inspiration. Cram's medievalism was motivated by his belief in the material corruption and symbolic impoverishment of contemporary American culture. In 1892, Cram and his partner Bertram Grosvenor Goodhue produced a short-lived quarterly, *The Knight Errant*, whose goal was "to assail the dragon of materialism."[29] Religious ceremony, particularly as practiced in pre-Reformation England, was the key to rejuvenation, and the architects' neo-Gothic churches provided spiritually rich environments in which worshipers could escape from the secular and noisy world of the surrounding city. They offered what historian Jackson Lears has called "gardens of cool repose—therapeutic antidotes to feverish modern haste." As Cram himself put it, "Within a church, whatever its environment, the motorbus and the motorcycle, the moving picture and the electric sky-signs, the newspaper and the billboard and the radio cannot come, and here at least you may demand and receive, peace, harmony and beauty."[30]

Cram considered the Gothic style appropriate only for those institutions that could actually trace their origins back to the Gothic era, specifically, churches and universities. Yet, he recognized that these institutions did not survive unchanged in the modern world. The Latin chant of a medieval mass, which was only enhanced by the rich reverberations of the space in which it was intoned, had been replaced by the sermon, particularly in the High Church Protestant denominations that were his firm's best clients. These churches emphasized intellectual as well as spiritual engagement on the part of the congregation, and Cram sought to accommodate both of these requirements in his architectural designs. While he thus insisted that the primary function of a church remained that its inhabitants "be filled with the righteous sense of awe and mystery and devotion," he also recognized that an ideal church must be a place "where a congregation may conveniently listen to the instruction of its spiritual leaders."[31] Cram's medieval aesthetic conflicted with modern acoustical necessities. To promote religious mystery, acoustical mastery was required. Cram turned to Wallace Sabine and Raphael Guastavino to provide this control.

Raphael Guastavino was a Catalan immigrant whose father (also named Raphael) had revived a traditional but long-neglected technique for constructing thin-shelled or "timbrel" vaults. In contrast to Roman vaults, which are supported by compressive forces between massive stone forms, a thin-shelled vault derives its strength from the curvature of its surface—much as a piece of paper, unable to support even itself when limp, can sustain a load when held in a curved form. In Spain, the elder Guastavino developed a technique for constructing such vaults out of multiple layers of terra cotta tiles sandwiched between thick blankets of cement mortar. This technology of "cohesive construction," as he called it, produced vaults that were strong, lightweight, and fireproof. Guastavino exhibited his construction technique at the Philadelphia Centennial Exposition of 1876 and was awarded a Medal of Merit. Recognizing the opportunity presented by the rapidly expanding cities of America, he decided to move his business here permanently. He arrived in New York with his young son in 1881, and before long the Guastavino Fireproof Construction Company was working with many of America's finest architects. McKim, Mead & White's Boston Public Library and Pennsylvania Station in New York; Heins & LaFarge's Cathedral of St. John the Divine in New York; and Cass Gilbert's Minnesota State Capitol at St. Paul represent just a few of many notable examples of Guastavino construction.[32]

The vast spaces created by Guastavino construction were voluminous and were lined throughout with hard ceramic tile. Sabine's formula only confirmed what any visitor to them already knew: the reverberation was impressive to a degree considered excessive by people increasingly preoccupied with silencing the sounds around them. Guastavino himself knew this, and he recognized the commercial potential of a sound-absorbing tile. Having read several of Sabine's articles in architectural journals, he decided to meet the physicist to discuss the possibility of creating such a tile.[33] Cram provided the letter of introduction, and the subsequent meeting was a success. The two men drew up a contract whereby Sabine agreed "to conduct a series of experiments planned to improve the acoustic quality of tile."[34]

The collaboration began with Sabine measuring the sound-absorbing power of Guastavino's standard tiles, which were produced at the company's kilns just outside of Boston in the town of Woburn, Massachusetts. Sabine took samples of tile intended for Cram, Goodhue & Ferguson's Chapel at the West Point Military Academy, and determined that they absorbed approximately 3 percent of incident sound energy at a frequency of 512 cps. According to Sabine, "The investigation then widened its scope, and, through the skill and great knowledge of ceramic processes of Mr. Raphael Guastavino, led to really remarkable results in the way of improved acoustical efficiency." Guastavino himself more modestly characterized his own role as that of a "practical ceramic worker," and it is evident that both men contributed to the success of their collaboration.[35]

As Sabine recalled, "The first endeavors to improve the tile acoustically had very slight results, but such as they were they were incorporated in the tile of the ceiling of the First Baptist Church in Pittsburgh."[36] The "Pittsburgh Tile" had an absorptivity of about 5 percent at 512 cps, a small improvement over that of the West Point Tile. In August 1911, Sabine sent to William Blodgett (the Boston-based treasurer for the Guastavino Company) a graph indicating the frequency-dependent absorptivity of the West Point Tile and a "special tile" (probably the version used in the Pittsburgh church). He compared the absorption curves for these tiles to those for brick, wood sheathing, felt, and "what is most interesting of all, as showing the possibilities of tile, a curve showing the absorbing power of 5/8 of an inch of beach sand" (which was about 30 percent at 512 cps). The "special tile," was not nearly as absorbent as the sand, and, in Sabine's opinion, was "not nearly as absorbent as it can be made."[37]

Sabine was confident that the tile could somehow be made to achieve the high absorptivity of sand, and he urged Blodgett to press ahead: "I am ready for new tile anytime now and you can count on me to push the work as rapidly as possible."[38] He did not indicate specifically how the new tile should differ from the old, and it appears to have been Guastavino's task to determine how to accomplish this. It was evident to both men that the porosity of the sand was the key to its absorbing power, and the goal was to re-create that porosity in a ceramic tile. This goal was ultimately achieved by formulating a tile that was a mixture of 25 percent clay, 10 percent feldspar, and 65 percent vegetable bearing earth, or peat. During firing, the peat was consumed by combustion and it left behind pores on the surface and throughout the body of the tile. Sabine praised the Guastavino Company's "tireless willingness to burn kiln after kiln" in experimentation, suggesting that the final formula was the result of a long process of trial and error, guided by Sabine's measurements and an idea of the kind of surface they sought to achieve.[39]

The new tile was named "Rumford,"[40] and in their patent application, filed in February 1913, Sabine and Guastavino emphasized its "peculiar porosity." Rumford was not a "cellular" structure, filled with numerous tiny, isolated air bubbles. Such a material was, the inventors claimed, "without value" for the purpose of sound absorption. Rumford's porosity was instead characterized by interconnecting air spaces, "channels traversing the rigid structure of the porous layer, and reaching to and penetrating the interior surface." "It is desirable," they continued, "that these channels be irregular in form, expanding and contracting in cross-section, so that their action will be like the muffling action of a muffler on an engine exhaust."[41] While the example of a layer of sand may have initially stimulated their thoughts on what was possible, Sabine and Guastavino later understood and explained their achievement in terms of the technology of the automobile muffler. Ralph Adams Cram, in his battle against materialism, had sought an environment isolated from the din of internal combustion engines; Sabine and Guastavino made that environment possible by creating a material that was filled with tiny engine mufflers.

Rumford tile was first employed in Cram, Goodhue & Ferguson's St. Thomas's Church in New York, and the overall effect of the building was exactly what Cram had hoped for. "As the Woolworth tower is an admirable symbol of our restless and material side," one critic concluded in 1913, "this church may well stand as a fitting expression of that great spiritual impulse that is slowly leavening the lump of our vast material achievement."[42] "The straight, strong ribs rise from the pavement in aspiring lines that lead the soul of the worshipper heavenward with them in simplicity and truth," waxed another. "The rushing world is left without."[43] (See figures 5.3 and 5.4.)

St. Thomas's Church, located along Fifth Avenue at 53d Street, was an Episcopal house of worship that ministered to the spiritual needs of many of New York's wealthiest families. The original church had been destroyed by fire

5.3
St. Thomas Church, exterior, Fifth Avenue & 53rd St., New York (Cram, Goodhue & Ferguson), c. 1913. While it looks like a relic from the Middle Ages, St. Thomas's conservative exterior belies a technologically innovative interior composed of sound-absorbing tiles. Half-tone, n.d., Museum of the City of New York, Print Archives.

5.4
St. Thomas Church, interior, Fifth Avenue & 53rd St., c. 1913. The use of sound-absorbing Rumford tile on the inner lining of the vaults resulted in a reverberation time much less than would have been the case with a traditional masonry finish. St. Thomas's was designed acoustically to accommodate the modern sermon, not the medieval mass. Half-tone, n.d., Museum of the City of New York, Print Archives.

in 1905, and no expense was spared in its reconstruction, as the parish "demanded that everything be genuine, that there should be no shams."[44] Cram was credited with the overall design of the building, and skilled craftsmen—following Bertram Goodhue's designs—filled the church with rich, handmade ornamentation: stone statuary, detailed wood carving, ornate metal hardware.[45] The architects' design, while inspired by the Gothic spirit, was characterized as uniquely their own. No mere copy, St. Thomas's Church was perceived to be the equal of those architectural masterpieces of the past, the great Gothic cathedrals of Europe. "No one in the materialism of the present day, in the rush and efficiency of the modern architect's office, could be expected to hold his own with those wonderful creations," the editors of one architectural journal asserted. "But in our opinion this has been done by the architects of St. Thomas's."[46]

In their design, the architects sought to maintain a balance between past and present. The building celebrated and embodied what was considered best about the past: Christian spirituality, a communal society, the pride of skilled labor and the beauty of its accomplishments. It also embraced certain aspects of modern life that promised to improve upon that past, and it accommodated others that were simply unavoidable. The ornamentation of the church, for example, mixed heroes from past and present. Alongside statues and carvings of long-dead saints stood more contemporary icons, including Woodrow Wilson, the Brooklyn Bridge, and a Salvation Army donut girl.[47] The technology of Guastavino construction allowed the architects not simply to re-create, but to surpass medieval strivings for lightness and openness of form. And, the sound-absorbing Rumford tiles that lined the inner surface of those vaults were also seen as a distinct improvement, for "the acoustics of no great European church would satisfy an American congregation of today."[48]

Architectural critic Montgomery Schuyler vividly emphasized the difference between medieval and modern congregations:

> As to the layman, the requirement of the medieval Gothic church, so far from betraying any disposition to "accommodate" him, was that he should be put in his place and made to feel that he was a worm, blessed above his deserts in being permitted to gaze from afar, in the dim recesses of the vaulting of the nave or the aisles, upon the celebration of the "mysteries" which was going on in the full light of the choir. Since then the layman has reclaimed his rights and has refused to be relegated to the shadowy background of what is going on. He pays, and he has to be conciliated.[49]

A large part of this conciliation, according to Schuyler, was centered on the modern tenet that "preaching holds the first place in the attractions of the church."[50] While Schuyler himself did not discuss the use of Rumford tile (perhaps because Sabine's own description of it appeared alongside Schuyler's account in the architectural journal *Brickbuilder*), other reviews of the new church did describe the development of the new sound-absorbing tile, and concluded that the acoustical result was "eminently satisfactory."[51]

Indeed, Rumford was so effective, it exceeded Sabine's "most extreme expectations."[52] According to his own measurements, the tile absorbed 29 percent of incident sound at 512 cps, far more than the 3 percent absorbed by regular Guastavino tiles.[53] The use of Rumford dramatically reduced the amount of reverberation that otherwise would have been present in the church, and it solved the problem of rendering a sermon intelligible in a large space lined with hard surfaces.[54] For Cram, the development of Rumford was in keeping with the tradition of technological innovation that was a defining characteristic of the Middle Ages, and the use of the new tile in St. Thomas's was a direct result of his desire, not to re-create the Middle Ages, but to draw upon its spirit to meet the spiritual needs of modern times. While this forward-looking aspect of Cram's work has been identified by scholars, it has primarily been associated with the architect's writings, not his buildings.[55] St. Thomas's Church not only exemplifies this philosophy, it also embodies another kind of modernity, acoustical modernity.

The degree of control over sound that Cram, Goodhue & Ferguson sought for St. Thomas's Church, and that Sabine and Guastavino provided, was unprecedented. The degree of sound absorption provided by Rumford was equally unprecedented in a large structure with the look, feel, and texture of masonry construction. With Rumford, a Gothic-looking church need not be a Gothic-sounding church. Natural laws and materials no longer limited what was acoustically possible. Sabine's formula had provided the key to working within those limitations, allowing architects to manipulate traditional materials of construction in order to achieve a desired end. Rumford, more powerfully, opened up entirely new acoustical possibilities. It was, in Sabine's own words, "a new factor at the disposal of the architect."[56]

Rumford initiated a transformation of the traditional relationship between sound and space that had been in place for as long as civilization had been constructing buildings. Since reverberation is a means by which we perceive space through time, Rumford additionally heralded the transformation of the aural aspect of space–time relationships. Historians have long identified the reformulation of perceptions of space and time as one of the signposts of modernity,[57] thus Rumford—and the sound that it produced—should be recognized as a modern artifact. Rumford introduced a new malleability to the relationship between sound and space. Over the next two decades—as will be seen—scientists and engineers would develop new materials to render this relationship even more malleable, until, by 1930, with the assistance of electroacoustic devices, the connection would virtually cease to exist. Any size or type of space could, by then, possess any type of sound. Knowing this outcome, it is possible to recognize that St. Thomas's Church was, acoustically, at the forefront of cultural change.

The modernity of St. Thomas's Church was, however, admittedly hard to perceive at the time it was built. Subtly defined by the absence, not presence, of sound, it was a modernity that whispered rather than shouted, and it seemed to have little in common with the voluble cultural transformations then taking place in art, music, and literature. Cram despised those transformations, and he would later condemn modernism as a "nervous fad for abnormality."[58] Goodhue wasn't even sure what the term *modern* meant. Yet, tellingly, he knew that it had something to do with science and with Wallace Sabine's contribution to his own architecture. In 1914, Goodhue wrote to Sabine concerning one of his ongoing church commissions:

> Dr. Parks has asked me to design a new exterior based on the present plan—a new exterior that shall be "*modern*"—again whatever this term may mean; so as you probably have already gathered I am almost at my wit's end. I was, however, clever enough to say that I thought in such a matter you should be the designer as much if not more than I. Perhaps if you were here you could shed light on the whole subject, which I must admit now is the most vexed one that I have ever had to do with, so please arrange to see me as soon as you conveniently can. And forgive me for the fuss I am making.[59]

While Sabine's reply to Goodhue's plea is lost to history, his continued involvement with Goodhue and Guastavino provides another means by which to follow the dialogue between American architects and acousticians as they worked together to develop an architecture with a distinctly modern sound.

Even as the Rumford-lined vaults of St. Thomas's Church were under construction, Sabine and Guastavino were hard at work on a material that would be even more sound-absorbing, as well as easier to manufacture. Since Rumford was ceramic, it was produced in kiln-sized lots, and each lot invariably varied slightly in color, composition, and absorptivity. The desire for a more uniform product led Sabine and Guastavino to develop an "artificial stone" tile, which they patented in 1916 as "Akoustolith." (See figure 5.5.) Akoustolith was an aggregate of pumice particles loosely bonded with Portland cement. Like Rumford, it was porous on its exposed face and throughout its thickness, and its absorptivity (38 percent at 512 cps) far surpassed that of Rumford.[60]

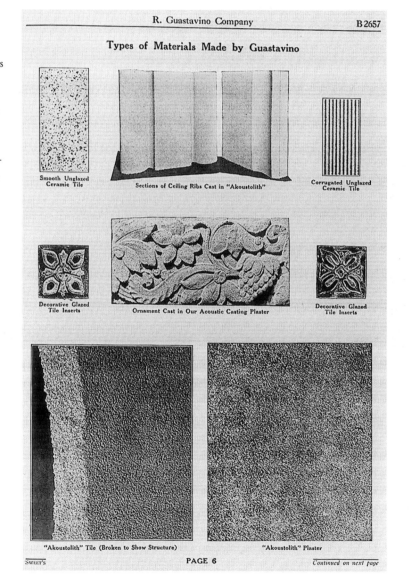

5.5
"Types of Material Made by Guastavino," 1931 advertisement for Akoustolith, a porous aggregate of pumice particles loosely bonded with Portland cement. Akoustolith could be cast in a variety of shapes, as well as cut into standard tiles. It absorbed 38 percent of incident sound energy at 512 cps. *Sweet's Architectural Trade Catalogue* (1931): B2657. Avery Architectural and Fine Arts Library, Columbia University in the City of New York.

Rumford and Akoustolith were installed in hundreds of churches and chapels, temples, and secular buildings across the United States. Rumford was employed in the auditorium of the museum of the University of Pennsylvania, and in Albert Gottlieb's B'Nai Jeshurun Synagogue in Newark, New Jersey.[61] (See figure 5.6.) Bertram Goodhue used it in the Church of St. Vincent Ferrer in New York, and in his First Congregational Church of Montclair, New Jersey. After attending opening services at the Montclair church, Goodhue wrote to Guastavino, "To the best of my knowledge and belief no such acoustical result has ever been achieved before except possibly by accident." "To you and Dr. Sabine," he continued, "all credit is due and it is difficult to express my satisfaction with the result of the years of patient effort spent by you both in the perfecting of this wholly new material."[62]

While the Guastavino Company continued to advertise Rumford well into the 1920s, Akoustolith clearly outsold its less absorbent predecessor. Examples of Akoustolith projects include Goodhue's National Academy of Sciences Building in Washington and his Nebraska State Capitol Building; Cram & Ferguson's Chapel at Princeton University; Albert Altschuler's Isaiah Temple in Chicago; and Fellheimer & Wagner's New York Central Railroad Terminal in Buffalo.[63]

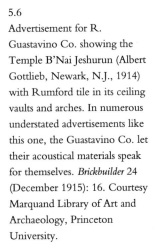

5.6
Advertisement for R. Guastavino Co. showing the Temple B'Nai Jeshurun (Albert Gottlieb, Newark, N.J., 1914) with Rumford tile in its ceiling vaults and arches. In numerous understated advertisements like this one, the Guastavino Co. let their acoustical materials speak for themselves. *Brickbuilder* 24 (December 1915): 16. Courtesy Marquand Library of Art and Archaeology, Princeton University.

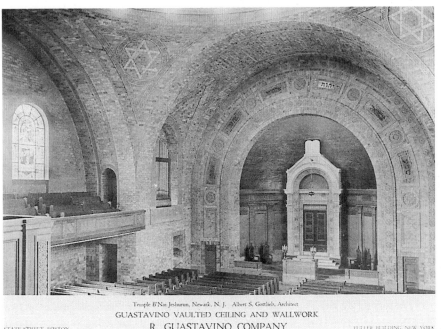

Guastavino's sound-absorbing tiles were most commonly used in monumental spaces that people visited but did not inhabit continually. As the teens gave way to the twenties, however, the range of application of architectural acoustics rapidly expanded. "Good sound" at church or in the lecture hall was no longer enough—people now sought to control sound throughout their daily lives; at home, at school, and especially at work. By 1923, the Guastavino Company noted that some architects were employing Akoustolith tiles independent of the Guastavino vaulting system.[64] The demand for the absorption of sound was clearly greater than the demand for monumental vaults, and a plethora of new sound-absorbing products were soon competing with Rumford and Akoustolith to quiet the spaces of everyday life.

New kinds of mass-produced materials and inexpensive systems of installation eventually rendered uneconomical the skilled manufacture and installation of Guastavino products.[65] The understated tone of the Guastavino advertisements, too, was soon overwhelmed by the modern techniques of marketing that the new competitors employed. Whereas Guastavino advertisements let the materials speak simply and quietly for themselves (see again figure 5.6), those of the new competitors clamored for attention in very different ways. What the ads sold so clamorously was not just acoustical building materials, but also the environment that those materials produced. Simply put, they sold silence.

IV Acoustical Materials and Modern Acoustics: The New York Life Insurance Company Building

By 1930, dozens of different companies were manufacturing a wide range of acoustical products that included not only felts and artificial masonry, but also ceiling tiles, rigid wallboards, plasters, floorings, soundproof doors, and mechanical devices that acoustically isolated floors, windows, walls, and ceilings. Architects could now choose from Audicoustone Plaster and Acoustifibrobloc, Insulite Acoustile, and Armstrong Corkoustic, to name but a few.[66]

Animal, vegetable, and mineral kingdoms alike were plundered to create these strange new materials for controlling sound. In addition to the old standby of felt made from animal hair, manufacturers now employed all sorts of plant fibers, including licorice, sugarcane, jute, flax, and cornstalks. They harvested cork and spun mineral wool. They mined asbestos, pumice, gypsum, lime, and volcanic silica. They developed "triple acting mechanical-aero-chemical processes"[67] to effervesce plaster into porous, sound-absorbing surfaces, and they devised pressurized guns to spray acoustical insulation onto and into walls. (See figure 5.7.) All of this innovation was dedicated to the mastery of sound, or rather to its elimination, as the overall goal was always to obtain "Control Through Absorption."⁶⁸ By 1932, so many products were available, the Absorbex Company declared that the problem of architectural acoustics was no longer the challenge of controlling sound, but rather, the dilemma of deciding which commercial product one should use to obtain that control.⁶⁹

Some of the new products were little more than familiar old building materials modernized with the adjective "acoustical." Tuckahoe Colored Interior Plaster, for example, was advertised rather vaguely as possessing "definite acoustical properties." "This, in itself," the manufacturer asserted, "makes Tuckahoe a highly desirable material."⁷⁰ In 1927, Floyd Watson warned of "commercial companies who have developed various products that have acoustic merit in greater or less degree and who present the matter by modern sales methods to the parties involved."⁷¹ The United States Gypsum Company may have tested the limits of modern sales methods when it suggested that their plaster not only solved problems of architectural acoustics, but also improved the romantic

5.7
"Sprayed on with Guns," 1931 advertisement for Sprayo-Flake Acoustical Plaster. Throughout the 1920s, the techniques of architectural acoustics were deployed in an ever-expanding field of battle against noise. Sprayo-Flake invited consumers to enlist this well-armed guard to protect them with a blanket of acoustical security. *Sweet's Architectural Trade Catalogue* (1931): B2513.

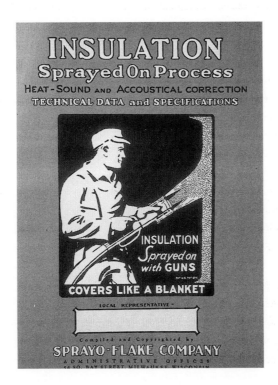

prospects of the plasterers who applied it.[72] (See figure 5.8.) But if such blatant salesmanship occasionally took precedence over actual acoustic merit, it was more often the case that manufacturers combined the two, presenting relevant information about legitimate acoustical products to architects and their clients through increasingly compelling sales techniques.

5.8

"Little Stories of the Job," a tale of romance and acoustical plaster, from a trade journal published by the United States Gypsum Co. It is not evident whether tales such as this actually enticed independent plasterers to promote USG's acoustical products. *The Gypsumist* (August 1926): 21. Series I, Box 13, Folder 11, Guastavino/Collins Collection, Avery Architectural and Fine Arts Library, Columbia University in the City of New York.

The Next Four Pages Are a Reprint from the Eight Page Magazine for Plasterers. Have You Sent Us a List of Yours?

red topics
for plasterers
PUBLISHED BY THE UNITED STATES GYPSUM COMPANY, CHICAGO

Vol. 2   AUGUST, 1926   No. 7

Little Stories of the Job
LESLIE LANDS A NIGHT JOB

WHEN Marian drove her car out of the garage at dusk, little did she expect to find it stalled on a country side-road an hour later. She wondered what she should do.

"May I help you?" asked a young man.

"Well I do believe I need some assistance," Marian smiled, "and if you don't mind—"

With the aid of a pair of pliers and Marian's flashlight the trouble was soon righted and the engine was humming beautifully.

"I hope you don't mind my complimenting you on your cleverness," said Marian, "and how about my driving you back to town?"

"Thanks, that will be fine," he answered, as he sat down beside Marian. "My name is Baker—Leslie Baker. I just strolled out to get the air, not for the exercise, for a plasterer gets plenty of that. I'm just starting in the game, and I do find more time than I need to practice on my banjo. By the way, you don't happen to know of anyone needing plastering done, do you?"

"I do know of two," Marian answered. "The old Methodist Church, of which my father is deacon, needs plaster badly and so does the Isis movie, which is owned by my beau."

They had reached town before they could realize it and said good-night.

Leslie wasted no time in seeing Marian's father. In the course of conversation the deacon said: "The church wasn't constructed right in the first place, because there are so many echoes that it requires ear-strain to hear the preacher, and the music doesn't sound well, either. The

"Well, I Do Believe I Need Some Assistance"

architect has suggested hanging curtains on the walls, but there's the price of the curtains, expense of cleaning and, besides, who wants curtains in a church?"

"What is the name of the architect?" Leslie queried, "I may be able to help out."

"It's Alfred Bruce, down in the National Bank Building," the deacon replied.

It was a week or two later when Marian greeted Leslie on the street with, "Say, Dad says you're a wizard to have plastered the church and stopped the echoes at the same time."

"I used 'acoustical plaster' that's why," came the answer with a little pride.

"What's *that*? An ailment of some kind," she laughed.

"Say your beau used acoustical plaster on his movie too, but another plasterer underbid me and got the job. I'm sorry, too, because, while it is a good job of plastering, it won't be worth a cent acoustically, for the plasterer used a carpet float where he should have used a cork float."

"Why, couldn't he swim?"

"Well, you see," Leslie explained, "a cork float is what is used to rub over the plaster to make it rough, and acoustical plaster requires that finish. I feel sorry for your friend, really I do."

"O him! A cork float won't do him any good because he's already sunk with me."

Leslie's eyes glistened with the light of hope, "May I call on you this evening?" he asked.

"Don't forget to bring your banjo," said Marian.

21

The best measure of any material's acoustical merit was its absorption coefficient, and the sales literature featured the escalating values of those coefficients as manufacturers engaged in a "coefficient war" to see who could offer the most absorbent product.[73] For architects who remained ignorant of what an absorption coefficient was and why it was important, sales brochures offered tutorials that conveniently educated their readers in the basic science and techniques of architectural acoustics. The manufacturer of Kalite Sound Absorbing Plaster, for example, presented an easy-to-understand graphic explanation of the effect of acoustical materials like Kalite upon the decay of sound in a room. (See figure 5.9.)

While architects were introduced to the basic principles of acoustical design, the trade literature stopped short of turning them into actual acousticians. Instead, they were encouraged to turn to the experts to ensure success, and the manufacturers themselves increasingly took on the role of providing this expertise. Following Johns-Manville's precedent, numerous companies established engineering departments that offered complimentary consulting services to help architects determine how best to use their products. "The Insulite Acoustile Engineering Staff is at your disposal," one ad graciously informed its readers. "Consult our experts on any problem in Architectural Acoustics," invited another.[74] Manufacturers encouraged such inquiries, promoting a general awareness of the need for "acoustically correct buildings,"[75] and presenting their own products as the means to achieve them.

Such aggressive promotional campaigns certainly contributed to the proliferation of acoustical materials in the 1920s. If the larger culture in which these sales efforts took place had not already perceived some need for what was being sold, however, all the marketing in the world would not have succeeded. In a world increasingly concerned with the problem of noise, the need for sound-

5.9
Cartoons demonstrating the effect of reflective and absorptive plaster surfaces upon sound, from a 1934 trade brochure. Promotional literature for acoustical building materials educated consumers on the basic principles of sound design, but always encouraged them to turn to professionals for guidance. "Kalite Sound Absorbing Plasters," (sales pamphlet, Certain-Teed Products, 1934), p. 2. Series I, Box 13, Folder 11, Guastavino/Collins Collection, Avery Architectural and Fine Arts Library, Columbia University in the City of New York.

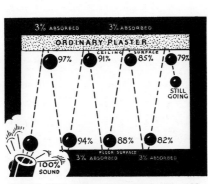
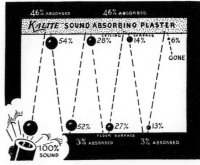

absorbing materials was already present. The advertisers had only to strike that resonant chord, or rather, to evoke the blare of a neighbor's saxophone, in order to make their message heard. (See figure 5.10.) The result was that an increasing number and variety of architectural spaces became increasingly absorbent.

Johns-Manville had been one of the first to enter the field of acoustical products and services, and it remained a leader in the industry throughout the

5.10
"Fire-and-Sound-Proofed by Herringbone," 1923 advertisement for Herringbone Rigid Metal Lath. The archetypically modern noisemaker, the saxophone, signifies everything from which the sober gentleman on the right must insulate himself. Herringbone offered "an effective barrier to the most penetrating sound." *Architectural Forum* 39 (July 1923): 25. Courtesy Marquand Library of Art and Archaeology, Princeton University.

teens and twenties. By 1931, the company offered over a dozen acoustical products with sound-absorption coefficients ranging from a low of 0.31 to a high of 0.82 (at 512 cps). (See figure 5.11.) While the company's earliest work in sound control had focused upon improving the acoustical quality of auditoriums, lecture halls, and churches, the problem of noise—especially office noise—soon attracted increased attention. In the office, acoustical materials were valued not

5.11
"Various Types of Johns Manville Acoustical Materials." In 1931, Johns-Manville offered a dozen different acoustical products, including Nashkote asbestos felt (available with a variety of surface finishes), Rockoustile, and Sanacoustic tile. Absorption coefficients ranged from 0.31 to 0.82 at 512 cps. "Johns-Manville," *Sweet's Architectural Trade Catalogue* (1931): B2668.

TYPE DESIGNATION	FOUNDATION REQUIRED	SIZE OF UNITS	LIMITATIONS OF SURFACE CONTOUR	CHARACTER OF FINISHED SURFACE	SOUND ABSORBING MEDIUM	FIRE RISK	UPKEEP REQUIRED	OVER ALL THICKNESS	WEIGHT PER SQUARE FOOT	SOUND ABSORBING EFFICIENCY AT 512 CYCLES	ADAPTED FOR
NASHKOTE TYPE·A· PERFORATED AFTER ERECTION	BROWN OR PUTTY COAT OF PLASTER. FLOATED SMOOTH. CONCRETE WITH LEVELING COAT OF PLASTER. WOOD SHEATHING. PLASTER BOARD. INSULATING BOARD. FLAT SHEET STEEL. WOOD STUDS OR JOIST 12-16 BC.		NO LIMITATIONS	SMOOTH OIL PAINT	ASBESTOS AKOUSTIKOS FELT	FIRE RESISTANCE APPROVED BY BUILDING DEPARTMENTS UNIVERSALLY	WASH OR REPAINT AND PERFORATE	1/2" 3/4" 7/8"	7 oz 10 oz 13 oz	.43 .53 .67	AUDITORIUMS·CHURCHES THEATRES·COURT ROOMS· WHEREVER A HIGH QUALITY FINISH IS REQUIRED TO MATCH PAINTED PLASTER
NASHKOTE ·TYPE·AIS·				PAINTED SAND FINISH PLASTER				1/2" 3/4" 7/8"	8 oz 11 oz 14 oz	.31 .38 .46	
NASHKOTE ·B·316·				PERFORATED OIL CLOTH 3/16" DIA. HOLES			WASH OR PAINT	1/2" 3/4" 7/8"	8 oz 11 oz 14 oz	.40 .53 .70	OFFICE HOSPITAL RESTAURANT AND GENERAL QUIETING
NASHKOTE ·B·332·				PERFORATED OIL CLOTH 3/32" DIA. HOLES				1/2" 3/4" 7/8"	8 oz 11 oz 14 oz	.43 .53 .67	
NASHKOTE ·B·085·				PERFORATED OIL CLOTH .085 DIA. HOLES				1/2" 3/4" 7/8"	8 oz 11 oz 14 oz	.36 .47 .60	
NASHKOTE ·B·068·				PERFORATED OIL CLOTH .068 DIA. HOLES				1/2" 3/4" 7/8"	8 oz 11 oz 14 oz	.39 .48 .63	
NASHKOTE ·B·045·				PERFORATED OIL CLOTH .045 DIA. HOLES				1/2" 3/4" 7/8"	8 oz 11 oz 14 oz	.39 .49 .64	SAME AS ABOVE AND ALSO FOR AUDITORIUM WORK.
NASHKOTE ·TYPE·C·				WHITE FACED FELT SIZED UNIQUE FINISH TEXTURE			WASH OR PAINT WITH SPRAY	1/2" 3/4"	7 oz 10 oz	.31 .42	HIGH AUDITORIUM CHURCH OR ARMORY CEILINGS.
NASHKOTE ·TYPE·F·				BURLAP REP BROCADE AWNING CLOTH OR ANY DYED FABRIC.			DRY OR VACUUM CLEAN	1/2" 3/4" 7/8" 1" 1 3/4" 3"	8 oz 11 oz 14 oz 21 oz 28 oz 42 oz	.35 .49 .65 .72 .76 .77	ALL TYPES OF AUDITORIUM CHURCH THEATRE WORK WHERE FABRIC FINISH IS DESIRABLE
NASHTILE		6"x12" 12"x12" 12"x24" 9"x18" 18"x18" SPECIAL SIZES		AN INDIVIDUAL TEXTURE NOT UNLIKE TRAVERTINE			WASH OR SPRAY WITH LACQUER	3/4"	13 oz	.38	ALL TYPES OF AUDITORIUMS · FOR QUIETING WHERE TILE PATTERNS ARE DESIRED
SANACOUSTIC TILE		12"x12" 12"x24" 8"x16" 16"x16"	NO LIMITATIONS FLAT WALLS OR CEILINGS· VAULTS OF LARGE RADIUS	ENAMELED PERFORATED METAL TYPE 068 PERFORATIONS	ROCK WOOL	FIREPROOF	WASH OR PAINT ANY MEDIA	1 1/4" 1 1/4" 4" 4"	STEEL 2.5 LB ALUM. 2.0 LB STEEL 5.2 LB ALUM. 4.7 LB	.82	·QUIETING OF EVERY TYPE· PROCURABLE IN ALUMINUM BEING PARTICULARLY ADAPTED FOR USE IN NATATORIUMS KITCHENS DISHWASHING ROOMS (CORROSION PROOF) THE 4" THICKNESS FOR RADIO OR SOUND FILM STUDIOS
ROCKOUSTILE	SAME AS NASHKOTE	6"x12" 12"x12" 12"x24" 9"x18" 18"x18" SPECIAL SIZES	NONE	AN INDIVIDUAL TEXTURE NOT UNLIKE TRAVERTINE	ROCK WOOL	FIRE RESISTANCE APPROVED BY BUILDING DEPTS UNIVERSALLY	CLEAN WITH VACUUM CLEANER AND SAND PAPER	1"	2.0 LB	.62	ALL TYPES OF AUDITORIUMS FOR QUIETING WHERE TILE PATTERNS ARE DESIRED

Table title: TABULATION OF VARIOUS TYPES OF JOHNS MANVILLE ACOUSTICAL MATERIALS just for the quality of sound that they produced, but also for the effect that this sound had on the people who worked there. Sound-absorbing materials reduced noise and thereby enhanced the productive efficiency of those who worked within those quieted spaces.

In 1913, *System: The Magazine of Business*—a primary resource for advocates of scientific management and industrial efficiency—interviewed Wallace Sabine on the problem of noise in offices. Soon thereafter, Johns-Manville began to present its acoustical materials as a solution to this new problem, and within a few years 40 percent of their sound-controlling business came from this new field.[76] In a 1920 sales brochure, Johns-Manville noted that the evolution of the "modern office building" had led to "an unbearable increase in unnecessary noise, confusion and nervous excitement, which has had a marked effect on the normal efficiency of both executives and office workers." "The effect of reverberation," the pamphlet continued, "upon the noise of typewriters, adding machines, telephone bells, conversation and all of the ordinary office disturbances as well as on the street noises entering from without, is to magnify them and cause a din that is fatal to proper concentration and nervous repose."[77] Confronted with this situation, the "business man of today" was led to ask: "What is the easiest, quickest and best way for me to eliminate confusing noises and loss of efficiency in my office?" Not surprisingly, "The Answer to Your Noise Problem" was provided by Johns-Manville: "The effect produced by Johns-Manville Acoustical Correction is remarkable. It produces a sense of comfort and quiet and of relaxed nerve tension which is hard to describe in words. There is a new sensation to be found in entering a room after the installation of Johns-Manville Acoustical Correction. . . . The office assumes a mien of order and dignity always impressive to the visitor and helpfully restful to the worker."[78]

Readers who resisted the quiet seduction of this description might instead have been convinced by some cold, hard facts. The pamphlet projected that an increase in working efficiency of as little as three-quarters of one percent would pay for the cost of acoustical correction, with interest, in just five years. Some businesses had claimed that the efficiency of their offices increased as much as 10 to 15 percent after acoustical materials had been installed and in these cases, Johns-Manville reported, the quieting treatment paid for itself in just a few months.[79]

Advertisements in architectural trade journals necessarily condensed this kind of presentation yet were equally effective in selling, not just acoustical products, but the quiet that resulted from their use. (See figure 5.12.) These

5.12
"Quiet—A Specification for Banks and Offices." This 1923 Johns-Manville advertisement cites over 125 "acoustical corrections" in banks where Akoustikos felt was employed to eliminate the "noise nuisance." *Architectural Forum* 38 (June 1923): ad sect., p. 136. Avery Architectural and Fine Arts Library, Columbia University in the City of New York.

advertisements appeared in journals that were devoting increased attention to acoustical design; thus their impact was amplified by their proximity to editorial accounts of acoustical problems and solutions. Special issues dedicated to banks, schools, churches, hospitals, and other building types included articles that detailed the particular acoustical challenge posed by each type of building, and described how best to meet that challenge.[80] These articles, often written by acousticians, reiterated much of what the companies asserted in their ads, and legitimized that message with their scientific authority and appearance of objectivity. The June 1923 issue of the *Architectural Forum*, for example, was dedicated to the design of banks. In addition to the Johns-Manville advertisement depicted in figure 5.12, ads for the Guastavino Company, Armstrong Linoleum, and Gold Seal Battleship Linoleum similarly described the beneficial acoustical properties of their products and illustrated their use with photographs of actual installations. Armstrong offered "Beautiful, Quiet Floors for the Bank," while Gold Seal asserted that "Modern Efficiency Demands Comfortable and Quiet Floors."[81] An article by Clifford Swan described "The Reduction of Noise in Banks and Offices" in language virtually indistinguishable from the most persuasive advertising copy:

> Two officials of a well known institution in New York were recently walking in earnest converse through the corridors of their new building. They were surrounded by the usual babel of echoing sound common to such public spaces, and were, moreover, unpleasantly conscious of their own voices reflected back upon them, making both speaking and hearing difficult and uncomfortable. They stepped through a door into a department on a large open floor, when suddenly it seemed as if a blanket had been thrown over their heads shutting off all sound. They stopped in sheer amazement until it dawned upon them that the ceiling of this department had been covered with material intended to absorb the sound and produce just the condition of quiet noted.[82]

Swan described and evaluated the different types of materials available, including felts, tiles, and artificial stone products, and he noted that the acoustical treatment of offices was "welcomed alike by office managers, efficiency experts, welfare workers, and physicians as a necessary factor in the conservation of human energy."[83] Five years later, the office managers, efficiency experts, welfare workers, and physicians of the New York Life Insurance Company would welcome the world's largest single installation of sound-absorbing materials when they moved into their new corporate headquarters at Madison Square.[84]

Insurance companies knew particularly well the value of protecting the health of workers, and the attention to acoustical design paid by New York Life was an important component of a larger program to provide "healthful working conditions" for its 3,500 home office employees.[85] The acoustical design was additionally part of an extensive effort to render the new building an exemplary embodiment of the efficiency of modern business. Sound-absorbing materials worked in tandem with high-speed passenger elevators and the world's largest pneumatic mail-delivery system to maximize workers' productivity while minimizing their fatigue. The incorporation of such modern business principles into the very fabric of the structure rendered the New York Life Insurance Company Building state-of-the-art corporate architecture in 1928, and the skyscraper was celebrated as the "epitome of modern civilization."[86]

The land on which the building rose, along Madison Avenue between 26th and 27th Streets, was rich with history, and the sounds of bygone eras echoed amid the din of modern construction as the building went up. Once the site of a busy terminal for the New York & Harlem Railroad, the station had been remodeled and reborn as an amusement palace in the 1870s. P. T. Barnum's circus, concerts by Theodore Thomas, religious revivals, and sporting events had all been held at Madison Square. In 1887, a group of wealthy New Yorkers purchased the property and commissioned Stanford White to design a lavish new playhouse in which they could enjoy everything from concerts and theater to fine dining and equestrian events. Madison Square Garden was White's masterpiece, and he was enjoying a meal at its rooftop restaurant when, in 1906, he was gunned down by the aggrieved husband of White's lover, the former showgirl Evelyn Nesbit.

In the years after White's death, "Society" migrated further uptown and the Garden gradually fell into disuse. It became the property of the New York Life Insurance Company in 1917 through a mortgage foreclosure. At the time, the property was considered a liability, but as business expanded rapidly in the 1920s, the company realized the value of the large site and began to envision a new, expanded home office there. A building committee was appointed in 1923. By 1925, the old Garden had been demolished and excavation for the new structure had begun. Cass Gilbert was selected as architect, with the Starrett Brothers' construction company to serve as general contractors.[87]

Thirty years had passed since Cass Gilbert's inquiry to the editors of the *American Architect and Building News* soliciting information about acoustical design. During those years, the field of architectural acoustics had flourished, and so, too, had Gilbert. The young midwestern architect had won the 1895 competition for the Minnesota State Capitol, and this project launched him on a successful career. His practice grew, he moved east, and his rising reputation reached its pinnacle with the completion of the Woolworth Building. Subsequent to that literal high point, however, Gilbert became increasingly uncomfortable with the noise and congestion that his buildings helped to create. Whereas in 1912 he had described New York as a "dream city of towers and pinnacles," in 1925 he concluded that "the noise, confusion and rapid traffic have changed it so that it is not a suitable place to live."[88] Gilbert continued to design large structures for that noisy city, but he now began to consider how to isolate and insulate those structures from the noise and confusion.

Gilbert's initial design for the New York Life Insurance Company Building was a "crushingly massive" sixteen-story tower rising from a five-story base.[89] The interior spaces of the tower—few of which would have been proximate to exterior windows—were all to be served with artificial light and ventilation. Gilbert wanted to create a drastically inward-looking environment that was isolated from the noisy world without. Builder Paul Starrett questioned this approach, however, and he objected to Gilbert's plan when his opinion was solicited by the building committee. "He's building a monument to himself, but a mausoleum for you fellows," Starrett candidly concluded. Artificial light was inferior to natural light, he explained. Artificial ventilation sufficient for such a massive structure would require uneconomically large air ducts running through the building. Starrett collected data to back up his assertions, and he assigned the task of developing an alternative design to one of his own draftsmen, Yasuo Matsui, who rapidly drew up a new plan. The building committee was convinced; Gilbert's original plan was rejected and his firm was instructed to adopt Matsui's plan and develop it as their own.[90] The cornerstone of the building was laid in June 1927, and the finished structure was dedicated on 12 December 1928. (See figure 5.13.)

While the new design may have compromised Gilbert's plans for a fully isolated tower, in fact, the building that resulted was still remarkably independent of its surrounding environment. "The building is a self-contained city," the company's historian declared, "providing its inhabitants with shelter, power, light, water, food and transportation."[91] The corporation that governed this city-in-a-building not only offered its citizen-workers all the necessities of daily life, it also promised to protect their health. "The attention given to the welfare of employees is particularly noticeable in the case of insurance companies," one commen-

5.13

New York Life Insurance Company Building, 51 Madison Ave., New York (Cass Gilbert, Inc., 1929), c. 1929. While the building's Gothicized facade evoked the distant past, the interior was celebrated as the epitome of modern civilization. That modernity was, in part, constituted of over ten acres of sound-absorbing felt, the largest single installation of acoustical materials in its day. Courtesy of New York Life Insurance Company, Corporate Communications Department.

tator observed, and the new headquarters for New York Life manifested this concern throughout its structure. Extensive medical facilities were provided, the heating of the building was automatically monitored and controlled, drinking water was sterilized, and the cafeterias were designed to handle food and to dispose of waste in a sanitary manner to prevent the spread of disease.[92]

But the most pervasive measures to promote employee welfare were the provisions for quiet, and the private realm of this city-in-a-building succeeded in its campaign for noise abatement in a way that the public city that surrounded it would not. Artificial ventilation discouraged the opening of the building's many windows, and the windows themselves were made of extra thick glass supported in heavy frames to constitute effective acoustical barriers against the noise of modern city life. Efforts to control noises generated within the building were even more extensive. The location of all motors and machinery in the building was carefully planned to prevent any disturbing effect from the noises that they generated. Interior walls were built of solid masonry, and office partitions were constructed of heavy metal and glass braced in rigid frames, to prevent the transmission of noise between rooms. Cork flooring minimized the sound of footfalls, while plumbing, hardware, and other equipment were selected for quiet operation.[93] Most significantly, offices, cafeterias, lounges, corridors, medical rooms, and even the pneumatic mail-tube system were all wrapped with a thick sound-absorbing felt to prevent the accumulation and transmission of unwanted and unnecessary sound. The acoustical treatment was used "in all spaces where excessive noise might originate or where it is desirable or essential that quietness should prevail." More than 450,000 square feet (over ten acres) of acoustical material made this the largest such installation in the world.[94]

The sound-absorbing felt was made of a combination of sanitized cattle hair and asbestos, cemented directly to walls and ceilings, then covered with a fabric chosen to suit the particular location.[95] Office ceilings were covered with a perforated oilcloth painted bright white to reflect light. (See figures 5.14 and 5.15.) In cafeterias and lounges, the material was decorated with painted murals. (See figures 5.16 and 5.17.) Whether the visual effect was functionally utilitarian or charmingly decorative, the acoustical effect was uniformly remarkable:

> Imagine, if you can, a large office with typewriters and adding machines clicking away, telephones ringing, filing cabinets being opened and closed, doors shutting, clerks coming and going—but with not a sound above a murmur reaching the ears. Even the sound of the steel worker riveting outside is subdued. Such a condition, which seems almost unbelieveable at first, is actually typical of the work rooms of the building, and is made possible only by an extensive installation of sound-absorbing materials. . . .
>
> Although only a few months have elapsed since the opening of the building, the study and precaution taken in eliminating noise have already improved working conditions very noticeably.[96]

5.14
Women working at teletype machines in the New York Life Insurance Company Building, c. 1929. The acoustical treatment of the ceiling consisted of an absorbent felt made of asbestos and cattle hair, covered with strips of perforated white oilcloth. The woman at the far right is placing a capsule into the pneumatic tube document-delivery system, which was also acoustically treated to ensure noise-free operation. Courtesy of New York Life Insurance Company, Corporate Communications Department.

5.15
Women working in an acoustically treated office in the New York Life Insurance Company Building, c. 1929. The seams and perforations of the fabric covering the sound-absorbing felt are visible on the ceiling. By absorbing sound and reducing the level of noise in the room, this ceiling protected the physical and mental health of the workers beneath it, and also increased their working efficiency, rendering them more productive for the company. Courtesy of New York Life Insurance Company, Corporate Communications Department.

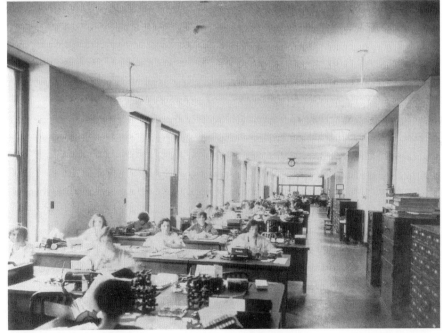

5.16
Men's Dining Room in the New York Life Insurance Company Building, c. 1930. The murals concealed a sound-absorbing wall treatment that reduced the din of the dining hall. A public address loudspeaker is also visible at the top of the image, on the support pillar that interrupts the mural. Courtesy of New York Life Insurance Company, Corporate Communications Department.

5.17
Ladies' Dining Room in the New York Life Insurance Company Building, 1929, showing a mural concealing a sound-absorbing wall treatment. This image is a frame from a scene filmed on location in the dining room for a Fox Movietone sound newsreel. Courtesy of New York Life Insurance Company, Corporate Communications Department.

By "improving" working conditions in this way, the New York Life Insurance Company not only protected the health of its workers, it also increased their working efficiency, a goal equally—if not more—important to any modern corporation.

The use of sound-absorbing materials was just one of numerous "mechanical and scientific improvements" that resulted in "an efficiency of operation and economy of motion" that was "little short of amazing."[97] High-speed elevators rapidly moved workers from floor to floor. Messages traveled even faster via teletype machines. When the transfer of physical documents was required—which was often, since the bulk of the company's daily operations consisted of the processing of millions of applications, approvals, denials, and claims—an extensive pneumatic mail-delivery system was relied upon. As many as 10,000 capsules per day moved through eight miles of tubing at 30 feet per second in the world's largest such system, which was, of course, muffled throughout with sound-absorbing materials. The capsules were sent to a central station in the basement, "a truly remarkable place, seeming, with its long rows of twisted tubes, and slides, and chutes, more like a scene from a futuristic 'movie' than an actual reality of the present."[98] Here, the capsules were sorted and then directed, via automatic conveyor belts, to the appropriate chutes that delivered them back up to their final destination.

Also located in the basement of the building was another marvel of modern efficiency, the kitchen. Operated by Savarin, Inc., it was well equipped to handle the "complicated engineering problem"[99] of expeditiously preparing nutritious lunches for the 6,000 people who worked in the building every day. A centralized food-preparation area served the four employee dining rooms (for men, women, department heads, and executive officers) as well as several public restaurants that Savarin operated for use by the tenants in the building. Conveyor belts and elevators distributed food from the kitchen to the different pantries, all "arranged for the rapid and convenient movement of food and personnel."[100] The efficiency of this kitchen was embodied in the smooth, flowing surfaces of its stainless steel counters, rails, and ventilation hoods. (See figures 5.18 and 5.19.) Harder to see are the acoustically treated ceilings, but they are there, absorbing the din of clattering dishes and contributing to the overall efficiency of the space. Impossible to see, of course, is the sound itself; such everyday sounds are virtually always lost to the historian, who must necessarily turn to textual descriptions and silent photographs to elicit the lost reverberations of the past. In the acoustically treated, sound-absorbing spaces of New York Life's new skyscraper, those reverberations were lost more quickly than ever before.

5.18
Kitchen of the New York Life Insurance Company Building, 1928. The kitchen, with its sleek surfaces of chrome and steel, was not considered an architecturally aestheticized space, but was instead characterized as the solution to the "complicated engineering problem" of efficiently providing lunch for the thousands of people who worked in the building. Courtesy of New York Life Insurance Company, Corporate Communications Department.

5.19
Dish Washing Room of the New York Life Insurance Company Building, 1929. The beauty of the engineered efficiency of spaces like this, while generally unacknowledged in America, was increasingly inspiring a new generation of architects in Europe. These modern architects would adopt the principles of engineering design as their own. Courtesy of New York Life Insurance Company, Corporate Communications Department.

The acoustical design of the New York Life Insurance Company Building demonstrates well how both the locus and the goals of sound control had changed in the years since St. Thomas's Church was built. Acoustical materials were no longer sequestered in churches and concert halls, devoted only to protecting and improving the sacred tones intoned within. Now, there was far more work to be done. As the world outside those sheltered spaces was perceived to become ever noisier, and as the deleterious effect of that noise upon human health and productivity was proven more convincingly, sound-absorbing materials were put to work on working people. Acoustical design came to be seen as "sound" economic practice, and the practice proliferated. Whereas people had previously only visited acoustically designed spaces, they now began to inhabit them. As a result, they gradually become accustomed to the sound—or lack thereof—therein.

V Modern Architecture and Modern Acoustics: The Philadelphia Saving Fund Society Building

While the New York Life Insurance Company Building was acclaimed as the epitome of modern business efficiency, and while the acoustical dimension of that modernity was celebrated, the visual modernity of its subterranean kitchen passed unnoticed. Its sleek surfaces were admired as engineering, not architecture. The beauty of the building was associated exclusively with its exterior, and that external beauty cloaked, concealed, even denied the modern business activity housed within its walls.[101] The tower presented to the world a Gothic-styled facade that conjured up images of a long-distant and slow-to-change past. Adorned with gargoyles and medieval tracery, its solid-looking exterior was patterned after centuries-old ideals of architectural beauty, evoking a dignified sense of permanence and stability. That image was certainly comforting to the many policy holders who hoped that the company would prevail come their time of need.[102] It also reflected the aesthetic and cultural conservatism of the architect, for Cass Gilbert shared Ralph Adams Cram's low estimation of modernism in art and architecture. Circa 1920, Gilbert had characterized the movement as nothing but "futile, rash experiments for novelty and sensation." The "jaded nerves" of Europeans, Gilbert argued, "may demand the vibrant screech of discordant sound or color or exaggerated and eccentric form to arouse sensation and excite a passing interest," but he felt that Americans would do best to reject these trends and instead practice "well ordered self restraint."[103]

The facade of the New York Life Insurance Company Building embodied this ideal of order and restraint, yet, the disjuncture of this placid exterior with the dynamic activity taking place within was striking, and it did not escape notice. The interior designer of the executive offices was just one who noted the incongruity:

> The problem of furnishing the special rooms of the New York Life Insurance Company Building presented many interesting aspects. To begin with, the period of the building designed by Cass Gilbert was, to use his own term, "American Perpendicular," but on examination one could see Mr. Gilbert's interpretation of Tudor or Gothic motives, connected with English tradition, adapted to our skyscraper form. In this twentieth century . . . it is necessary to select a period of interiors and furnishings wholly different from those used by our mediaeval ancestors. To be more in accord with our present day business life, then, it was decided that the English period of the eighteenth century would be generally featured in the principal rooms.[104]

The designer, while leaping forward several centuries by foregoing Gothic for Georgian, still landed far short of his own era. His modernizing tendencies were not strong enough to allow him to recognize the engineered look of "present day business life" itself as an appropriate design model. But that very look—the look of technological efficiency—while largely ignored by American architects like Gilbert and Cram, was simultaneously stimulating a transformation of architectural design in Europe. In much the same way that European composers had engaged with technology to create a new, modern music, so, too, did Europe's architects turn to technology—American technology—to construct a modern architecture.[105]

In 1923, the Swiss architect Charles-Eduard Jeanneret, better known as Le Corbusier, brought forth a manifesto in which he declared that "The machinery of Society" was "profoundly out of gear."[106] If society was a mechanism, who better than engineers to repair it? "Let us listen to the counsels of American engineers," Le Corbusier proclaimed, and he turned to the world of engineered objects for inspiration as he worked to develop a new architecture that would both shelter and engender a new society.[107] In Germany, Walter Gropius and his colleagues at the Bauhaus were similarly reformulating the art of building. "Architecture during the last few generations has become weakly sentimental, aesthetic and decorative," Gropius declared.

> This kind of architecture we disown. We aim to create a clear, organic architecture whose inner logic will be radiant and naked, unencumbered by lying facings and

trickery; we want an architecture adapted to our world of machines, radios and fast cars . . . with the increasing strength and solidity of the new materials—steel, concrete, glass—and with the new audacity of engineering, the ponderousness of the old methods of building is giving way to a new lightness and airiness.[108]

The crisp white villas of Le Corbusier (which he called *machines à habiter*, or "machines for living"), and Ludwig Mies van der Rohe's project for a glass-walled skyscraper (see figure 5.20) epitomize both the small- and large-scale aspirations of modern architects.

The technological purity of such modern spaces was, however, at times apparently achieved at the cost of comfort. "More than one loyal supporter of the Bauhaus," Reyner Banham has claimed, was "prepared to admit that the glaring lighting and the ringing acoustics were distressing."[109] Walls of expansive glass and hard, thin plaster partitions resulted in uncomfortably reverberant spaces that easily transmitted sound. In 1931, Le Corbusier attempted to alleviate these problems in his *Pavillon Suisse* dormitory by suspending sheets of lead within the lightweight partitions that separated the building's rooms. While the experiment was unsuccessful (an "ear-witness" complained that you could "hear an electric razor three rooms away"[110]), it demonstrates that the architect was well aware of the acoustical shortcomings of his structure.

In fact, Le Corbusier had attempted to control sound several years before he built his *Pavillon Suisse*. For his entry in the 1927 competition for a large assembly hall for the League of Nations, the architect collaborated with the French acoustical engineer Gustave Lyon.[111] They designed an auditorium with a roof composed of double-plated glass arranged in parabolic sections. Their intent was to allow the form of the room to direct and disperse reflected sound throughout the body of the hall. One critic of the design, Swiss engineer F. M. Osswald, called attention to the extensive use of glass and suggested that "the low sound absorption of such a material, together with the rather large volume (1,400,000 cubic feet), presents a combination that is practically sure to yield a room that is too reverberant for speech."[112]

Osswald also remarked that, while hundreds of designs had been entered into the League of Nations competition, only a few considered acoustics at all. "The majority of competitors," he concluded, "had an insufficient knowledge of the principles which govern hearing conditions in large enclosed spaces."[113] A small number of entrants had considered the reflection of sound, but even fewer considered its absorption by architectural materials. Acoustical materials simply were not as prevalent in Europe as they were in America, nor were their tech-

5.20

Mies van der Rohe's drawing of a glass skyscraper, 1921. Modern architects like Mies abandoned past models of architectural beauty. In its place, they embraced the efficient essentials of engineered structure and modern materials. Ludwig Mies van der Rohe. Friedrichstrasse Skyscraper. 1921. Presentation perspective (north and east sides). Charcoal, pencil on brown paper, 68 1/4 x 48" (173.5 x 122 cm). The Mies van der Rohe Archive, The Museum of Modern Art, New York. Gift of the architect. © 2001 The Museum of Modern Art, New York.

niques of employment as widely known.[114] Mies van der Rohe had called for such materials in 1924 when he announced: "Our technology must and will succeed in inventing a building material that can be manufactured technologically and utilized industrially, that is solid, weather-resistant, soundproof, and possessed of good insulating properties."[115]

Mies's call went largely unheeded abroad, but when modern architecture arrived in America, its architects encountered a market full of the very kinds of materials that Mies had sought. The existence of those materials—including acoustical materials—and the tradition of their employment in American buildings helped establish the new style in this country. As Reyner Banham put it, "while European modern architects had been trying to devise a style that would 'civilise technology,' U.S. engineers had devised a technology that would make the modern style of architecture habitable by civilised human beings."[116]

The affiliation between the technology of sound control and modern architecture was suggested in 1931 when the United States Gypsum Co. evoked a Miesian tower in advertising its System of Sound Insulation. (Compare figures 5.20 and 5.21.) The Celotex Company's slogan, "Less Noise . . . Better Hearing" also echoed Mies's classic dictum, "Less Is More," and highlighted the enthusiasm for efficiency shared by modern architects and acousticians alike.[117] This implicit affiliation was made explicit in 1932, when the new architecture formally arrived in America.

In 1932, the Museum of Modern Art presented an exhibit on "The International Style," surveying and summarizing the new trend in architecture that had been developing in Europe over the past decade. Simultaneously, the first significant example of that architecture to be built in America was taking shape on the streets of Philadelphia. The MoMA exhibit elucidated the principles that the Philadelphia Saving Fund Society Building of Howe & Lescaze exemplified; a new emphasis on architectonic volume rather than mass, regularity rather than symmetry, and a vehement proscription of "arbitrary" applied decoration.[118] The functionally differentiated spaces of the PSFS Building—the ground level shops, the second-floor banking room, the office block that rose above, and the elevator block that served those offices—were all distinguished by the different volumes that constituted the structure. The asymmetrical arrangement was orderly, and ornament was "conspicuous by its absence." "The surfaces of machine production are inherently beautiful," one reviewer wrote, and they produced "a natural aesthetic movement supplied in other buildings by sculptured swags and terra cotta gargoyles."[119] (See figure 5.22.)

5.21

"USG System of Sound Insulation," 1931 advertisement for the United States Gypsum Co. The skyscraper depicted draws upon the visual iconography of modern architecture and evokes the drawing by Mies van der Rohe reproduced in figure 5.20. The product, a system of interior construction that prevented the passage of sound between rooms, offered the isolation and control that characterized modern acoustics. *American Architect* (October 1930): 11.

A MESSAGE TO ARCHITECTS FROM THE
UNITED STATES GYPSUM COMPANY

The left illustration shows noise vibrations crashing against the exterior of a building, like waves breaking on the seashore. At the right is a USG sound insulated "floating" partition which prevents similar sounds created within the building from being transmitted from one room to another.

You Are Invited to Use This Service in Architectural Acoustics

THE United States Gypsum Company has undertaken to supply a new and comprehensive service in the field of Architectural Acoustics. For this purpose we maintain a complete sound research laboratory and an extensive department devoted exclusively to the solution of problems in the field of sound control.

In order to handle all assignments in Architectural Acoustics it has been necessary for us to develop Acoustone, the USG Acoustical Tile for sound absorption, and in addition a complete System of Sound Insulation for preventing the transmission of sound from one room to another or from one floor to another.

The USG System of Sound Insulation is a scientific method of floor, wall, ceiling and door construction so designed as to prevent the force of sound striking on one side of the construction from carrying through to the other side. The United States Gypsum Company furnishes all the special materials required, supervises the entire installation and assumes full responsibility for the predicted performance of the in-

Detail of USG Sound Insulative Door which prevents the transmission of sound from room to room. It is used in connection with the USG System of Sound Insulation.

stallation. This system has been used with highly satisfactory results in hotel, apartment and office buildings, industrial plants, schools, studios, etc. It may be used in any construction where noise abatement is desirable.

Architects are invited to write for particulars about the USG System of Sound Insulation and to avail themselves of our services on any problem in Architectural Acoustics. No obligation is involved. Address the United States Gypsum Company, Dept. 26K, 300 W. Adams St., Chicago, Ill.

USG SYSTEM *of* SOUND INSULATION

FOR OCTOBER 1930

5.22

Philadelphia Saving Fund Society Building (Howe & Lescaze, 1932). The PSFS Building was the first large-scale example of modern architecture to appear in America. It embodied aesthetic principles that emphasized volume over mass and regularity over symmetry, and rejected ornamentation of any kind. PSFS Archive, Box 6, Folder: PFSF Building Exterior Views. Courtesy of Hagley Museum and Library.

Customers were conveyed by sleek steel escalators to a banking room filled with gleaming surfaces of chrome, glass, and polished marble. (See figures 5.23 and 5.24.) The society, usually conservative in matters of style as well as finance, had "gone Gershwin,"[120] and it was Ira's brother, not some long-dead king of England, who inspired the Georgian interior of this office building. With the PSFS Building, engineered objects were now celebrated for their aesthetic appeal as well as for their inherent efficiency. Whereas the smooth flowing efficiency of the New

5.23
Escalator and stairs leading from the street entrance to the Main Banking Room of the PSFS Building, 1932. The gleaming surfaces that had characterized the engineered kitchen spaces in the basement of the New York Life Insurance Company Building were brought upstairs and legitimated as architecture in the PSFS Building. PSFS Archive, Box 3, Folder: Escalators and Stairways. Courtesy of Hagley Museum and Library.

5.24
Main Banking Room of the PSFS Building, 1932. The chrome, steel, and glass surfaces, as well as the functionally designed furniture and accessories that constituted the Main Banking Room, were celebrated as "inherently beautiful." An acoustical tile ceiling, whose grid is faintly visible in this photograph, kept the hubbub of commerce to a minimum. PSFS Archive, Box 3, Folder: PSFS Building, Banking Floor, Dooner Photos. Courtesy of Hagley Museum and Library.

York Life Insurance Company Building kitchen had been concealed deep within the basement, the main banking room of the PFSF Building—which looked strikingly similar to the kitchen—constituted the architectural showpiece of that tower, and it was heralded for ushering in "a new architectural era."[121]

That the debut of modern American architecture should occur on the staid streets of Philadelphia was an irony well noted at the time. "The closing of Wanamaker's," *Fortune* magazine reported, "the razing of Independence Hall, and a Democratic victory in the city would have been plausible and likely happenings in comparison."[122] And that such a building should come from George Howe seemed equally implausible.

As a partner of Mellor, Meigs & Howe, the architect had spent a long and productive career designing Renaissance palaces, Georgian mansions, and Elizabethan manors for the residents of Chestnut Hill, the Main Line, and other wealthy Philadelphia suburbs. But over the course of the 1920s Howe became increasingly dissatisfied with this kind of architecture, and he began to search for a new approach to design. "I have been looking," he wrote in 1930, "for a means of architectural expression which should not be in conflict with any form of modern activity outside the field of architecture. I felt I had failed either to evolve or discover such an expression until I became conscious of the meaning of the so-called modern system of design."[123]

In May 1929, Howe formed a new partnership with William Lescaze, a young Swiss emigré who brought training in and a commitment to the new modern system of design, if not many executed commissions. Howe made it possible for Lescaze to build buildings, and Lescaze helped his partner to build new kinds of buildings, to achieve an architecture "far superior to the masking beauty of Classic and Gothic derivatives."[124] Together, they created for the PSFS "a sound economic working building which acknowledges itself frankly as such instead of pretending to be a temple or a cathedral."[125]

Howe & Lescaze emphasized the practical efficiency rather than the stylistic novelty of their design to the PSFS building committee, which was controlled by the imposing and conservative president of the society, James Willcox. Howe had previously designed several traditional-looking suburban branch offices for the bank, and he gradually convinced Willcox to share his bold vision for this new building. By the time the society was ready to advertise rental space in the new office tower, it proudly proclaimed that there was "Nothing More Modern."[126] Philadelphia's resident Modern, Leopold Stokowski, agreed, declaring the new building "a marvelous pile of architecture." "I am so happy that we have a wonderful example of modern art growing in Philadelphia," the conductor exclaimed. "It will help us to develop other forms and it will be a delight to everybody who is alive to the things of today."[127]

Many Philadelphians were about as enthusiastic for their new modern skyscraper as they were for Stokowski's orchestral programs of modern music, which is to say, not very enthusiastic at all. A survey of the local citizenry indicated that "almost 50 per cent think the building a 'botch.'" "What a hideous thing that building is," the *Sunday Dispatch* declared, "utterly destitute of the faintest claim to comeliness, an affront to public taste. . . . It's barbaric, repellent, epically stupid." The *Sunday Transcript* agreed, asserting that "never has such an ugly building been perpetrated."[128] Others, however, shared Stokowski's enthusiasm, and a lengthy paean to the new building appeared in the Philadelphia *Record*:

> Sheer out of the earth
> The black edifice
> The crowning point
> Of Man's climb from the caves
> This black and silver stem
> Shivering in the sun
> A frozen exclamation mark
> In the march of time
> Regular
> Angular
> Precise
> Perfect
> Calm
> Cold
> Frozen
> Up from the city streets
> In unbroken perfection
> Up from the streets
> To the clean air
> Up from the noisy streets
> The hideous clamor
> To the secrecy of the upper air
> The silver stillness
> The quiet beauty
> The fulfilled promise
> Of the upper air.[129]

Whether one loved it or hated it, all agreed that the PSFS Building looked like no other skyscraper in America. The building sounded as modern as it looked, and the acoustical technology employed within it was both extensive and extraordinarily absorptive. This modern sound was not nearly as startling as was the modern look, however; indeed, in the PSFS Building, the modern look was only catching up to the modern sound that had evolved over the course of the past two decades in traditional-looking structures like St. Thomas's Church and the New York Life Insurance Company Building. What was most innovative about the acoustical technology in the PSFS Building was how well it was integrated with the other technological systems of the building, the systems for artificial ventilation and illumination. In addition, the fact that all these technologies were accepted as contributing directly to the visual impact, the aesthetic beauty of the structure, was something strikingly new.

In the PSFS Building, standard as well as customized acoustical materials and products were used throughout. The ventilating ducts located within the columns of the main banking room were damped with sound-absorbing materials; the dishwashing room was structurally isolated and insulated with felt; Armstrong sound-absorbing corkboard lined the walls and ceilings of the machine rooms; and rubber flooring was installed in areas of heavy traffic.[130] Custom-designed soundproof office partitions were designated "an important contribution to building science." "Neither typewriter clatter nor telephone jingling can penetrate the two layers of one-half-inch Insulite, separated by a two-inch air space, which form their core."[131] Most prevalent, however, were the sound-absorbing ceiling tiles installed throughout the building.

By 1932, it had become standard practice to concentrate sound-absorbing materials on the ceiling, rather than the walls, of a room.[132] Throughout the 1920s, it had been common for such treatment to be built up of long strips of felt, cemented or tacked into place. As in the New York Life Insurance Company Building, such treatment was typically covered with a sound-permeable cloth for a more finished look. In addition to the perforated oilcloth used by Cass Gilbert for New York Life, Johns-Manville offered coated fabric finishes that imitated the look of more substantial materials such as plaster and travertine.[133] These felt treatments offered very high levels of sound absorption, but as Sabine himself had long ago recognized, architects sought a more inherently structural solution, one without obvious seams and covers, and they turned increasingly to new kinds of sound-absorbing materials to achieve this.

Acoustical plasters were seldom as absorbent as felt, but they offered a smooth, seamless finish that many architects preferred. In 1922, the Mechanically

5.25
Installation of Sabinite Acoustical Plaster at 300 West Adams St., Chicago. An acoustical plaster ceiling has been deployed over an office typing pool to reduce the noise of the typists in the room and to improve their working efficiency. This room is additionally isolated acoustically from the surrounding spaces, which are visible through the interior windows. "U.S.G. Sound Control Service," (sales pamphlet, United States Gypsum Co., 1931), p. 22. Series I, Box 13, Folder 11, Guastavino/Collins Collection, Avery Architectural and Fine Arts Library, Columbia University in the City of New York.

Applied Products Co. advertised Macoustic Plaster, and in 1926, the U.S. Gypsum Company introduced "Sabinite," a sound-absorbing plaster developed by Paul Sabine at the Riverbank Acoustical Laboratory.[134] (See figure 5.25.) By 1932, Kalite, Wyodak, Old Newark, Sprayo-Flake, and numerous other brands were competing with Macoustic and Sabinite for customers.[135] While these plasters were widely employed, there were drawbacks to their use and, like the felts, they were not perceived as an ideal means for controlling sound. Acoustical plasters were specially formulated to create a porous, sound-absorbing surface, but while manufacturers claimed that no extraordinary techniques of application were required, the final result was, in fact, dependent on following to the letter complicated installation instructions.[136] Without such skilled installation, the absorption coefficient of the plaster could differ significantly from that promised by the manufacturer. What was needed was a standardized, easy-to-install product that offered a dependably constant and high level of absorption. Acoustical tiles offered all of the above, and architects increasingly turned to tiles as the solution to their problems of sound control.

Rumford, and especially Akoustolith, had constituted the original standardized acoustical tiles, but when the demand for tile increased in the 1920s, a host of competitors appeared. In 1921, Sabine's old nemesis Jacob Mazer introduced the Mazer Acoustile Sound Controlling System ("fully covered by letters

patent"). Mazer depended on flexible felt to provide the sound absorption for his system, but he deployed the sheets of felt in "rigid self-sustaining panels built complete before erection." These panels were custom-designed to fit each particular application, and each felted frame was covered with a permeable membrane. While Mazer claimed hundreds of installations in his advertisements, this type of customized acoustical treatment did not really catch on.[137] Far more typical were standardized rigid products like "Acoustibloc," a tile molded from a combination of flax fiber and rock wool that the Union Fibre Company offered in 1924. Four years later, the Boston Acoustical Engineering Company introduced "Silen-Stone," an artificial stone tile made of sand and Portland cement, and "Acoustex," a "sound-absorbing slab" made of wood fiber and cement.[138] In 1931, the U.S. Gypsum Company advertised "Acoustone," a tile made of "specially processed stone" available in standard sizes and thicknesses, with absorption coefficients (at 512 cps) ranging from 0.46 to 0.62.[139] Acoustone, like Silen-Stone and Akoustolith, was intended to mimic the look of traditional masonry construction. The most widely used tile product of the twenties, in contrast, looked like nothing else, and the manufacturers of Acousti-Celotex made little effort to disguise their product's distinctive appearance. (See figure 5.26.)

The Celotex Company was founded in 1920 by Bror Dahlberg and Carl Muench. Dahlberg was a Swedish immigrant who had worked his way up from a clerkship with the Great Northern Railroad to management of the Minneapolis and Ontario Paper Company. Muench joined him there in 1914, and the two men developed a means to turn cellulose waste fibers into a rigid building board. The resulting product, Insulite, was a commercial success, and the Insulite Company was established as a subsidiary of M&O in 1916. In 1919, Muench and Dahlberg hoped to repeat their success with bagasse, the fibrous waste of sugarcane refineries. Their employer was unwilling to support this endeavor, however, so the men left to form their own company. Just one year later, the Celotex plant was turning out 200,000 square feet of tile per day, and the product "found ready acceptance as a standard building material."[140]

An excellent heat insulator, Celotex was used in home construction as well as in the manufacture of refrigerated railroad cars. Early advertisements also noted that Celotex was an "efficient sound deadener."[141] In 1925, the company introduced a new product, Acousti-Celotex, which was vigorously promoted in sales pamphlets and with distinct coverage in *Sweet's*. Acousti-Celotex was made of the same material as the regular building board; in addition, it had numerous holes drilled into its surface. These perforations increased significantly the

5.26
"Acousti-Celotex Types and Sizes." Illustration from a 1927 sales brochure depicting the variety of acoustical tile products offered by the Celotex Co., with absorption coefficients ranging from 0.25 to 0.70 at 512 cps. Acousti-Celotex tiles were made of bagasse, the fibrous waste product of sugarcane refineries. "Less Noise—Better Hearing," (sales brochure, Celotex Co., 1927), p. 29. Series I, Box 13, Folder 11, Guastavino/Collins Collection, Avery Architectural and Fine Arts Library, Columbia University in the City of New York.

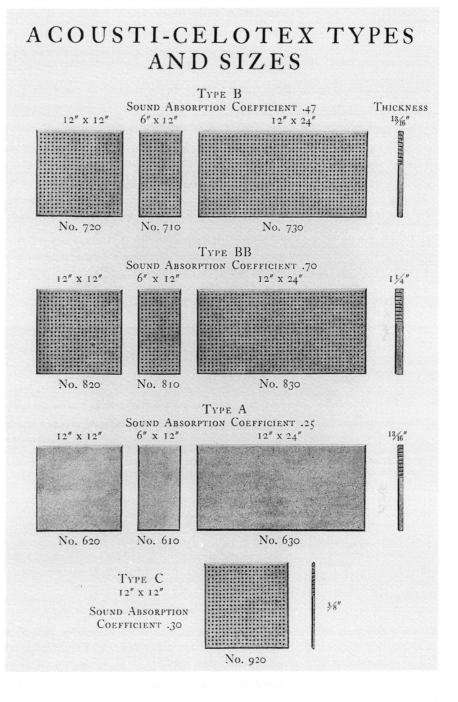

absorptivity of the material, which was rated as high as 0.70 (at 512 cps) for tiles that were 1.25 inches thick.[142]

Regularly perforated acoustical materials had been patented in 1924 by Jacob Mazer. In 1925, Celotex obtained a license from Mazer for the manufacture and sale of perforated acoustical materials,[143] and soon thereafter, the company introduced Acousti-Celotex, advertising its employment in "Armories, Auditoriums, Ballrooms, Banking rooms, Banquet rooms, Bowling alleys, Churches, Composing rooms, Computing rooms, Corridors, Courtrooms, Dining rooms, Factories, Halls, Hospitals, Lecture rooms, Libraries, Lodge rooms, Music rooms, Mailing rooms, Offices, Printing plants, Radio broadcasting stations, Radio receiving stations, Railway stations, Reading rooms, Residences, Restaurants, Schoolrooms, Telegraph rooms, Telephone rooms, Temples, Theaters and Typewriting rooms." By 1927, Acousti-Celotex was "bringing relieving quiet into the nerve-worn world of commerce, industry and education. It subdues irritating noises . . . deadens the roar of traffic . . . increases working efficiency."[144]

Not only did Acousti-Celotex help create efficient workers, the tiles themselves were the epitome of efficient production. Dahlberg and Meunch had taken industrial dross and spun it into gold as the waste product of sugar refineries was transformed into a valuable commercial commodity by "high-speed continuous fabricating lines."[145] The sound that those tiles produced, too, was itself efficient and advertised for its inherent qualities as well as its effects. The manufacturer proclaimed that Acousti-Celotex "swallows up all distracting noises, clears the air of echoes and reverberations . . . allows only the true, intended sound to strike your ear."[146] By 1933, this true, intended sound—the modern sound—was heard in over 6,000 locations.[147]

As acoustical tile technology developed over the course of the 1920s, new methods of ceiling construction, particularly suspended ceilings, began to appear. New types and brands of acoustical tiles were designed to fit into systems for suspended ceiling construction, as well as to avoid the proprietary aspects of the Celotex-controlled patent on integrally perforated materials. In 1929, the Burgess Laboratories introduced "Sanacoustic," a system of "perforated metal forms, back of which is placed a highly sound absorbent material such as Mineral Wool, Balsam-Wool, Flaxlinum or Hairfelt. These forms in turn are locked definitely into light structural T-sections which have been bolted or screwed to wall or ceiling surfaces." In 1929, Burgess listed twelve installations of Sanacoustic Tile, including the auditorium of the University of Minnesota and the Boston offices of the Sears Roebuck Company.[148] Johns-Manville began to distribute the product in 1930, and within a year they offered four sizes of Sanacoustic Tile, all of which were now filled with their own sound-absorbing products, such as Asbestos Akoustikos Felt or Banroc Wool. The absorption coefficient (at 512 cps) was advertised as 0.82.[149]

Other suspended-ceiling systems soon appeared to compete with Sanacoustic. The Truscon Steel Company's "Ferrocoustic" metal ceiling framework was designed to hold any type of acoustical tile from any manufacturer to create a suspended sound-absorbing ceiling.[150] The Acoustical Corporation of America advertised "Silent-Ceal," a "complete suspended ceiling construction" developed by the acoustical engineer M. C. Rosenblatt. The Silent-Ceal system consisted of perforated metal trays suspended from metal furring and filled with rock wool for sound-absorption. It offered "the highest sound absorption possible—above 70 per cent for four principal octaves."[151] Silent-Ceal was soon superseded by the "Mutetile" system, in which perforated, cast plaster tiles were filled with "nodulated rock wool" and then "spring suspended" from the metal framework. The absorption of Mutetile was advertised as even greater than that of Silent-Ceal, as high as 94 percent at a frequency of 1,024 cps,[152] and it was Mutetile that was used throughout the PSFS Building.

The architects' specifications for the main banking room, the conference rooms and school department on the second mezzanine, and the dishwashing room on the thirty-third floor of the PSFS Building had originally called for Johns-Manville Sanacoustic Tile in a suspended ceiling construction. For what appear to be financial reasons, the contractors turned instead to the Acoustical Corporation of America to supply the tiles for all these locations.[153] Mutetile ceilings were suspended throughout, absorbing unprecedented amounts of sound and creating the "silver stillness" and "quiet beauty" that the Philadelphia *Record* had poetized upon. (See again figure 5.24, and see figure 5.27.)

Press accounts of the new building uniformly referred to its state-of-the-art acoustical design, but what really attracted the attention of journalists, architectural critics, savings-account holders, and prospective tenants was the building's overall accomplishment of environmental control. Sound, light, air, and temperature were all integrated and regulated by a complex technological system that created a complete, and completely comfortable, artificial environment within the building. The most innovative and obvious element was not the acoustical tiling, but the air conditioning.

Air conditioning, developed by Willis Carrier and others around the turn of the century, was initially applied to sites of industrial manufacture to provide

5.27
Conference Room with Mutetile ceiling on the second floor mezzanine of the PSFS Building, 1932. Mutetile was advertised to absorb 94 percent of incident sound energy, constituting one of the most absorbent materials available. Note the incongruously Gothicized radio receiver on the windowsill. PSFS Archive, Box 3, Folder: Executive Row. Courtesy of Hagley Museum and Library.

humidity control rather than temperature control. Only around 1920 did the application to cooling for personal comfort begin to develop. By the end of the 1920s, countless movie theaters enticed patrons with chilled air, but in 1932 the PSFS Building was only the second office tower in America to provide what was then called "manufactured weather."[154] It was this feature that most dramatically distinguished the PSFS Building from its neighbors and filled it with new tenants, even in the midst of suddenly difficult economic times.

While the air-conditioning system captured people's attention, it is best thought of as just one component of a complete package of environmental control offered by the PSFS Building. Temperature and humidity, light and sound were all controlled by the tightly integrated technological systems of the building. Most simply, air conditioning meant that windows could remain closed year-round, keeping out the dirt and noise of the city streets below.[155] For those who could not afford air-conditioned office space like that in the PSFS Building, the same benefit could be achieved with less expensive ventilation units like the "Silentaire," which vented fresh air into a room while filtering out the unwanted external noise.[156] (See figure 5.28.)

In the main banking room of the PSFS Building, and in the interior offices of the school banking department, the integration of the different technologies

5.28 "Silentaire," 1932 advertisement for a sound-absorbing ventilation unit for installation in double-hung windows. For those who couldn't afford air conditioning, units like the Silentaire allowed "non-draft circulation" via open windows while maintaining the "restful and efficient quiet of a closed room." *Architectural Forum* 57 (October 1932): 29. Courtesy Marquand Library of Art and Archaeology, Princeton University.

5.29
School Department office space on the second floor mezzanine of the PSFS Building, 1932. The acoustical tile ceiling, which includes custom-designed fixtures for illumination and ventilation, constitutes a technological system for complete environmental control. Air, light, and sound were all regulated to create the ideal working environment. PFSF Archive, Box 3, Folder: Open Office Space. Courtesy of Hagley Museum and Library.

was even tighter. (See figure 5.29.) Here, custom-designed fixtures that provided both artificial lighting and air-conditioned air fit neatly into the network of suspended acoustical tiles that constituted the ceiling. The gridded surface was regular and orderly—an exemplar of good modern design—and the suspended ceiling additionally created a hidden space above it, in which to conceal the apparently less orderly electrical conduits and ventilation ducts.[157] The integration of the different systems into a total package of environmental control offered a "New Deal" to Philadelphians; a winning hand whose whole was far greater than the sum of its parts. (See figure 5.30.) Prospective tenants were additionally sold on the resultant efficiency of this total environment. "Your office will be more efficient," the ads promised, "because of Manufactured Weather, thermostatic heat control, and absolute quiet."[158]

Less than a year earlier, the editors of the *New York Times* had, somewhat playfully, imagined the outcome of Benjamin Betts's call for a new business of

5.30
"A New Deal," advertisement for the PSFS Building from the Philadelphia *Public Ledger,* c. 1933. The winning hand of complete environmental control over air, light, and sound enticed prospective tenants to come "see today what will be called modern tomorrow." MSS Acc 2062 (Box 90). Courtesy of Hagley Museum and Library.

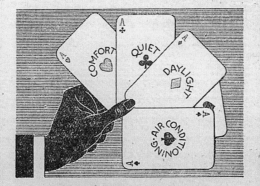

A NEW DEAL

There are office buildings and an office building. In other words, there are many buildings in which you may locate your office — buildings old and buildings new — *and* there is *Twelve South Twelfth.* You may feel that your present office is modern in every particular, so up-to-date that it is worth the premium which, if it is in a new building, you may be paying for it. Visit *Twelve South Twelfth* and see today what will be called modern tomorrow. Feel the comfort of genuine air conditioning. See what true daylighting can mean to you. Ask about the rentals. Make a personal new deal.

A new and well equipped garage is located at 12th & Filbert Streets

THE PHILADELPHIA SAVING FUND BUILDING
Twelve South Twelfth

Rental Agent: RICHARD J. SELTZER
225 So. 15th St., Phila., PENnypacker 7532

5.31

"Quiet!" advertisement for the PSFS Building from the Philadelphia *Public Ledger,* c. 1932. The building is represented as the epitome in acoustical isolation, hermetically sealed off from all external noises by the vacuum of a bell jar. With the PSFS Building, "The silence you have wished for is available." MSS Acc 2062 (Box 90). Courtesy of Hagley Museum and Library.

sound control. Might the day come, they asked, when "The home hunter consulting an agent will ask: 'Is it soundproof? Is the air conditioner in good working order? Are the imitation windows fitted with both sunlight and twilight lamps?'"[159] Like Betts, they seem not to have realized that such a day was already at hand. With the PSFS Building, their fanciful prediction of a "hermetically sealed"[160] architecture, an architecture of total environmental control, came true. (See figure 5.31.)

VI Conclusion

The unprecedented degree of technological control provided by the PSFS Building was not concealed behind tapestry or tracery. It was celebrated for what it was, a modern building, "inside and out."[161] With Corbusian enthusiasm, the PSFS proclaimed "this building is a working machine!" and when Le Corbusier himself first saw the building in 1935, he declared, "C'est Magnifique!"[162] The suspended acoustical tile ceilings and the sound that they created were significant elements of this magnificent modern machine.

By absorbing an extraordinary amount of sound energy, the acoustical tile ceilings in the PSFS Building rendered its spaces virtually free of all reverberation. The sound that remained was clear and direct, efficiently stripped of all aspects unnecessary for communication. The ceilings also ensured that the workers who worked below them were able to do so with maximum efficiency. They demonstrated the total control over the environment that was possible with modern technology, and that environment was unapologetically offered as a commodity for sale to prospective tenants. The sound-absorbing tiles removed all spatial characteristics from the sound within the building, and the building itself was dissociated from the urban space around it by the "hermetic seal" of environmental control. There was "Nothing More Modern" than the sound of the PSFS Building, and this acoustical modernity was, at last, celebrated with a new visual style that drew upon the tiles to accomplish its modern look.

The mass-produced tiles of the PSFS ceilings created the regular, modular patterns that modern purists had called for. Indeed, in their definitive catalog of the new style, the curators of the MoMA exhibit on modern architecture had referred to the "geometrical web of imaginary lines" that "integrates and informs a thoroughly designed modern building."[163] In the PSFS Building, this imaginary web became real, stretching out across the ceilings of the various rooms and swallowing up the wayward sounds that impinged upon its geometrically ordered surface.

The PSFS Building offered unprecedented degrees of control over sound. That control was exercised in ways that acoustically denied the existence of space, by minimizing the reverberation within and by acoustically isolating inhabitants from the soundscape without. Ultimately, however, this silent antispace threatened to be a bit too quiet for comfort. Plans were made to allow sound back into those architecturally silenced spaces, but only sounds of a particular sort. The PSFS Building was wired for radio reception so that each office could be filled with the electroacoustically controlled sounds of radio signals. Those who worked within plugged their incongruously Gothicized radios into special outlets. (See again figure 5.27.) The signals so emitted filled the silence of the vacuum with the sounds of life, and connected those workers to the wider world without.[164]

Anyone working late in the PSFS Building on the night of 27 December 1932 might have listened in on a live radio broadcast of opening night at Radio City Music Hall in New York. From within the silent stillness of the glass tower, this lonely listener would have heard the noise and excitement of a crowd of thousands that had gathered at the new theater to behold "the first great modern show of shows."[165] In Radio City, as in the PSFS Building, the silence of architectural acoustics combined with the sounds of electroacoustics to complete the modern soundscape.

CHAPTER 6   ELECTROACOUSTICS AND MODERN SOUND, 1900–1933

> [A] new factor has come strongly into the picture, and I believe that it will call for some radical revisions of our criteria for best acoustics. I refer to the electrical reproduction of sound.[1]
>
> Edward W. Kellogg, General Electric Research Lab, 1930

I INTRODUCTION: OPENING NIGHT AT RADIO CITY

It was cold and rainy in New York on the night of 27 December 1932, but that didn't prevent a large crowd from gathering at the corner of 50th Street and 6th Avenue. Six thousand had come to witness the grand opening of Radio City Music Hall, and many others turned out hoping to catch a glimpse of the rich and famous as they entered the building. The doors opened at 7:30 P.M., and those fortunate enough to hold tickets entered the theater through a narrow hallway, then emerged into the foyer, which stretched 140 feet toward the grand staircase at its far end. (See figure 6.1.) None of the austere, technologically pure modernism of the PSFS Building was to be found here. Instead, Radio City Music Hall was flamboyantly Moderne, an Art Deco dream in which "Beaux-Arts monumentality is wedded to jazz cubism and the Hollywood stage set."[2] The Music Hall offered its guests a glimpse of "sophisticated life lived among skyscrapers,"[3] and on opening night the sophisticates themselves were out in force.

An NBC radio announcer was stationed in the lobby and he described to distant listeners the arrival of John D. Rockefeller, Jr., whose wealth had funded the new Music Hall as part of Rockefeller Center. Former New York governor Al Smith soon followed, as did aviatrix Amelia Earhart, comedian Charlie Chaplin, prize-fighter Gene Tunney, conductor Leopold Stokowski, and thousands of others. Some of the stars stopped by "to say a word to the radio audi-

6.1
Rockefeller Center, Radio City Music Hall Lobby, view from balcony, c. 1934. The ornate interior of Radio City Music Hall, which contrasts sharply with the austere modernism of the PSFS Building, was characterized as an Art Deco dream in which "Beaux-Arts monumentality is wedded to jazz cubism and the Hollywood stage set." Photograph, n.d., Museum of the City of New York, The Wurts Collection.

ence" on their way in, and Mayor-elect John O'Brien went on for so long he had to be pulled away from the microphone.[4] The hubbub of arriving guests, the noise and confusion of the traffic outside, the crowd of onlookers, and the police overseeing them were described to millions of listeners far removed from the event. Those distant listeners were vicariously present through the modern machinations of electroacoustic technology.

The show itself was not broadcast, so the radio audience was left behind in the lobby as the guests moved into the auditorium and found their seats. Their attention was immediately drawn to the series of immense, telescoping arches that made up the walls and ceiling of the auditorium. (See figures 6.2 and 6.3.) "The hall has a mighty, swift sweep," architectural critic Douglas Haskell explained. "It has focus and energy. The focus is the great proscenium arch, over sixty feet high and one hundred feet wide, a huge semicircular void, filled, at the moment, by the folds of a golden curtain. From that the energy disperses."[5]

The golden curtain finally rose at 9:00 P.M., or, rather, it danced. Thirteen motors controlled its folds and contours as the fabric undulated to the music of Rimsky-Korsakov in a "Symphony of the Curtains." Patriotic music from the mighty Wurlitzer organ followed; the acrobatic Wallenda Troupe tumbled; Fräulein Vera Schwartz sang Johann Strauss's "Liebeswalzer"; the Tuskegee Institute Choir offered gospel tunes amidst "clouds of Wagnerian steam"[6]; Ray Bolger clowned; forty-eight nimble-legged "Roxyettes" kicked; the Martha Graham Ballet interpreted a Greek tragedy; and five hours after the curtain had risen, the classic schtick of "old-time"[7] vaudeville comedians Joe Weber and Lew Fields finally brought the inaugural program to a close.

Critics subsequently panned the show for being long and dull, and the *New York Times* condemned it as the "product of a radio and motion-picture mind."[8] The remark was a gibe at the show's producer, Samuel "Roxy" Rothafel, who was renowned for managing deluxe motion picture palaces in which elaborate live stage shows (regularly broadcast on radio) preceded the presentation of the films.[9] While the *Times* blamed radio and motion pictures for the dramatic failure of the spectacle, those same technologies were equally responsible for its acoustical success. For, in spite of the unprecedented size of the Music Hall, reviewers unanimously concluded that everyone could hear "quite well, even from the seats furthest from the stage."[10] What the audience heard, however, was not the natural voices of the performers, but their reproduction as rendered by loudspeakers concealed behind the golden grilles of the magnificent ceiling arches. Radio City Music Hall was wired for sound, and no one seemed to mind.

6.2
Radio City Music Hall, view of stage, c. 1933. Although the shape of the proscenium suggests expanding waves of sound, the huge arches were actually made of sound-absorbing plaster. Loudspeakers were hidden behind the grilles that were integrated into the arches. Photograph, n.d., Museum of the City of New York, Theater Collection.

6.3
View of Radio City Music Hall 6,200-seat auditorium, c. 1933. The reverberation of the vast auditorium was minimized by the use of sound-absorbing plaster for the ceiling arches and by a highly absorptive covering on the rear wall, ensuring clear and distinct reception of sound throughout the hall. Photograph, n.d., Museum of the City of New York, Gift of Charles B. MacDonald, 50.326.44.

The deployment of microphones and loudspeakers into the soundscape occurred gradually but persistently over the course of the 1920s. Devices first developed in scientific laboratories as tools to study sound now became mass-marketed products that provided listeners with an expanding array of new acoustical commodities. In the home, electrically amplified phonographs and radio loudspeakers became increasingly popular sources of aural entertainment. Public address systems and talking motion pictures transformed public spaces for listening. By 1932, it was customary for people to gather and listen to loudspeakers broadcasting reproduced sound; this is why the electrically generated sound in Radio City Music Hall was so unremarkable.

That sound would not have been satisfactory, however, if the new technology had not been deployed in tandem with that more traditional tool of acoustical control, sound-absorbing building materials. The dramatic arches that constituted the envelope of the auditorium may have looked like expanding waves of sound energy, but they were, in fact, constructed of sound-absorbing plaster. That plaster minimized the reverberation in the hall and ensured that each member of the audience enjoyed distinct and direct reception of the sound signals emanating from the loudspeakers.

In its powerful combination of architectural and electrical control over sound, Radio City Music Hall represents a culmination of the modern soundscape. Within its walls, the age-old "mysteries of the acoustic" were finally and fully revealed by modern acoustical technologies. A forlorn architect had evoked those mysteries in a letter to Wallace Sabine many years before, but that frustration was now replaced by a pervasive sense of mastery. Roxy, who ruled over Radio City as absolutely as Henry Higginson had over Symphony Hall, predicted that the acoustics of his hall would be "perfect,"[11] and no one questioned his confidence in this result. Just as Roxy's confidence contrasted with the tentative attitude of those who first gathered to listen in Symphony Hall, so, too, did the sound of Roxy's hall differ from its turn-of-the-century predecessor. By returning to performance spaces, and by charting the transformations that occurred within them, the architectural and electrical construction of this new modern sound will be fully elaborated.

Radios, electrically amplified phonographs, public address systems, and sound motion pictures transformed the soundscape by introducing auditors not only to electrically reproduced sound but also to new ways of listening. As people self-consciously consumed these new products they became increasingly "sound conscious,"[12] and the sound that they sought was of a particular type. Clear and focused, it issued directly toward them with little opportunity to reflect and reverberate off the surfaces of the room in which it was generated. Indeed, the sound of space was effectively eliminated from the new modern sound as reverberation came to be considered an impediment, a noise that only interfered with the successful transmission and reception of the desired sound signal.

But this modern sound was not simply the outcome, or output, of new electroacoustic technologies; it was also heard in rooms for live performance that were not wired for sound. Well before application of the new electrical technologies had become widespread, acousticians had begun to promote new acoustical criteria that minimized the significance of reverberation and emphasized the direct transmission and clear reception of sound. The modern spaces that embodied these new standards—from the Eastman Theatre to the Hollywood Bowl—thus produced sounds much like those increasingly being reproduced via microphones and loudspeakers.

Most Americans encountered this modern sound most frequently, however, in auditoriums that were wired for sound, particularly in the sound motion picture theaters that proliferated after 1927. The motion picture industry played a crucial role in defining and disseminating the new sound, and the evolution of acoustical technologies in theaters and studios demonstrates how architectural acoustics and electroacoustics gradually merged. Physically as well as conceptually, the distinction between sound in space and sound signals in circuits fell away, as acousticians and sound engineers sought to achieve ever greater degrees of control.

As sound engineers grew adept in the new techniques of electrical recording, they learned to employ those techniques to create artificially the sound of space that had been banished from the studio itself. The "virtual space" (as we might call it today) that they created was not, however, associated with the real architecture of studio or theater, but instead represented the fictional space inhabited by the characters in the program being broadcast or filmed. The sound track itself constituted a new site in which the sound of space could be constructed and manipulated to a degree not fully attainable in the architectural world. Even so, the desire for direct, nonreverberant sound was pervasive, and sound engineers exercised their new power with discretion, creating distinctive virtual spaces only occasionally as "sound effects."

The modern soundscape that resulted from all these developments in the science and practice of architectural acoustics and electroacoustics was, by 1930, ubiquitous. It differed so significantly from its predecessor that the very foundation of architectural acoustics had to be reformulated in order to characterize accurately the new aural environment. Wallace Sabine's reverberation equation had constituted the first significant and successful effort to control the behavior of sound in rooms, and it had stimulated an extensive development of the science and technology of architectural acoustics in the decades that followed. By 1930, the success and extent of that development were such that Sabine's equation no longer described the modern world of rooms filled with modern sound. Sabine's formula was revised, and with this revision, the transformation of the soundscape was complete.

II Listening to Loudspeakers: The Electroacoustic Soundscape

In 1876, Alexander Graham Bell's telephone announced the arrival of electrically reproduced sound. This new, technologically mediated sound immediately reconfigured traditional relationships between sound and space.[13] The telephone—like the telegraph before it—was heralded for "annihilating" space and time, by effectively eradicating the physical distance between people who wished to communicate, and by transmitting their communications across space virtually instantaneously.[14] Yet, geographic space was not the only kind of space annihilated by the telephone.

When two people converse face-to-face, the sound is modified as it passes from speaker to listener. This modification is the result not only of the distance between them (which affects the volume or loudness of sound), but also by the acoustical character of the space that they inhabit (which affects the quality of sound). Little such spatial modification occurred when people began to converse over the telephone. In order for a telephone conversation to be audible, the transmitter had to be held close to the speaker's mouth and the receiver adjacent to the listener's ear; thus telephonic sounds did not fully occupy architectural space as did the sounds of an ordinary conversation.[15] It was as if the telephonic conversants were speaking directly and intimately into each others' ears, oblivious to not only the distance between them, but also the space around them.[16] When the sound of that space did intrude (for example, with a public telephone in a reverberant location), it was perceived as unwanted noise, much like the electrically generated disturbances and distortions that similarly interfered with the intelligibility of the speech signal. Telephone engineers modified their circuits to eradicate the electrical noise; spatial noise was eliminated by the construction of the soundproof and nonreverberant space of the telephone booth.[17]

Thomas Edison's phonograph appeared just a year after Bell's telephone. Like the telephone, the phonograph introduced people to sounds that had been severed from architectural space, and it taught them to distinguish between desired sound signals and unwanted sounds or noises.[18] Early phonograph recordings were made by speaking directly into the large end of a conical horn. The sound vibrations set in motion a diaphragm positioned at the apex of the horn, and a stylus mounted on the vibrating diaphragm cut an undulating groove into a wax cylinder that revolved beneath it.[19] Since the sound of the voice was channeled directly into the horn, there was little opportunity for the surrounding space to modify that sound before it was recorded onto the record. For phonographic reproduction, the undulating groove of the record was passed under a stylus whose motions were transmitted to a reproducing diaphragm. The moving diaphragm set the surrounding air in motion, re-creating the sound of the original source. The acoustical output of the earliest phonographs—like the electroacoustic output of the telephone—was weak, and listeners often listened through narrow tubes that carried the sound directly into their ears. Thus, here, too, the room in which the listener listened played little role in shaping the character of the sound heard. From start to finish, phonographic sound was isolated as much as possible from any spatial context.

As Bell, Edison, and their colleagues and competitors worked to improve the quality of telephonic and phonographic sound signals and to minimize the interfering effects of noise, others were exploring the technology of radio.[20] At the turn of the century, those who listened to radio transmissions relied upon electroacoustic headsets to render audible the faint signals captured by their homemade receiving apparatus. These headsets, like telephone receivers, converted the electrical signal into sound vibrations and transmitted that sound directly into the listeners' ears. The headsets were identical to those worn by telegraph operators; indeed, early radio was known as "wireless telegraphy" and the signals received were simply the dots and dashes of Morse code. But when continuous wave transmission became possible, the sounds of speech and music were soon being transmitted across the ether and into the ears of eager listeners.[21]

Susan Douglas has examined the different modes of listening associated with radio technology as it evolved over the course of the twentieth century. From the turn of the century until around 1925, the mode was known as "DX-ing," or listening for distant transmissions. The goal was to see "how far" one could hear. Radio listeners, typically boys or young men, designed and manipulated their homemade wireless sets to tune in to distant transmissions. By listening carefully through their headsets, they learned to detect the faint radio signals amid the ever-present static or electromagnetic noise. This kind of listening celebrated the same annihilation of distance that had been heralded with the telephone, and, as with the telephone and phonograph, the sound of the space occupied by the listener played little if any role in the experience. DX-ing also required a mode of listening that kept the distinction between signal and noise constantly in mind.

While the telephone remained a device for person-to-person conversation and therefore maintained its intimate contact with users' mouths and ears, radio and the acoustical phonograph were soon modified to allow their re-created sounds to fill the rooms in which they were heard, enabling communal listening. For the phonograph, this was accomplished by the use of a reproducing horn. Inverting the function of the recording horn, the reproducing horn picked up the faint sound vibrations given off by the reproducing diaphragm and effectively amplified those vibrations so that the resulting volume was sufficient for a number of people to listen together. The flowery horn of the phonograph soon became its most recognizable feature, until a new concealed-horn style of cabinet, introduced in 1906 as the Victor "Victrola," became standard.[22]

In 1907, *Littel's Living Age* described a collector who endeavored "to possess perfect specimens of the recording art. To this man the class of record is immaterial, his aim being only records which for clearness, volume, and quality of tone are absolutely faultless."[23] To this man and others like him, consuming sound quality was more compelling than listening to music. He derived pleasure from knowing that he had obtained the clearest and best-sounding reproduction possible, and his consummate taste enabled him to avoid the noises that characterized the inferior records that he had rejected. Competition among phonograph manufacturers was intense, and advertising campaigns encouraged all consumers to engage in such critical listening to determine which brand of phonograph offered the best sound.

The Edison Company preferred to compare its sound, not to that of competing machines, but rather to the sound of live music itself. From 1915 to 1926, the company sponsored Tone Tests, recitals in which phonographic "re-creations" of musicians, as reproduced by the Edison Diamond Disc Phonograph, were compared directly to live performances by those same musicians. In auditoriums and concert halls across the nation, curious crowds gathered to engage in a very public kind of critical listening. (See figure 6.4.) Opinions may have varied as to whether or not the Diamond Disc re-creation was truly indistinguishable from the original, but more important, Tone Test audiences universally accepted the premise of comparison. The act of listening to reproductions was implicitly accepted as culturally equivalent to the act of listening to live performers.[24] The establishment of this equivalence was no small accomplishment; for years, the reproduced melodies of the phonograph had been disparaged as "canned music," mechanically preserved products that had more in common with a tin of sardines than with live music.[25] Tone Tests demonstrated, and perhaps helped bring about, a new willingness to accept these reproductions as an authentic aspect of musical culture. The tests also emphasized the importance of critical listening; an inattentive auditor who was not committed to careful, evaluative listening would not be able to distinguish, then obtain, the best possible sound. As countless phonograph ads made clear, such persons were bound to suffer—musically and socially—for their neglect.[26]

Tone Testing reached its peak of popularity around 1920, when over two thousand recitals were presented across the nation, including one at Carnegie Hall in New York. Subsequently, the number of events, as well as the attention paid to them, declined, and in 1926 the campaign was discontinued. By then,

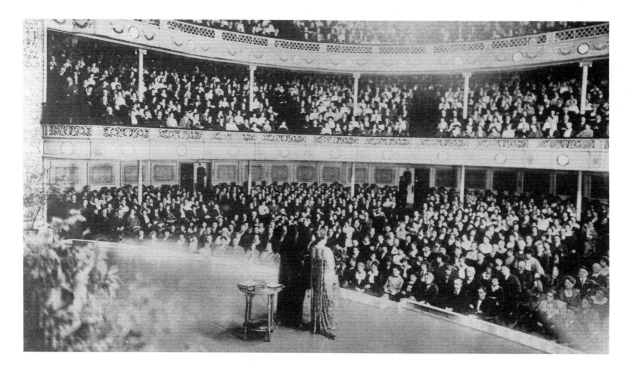

6.4
Operatic soprano Marie Rappold performing a Tone Test Recital with the Edison Diamond Disc Phonograph at Carnegie Music Hall, Pittsburgh, 1919. The audience was challenged by the Edison Company to distinguish Rappold's live voice from its reproduction by the Diamond Disc. United States Department of the Interior, National Park Service, Edison National Historic Site.

the novelty had worn off. More significantly, consumers were now far more interested in listening to the electrically generated sound of radio.

By 1925, radio receivers were no longer complicated contraptions whereby solitary auditors listened through headsets to intermittently broadcast signals. A rapidly growing industry now mass-produced products that any consumer—even the most technologically uninformed—could purchase, take home, and enjoy. A handsome cabinet concealed the tubes, wires, and other technological trappings. Tuning was still a skill that had to be acquired, but innovations in vacuum-tube technology and circuitry made this task easier and additionally improved the quality of the sound signals received.[27]

These improved receivers were accompanied by new sources of transmission. As commercial radio stations were established, beginning with Pittsburgh's KDKA in 1920, regularly programmed entertainment was broadcast to listeners across the nation. The first programmers simply played phonograph records into telephonic transmitters, but soon live musicians were being brought into the studio to perform into high-quality carbon and condenser microphones. The equipment was also taken out of the studio and set up in hotel ballrooms and nightclubs to broadcast the performances of jazz bands and dance orchestras. The result was that listeners at home heard a reproduced but "live" signal that offered a currency and connectedness to other listeners that even the most up-to-date phonograph record was perceived to lack.

Radio listeners were not only acoustically connected to distant companions simultaneously enjoying the same program, they were now also able to share that program with others in the immediacy of their own home. The old headsets were replaced by electroacoustic loudspeakers that projected the sound out into the room, enabling an entire family to listen together.[28] The earliest type of loudspeaker appeared around 1921 and consisted of a small electromagnetic receiver, like that found in a telephone earpiece, attached to a gooseknecked horn. This model was soon accompanied by the "cone-type" loudspeaker, an electromagnetically driven paper diaphragm that was capable of filling a room with sound without the assistance of any horn.[29] (See figure 6.5.) Although the new loudspeakers now projected the sound out into the space of a living room or parlor, listeners preferred to sit close to their speakers, in order to receive as much of the direct sound output as possible. In doing so, they minimized the effect of the architectural locale upon their listening experience.

Loudspeakers did not simply amplify reproduced sound; they also added their own characteristic to the reproduction, and people generally enjoyed this

6.5

Radio shop in Peekskill, N.Y., c. 1925. Horn and diaphragm models of radio loudspeakers, as well as headsets, were sold here. The gooseneck horn sits on top of the receiver on the counter at the center of the image. A moving diaphragm, or cone, speaker is visible on top of the glass case to the right, with another inside the case on the bottom shelf. A headset is displayed on the fashionably modern mannequin head. George H. Clark Collection, Archives Center, National Museum of American History, Smithsonian Institution, SI negative #92-16437.

new kind of sound. The phonograph industry was inundated by a "flood of radio-generated public demand for more bass, more volume,"[30] and it responded by applying electroacoustic technologies to its own products. The techniques of electrical recording and reproduction developed at Bell Laboratories in 1925 were licensed by the Victor, Columbia, and Brunswick phonograph companies, and microphones replaced the recording horn in the studio. In the home, an electromagnetic pick-up replaced the reproducing diaphragm, a loudspeaker took the place of the horn, and the phonograph now offered the same "smooth, uninterrupted flow of sound" that radio listeners had come to love.[31] A 1927 advertisement for the Orthophonic Victrola described the new sound as "Vivid! Lifelike! As radically different as the modern motor-car in comparison to the 'horseless carriage.' And the new Orthophonic Victor Records, recorded by microphone, have a *character of tone* that is pleasing beyond description. Rich. Round. Mellow."[32] Edison had earlier boasted that his Diamond Disc phonograph had no tone of its own to distort the sound of the music recorded on its records.[33] With the new electrical phonographs, the characteristic qualities of electroacoustic reproduction became a desired feature, a commodity to be experienced and enjoyed.

Even as they transformed the habits and goals of domestic listening, loudspeakers were increasingly employed at sites for public listening. On 27 August 1928, for example, when Leon Theremin and his students performed before 12,000 people at Lewisohn Stadium in New York, the Theremin-Voxes on which they performed were equipped with "massive" loudspeakers. While music critics were wary of the potential of these new instruments for "practically unlimited volume," the *New York Times* indicated that the audience responded enthusiastically to the "loud full tones with a radio sound similar to a movie theatre vitaphone."[34] By 1928, stadium audiences were accustomed to hearing "radio sound" emitted from loudspeakers, as the use of public address systems for large gatherings of all sorts was now well established. And, as the *Times* acknowledged, movie audiences were also now encountering the sound of loudspeakers as Vitaphone, a new sound motion picture system, was transforming the movie-going experience.

Public address, or P.A., systems and Vitaphone sound movies were developed by scientists and engineers at AT&T as part of a strategy to expand the corporation's product line beyond telephony to encompass as many new electroacoustical sound products as possible. P.A. systems employed the same vacuum-tube amplifier that AT&T researchers had devised for use in radio and long-distance telephone transmission. Military applications of P.A. systems were explored during the First World War, and civilian uses for the technology were promoted soon after the war's end. Newsworthy events, including Warren Harding's presidential inauguration, were captured by microphonic receivers at their source, transmitted electrically over long-distance telephone lines, and then broadcast via loudspeakers to large crowds gathered at public sites in distant cities. The systems found numerous other more local applications and, by 1922, Western Electric was selling and installing P.A. systems anywhere that sound amplification was desired, including sports stadiums and ball parks, racetracks, convention halls, hotels, department stores, and large churches.[35]

Theater directors also found the systems useful. In 1922, Roxy Rothafel used a Western Electric P.A. system to direct rehearsals of his famous musical reviews.[36] Three years later, the now-improved system sounded good enough for the director to consider employing it during the show itself. With customary hyperbole Roxy proclaimed, "Acoustics no longer present a problem, since the amplification system, with which we are now experimenting, will carry the voice and will send it perfectly almost any distance within reason, and certainly a distance greater than could be found in any theater."[37] Hyperbole soon became reality, and by 1929 Roxy was using the system to manipulate the balance between the string sections of his orchestra during the performance, as well as to enhance reception by the audience throughout the vast auditorium of the Roxy Theatre.[38] In 1932, Roxy's shows in the even larger Radio City Music Hall depended on a similar kind of sound system to broadcast their sounds to the huge audience assembled in the hall.

P.A. systems were also used by motion picture directors to instruct large crowds of extras during the filming of silent films. Previously, directors had shouted into enormous megaphones or created elaborate chains of command whereby instructions were transmitted, by gunshot, semaphore, or telegraph, to cadres of assistant directors scattered throughout the field of action. D. W. Griffith turned to signalmen from the United States Signal Corps to coordinate the large battle scenes in his 1915 epic, *The Birth of a Nation*. In 1923, Wallace Worsley became the first motion picture director to put the new Western Electric P.A. system to use as he shot *The Hunchback of Notre Dame*. Curiosity about the new system attracted other directors to the *Hunchback* shoot, and the visitors were impressed by what they heard there. Before long, the amplified commands of dictatorial directors were echoing across studio backlots all over Hollywood.[39] But the telephone company had far greater ambitions for transforming moviemaking, and its engineers now turned to the long-standing challenge of making the movies themselves talk.

Thomas Edison's earliest ideas for creating moving pictures had been stimulated by his invention of the phonograph, and he had intended from the very start to synchronize his images with recorded sounds.[40] Turning this idea into a working technology proved difficult, however. Only after years of work, with the considerable input of his assistant William K. L. Dickson, and with the abandonment of the idea of synchronized sound, was Edison able to achieve his goal of making pictures move.[41]

In April 1894, the world's first Kinetoscope Parlor opened. A former shoe store at 1155 Broadway in New York was now outfitted with ten of Edison's new motion picture machines. Each "peep show" Kinetoscope contained a twenty-second loop of film that customers viewed individually for a nickel a shot. Strongman Eugene Sandow flexed his muscles in one machine; in another, blacksmiths (Edison's own machinists) hammered a piece of iron and shared a bottle of beer. Other fare included a barber shaving a bearded customer, the contortions of Madame Bartholdi, and a pair of fighting roosters.

The novelty was a tremendous success, and exhibitors were soon placing the machines in bars, amusement parks, and arcades across the nation. Rival devices appeared, too, including the peep-show Mutoscope, in which the customer turned a crank to flip rapidly through a series of postcardlike photographs. The public developed a voracious appetite for moving images, and a new industry was born as producers photographed virtually anything that moved—from famous actors to risqué dancers to boxing cats—to meet the seemingly incessant demand.

Within a year, however, the novelty had worn off. Edison attempted to reinvigorate the business by returning to his idea of pairing the picture with sound. With the Kinetophone, a customer peered through the standard viewfinder and listened to the sound of an accompanying phonograph through a set of ear tubes. No synchronization was attempted, and the sound consisted of little more than background music. The films themselves were no different from the standard Kinetoscope fare, and the public not surprisingly failed to respond with enthusiasm to the new device.[42]

The nascent industry was rejuvenated not by sound, but by projection. In France, Louis and Auguste Lumière developed a means by which to project motion pictures onto a large screen, and by the end of 1895 they were offering regular screenings to paying customers in the basement of a Parisian café. The Edison Company's new Vitascope presented the first commercial projection of motion pictures in America in New York on 23 April 1896.[43] Moving images projected onto a large screen, and viewed in the company of others, left a far greater impression upon an audience than did the tiny, individually experienced peep-show images, and with projection, a new and permanent class of popular entertainment was established.

With projection, however, the challenge of providing synchronized sound became even more challenging. Now, there was not only the difficulty of maintaining synchronization between sound and image, but also the problem of providing sound loud enough for everyone in the theater to hear. Some enterprising impresarios avoided these problems by concealing behind the screen live actors who spoke and sang along with the characters projected onto it.[44] But numerous other inventors in Europe and America confronted the dual challenges of synchronization and amplification, and a variety of sound motion picture systems appeared in the first two decades of the century. None was a commercial success.

As early as 1902, Leon Gaumont's Chronophone presented films of French music hall performers who declaimed very loudly into a recording phonograph that was located just out of camera range. Gaumont initially depended on two phonographs to provide sufficient sound in the theater, but in 1913 he turned instead to a phonograph whose output was magnified by a compressed-air amplifier. Early Chronophone demonstrations were generally well received, but the system was not economically viable for exhibitors. A trained operator was required to maintain synchronization between sound and image by constantly manipulating the speed of the projector to match the record. This labor was expensive and seldom up to the task, and the few exhibitors who tried the Chronophone soon dropped it from their programs.[45]

A similar system, the Cameraphone, was developed in America around 1906. The Cameraphone technique used phonographic recordings made in advance of the cinematography. During filming, the performers lip-synched their performance to match the record. Large-horned phonographs were employed in the theater to achieve maximum volume, but, as with the Chronophone, it was difficult and expensive to keep the sound in sync with the image, and the Cameraphone company went out of business in 1910.[46]

Edison himself tried one last time to marry his two inventions. A mechanically amplified phonograph playing large-diameter cylinders was tenuously linked to a projector via belts and pulleys; while initially impressive, Edison's system ultimately proved as vulnerable as others to the loss of synchronization. At the Kinetophone's debut in February 1913, the audience was "literally spellbound," but subsequent screenings were far less successful. Synchronization came and went, the amplifier amplified the surface noise of the record as well as the voices recorded upon it, and within a month *Variety* branded the Kinetophone "The Sensation That Failed."[47]

After this failure, the motion picture industry basically gave up on the idea of synchronized sound. If Edison himself couldn't make the movies talk, who could? Besides, the public clamored for silent films; why change an already successful product? The impetus to continue experiments now came, not from the industry itself, but from outsiders, electrical inventors and manufacturers who were not already benefitting from the success of silent films, and who had not been discouraged by previous attempts to add sound to them. These men realized that the vacuum-tube amplifiers and loudspeakers currently being used in long-distance telephony, radio, and public address could provide high-quality amplification of sound in a motion picture theater. All that was required was to find a means of maintaining synchronization between the image and the medium on which the sound was recorded.

Lee de Forest, whose audion tube was the basis for all forms of electroacoustic amplification, began experimenting around 1913 with a means to record sound onto photographic film. He developed a variant of his audion amplifier called the photion, which enabled him to generate an optical image of an electroacoustical signal. Inventor Theodore Case improved upon de Forest's design and devised a means by which to reverse the process, thereby re-creating the sound that had originally been recorded on film.[48]

Case and de Forest ultimately developed a system that provided synchronized and amplified sound, and the De Forest Phonofilm Corporation was formed in 1924 with Case as a partner. Several dozen theater owners were persuaded by de Forest to install his equipment and present the short sound films that Phonofilm produced. These films—typically musical numbers by vaudeville performers—met with mixed reviews, but cranky critics were soon the least of the inventors' worries. De Forest pursued creative financial strategies to generate operating income for Phonofilm, and he soon ran afoul of the United States Department of Justice. Case left the organization, taking with him the patents for his own contributions to the system, and de Forest's company went bankrupt in 1926.[49]

Simultaneous with the efforts of de Forest and Case, AT&T and General Electric—both of whom shared legal access to the technology of vacuum-tube amplification—also began to explore the development of sound pictures. GE researcher Charles Hoxie devised his own version of an optical sound recording system and euphoniously dubbed it the Pallophotophone. When the Radio Corporation of America was created in 1919 by merging the radio-related resources of GE and Westinghouse, the Pallophotophone was put to use to record music and speech for delayed radio broadcast. The company chose not to pursue its application to motion pictures.[50]

Unlike RCA, the telephone company was interested in moving into the movie business. Even as Western Electric's P.A. systems were finding their way onto Hollywood back lots, the company had begun to explore how best to make sound motion pictures. Experiments were made with both sound-on-film and sound-on-disc, but the Western Electric engineers chose to focus on discs, taking advantage of the recording skills they had developed when they electrified the phonograph. A means of maintaining synchronization between camera, phonograph, and projector was devised, and by 1924, salesmen were demonstrating the system to Hollywood's biggest players. But in 1924 no one was interested. Virtually all of the leaders in the industry had long since dismissed the viability of sound pictures, and the phone company was not about to change their minds. While Paramount, Metro-Goldwyn-Mayer, and other first-tier studios all closed their ears to the new technology, a second-tier outfit run by four brothers named Warner chose instead to listen.

In 1924, Warner Brothers was a small but ambitious studio whose biggest asset was the canine action hero Rin Tin Tin. The studio had, however, recently initiated an aggressive campaign to become a dominant player in the production, distribution, and exhibition of films. As a part of this campaign, Warner Brothers purchased a radio station in Los Angeles, and Sam, the most technically minded of the brothers, supervised its operation as a medium of publicity for the studio. When shown the Western Electric sound film system, Sam liked what he heard and convinced his brothers that this was how the studio could make a name for itself. Sam proposed that they use recorded sound to replace the live music heard in their theaters. Short films of Broadway's best vaudevillians could replace the less-than-stellar local fare offered in provincial theaters, and recorded orchestral scores for feature films could similarly replace the variable quality of musical accompaniment that was rendered in each individual house. By offering a standardized and high-quality musical program, Warner Brothers could transform every Warner theater—no matter how small—into the equivalent of a "first-run" house and thus make their mark on the industry.

Warner Brothers and Western Electric joined forces in 1925 to form the Vitaphone Corporation, and on 6 August 1926, Vitaphone presented its first program at the Warner Theatre in New York. A brief address by Will Hays, president of the Motion Picture Producers and Distributors of America, opened the show. The image of the motion picture czar appeared on screen, and when his image audibly rapped its knuckles on the table in front of him, he immediately captured the audience's attention. Hays's talking image described how Vitaphone would inaugurate "a new era in music and motion pictures,"[51] and his address was followed by a series of "high-class" musical shorts. The New York Philharmonic performed Wagner's Overture to *Tannhäuser*, violinist Efrem Zimbalist and pianist Harold Bauer performed Beethoven's *Kreutzer* Sonata, and numerous other stars performed on screen and synchronized disc for the audience. Best received by far was tenor Giovanni Martinelli's dynamic rendition of "Vesti la giubba." The Vitaphone shorts were followed by the feature attraction, John Barrymore's *Don Juan*, a silent swashbuckler that was accompanied by a recorded, synchronized score of symphonic music with sound effects.[52]

Musical shorts followed by a sync-scored feature also made up the second Vitaphone program, and this time the performances of George Jessel and Al Jolson stole the show. Warner's competitors took note of the growing success of these films, but most producers remained convinced that Vitaphone was nothing more than a fad. Al Jolson's subsequent Vitaphone feature, *The Jazz Singer* (1927), would force them to reevaluate this opinion.

In *The Jazz Singer*, musical shorts by Jolson himself were effectively inserted into a nontalking, sync-scored melodramatic feature. But when Jolson's character briefly conversed with his mother before bursting into song in one such segment, the possibilities of truly talking films suddenly became obvious.[53] Over the next year, Warner Brothers released a series of "part-talking" films, and the percentage of talking footage gradually increased until, in October 1928, they could advertise *The Lights of New York* as the first "100% talking" feature film.

By 1928, Hollywood finally realized that this new sound technology would not fade away like its predecessors. RCA offered a sound-on-film system called Photophone to compete with Western Electric's sound-on-disc, and producer William Fox was turning out newsreels and feature films with synchronized sound provided by Theodore Case. Production of talking films increased dramatically during 1928 as studios frantically raced to build new soundstages, install new sound equipment, and learn how to operate it. The number of theaters wired for sound grew, too, as exhibitors were now eager to present the popular new films. By 1932, only 2 percent of America's theaters remained silent.[54]

Western Electric emphasized the connection between sound pictures and its older electroacoustic technologies by proclaiming the new technology "a product of the Telephone." RCA similarly designated its sound films as "Radio Pictures" to highlight their connection to its own electroacoustic products of the past.[55] But the transition to sound in the movies was strikingly abrupt, and it focused peoples' attention in a way that these earlier technologies had not. The celebratory publicity and intense competition surrounding the different systems led listeners to listen more closely than ever before. Audiences critically consumed these new products as they developed "the listening habit" as an important new element of their "modern life."[56]

The new habits of modern listeners were not simply a response to new technologies, however, and the sounds that they so carefully evaluated were not exclusively the output of electroacoustic devices. The same kind of sound to which they listened intently in the cinema was also encountered in places where no microphones, amplifiers, or loudspeakers could be found. Here, in modern auditoriums and concert halls, that same clear, direct, and nonreverberant sound was strictly the result of architectural construction.

III THE MODERN AUDITORIUM

Acousticians began to promote a new "ideal"[57] type of auditorium in the 1920s, and architects simultaneously made that ideal a reality. The new auditorium was low and wide, "spatulate"[58] or fan-shaped, with diverging side walls spreading out from a small stage area to form an increasingly wide seating area. The ceiling rose toward the rear to accommodate a balcony or two. The stage area was constructed of reflective materials, but the auditorium itself was highly absorbent. The acoustical result was that performers on stage effectively occupied the apex of a large horn. The sound that the audience received issued directly from the horn, or was perhaps once-reflected off the side walls. There was little opportunity for reverberation to develop, as the shape and material constitution of the new auditorium were designed "to blend and unify the music at its source and then transmit this music efficiently and uniformly throughout the extended seating area."[59] Efficient transmission—a primary goal in electroacoustical design—was equally valued in the realm of auditorium design.

Real examples of this ideal type include the Eastman Theatre in Rochester (Gordon & Kaelber, 1923); the Chicago Civic Opera Auditorium (Graham, Anderson, Probst & White, 1930); Severance Hall in Cleveland (Walter & Weeks, 1930); and the Kleinhans Music Hall in Buffalo (F. J. & W. A. Kidd with Eliel Saarinen, 1940). Numerous college and innumerable high school auditoriums also followed the trend. Describing the Kleinhans Music Hall in 1962, Leo Beranek wrote, "Listening to music there is rather like listening to a very fine FM-stereophonic reproducing system in a carpeted living room."[60] Historian Michael Forsyth has developed Beranek's characterization, identifying auditoriums built in America after 1925 as "Hi-Fi Concert Halls." Their sound, according to Forsyth, is sharp and lucid, much like a "'front-row' close-to-microphone recording."[61]

But the modern auditorium was more than a conscious or unconscious attempt to simulate architecturally the sound of electrically reproduced music. While the popularity of this type of auditorium was certainly reinforced by the similarity of its sound to that of the new electroacoustic technologies, its origins preceded the diffusion of those technologies. The historical development of this auditorium was the result of other factors, including the widespread use of sound-absorbing materials, the desire to eliminate noise and reverberation, scientific research on the intelligibility of speech, and enthusiasm for outdoor sound. These factors simultaneously influenced developments in architectural acoustics and in electroacoustics; thus it should not be surprising that the results should sound so similar.

The Eastman Theatre in Rochester, New York, was one of the first notable examples of a modern auditorium. Built in 1923 to serve as a concert hall for the Eastman School of Music, the theater was also intended to operate as a luxury cinema whose income would help support the new music school.[62] The irregularity of the corner lot led the architects to develop a fan-shaped plan, unusual for a concert hall but common for cinemas in 1923.[63] (See figure 6.6.)

Floyd Watson served as acoustical consultant for the project, and he initially expressed concern that seats in the mezzanine balcony would receive insufficient sound. The opening to the mezzanine was restricted by the gallery balcony above, and he predicted that the small absorbent space would suffer acoustically. "On completion of the theater, however," Watson discovered, "the reception of music on this floor was thought superior to other locations."[64] Auditors seated there liked the effect of the direct and once-reflected sound as it projected

6.6
Plan of the Eastman Theatre and School of Music, Rochester, N.Y. (Gordon & Kaelber, 1923). The fan-shaped form of the auditorium, while unusual for a concert hall, was typical for motion picture theaters at this time. *American Architect and Architectural Review* 123 (28 February 1923): 197.

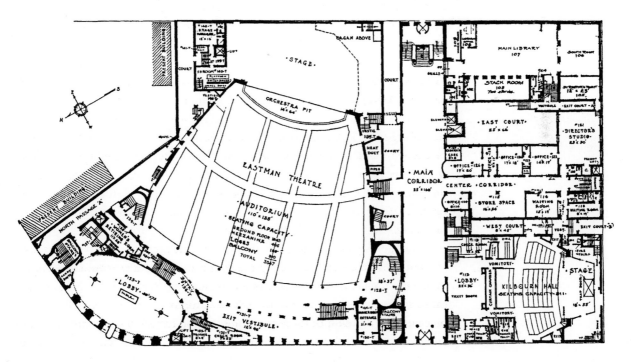

6.7
Eastman Theatre, Rochester, N.Y. (Gordon & Kaelber, 1923), rear balconies as seen from the stage. Acoustical consultant Floyd Watson initially feared that the narrow opening to the mezzanine balcony would cause those seats to suffer acoustically, but they were soon judged to be the best in the house. *American Architect and Architectural Review* 123 (28 February 1923): Pl. 53. Courtesy Marquand Library of Art and Archaeology, Princeton University.

through the narrow opening into the highly absorbent space, and these seats soon became known as the best in the house.[65] (See figures 6.7 and 6.8.)

The acoustical success of the mezzanine balcony at the Eastman Theatre was unexpected and cannot be attributed to the influence of electroacoustic technologies. In 1923, loudspeakers for radio and public address were just beginning to be heard, and sound movies were little more than failed experiments. Vern Knudsen later suggested that the new criteria exemplified in the Eastman Theatre were the result of "the enormous increase in the use of absorptive materials in all sorts of rooms, thus conditioning or predisposing people to non-reverberant rooms."[66] It is also conceivable, if difficult to document, that people increasingly surrounded by a sea of city noise would seek out a different kind of sound in places where they chose to listen, as in a concert hall. The clear, directional flow of sound experienced in the Eastman Theatre differed remarkably from the omnidirectional aural chaos of city streets. Perhaps people took comfort in that highly controlled sound. For whatever reason, people liked the sound of this space and Watson began to promote a new type of auditorium that would create this sound throughout the house.

6.8 Subscribers' Mezzanine Balcony, Eastman Theatre, Rochester, N.Y. (Gordon & Kaelber, 1923). Audience members seated at this location enjoyed the quality of sound in this highly absorbent space, and George Eastman himself sat at the center of this balcony when he attended performances in the theater. *American Architect and Architectural Review* 123 (28 February 1923): Pl. 60. Courtesy Marquand Library of Art and Archaeology, Princeton University.

The relatively small opening to the Eastman mezzanine balcony rendered it not unlike a separate but acoustically coupled room, and Watson began to conceptualize the concert hall as a combination of two different rooms, one for performers and another for auditors. He compared the acoustical conditions preferred by performing musicians to those preferred by listeners, and concluded that "conditions in the same room must be quite different for playing and listening."[67] Whereas performers worked best in a reverberant environment that blended and reinforced their efforts, listeners in the 1920s were happiest in a far more absorptive environment. Watson thus suggested that "the generation of sound should be done in a room more or less separated from the main auditorium, while the listening is best in the latter room with a sound deadened interior."[68] He proposed effecting this acoustical separation by leaving the stage end of the hall "live" or reflective, and concentrating sound-absorbing materials toward the rear, where they would surround the audience in a reverberation-muffling blanket.

The result was a concert hall that provided a sound similar to that beginning to be heard via electroacoustic technologies in the home. When Watson considered the performance of live music in the home itself, the parallel was even more striking. Here, he proposed an actual physical separation between musicians and auditors. Recommending how best to create an arrangement for domestic music, he suggested that "the listeners would find a better effect in adjacent rooms connected with the studio by an open door."[69] By removing the live performers from view, and by channeling their sound through a restricted opening (a doorway) prior to its reception by the audience, Watson's prescription neglected the shared pleasures of live performance and listening, and instead emphasized the attentive but detached mode of listening associated with sound-reproducing technologies like phonographs and radios. The physical separation of music producers from consumers, a separation first made possible with sound recording devices, was re-created in this domestic proposal. It was equally present, if less visually explicit, in his recommendations for auditorium design.

In addition to reconfiguring the acoustical relationship between performers and listeners, Watson also revised his recommendations for the optimal reverberation time of an auditorium. When he first published his textbook *Acoustics of Buildings* in 1923, Watson had recommended an average reverberation time of over three seconds for a room whose volume was one million cubic feet. In the second edition of his text, published in 1930, he lowered that figure to two seconds. (See figure 6.9.) "Some years ago," Watson wrote in 1930, "the author published the curves shown in Fig. 15, and used them extensively in the correction of acoustics. Experience indicates, however, that shorter times of reverberation than given in Fig. 15 produce better results, and a later graph of 'optimum' times was advocated, as shown in Fig. 16."[70]

Vern Knudsen later recalled the "unmistakable trend toward shorter reverberation times" that began in the 1920s, and he contributed to that trend along with his colleague Watson. Referring to a room of 500,000 cubic feet used for both speech and music, Knudsen noted that Watson, circa 1920, had recommended 2.6 seconds as optimal reverberation. In 1923, Watson lowered his recommendation to 1.9 seconds, and in 1932, Knudsen himself advocated a reverberation time as low as 1.5 seconds.[71] Knudsen's recommendations drew not only on his own experience as an acoustical consultant, but also on research that he was carrying out on the effect of reverberation upon the intelligibility of speech.

Knudsen's experiments focused on the determination of "percent articulation," a measure of the degree to which speech is understood by a listener, and he investigated how intelligibility rose as reverberation decreased. The technique for measuring it was borrowed by Knudsen from the telephone industry, where the procedure had been developed to rate the quality of telephone lines and other electroacoustic transmission systems. Speakers recited nonsensical (but phonetically significant) sequences of syllables into the system being tested. Auditors at

6.9
Charts indicating the drop in optimum reverberation times recommended by Floyd Watson between 1923 and 1930. According to his figure 15, originally published in 1923, the average acceptable reverberation time for a room of volume 1,000,000 cubic feet was just over three seconds. In 1930, he reduced this value to slightly over two seconds, as indicated in his figure 16. Floyd Watson, *Acoustics of Buildings* (New York: John Wiley and Sons, 1930), pp. 35, 36.

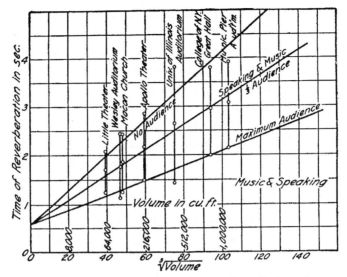

FIG. 15. Acceptable time of reverberation for auditoriums of different volume.

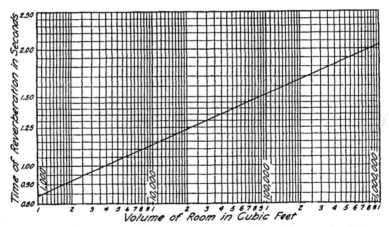

FIG. 16. Plot showing optimum time of reverberation for different auditoriums.

the receiving end of that system wrote down what they heard, and their results were compared against the original to calculate the percentage of correctly perceived articulations. The tests thus measured the effect of electrical noise and distortion upon the transmission and reception of a telephonic speech signal.[72]

Knudsen transferred this test to the realm of architectural acoustics, and in this context, the noise or distortion was simply the reverberation of the room.

He determined that the percent articulation in a room decreased 6 percent for every added second of reverberation, and concluded that the optimum time of reverberation for speech was considerably less than current practice acknowledged. Knudsen proposed that, in rooms for speech, reverberation should always be kept below one second. "Even for music," he claimed, "there seem to be no physical factors which would warrant a time of reverberation much in excess of 1.0 second."[73]

Knudsen's research led him to argue in 1926 that the best environment for listening was the reverberation-free outdoors. *Science* reported upon this "contradiction of the widespread idea that a properly constructed auditorium reinforces and improves audibility," and the journal declared that auditorium walls were nothing but "a necessary nuisance."[74] Floyd Watson echoed this sentiment, asserting that "reflected sound could be omitted entirely without vital consequence—a conclusion that is quite contrary to the usual conception of auditorium acoustics, where the reflecting walls are supposed to be quite beneficial."[75]

This surprising conclusion, that "the auditorium should be made as dead as outdoors for the benefit of the auditors,"[76] stimulated a growing interest in outdoor acoustics. Watson opened his 1928 article, "Ideal Auditorium Acoustics," with a drawing of the ancient amphitheater at Ostia, and Vern Knudsen also included a section on Greek and Roman theaters in his 1932 textbook.[77] A 1929 article in the *Architectural Forum* noted that a "rapidly increasing number of outdoor auditoriums" were being built, and Knudsen himself helped build the most famous of all.[78]

The Hollywood Bowl was a natural amphitheater located in the Bolton Canyon amid the hills of Hollywood outside Los Angeles. The site housed its first concert event in 1920, with musicians performing from a simple wooden platform and listeners gathered on crude wooden benches. A series of temporary orchestra shells were constructed over the next seven years, including a striking design by the architect Lloyd Wright made up of a series of concentric elliptical arches. Wright's ellipses were replaced in 1929 with a permanent structure designed by the engineering firm of Elliot, Bowen & Walz, with Vern Knudsen serving as acoustical consultant. The new shell was similar to Wright's, but its nine concentric arches were now semicircular and were made not of wood but of transite, a mixture of cement and asbestos, formed over a steel frame.[79] (See figures 6.10 and 6.11.)

The transite that composed the arches was hard and reflective, "to add greatly to the sound projection qualities of the structure,"[80] and each arch was

6.10
Hollywood Bowl Orchestra Shell and Grounds, 1929. The natural amphitheater formed by the Bolton Canyon in the Hollywood hills was first used for concerts in 1920. Over the next decade, a series of temporary stage shells were constructed, and the permanent shell shown here was built in 1929, with Vern Knudsen serving as the acoustical consultant. Courtesy Edmund D. Edelman Hollywood Bowl Museum.

6.11
Graduation Ceremony (probably Hollywood High School) in the orchestra shell of the Hollywood Bowl, c. 1930. The curves of the shell were formed of transite, a mixture of cement and asbestos, over a steel frame. Each curve was angled to direct the sound out toward the vast audience gathered on the surrounding hillside. Courtesy Edmund D. Edelman Hollywood Bowl Museum.

inclined at just the right angle to reflect the sound out toward the audience on the upwardly sloping grounds. Knudsen's goal was to provide "a pronounced directional flow of sound toward the audience," to ensure that the "myriads of attentive people" gathered in the bowl could all hear clearly and distinctly.[81] Considering that the bowl held as many as twenty thousand attentive listeners, and that those in the most remote seats sat over five hundred feet away from the musicians on stage, achieving this goal was a considerable challenge.

While some criticized the new shell, complaining that certain seats still received insufficient or unbalanced sound and that the transite and steel arches resulted in "metallic and strident" tones, it was nonetheless celebrated for its "utterly echo-less and amplifying traits."[82] "The faintest tones of the violin are clearly audible in the most remote seats," one reviewer claimed, adding, "The acoustics of the Bowl are enthusiastically praised by musical critics."[83]

The sound of the Hollywood Bowl, with its pronounced directional flow and its echoless and amplifying traits, constitutes another example of the modern sound that was now being presented to auditors by auditoriums, amphitheaters, and loudspeakers alike. Hollywood would increasingly be associated with this new sound, but not via its connection to the Hollywood Bowl, nor to any other venue for live music. Hollywood was the headquarters of the motion picture industry and it was within the walls of the motion picture theater that most Americans were exposed to the new sound.

IV Architectural Electroacoustics: Theater and Studio Design

The silent cinema had never really been silent; it had always been filled with sound.[84] Kinetoscopes were viewed amid the clatter and din of the amusement parlor, and the earliest theaters for projected motion pictures were just as noisy and chaotic. Many people saw their first projected films as part of the bill of fare at a local vaudeville theater. Others viewed them in makeshift storefront theaters leased by itinerant showmen. A draped sheet served as the screen, and folding chairs were the extent of accommodation. In spite of such spartan surroundings, the excitement of seeing something new packed these houses night after night. By the turn of the century, when it was clear that the movies were more than a fad, exhibitors began to set up more permanent facilities and accommodations gradually began to improve. Nondescript storefronts were transformed into alluring portals to paradise with prefabricated facades of sculpted terra cotta or stamped tin, gaudily festooned with electric lights. Interiors were enhanced, and architectural journals began to publish guidelines for seating arrangements, sight lines, ventilation, projection booth layout, and other aspects of motion picture theater design.[85]

As attention to architectural accommodations increased, so, too, did that paid to the provision of music in the theater. The monotonous din of a player piano had filled the storefront cinemas with sound, but patrons now expected more for their price of admission. Some exhibitors included live musical acts as part of the show. The audience itself contributed, too, when illustrated song-slides were projected to guide sing-alongs of sentimental favorites.[86] Finally, music to accompany and enhance the material depicted on screen became an integral part of the program. Local pianists improvised scores in the smaller houses, while larger theaters employed an organist or a small orchestra. Musicians fortunate enough to work in houses featuring powerful Wurlitzer or Marr & Colton organs could create different sounds and moods with the push of a button or the pull of a stop. These instruments additionally provided an arsenal of sound effects, from bells and sirens to gunshots.[87]

Still, the early-twentieth-century motion picture theater hardly encouraged rapt, attentive listening. The program of numerous short films ran continuously, and audience members came and went, constantly and noisily, throughout the program.[88] The music was often raucous, inappropriate, or both.[89] As the creative ambitions of producers and exhibitors grew, however, this situation would change. On the production side, the short one-reelers grew into multireeled features that could last two hours or longer. Rich character portrayals and complex stories now unwound along with the celluloid, drawing the viewer into an increasingly compelling world of fantasy. Producers began to invest heavily in elaborate stage sets and exotic on-location shooting to achieve an unprecedented degree of spectacle on screen.

As the films themselves became more sensational, so, too, did the theaters in which they appeared. "Picture palaces" in the larger cities rivaled the on-screen spectacles for extravagance, offering their patrons richly upholstered seats, smoking lounges, and liveried attendants in addition to the entertainment that appeared on screen. Throughout the teens and twenties, architects like Thomas Lamb, Rapp & Rapp, Meyer & Holler, and John Eberson created Chinese pagodas, Egyptian temples, and Italian villas out of stucco, plaster, velvet, and gilt. Perhaps the most fantastic were the "atmospheric" theaters of Eberson. Here, the screen was surrounded by a stage-set-like construction that created the effect of a Mediterranean garden, a Middle Eastern village, or some other exotic outdoor locale. The theater was surmounted by a smoothly curving plaster ceiling that, while plain in itself, was illuminated during the show with rich blue hues to effect a night sky. Special light projectors wafted clouds and twinkling stars across the heavens to complete the illusion.

The managers of these picture palaces, men like Sid Grauman in Los Angeles and Roxy Rothafel in New York, took pride not just in their architectural surroundings, but also in the elaborate live productions that showcased the films they exhibited. Organ preludes, orchestral overtures, guest soloists, and elaborate "ballets" opened each night's program. Regal musical directors like Rouben Mamoulian and Hugo Reisenfeld not only led large orchestras of talented musicians, but also composed and compiled unique scores to accompany each new feature film.

Of course, only a small number of theaters in large cities could offer such musical amenities. Still, there was a "trickle-down" effect that improved the quality of music offered in more typical neighborhood theaters. Famed music directors published guidebooks that helped less-talented musicians create effective accompaniments to films.[90] As individual exhibitors expanded their theatrical empires into regional and national chains, the musical resources of their first-run flagship theaters became available to their less urbane second-run houses. Film scores were passed along, and in a few cases, the live productions actually became road shows that traveled into the hinterlands along with the feature films they showcased.[91]

Producers as well as exhibitors worked to improve the quality of music in the theaters. As early as 1909, the Edison and Vitagraph Companies had offered suggestions for music to accompany their films. With the rise of the feature film, production companies began to provide detailed cue-sheets, not only suggesting songs or themes, but also indicating the precise points in the film at which these themes should enter and exit. For the most significant features, a complete and original musical score was commissioned and distributed to exhibitors.[92]

In spite of the increasing attention paid to music, little such attention was paid to the acoustics of theaters until the arrival of sound film in the late 1920s. Floyd Watson noted that "the necessity for adjusting the acoustics of theaters has not arisen so often nor so seriously as in the case of churches and other auditoriums," and the published record confirms his conclusion.[93] The earliest articles on theater design said little, if anything, about acoustics.[94] Roxy Rothafel's experiments with his P.A. system in the early 1920s suggest that, in the largest houses, it may have been difficult to produce a volume of sound sufficient to reach all seats, but this problem was not significant enough to provoke discussion in articles and guidebooks on theater architecture. With the advent of sound movies, however, all this would change.

"The telephone rings. 'Long distance calling. Smithtown, Palatial Theatre. New installation of talking picture a failure owing to bad acoustics. Advice necessary at once or house must close.'" "Such," declared acoustical consultant Clifford Swan, "is the typical S.O.S. call for help."[95] By 1929, according to Swan, the problem of acoustics had become "insistent," as previously silent theaters, in which "the question of hearing was not a matter to consider," were wired for sound.[96] Countless theaters across the nation were suddenly discovered to be acoustically deficient, and consultants like Swan found a wide new field in which to exercise their expertise.[97] As one observer put it, "The film being no longer silent, the acoustic expert must be heard."[98]

The film industry initially turned to academic consultants like Watson and Knudsen, or to men associated with the acoustical materials industry like Swan, to provide that expertise. But the companies that manufactured and sold the electroacoustic sound film systems soon undertook to obtain that knowledge for themselves, and before long, acoustical consulting became yet another of the many sound products they offered for sale.

The first and foremost of these organizations was Electrical Research Products Incorporated (ERPI), the company that leased, installed, and serviced the Western Electric sound picture systems. ERPI was a wholly owned subsidiary of Western Electric established in January 1927 to handle this new business. The original personnel was recruited largely from Western Electric and AT&T, but the company grew rapidly and incoming classes of sales and service engineers came not only from the Bell System, but also from radio manufacturers, power and light companies, and other related industries.[99]

Initially, the electrically minded ERPI engineers focused their attention upon the sound equipment itself, and architectural acoustics was mentioned only briefly in the instruction provided to new installation engineers.[100] But it quickly became clear that acoustical expertise was required to ensure a successful installation, and in February 1929, training instructor S.K. Wolf was reassigned to the theater engineering group to lead research in architectural acoustics. Wolf traveled to the Riverbank and Burgess Laboratories to study firsthand the "latest developments in the field of acoustical research."[101] He hired academic acousticians as well as experts from the building materials industry, and by October his technical staff of nine had not only begun numerous fundamental investigations but had also analyzed, and recommended alterations to, over 300 theaters. Several of the large theater chains arranged to have Wolf's staff examine all of their plans for new theaters, and prominent theater architects also availed themselves of this ERPI service. By December, Wolf's men were reviewing the acoustics of 75 theaters, old and new, every week.[102]

Most of ERPI's work on theater acoustics was dedicated to rendering the old, so-called silent theaters suitable for the new sound equipment. The acoustical survey became an integral aspect of the work of ERPI installation engineers stationed across the nation and around the world. The company newsletter, *Erpigram*, explained the procedure:

> A complete acoustical survey of the theatre is first made by the Installation engineer who is assigned to make the regular survey. Written reports of this survey are then sent to the acoustic engineers in the home office [Wolf's group] who analyze them to determine the acoustic values of the house, and to draw up recommendations for treatment when needed.
>
> In making the surveys, engineers are required to determine the exact volume and seating capacity, nature and thickness and amount of draping and decorating material used in the theatre, exact nature of all seats and furniture, etc. Also included is a noise survey and recommendations for eliminating all noises in the house. So complete is this survey, the report covers five pages and either accurate sketches or architects' drawings must be included in the survey reports.[103]

ERPI engineers, outfitted like big game hunters or members of some expeditionary force, "went on the warpath with a full complement of weapons to banish the bogy Silence and his near relation General Reverberation." "Each man," the *Erpigram* explained, "has been supplied with a large fibre knapsack in which to carry his equipment. Among other things, it contains a steel tape so that he may measure a house, and the structure with which he comes in contact will have to be analyzed for hidden horrors, such as 'plaster backed by brick,' [and] 'leather covered seats, filled with straw.'" The kit also contained a cap pistol, to "hunt out Reverberation, and his Echoes, and banish him from the theater."[104]

It was immediately evident that the problem in the old theaters was too much reverberation. The metropolitan movie palaces may have suffered less in spite of their large size, as their drapes, carpets, and well-stuffed upholstery would have created an absorbent environment, but, as Clifford Swan noted, the majority of theaters were "mere barren halls with plaster walls and ceiling, wood or concrete floors, and bare wood seats."[105] The audience itself provided the only significantly absorptive surface, and the ERPI engineer called upon to correct such houses would "eagerly watch the door, and every time an additional person enters optimistically mutter to himself, 'Here comes four and seven-tenths units more.'"[106]

If it was clear that too much reverberation was plaguing sound movie theaters, it was not immediately obvious what constituted an optimal reverberation time for these rooms. All previous research, dating back to Wallace Sabine's experiments at the New England Conservatory of Music, had considered only live music and speech. The technology of sound reproduction fundamentally changed the situation, and one of first tasks undertaken by Wolf's group, as well as by others, was to reevaluate the role of reverberation and to determine new optimum reverberation times for rooms that were wired for sound.

In 1930, a researcher at the acoustical laboratory of the General Electric Company discussed "Some New Aspects of Reverberation" before the Society of Motion Picture Engineers. Edward Kellogg identified three primary contributions of reverberation to the acoustics of live-performance spaces: It served to build up and thus increase the total volume of sound in the room; it mixed the elements of sound present at any given instant (for example, the various instruments of an orchestra); and it caused sounds produced sequentially in time to overlap with each other. The first two functions were beneficial and the last, Kellogg asserted, was strictly detrimental. Traditional prescriptions for optimal reverberation times for auditoriums thus were a compromise between the good and bad roles that reverberation played.[107]

In auditoriums that were wired for sound reproduction, however, no such compromise was required. An appropriate level of loudness could be achieved simply by adjusting the gain of the amplifiers. Proper positioning of the highly directional loud speakers further ensured that listeners located in even the most distant seats would receive a sufficient volume of sound. Nor was reverberation required to mix or blend the sounds; this mixing, Kellogg pointed out, was already accomplished during the recording process. The only role left for reverberation was to cause the overlapping of sounds, a role best eliminated. "So far as we can see, then," Kellogg concluded, "there is practically nothing which auditorium reverberation accomplishes which cannot be secured in a highly damped auditorium by other means," and Kellogg recommended that auditoriums for reproduced sound should be designed with maximum possible absorption. In other words, his optimum reverberation time was zero.[108]

Kellogg's pronouncements were extreme, but his conclusions differed from others' only in degree, not substance. S. K. Wolf, too, emphasized that the electrical amplification of sound rendered unnecessary any dependence on reverberation to achieve sufficient loudness. He also noted that the presence of studio reverberation on the recording itself decreased the need for theater-generated reverberation. Both factors indicated that optimum reverberation time in theaters for sound reproduction be considerably lower than that for live performance spaces, and Wolf recommended a difference of about 0.25 seconds.[109] Actual recommendations ranged from about 1.25 seconds for a theater of 175,000 cubic feet (a seating capacity of around 1,200 people) to 1.75 seconds for a theater of 1,000,000 cubic feet (the very largest, with a capacity of about 6,000 people).[110]

Even if reverberation in theaters were only to be reduced and not eliminated, that reduction was still significant. The rooms were generally overreverberant to begin with, and the goal was now to reduce the reverberation time to as little as one second.[111] To bring about this transformation, large quantities of sound-absorbing materials were introduced. Upholstered seats were chosen to effect the same absorption as the people who filled them, so that reverberation would remain constant whether the house was full or not. Drapes and tapestries were hung in some theaters, acoustical plasters were applied in others, and sound-absorbing materials like Celotex were installed on walls and ceilings just about everywhere.[112] The cost of "correcting" a motion picture theater could be considerable, and this cost was in addition to the expense of acquiring the sound equipment itself. While large theater chains could absorb these expenses, independent operators were hard-pressed to finance such expenditures. As a result, the already declining role of the independent exhibitor in the motion picture industry declined even further.[113]

In theaters that were successfully altered, the sound was "beamed"[114] directly out at the audience by highly directional loudspeakers located up front, typically behind the screen. This sound had much in common with the electrical signal that was its source. As theaters were wired for sound, the distinction between the architectural space of the auditorium and the electrical circuitry that transmitted the signal into that space began to fall away, until it ultimately became difficult to determine where the signals ended and the sounds began.

ERPI engineer G. T. Stanton, for example, defined the motion picture auditorium as "a system for transmission of sound." As he described it, the theater was fundamentally no different from the telephone, radio, or any other such system, and his criteria for evaluating auditorium performance were the same as for those electrical systems. To Stanton, the sound—whether in the circuits or in the architectural space of the theater—was a signal, a carrier of information whose goal was to arrive efficiently and accurately at its final destination.[115] For Edward Kellogg, the architectural and electrical systems merged in ways not just conceptual. In 1931, Kellogg proposed a new type of theater loudspeaker in which the speaker driver was to be mounted in the corner of a room, and the three surfaces of the room that emerged from that corner would serve as the horn of the speaker. The room itself thus became the loudspeaker's horn, as architecture and electroacoustic technology merged seamlessly into one continuous system of transmission.[116] (See figure 6.12.) While architectural acoustics and electroacoustics began to merge, physically and conceptually, in the sound motion picture theater, that merger would occur even more dramatically in the sound studios.

Recording studios date back to the origins of phonographic recording, but here, too, little attention was paid to room acoustics until electricity entered the scene. With preelectric, or acoustic, recording, musicians were placed as close as possible to the horn that collected their sounds and channeled them to the

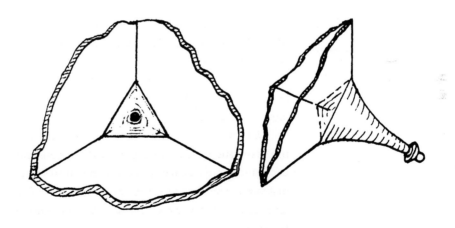

6.12
The convergence of architectural acoustics and electroacoustics is demonstrated in Edward Kellogg's design for a loudspeaker whose horn was to merge with the walls of the room in which it was installed. Reprinted with permission from Edward W. Kellogg, "Means for Radiating Large Amounts of Low Frequency Sound," *Journal of the Acoustical Society of America* 3 (July 1931): 106. © 1931, Acoustical Society of America.

recording apparatus. Solo performers stood directly in front of the horn. Small ensembles of musicians were necessarily further away, but still crowded as proximately as possible. Larger groups, such as symphonic orchestras, were virtually impossible to record successfully.[117] The goal of acoustic recording was to capture as much of the direct sound energy of the performance as possible, and there was little discussion about capturing (or eliminating) the sound of the studio itself. Extant photos and drawings of early recording studios offer little evidence of any significant effort to control the acoustic character of the rooms. Recording quality was controlled primarily through the selection of different sizes and shapes of horns, and through the arrangement of musicians with respect to the horn.[118] (See figure 6.13.) With the advent of radio broadcasting in the early 1920s, however, and with the electrification of phonographic recording, the acoustic properties of the studio suddenly became significant.

Microphones immediately freed the musicians in the studio from the cramped spatial arrangements that acoustic recording had necessitated. Now, electrical amplifiers ensured adequate sound intensity. An appropriate balance between instruments was achieved not through the awkward placement of musicians, but through the use of multiple microphones and mixing consoles in which the signals from those microphones were blended and balanced electrically. If the physical space of elbow room was no longer a problem in the electrified studio, however, acoustical space was. The earliest microphones were omnidirectional, "listening" in all directions at once. They thus captured the reflected as well as the direct sounds of the musicians, and electrical recordings therefore included the reverberatory character of the studio to a degree that acoustic recordings had not.

Some perceived this new characteristic as a move toward greater realism and fidelity; it made a record sound more like a live performance heard in a concert hall. Others were troubled by the layering of different acoustical spaces that occurred when recorded reverberation was reproduced in a room that additionally contributed its own acoustical character. Even proponents of recorded room sound realized that a little reverberation went a long way, however, and electrified studios were soon swaddled with sound-absorbing materials.[119] (See figure 6.14.)

In 1928, Paul Sabine recalled that the "early practice" in electroacoustic studio design had been "to cut down sound reflection to the limit." "Gradually," he noted, "the tendency toward less deadening and longer reverberation times has grown up." But "longer" was clearly a relative term here; Sabine described an

6.13
Acoustic recording session at the Edison studio in New York City, 1912. The differently shaped horns on the wall and floor were used to control the quality of the recording. The recording phonograph, not visible here, was located behind the barrier at the far left of the image. Musicians were arranged in space to balance their sounds on the recording. The black partitions may have been covered with sound-absorbing material, but such materials were not widely used in acoustic recording studios. United States Department of the Interior, National Park Service, Edison National Historic Site.

6.14
The KDKA broadcast studio in Pittsburgh, heavily draped for sound absorption. A microphone hangs from a boom to the right of the piano. One wonders if the creaking of wicker rocking chairs created problems for the sound engineers who worked here. D.G. Little, "KDKA: The Radio Telephone Broadcasting Station of the Westinghouse Electric and Manufacturing Company at East Pittsburgh, Pennsylvania," *Proceedings of the Institute of Radio Engineers* 12 (June 1924): 273. © 1924 IRE, now IEEE.

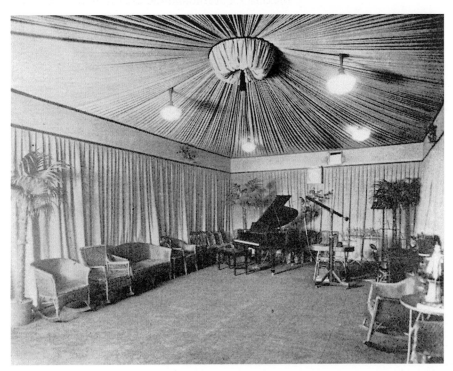

experiment he had carried out for radio station WLS in Chicago in 1926, to find out what conditions were preferred by listeners. An identical program was broadcast three times from a studio whose reverberation was varied from 0.25 to 0.64 seconds. While the listeners indicated a preference for the program with the greatest amount of reverberation, that amount—just 0.64 seconds—could hardly be considered "live."[120]

Joseph Maxfield, too, was a proponent of recorded room sound, particularly for recordings of orchestras, where reverberation constituted part of the "musical and artistic effect."[121] Nonetheless, Maxfield argued that a studio for recording should still be considerably less reverberant than a room intended for listening to live performances. He explained that extra damping was required to compensate for the fact that the monaural microphone in the studio detected sound differently from the binaural human listener. The "one-eared" microphone perceived more reverberation in a given space than did a two-eared person; thus the absorptivity of a space had to be increased so that the recorded signal would not sound excessively reverberant when later heard by human listeners.[122] As acousticians like Watson and Knudsen were lowering their recommendations for optimum reverberation in live performance spaces, and as theater consultants like Wolf were recommending reduced reverberation in spaces for the reproduction of sound, studio consultants were recommending even less reverberation for the spaces in which sound was recorded. Other than the soundproof, anechoic laboratories that were constructed for scientific research, these studios were the most absorptive spaces around, with recommended reverberation times falling well below one second.[123]

In addition to eliminating virtually all of the reverberatory sounds within the studio, it was just as critical to keep extraneous noise out. In 1928, the new NBC studios in New York exemplified state-of-the-art design for sound absorption and isolation. The problem of broadcast studio design was, as architect Raymond Hood put it, "as modern as a problem could be." "About the technical side there could be no discussion. We were to work with their engineers to make the studio as sound-proof and as acoustically perfect as possible."[124] The NBC studios employed floating construction in which the walls, ceilings, and floors were all mechanically isolated from the surrounding structure to prevent the transmission of sound.[125] Observation windows were double- and triple-glazed, and heavy doors were lined with airtight rubber gaskets to create a "hermetically sealed"[126] environment. If the hermetic seal evoked in advertising for the PSFS Building had been metaphoric, the term was applied far more literally to the new electroacoustic studios. In such airtight surroundings, artificial ventilation was a necessity, and the requisite air-conditioning systems were carefully designed for silent operation. All machinery was kept distant from the studio site and mechanically isolated, and air ducts were lined, inside and out, with sound-absorbing materials so that noise would not travel into the studios along with the cool air.

By 1928, just as the challenge of broadcast studio design appeared to have been successfully met, an even greater challenge arose. Studios for sound motion pictures required an even greater degree of acoustical control. They had to provide this control in a much larger space, and they had to do so in a way that did not interfere with the visual aspects of film production. While soundstage designers could thus draw upon the principles of design developed for radio and phonograph studios, distinctly new problems had to be addressed.

The first Vitaphone production facility was the old Vitagraph motion picture studio in Brooklyn. Warner Brothers had acquired the property in 1925 and they chose to begin their experiments with sound here, close by the scientists and engineers at Bell Laboratories. The need for a soundproof location, isolated from the noises of the city, was quickly made evident, but little could be done here except to record at times when such noise was at a minimum. The first sound recordings made in the Vitagraph studio also suffered from distinct echoes and excessive reverberation, so carpets were taken out of the prop room and heavy cloth was draped around the set to absorb as much sound as possible.[127]

Vitaphone soon relocated to the Manhattan Opera House, Oscar Hammerstein's old theater on 34th Street at 7th Avenue, and the musical shorts that premiered with *Don Juan* were produced here, as was *The Voice from the Screen,* a documentary produced for the New York Electrical Society by Bell Labs to explain and demonstrate the new sound pictures. As at the Vitagraph studio, city noises intruded and the theater was draped to reduce its reverberation.[128] (See figure 6.15.) When other motion picture producers entered the sound scene, they, too, established facilities in or near New York, to be close to the voices of Broadway and the Metropolitan Opera, as well as the sound engineers in Manhattan and New Jersey. By 1927, however, Warner Brothers had already begun to relocate its operations to new soundstages in Hollywood, and the other studios soon followed. By 1929, virtually all of the major producers were building new studios in and around Los Angeles, and they depended on acoustical experts to ensure that these structures were both soundproof and nonreverberant.[129]

6.15

Recording a Vitaphone sound motion picture in the Manhattan Opera House, New York, 1926. This photo was taken during the making of the short film *The Voice from the Screen* produced by Bell Laboratories to demonstrate the new technology. The soundproof camera booth was left open to show the camera's operation, and three suspended microphones recorded the process of recording as well as the performance of musicians. Drapes reduced the reverberation, and the megaphone at the feet of Bell Labs vice president Edward B. Craft could have been used only to command silence on the set. Photo #W4991. Property of AT&T Archives. Reprinted with permission of AT&T.

Vern Knudsen—fortuitously located at UCLA—recalled being called in to the executive offices of Metro-Goldwyn-Mayer in 1928 to consult upon the design of their first soundstages. "We want these two stages, stages A and B," Louis B. Mayer explained, "to be insulated from each other so well that you can have gunfire on one stage and record chamber music on the other stage." "Well," Knudsen replied, "This calls for a very costly type of building." "We don't care," Mayer responded. "We want that; that's the requirement. That *must* be the requirement."[130]

Knudsen supervised the construction of MGM's first two soundstages. They were heavy, rigid structures with ten-inch-thick concrete walls and a concrete slab ceiling to keep out external noise. The studios themselves were located within, but structurally isolated from, this outer shell, and were lined with thick layers of sound-absorbing material. The expense of this design led the studios to search for a cheaper method of construction that would provide the same degree of acoustical control. The use of multiple layers of building materials such as plaster- and fiber-board, separated by air spaces lined with sound-absorbing materials and mechanically isolated from each other, proved equally effective, and this type of building became the industry standard.[131]

There was no debate about optimum reverberation for a soundstage; the goal was to eliminate it entirely. Even Joseph Maxfield agreed that the motion picture studio should be "as dead as possible."[132] While the complete elimination of reverberation was physically impossible, times well below 0.50 seconds were recommended and obtained, even in very large studios.[133] Such low reverberation times were effected by lining the entire stage with a thick blanket—as much as four inches—of sound-absorbing materials.[134] (See figure 6.16.)

Far above the silenced soundstage loomed the monitor booth, a glass-enclosed bay that housed the recording engineer at his mixing panel. (See figure 6.17.) Here, he adjusted and controlled the signals created by the microphones to ensure a high-quality recording. While little physical space was required to perform this task, the room in which he worked required a great deal of acoustical space. In order to create a recording that would sound good in a typical theater, the monitor room had to constitute an acoustical facsimile of such a theater. The room therefore had to be large, treated with acoustical materials to effect a typical theater reverberation time, and outfitted with loudspeakers identical to those used in theaters.[135]

Like the theater, the studio constituted a site where the distinction between architectural acoustics and electroacoustics was blurred, a place where sounds

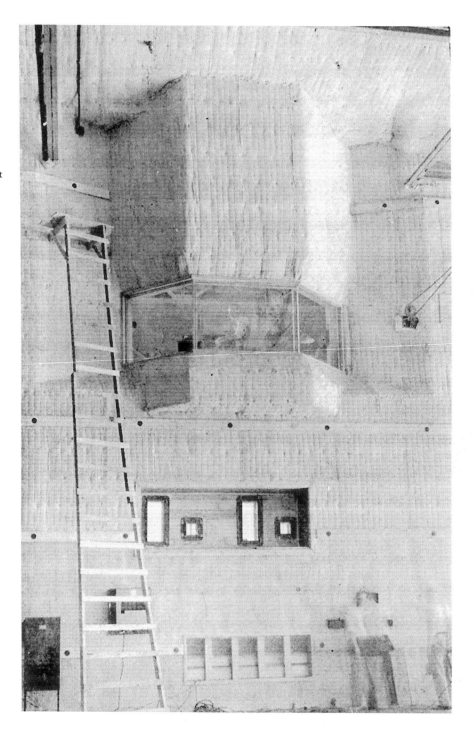

6.16
Metro-Goldwyn-Mayer sound stage, c. 1929. The extensive acoustical treatment evident here would have rendered this large room almost completely nonreverberant. The bay window allowed the sound engineers in the monitoring booth to observe the action on the set below. *Western Electric News* 18 (April 1929): 36. Property of AT&T Archives. Reprinted with permission of AT&T.

6.17

Sound engineer working in the monitoring balcony of an unidentified studio, c. 1929. The engineer balanced the signals from different microphones on the set by manipulating the dials on the mixing console. Telephones allowed him to communicate with people on the set below. Photo #W2085A. Property of AT&T Archives. Reprinted with permission of AT&T.

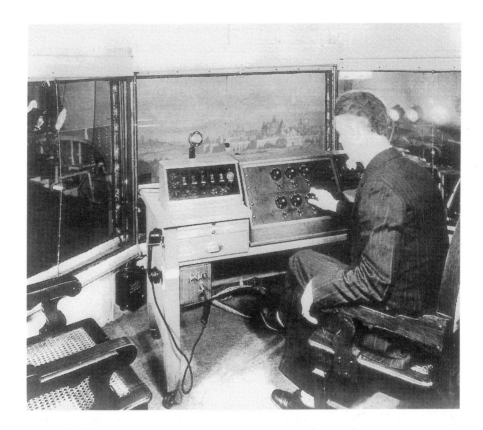

and signals combined and converged. As early as 1924, radio engineers had depicted the architectural space of the studio as a discrete element, like a rectifier or an amplifier, in their circuit schematics.[136] (See figure 6.18.) Multipaned monitor booth windows were compared to electrical filters, blocking the transmission of sound in the same way that those filters blocked the transmission of signals.[137] ERPI engineer H. C. Humphrey even suggested that a special monitoring headset could be designed to re-create, electrically, the acoustical characteristics of the average theater. A simple circuit could then replace the physical space of the monitoring room.[138]

There is no evidence that Humphrey's suggestion was carried out at this time. Still, studio technicians did manipulate electrical technology in other ways to create the effect of architectural space. When Edward Kellogg reevaluated the role of reverberation in motion picture theaters in 1930, his argument for eliminating it was based on the fact that "the desirable effects of reverberation can all be simulated by a high grade electrical system."[139] When Joseph Maxfield

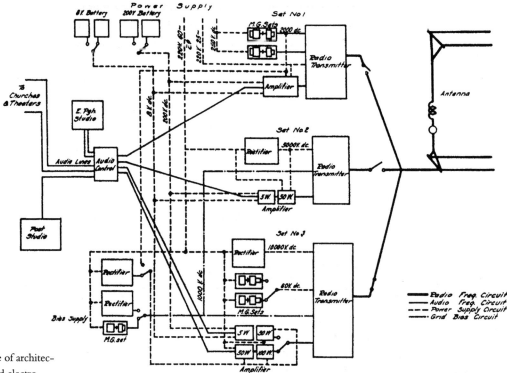

6.18

The convergence of architectural acoustics and electroacoustics is documented in this circuit diagram for the KDKA radio station in Pittsburgh, which represents studio architecture as just another element of the circuitry. See boxes labeled "Post Studio" and "E. Pgh. Studio" toward the left of the diagram. D. G. Little, "KDKA: The Radio Telephone Broadcasting Station of the Westinghouse Electric and Manufacturing Company at East Pittsburgh, Pennsylvania," *Proceedings of the Institute of Radio Engineers* 12 (June 1924): 256. © 1924 IRE, now IEEE.

declared that the soundstage should be as absorbent as possible, he, too, knew that there were other means—electrical means—to create the sound of space.[140]

V Electroacoustic Architecture: Sound Engineers and the Electrical Construction of Space

As acousticians worked to silence the architectural spaces of studios and theaters, sound engineers used their electroacoustic tools to fill that silence with a new kind of sound, the sound of the motion picture sound track. Just what a sound track should sound like, however, was not immediately evident, and the early years of sound film production were filled with debate over how best to answer this question. As MGM sound engineer Wesley Miller frankly admitted, the industry was "groping for an understanding of what is to be expected from the sound product itself."[141] The Society of Motion Picture Engineers, the American Society of Cinematographers, the Academy of Motion Picture Arts and Sciences, and the Acoustical Society of America all served as clearinghouses for ideas and sponsored educational forums to keep everyone abreast of the rapidly changing state of the industry.[142] Between 1926 and 1930, as the nature of sound film and the techniques for creating it rapidly evolved, so, too, did ideas about how best to constitute the sound product.[143]

From 1926 through early 1928, sound movies consisted primarily of filmed renditions of staged musical performances (most notably, the Vitaphone shorts of vaudeville and opera stars), or sync-scored features like *Don Juan*, silent films accompanied by a recorded orchestral score and sound effects. In either case, it was assumed that the goal of recording was simply to re-create the sound of live theater, an aural context appropriate for both the filmed theatrical performers in the shorts and the recorded theater orchestra in the features. Paul Sabine, speaking before the Society of Motion Picture Engineers in 1928, argued that the engineers should strive to achieve "acoustic conditions for recording which will produce a record that most nearly simulates music and speech as heard by an audience from an actual stage."[144] He confidently asserted the ability of acousticians like himself to create those conditions through the techniques of architectural acoustics. But even at this early date, an alternative goal for the sound track as well as alternative means for achieving it were being developed.

In 1928, musical shorts and sync-scored features were suddenly overwhelmed by a new demand for "talking films," as Al Jolson's performance in *The Jazz Singer* captivated audiences and left them eager to hear more. While the film is famous for the brief dialogue that occurs between Jolson's character and his mother, its historical significance also derives from the fact that it moved the sound movie out of the "virtual theater" inhabited by the performers in Vitaphone shorts and by the orchestra members who created the synchronized scores of earlier Vitaphone features. The voice of Jolson's character was indeed heard in a theater, but also in a temple, a restaurant, and his mother's front parlor. With *The Jazz Singer*, the sound track began to move through space, inhabiting the numerous and diverse places that had long been represented visually in silent films.

Warner Brothers' first "all talking" film, *The Lights of New York* (1928), moved its audience around even more, from a small town to the lights and lures of the big city, including Broadway, Central Park, a barber shop, an apartment, and a nightclub "where anything can happen and usually does."[145] Once talking films began to present this variety of acoustical spaces, the goal of simply creating an accurate reproduction of "theater sound" was no longer perceived to be adequate or appropriate. As one engineer now suggested, "The reproduction

6.19
Marquee at the Chaloner Theater, New York, advertising the Western Electric Sound System as "The Voice of Action," 1930. Current and coming attractions, all "100% Talking," include Chester Morris in *Alibi* and Mary Pickford in *Coquette,* her first talking film. Note advertisements for the sound system in the display cases at the left of the image and immediately to the right of the ticket booth. Note also the shadow cast by the tracks of the elevated train. Photo #W1953A. Property of AT&T Archives. Reprinted with permission of AT&T.

should sound the way you would expect the original to sound under the circumstances that are brought to your mind by the illusion created by the picture."[146] That is, a scene set in a large dance hall should sound different from a scene set in a small cottage, or one depicting people outdoors. Others, however, opposed this definition of the sound track, arguing instead that it was more important to maintain continuity of sound quality. According to this view, clarity and uniformity of sound were more important than spatial realism; if a person's voice sounded different in each scene, this would detract from, rather than enhance, the effect of the film.[147]

A theoretical debate about the fundamental role of the sound track was beginning to take shape, but in practice, sound engineers were initially preoccupied with the far more basic task of getting the new equipment to register the voices of the players on the set. Actors were required to stand still and speak directly into immobile microphones that were hidden in props or suspended above the players' heads just out of camera range. Carbon arc and mercury vapor lamps emitted audible and radio-frequency noises that were picked up by the recording equipment, so they had to be replaced with silent incandescent lamps. These new "inkies" were hot enough to melt makeup and to drench performers in perspiration, so the new soundstages now required powerful air-conditioning systems, which, if not properly designed, would themselves introduce mechanical noise. The camera, too, generated noise that was picked up by the microphones, so the camera and cameraman were encased within a small, soundproof booth equipped with a glass window out of which to shoot the image. Techniques for editing sound, on disc or film, were initially impractical, so scenes were shot and recorded in their entirety. If different camera angles of a given scene were required, multiple camera booths had to be set up to film simultaneously, so that each viewpoint would be synchronized to the recording. Cinematographers were thus forced to abandon their more creative lighting techniques, and instead provide flat, uniform lighting that generally served all camera angles at once.

The Lights of New York demonstrates well the many limitations imposed by the equipment and techniques of sound recording circa 1928; the film is infamous for its static camera work, flat lighting, and stolid pace. But it wasn't long before filmmakers and sound engineers found ways to transcend these limitations. Camera booths were placed on rubber wheels so they could be rolled around, and then were completely eliminated when quieter cameras, fitted with "blimps" or close-fitting sound-absorbing blankets, were introduced. By 1929, motion pictures were moving once again, and Western Electric began to advertise its sound system with a new slogan, "The Voice of Action." (See figure 6.19.)

Microphones also proved more mobile than had originally been assumed. The microphone boom appeared simultaneously in several studios, and by 1930 this portable, counterweighted support was standard equipment, allowing an operator to suspend a microphone immediately above the players and to follow them as they moved around the set. But even as the restored mobility of the camera and the newfound mobility of the microphone opened up new visual possibilities for sound films, the basic question remained of just what these films should sound like. The debate over "sound perspective,"[148] the relationship

between an image and its accompanying sound, grew louder as sound engineers gained control over their tools.

One fundamental question concerned how the volume level of the recorded sound should relate to the image on screen when a film cut between long shots, medium shots, and close-ups. If, for example, a woman were shown speaking to a man in a medium shot, and the film then cut to a close-up of her, still talking, should her voice suddenly get louder to match the increased size of her image on screen?[149] If a talking man were filmed gradually walking away from the camera, or if the camera pulled away as he spoke, should his voice level diminish as he receded into the distance? In each of these examples, the point-of-view presented to the audience moved through space; abruptly in the former, gradually in the latter. Whether the point-of-audition should similarly move was a question that had to be answered. The question of whether or not to represent aurally the particular kind of space depicted on screen was also reexamined, as sound engineers considered new means by which to control the amount of reverberation recorded on the sound track.

In 1928, Paul Sabine had confidently volunteered the services of architectural acousticians to control the quality of sound in the new sound films. The traditional, architectural means of control that Sabine proposed were indeed pursued by the motion picture industry, albeit in a slightly modified form. The "architecture" of motion pictures, like everything else associated with the medium, existed more as illusion than reality. Set designers used forced perspective and other tricks to create the visual effect of architectural construction out of flats made of paper, plaster, and two-by-fours. To control the sound quality of this illusory architecture, therefore, one had to control the acoustic properties of the sets out of which these virtual structures were made. Absorption coefficients previously determined by architectural acousticians were applied to the construction of stage sets, but it soon became clear that much of the acoustical data compiled by Wallace Sabine, Floyd Watson, and others was "useless for studio application."[150]

The problem was that these acousticians had measured the coefficients of materials employed in solid and substantial architectural constructions. What the studios required was data relating to how these materials functioned in the far less substantial construction of Hollywood stage sets. To determine these new coefficients, a special committee of the Academy of Motion Picture Arts and Sciences enlisted the services of Vern Knudsen and ERPI engineer F. L. Hopper. Working in Knudsen's new acoustical laboratory at UCLA, the men measured the acoustical performance of different kinds of materials as employed in actual set constructions that were donated by the various studios.[151]

This material approach to the control of sound on the soundstage was, however, soon overshadowed by a new and more powerful means of control. As early as 1928, sound engineers had begun to use the tools and techniques of sound recording itself to create the effect of space. Indeed, even as Paul Sabine was promoting the value of architectural acoustics to the Society of Motion Picture Engineers, motion picture engineer Edward Kellogg steered the discussion away from the material control of sound, citing instead the power of electroacoustic tools to effect this control. "The liveliness of the room can be compensated for," Kellogg proclaimed, "by the position of the microphone." As Joseph Maxfield explained, "If you record only the direct sound, you can get a sound track without reverberation, but with the microphone farther away you get a record with considerable reverberation." One sound engineer suggested even further that multiple microphones could be used simultaneously in a dead room, as "a substitute for the reflecting surfaces."[152]

Others, however, opposed this technique. RCA engineer John Cass objected that, "When a number of microphones are used, the resultant blend of sound may not be said to represent any given point of audition, but is the sound which would be heard by a man with five or six very long ears, said ears extending in various directions."[153] Cass's description brings to mind the technique of cubist painters like Pablo Picasso, in which multiple visual perspectives were simultaneously represented on a single canvas. While Cass clearly opposed the construction of a cubist sound track, something very much like this—a sound track simultaneously everywhere and nowhere—would eventually become the industry standard. Clearly, as Cass, Maxfield, and Kellogg all recognized, the technique of microphone placement constituted a powerful new means by which to create or efface the aural effect of space. The opportunities afforded by the use of multiple microphones were increased even further as techniques for sound mixing, editing, and dubbing, or rerecording, developed.

The role of the "mixer man" in the earliest years of sound film was simply to monitor and control the level of sound being picked up by the microphone. If a voice was too faint or too loud, a turn of the dial on the mixing console would amplify or diminish the strength of the signal to an appropriate level before it was recorded. On sets equipped with several microphones, the mixer additionally had to follow the action, opening up, or activating, the microphone closest to the speaking actors, then closing it off and opening another when the action moved to a different spot on the set. Background music also had to be added to the mix as the recording occurred. Often, a band or orchestra was written into the story so that it could appear on camera; otherwise, the musicians were located off stage and out of camera range. In either case, the mixer mixed the signals from the orchestra microphones with the dialogue signals of the actors as all performed at once. As long as scenes were shot and recorded in continuity, all of this manipulation and fine-tuning of the signal had to occur in real time, as the scene was played out before the cameras and microphones. As early as 1927, however, experiments in sound editing and rerecording had begun, and within a year or two these techniques were highly developed.[154]

It was relatively easy to cut and splice together different "takes" or recordings of sound on film; the challenge was to maintain synchronism with the separate strip of film that carried the image. Around 1930, special-purpose sound-editing consoles appeared to help editors meet this challenge. Soon thereafter, new kinds of film stock with sequentially numbered frames further expedited the process.[155] In addition to piecing together serially several recordings, the signals of multiple sound tracks could also be mixed together to create a new, combination track, as when a dialogue track was mixed with a track of synchronized sound effects or music. Here, the limitation was that, with each new generation, the level of noise inherent to the sound-on-film process increased. Finer-grained film stock helped alleviate this problem, until, in 1931, the aptly named ERPI engineer H. C. Silent designed a new "noiseless" system for sound-on-film recording.[156]

Although sound on disc could not be physically cut and pasted like sound on film, an extraordinarily complicated procedure was developed at Warner Brothers in 1928 to enable engineers to mix and edit disc-recorded sound.[157] But here, too, the noise level increased with each successive generation of reproduction. By the time Western Electric introduced their noiseless sound-on-film recording system, however, every studio in Hollywood—even Warner Brothers—had abandoned discs. Indeed, the increasing importance of editing played a strong role in the adoption of sound on film as the production standard for the industry.

Sound engineers developed techniques not only to add and layer dialogue, music, and sound effects, but also to manipulate the quality of these constituent sounds. They eliminated certain kinds of noise with electrical filters, created sound fades and dissolves to segue one scene into another, and controlled widely ranging volume levels with automatic limiting devices.[158] Dubbing was now dubbed of "supreme importance to the advancement of the art." "It makes possible," sound engineer Joe Coffman declared in 1930, "the improvement of voices and effects through changing their frequency content by use of the requisite filters; it permits almost any imaginable acoustic trick, and the inclusion of effects which occur as afterthoughts." "It is probable," Coffman predicted, "that within a year no original sound records will be used for the making of release prints of feature productions of high quality."[159] Indeed, as historian Donald Crafton has documented, by 1930 the sound track "came to be seen more as an ensemble constructed in postproduction rather than as a record of an acoustical performance."[160]

With this redefinition of the sound track, the task of studio recording was similarly redefined. Although some still argued for a recording technique that produced a "natural" representation of space that would necessarily vary from shot to shot, this approach was now seldom followed in practice. Instead, sound engineers focused almost exclusively on collecting a uniformly "close-up" sound signal. The goal was to capture the actors' voices clearly and directly, and this was accomplished by following the players closely with moving microphones suspended from booms. "When speech is picked up electrically with a microphone," RKO sound engineer Carl Dreher explained, "it is usually possible to secure high quality only by placing the pickup device relatively close to the source of sound." The best procedure, according to Dreher, was thus "to shoot close-up sound only, modifying the quality in re-recording when necessary to simulate more distant pickup for the long shot picture."[161]

When this technique proved impractical, for example, with extreme long shots in which a close microphone would fall within the camera's field of vision, new devices were devised to overcome the obstacles. "Sound concentrators" were developed at RKO in 1930 to enable engineers to obtain close-up sound from a distant source. These large, parabolic reflecting horns collected sound energy from the direction in which they were pointed, and focused that energy on a microphone mounted within the horn, effectively creating a highly directional and sensitive microphone. (See figure 6.20.) Concentrators allowed engineers to record physically distant sound with the desired close-up quality. Additionally, the directional characteristics of the concentrator contributed markedly to "overcoming the detrimental effects of reverberation or generally reflected sounds."[162]

Sound concentrators were used on a number of RKO films, including *Danger Lights* and *Cimarron*.[163] In 1931, RCA introduced a new type of micro-

6.20

Radio-Keith-Orpheum film crew shooting a scene with microphone concentrators. The parabolic reflectors directed sound to a microphone mounted at the focus of the curve. These devices picked up sounds from a much greater distance than was otherwise possible, and they also allowed highly directional recording. Carl Dreher, "Microphone Concentrators in Picture Production," *Journal of the Society of Motion Picture Engineers* 16 (January 1931): 27. Courtesy Society of Motion Picture and Television Engineers, and Princeton University Library.

phone that achieved the same effect in a much smaller package, and RKO engineers were soon using these new "ribbon microphones" on all of their sound pictures.[164] Unlike omnidirectional carbon and condenser microphones, which picked up sound equally in all directions, ribbon microphones possessed strongly directional characteristics. They "listened" acutely to sounds directly in front, and "ignored" sounds coming from other directions. As a result, ribbon microphones picked up actors' voices loudly and clearly, even from a distance. They also reduced the pickup of studio reverberation to approximately one third the level recorded by omnidirectional microphones.[165] Microphone booms were equipped with swivel controls that allowed engineers to pivot and point the ribbon microphone at actors as they spoke, and the goal of recording clear, direct, close-up, and nonreverberant sound was fully achieved.

As Carl Dreher had noted, the close-up recording that resulted from the use of these tools constituted only the first stage in the construction of the sound track. The sounds on this recording were modified and mixed with others before they were released to the public. Each stage of this process, and each element in the mix, was now fully under the control of sound engineers. Perhaps ironically, those engineers sometimes chose to reintroduce certain kinds of noise into their painstakingly wrought noise-free recordings. Numerous early sound films took place in and celebrated the urban environment, and they often included aural montages of city noises in which car horns, police whistles, trolley bells, subways, and shouting newsboys were all heard.[166] Sounds that city-dwellers were seeking to escape in real life were vicariously enjoyed when experienced within the artificial—and highly controlled—setting of a sound motion picture theater, and the noises themselves were artificially created and controlled by sound engineers in the studio. For example, a special "noise machine" was constructed at one studio to simulate the noise made by a subway train pulling out of a station.[167] (See figure 6.21.) Not just the noise of machines, but the sound of space, too, was created in equally artificial ways.

In 1930, Edward Kellogg described to the Society of Motion Picture Engineers how a British radio station had begun to add reverberatory effects to its program material through the use of a special "reverberant chamber," and American sound engineers were similarly experimenting with this new technique. In the broadcast studio, close-miking of the performers generated a non-reverberant signal that was subsequently directed to a distant loudspeaker that reproduced the sound in a small but hard-surfaced chamber. A microphone within this chamber picked up this sound, which now consisted of a highly reverberant reproduction of the original. The engineers then mixed this signal

6.21
"Noise Machine" at an unidentified Hollywood studio, 1929. Sound engineer Kenneth Morgan noted that devices for adding sound effects through rerecording were "both novel and elaborate as well as numerous." This device simulated the noise of a subway train pulling out of a station. K. F. Morgan, "Scoring, Synchronizing, and Re-recording Sound Pictures," *Transactions of the Society of Motion Picture Engineers* 13 (1929): 283. Courtesy Society of Motion Picture and Television Engineers, and Princeton University Library.

back into the original, varying the proportion of the two until just the desired degree of "space" was achieved, and this became the broadcast signal that listeners heard at home.[168] (See figure 6.22.)

While acoustical building materials had first introduced the possibility of transforming traditional relationships between sound and space, the new electroacoustic techniques associated with radio and sound motion picture production expanded these possibilities dramatically. As Edward Kellogg put it, "the

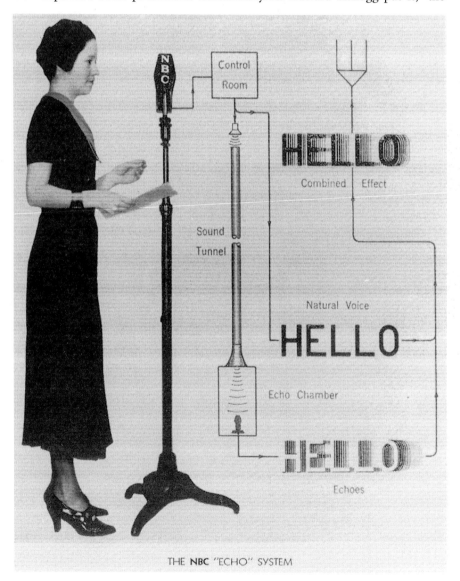

6.22
The NBC "Echo System." By the early 1930s, sound engineers for both radio and motion pictures depended upon systems like this to create artificially the sound of space by simulating echoes and reverberatory effects. "How Echoes Are Produced," *Broadcast News* 13 (December 1934): 26. Courtesy David Sarnoff Library, Princeton, New Jersey.

desirable effects of reverberation" could now be "simulated by a high grade electrical system," and these effects were now "subject to complete control."[169] The sound of space could now exist free of any architectural location in which a sound might be created; it was nothing but an effect, a quality that could be meted out at will and added in any quantity to any electrical signal.

By 1931, NBC had begun to add this "artificial reverberation" to radio broadcasts of the Roxy Theatre orchestra in order to "give to the listener a tone picture, corresponding to their impression of how the orchestra would sound to them were they present in the theater."[170] Filmmakers, too, began to explore the possibilities of simulated reverberatory effects. In John Ford's first sound film, *The Black Watch*, several scenes that occur within a "Cave of Echoes" have a distinctly reverberant quality that may have been achieved artificially. Two years later, in Frank Capra's *Platinum Blonde*, the character of Stew Smith, a hard-boiled journalist feeling increasingly trapped in his marriage to a wealthy socialite, shouts out his frustration in the cavernous foyer of their mansion. His voice echoes and reverberates, but when he subsequently turns and speaks to his butler, it is immediately close-sounding and nonreverberant, suggesting that the reverberant effect was achieved in postproduction.[171] Film historian Arthur Knight has noted that the strange mixture of sounds heard in Rouben Mamoulian's *Dr. Jekyll and Mr. Hyde* during the doctor's frightening transformation into the monster, includes "exaggerated heartbeats mingled with the reverberations of gongs played backwards, bells heard through echo chambers and completely artificial sounds created by photographing light frequencies directly onto the sound track."[172] By the mid–1930s, according to Rick Altman, devices for adding reverberation abounded.[173]

But if these new means for creating the sound of space were widely available, they were not widely employed. Nor, when used, was the goal to achieve an unobtrusively realistic representation of space, but rather to create discrete and highly irregular special effects. Sound engineers exercised their newfound ability to create the effect of space with remarkable discretion. The typical sound track of the early 1930s emphasized clarity and intelligibility, not spatial realism. Uniformity, not variation, was the norm, and a close-up, direct, and nonreverberant sound prevailed. Cuts between long shots and close-ups were seldom accompanied by volume level changes, and realistic representations of reverberatory spaces were presented even less frequently.[174] Donald Crafton has characterized the result as a "well-tempered sound track," and, as James Lastra has also established, the debate over how a sound track should sound was finally settled "by the adoption of the standard of close-miking and a certain 'frontality.'"[175] Lastra characterizes the sound that resulted as "'contextless' or spaceless," bringing to mind the cubist sound track described by the sound engineer John Cass, who complained—to little avail—of the "indefinite position" of the auditor that resulted.[176] Having thus settled the fundamental question of what a sound track should sound like, these engineers left the historian another problem to ponder: Why didn't they take fuller advantage of their ability to add a spatial dimension to their sound tracks?

Many of these men were originally trained as radio and telephone engineers.[177] These industries had long emphasized clear, intelligible voice signals as the criterion for "good sound" and their engineers perceived reverberation as just another form of noise. When these men moved into the motion picture business, they brought those aural standards with them.[178] Radio and telephone engineers had also been trained to think of the sound they produced as a product, an aural commodity, and Rick Altman has argued that the kind of sound track they ultimately constructed privileged the listener as a consumer of sound, offering "sound that is made for us."[179] This sound was indeed attractive, not only to the engineers who produced it, but also to the listeners who consumed it, and to understand fully the source of its attraction, one need only consider the lives of those listeners within the larger soundscape that they inhabited.

The sound of the modern sound track only echoed that being heard in countless other contexts in modern America. From the soundproofed offices of the PSFS Building to the pronounced directional flow of sound at the Eastman Theatre and the Hollywood Bowl, to the electroacoustic offerings of Radio City Music Hall, this kind of sound was everywhere. In its commodified nature, in its direct and nonreverberant quality, in its emphasis on the signal and its freedom from noise, and in its ability to transcend traditional constraints of time and space, the sound of the sound track was just another constituent of the modern soundscape. Indeed, the sound track epitomized the sound of modern America. The many changes in the soundscape that had occurred since the turn of the century—the development of new tools for studying sound, the crescendo of new kinds of noise and the deployment of sound-absorbing materials, the rise of radio, and the transformation of the concert hall—all these phenomena culminated just as sound cinema was finding its voice. The voice it found thus proclaimed these changes loudly and clearly.

When Edward Kellogg reevaluated reverberation for the Society of Motion Picture Engineers in 1930, he noted that, in spite of the tremendous changes wrought in the world of sound over the past thirty years, "the general conclusions reached in the pioneer work of Prof. Wallace Sabine have not been materially altered."[180] Kellogg did not realize that the very revision of Sabine's pioneer work that he subsequently called for was already under way.

VI Conclusion: Reformulating Reverberation

In 1929, Bell Laboratories opened a new facility at 151 Bank Street in New York for making experimental sound pictures "under conditions similar to those in practice."[181] The three-story building contained a soundstage, a monitoring room, film and disc recording rooms, developing and printing rooms, a small theater, dressing rooms, a film storage vault, and laboratories for research in optics and acoustics. The "central thought in the planning of the laboratory" was "to provide for experimental control of every factor influencing sound quality, from set and microphone to loud speaker and auditorium."[182] (See figure 6.23.)

The large monitoring room in the Sound Picture Lab was equipped with full-sized theater loudspeakers, and it was acoustically treated to simulate sound

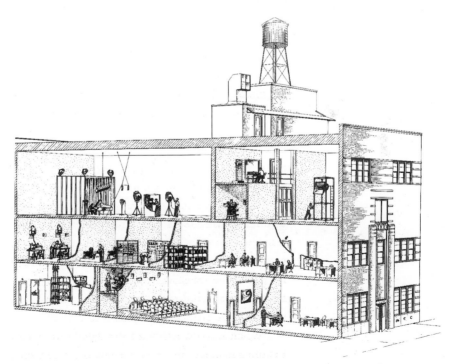

6.23
Sound Motion Picture Laboratory, Bell Telephone Laboratories, New York, 1929. Designed "to provide facilities for making experimental sound pictures under conditions similar to those in practice," the laboratory allowed researchers at Bell Labs to experiment on the processes of making sound films. Photo #W2003B. Property of AT&T Archives. Reprinted with permission of AT&T.

as heard in a typical theater. The sound engineers who worked in this room sat at a mixing desk located on a small balcony. From this perch, they observed the action taking place on the adjacent soundstage through a soundproof, double-glazed window. The stage was acoustically treated to provide the best possible conditions for recording. A thick layer of rock wool covered all walls and ceiling, and adjustable drapes provided further absorption. (See figures 6.24 and 6.25.)

The Sound Picture Lab was the highlight of tours of Bell Labs in the early 1930s, with visitors ranging from Hollywood royalty like movie stars Rod La Roque and Vilma Banky, to real royals like the king and queen of Siam.[183] Aside from the occasional distinguished guest, a technical staff of thirty men inhabited the lab; among these was Carl Eyring, a physicist from Brigham Young University. Perhaps lured by the excitement of the movies, or (more likely for a devout Mormon who had been a student of Harvey Fletcher) attracted by the opportunity to work with state-of-the-art electroacoustic technologies, Eyring had taken a leave of absence from his academic position to work in the Sound Picture Lab.[184] In the course of working in the extremely sound-absorbent environment of the Bell Labs soundstage, Eyring discovered that Sabine's reverberation equation did not accurately describe the behavior of sound in this room.

Wallace Sabine's equation had been a product of the soundscape in which he had worked. In deriving it, he had assumed that the sound energy in a room could be characterized as a homogeneous field, distributed uniformly through space, gradually absorbed by the surfaces to which it was exposed. This assumption was based on his experience working in the various rooms whose qualities he studied, reverberant rooms constructed of wood, plaster, and glass. The absorption coefficients that he ultimately derived for these materials ranged from .025 for plaster to .061 for hard pine sheathing. In other words, these materials absorbed just 2.5 percent to 6.1 percent of the impinging sound energy at each reflection.[185] The sound in these rooms was therefore reflected off the various surfaces hundreds of times before it died away to inaudibility, resulting in reverberation times ranging from 1.91 seconds to 7.04 seconds.[186]

The reverberant nature of Sabine's material environment not only shaped his conception of the physical process of reverberation, it was also embodied in his mathematical analysis. For example, he constructed an infinite series to represent the total sound energy in a room, with each element of the series representing the portion of sound that has suffered a given number of reflections off the surfaces of the room. This series took the form:

6.24

Sound engineer in the Sound Motion Picture Lab, Bell Labs, 1929. The engineer at his mixing desk observed though the window the action taking place on the large soundstage below. He listened to sound reproduced through full-sized theater loudspeakers broadcasting the signal into the large monitoring room (see figure 6.23), which was acoustically designed to imitate the acoustical characteristics of a typical theater. Photo #HM38172. Property of AT&T Archives. Reprinted with permission of AT&T.

6.25

Sound Stage of the Sound Motion Picture Lab, Bell Labs, 1929. The walls and ceiling were all treated for high levels of sound absorption, and the acoustics were further controlled through the use of adjustable drapes. The average reverberation time of this room was just 0.35 seconds. Photo #HM38162. Property of AT&T Archives. Reprinted with permission of AT&T.

$$\text{Energy} = \frac{p\,E}{V}\,[1 + (1 - a/s) + (1 - a/s)^2 + \ldots + (1 - a/s)^n],$$

where

p = mean free path of sound between reflections,
E = rate of emission of energy from the sound source,
a = absorbing power of the room,
s = surface area of the room,
V = volume of the room, and
n = number of reflections suffered by each component of sound energy.

General mathematical rules applying to series of this form allowed Sabine to write his series in the condensed form:

$$\text{Energy} = \frac{p\,E}{V}\,\frac{1 - (1 - a/s)^n}{1 - (1 - a/s)}.$$

Sabine next assumed that n was large; the room was reverberant enough that the sound in it would suffer many reflections before any individual contribution of reflected energy would become negligible. This assumption allowed him to simplify the series further, to:

$$\text{Energy} = \frac{p\,E\,s}{V\,a}.$$

Sabine used this quantity to represent the total energy in the room as he continued his analysis. In this way, the liveness of his rooms was embedded in his equations.

Carl Eyring's acoustical environment differed dramatically from that of Sabine, and it was this difference that drove him to reformulate Sabine's equation. Simply put, Eyring worked in a world swaddled in sound-absorbing materials. The absorption coefficient of the thick material that lined the walls of the Bell Labs soundstage was 0.77, much greater than any coefficient with which Sabine had worked. The resulting reverberation time of the soundstage was just 0.35 seconds, far less than any time that Sabine had ever measured.[187] In such an environment, sound energy was absorbed so quickly and completely that Sabine's assumptions about the gradual, diffuse absorption of sound no longer applied. Eyring's working environment constituted an extreme case that Sabine had neither encountered nor considered; thus, Sabine's equation failed Eyring in a way that it hadn't failed Sabine thirty years earlier.[188] Eyring's task was to modify Sabine's equation to fit the acoustically dead rooms of his world as well as the live ones that Sabine had inhabited. He chose to start at the very beginning, to reconceptualize the phenomenon of reverberation in a way that would have been inconceivable to Sabine.

To measure reverberation, Sabine had employed as his source an organ pipe sounded by a tank of compressed air. His detector was his own sense of hearing. He listened to the sound of the organ pipe as it gradually died away and recorded the moment at which it became inaudible. To Sabine, reverberation was defined by a human auditor located in architectural space, listening to the decay of a traditional musical sound as it was reflected off, and gradually absorbed by, the surfaces of that room.

Eyring's technique differed dramatically. He replaced the mechanically sounded musical tone of the organ pipe with an electrically driven oscillator whose pure signal was amplified and then projected from a loudspeaker. The human detector was replaced with an "electro-acoustical ear," a microphone that automatically triggered a recording chronograph to register the instant at which the received sound signal had attenuated by 60 decibels.[189] (See figure 6.26 and compare it to figure 2.12.) For Eyring, reverberation was dissociated from rooms filled with musical sounds, as well as from human listeners located in those rooms. To him, it was instead the time required for an electrical signal to suffer a standard degree of attenuation.

Just as Sabine's experimental technique shaped his understanding of the physical process of reverberation, Eyring's own technique helped him reconceptualize Sabine's understanding. Simply put, Eyring presented "an analysis based

6.26
Schematic diagram for an electrical means of measuring reverberation, 1930. The technique indicated here stood in sharp contrast to Wallace Sabine's technique of 1900. (See figure 2.12.) An electrically generated tone was amplified and projected from a loudspeaker into a room. After this signal was cut off, an "electro-acoustical ear" listened and automatically registered the instant at which the resultant sound signal had attenuated by 60 decibels. Reprinted with permission from E. C. Wente and E. H. Bedell, "A Chronographic Method of Measuring Reverberation Time," *Journal of the Acoustical Society of America* 1 (April 1930): 422. © 1930, Acoustical Society of America.

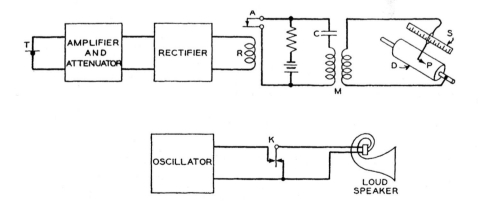

on the assumption that image sources may replace the walls of a room in calculating the rate of decay of sound intensity."[190] He imagined an abstract source located in free space, surrounded not by walls but by an infinite number of other sources located at increasing distances from the original, all simultaneously emitting sound back toward that original source. (See figure 6.27.)

Carl Eyring was not the first to envision sound reflections as emissions of sound from image sources. Textbooks on physics and sound had long and regularly portrayed and explained reflections of sound by the method of images. Floyd Watson had used this approach in 1928. But Watson *compared* walls to "acoustical mirrors" that created images of sources of sound, and, he interjected, "of course this image is imaginary, and its speech is nothing more than the reflected sound."[191] His illustration of this way of thinking about sound emphasized the architectural reality of the room over the imaginary sources of sound. (See figure 6.28.)

Eyring, in contrast, *replaced* the walls of the room with "image sources located in evenly spaced discrete zones."[192] No walls at all were depicted in his illustration. Eyring transformed the acoustical phenomena within a room into an

6.27
Carl Eyring's representation of the behavior of sound in a room, 1930. Eyring conceptually replaced reflections of sound off architectural surfaces with emissions from imagined sound sources located at increasing distances from the original source, all emitting sound back toward the source. The original source of sound is represented by the black dot at the top, front corner of the diagram, and the sequentially numbered dots represent some of the image sources. Reprinted with permission from Carl Eyring, "Reverberation Time in 'Dead' Rooms," *Journal of the Acoustical Society of America* 1 (January 1930): 223. © 1930, Acoustical Society of America.

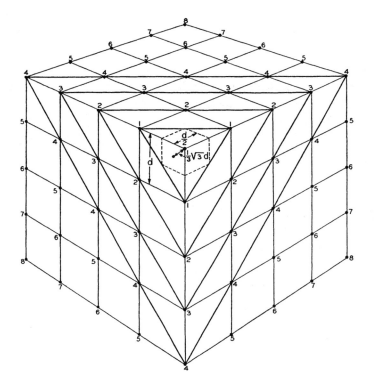

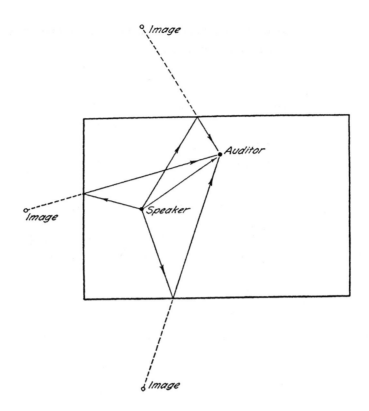

6.28
Floyd Watson's representation of the reflection of sound in a room by means of image sources. In this drawing, unlike in figure 6.27, the physical space of the room is clearly delineated and is represented as being inhabited by a human speaker and auditor. The images are also visually distinguished from the original source. Floyd Watson, "Ideal Auditorium Acoustics," *Journal of the American Institute of Architects* 16 (July 1928): 260.

abstract array of sources existing in unbounded space—an array that one might easily imagine as a network of loudspeakers. He not only studied sound with electroacoustic tools, these tools additionally provided the very means for him to reconceptualize its behavior. Sounds and signals had physically and intellectually commingled and coalesced to the point where not only sound engineers, but physicists, were uninterested, perhaps even unable, to separate the two.

The analysis that followed from Eyring's reconceptualization of reverberation led him to understand the absorption of sound energy in a way that differed distinctly from Sabine's earlier characterization. Where Sabine had supposed a smooth, gradual, and continuous decay of energy, Eyring described a discontinuous process whereby the flow of energy suffered abrupt drops. "This constant energy flow followed by an abrupt drop, rather than a continuous drop to this same level," Eyring observed, "means a greater absorption during the same interval of time and hence a more rapid decay of the sound."[193] With this new understanding of the decay of sound energy, Eyring ultimately derived a new equation for calculating reverberation time:

Sabine equation	Eyring equation
$t = \dfrac{.164\ V}{S\, \alpha_a}$	$t = \dfrac{.164\ V}{-S\ \ln(1-\alpha_a)},$

where:

t = reverberation time (in seconds),
.164 = hyperbolic constant,
V = volume of room (in meters),
S = surface area of room (in square meters),
α_a = average coefficient of absorption for the room,
$\quad = \sum \dfrac{S_n\, \alpha_n}{S}$, where S_n = surface area of material n,
$\qquad\qquad \alpha_n$ = absorption coefficient for material n.

With the new equation, the reverberation time of an infinitely absorbent room calculated out at zero seconds, a mathematical criterion that had not been met by Sabine's original equation. Sabine and other acousticians circa 1900 had not been concerned with this limitation, as the existence of such an absorbent room was virtually inconceivable at that time. In 1930, such a space was fast becoming an architectural reality, and the limitation now became a problem that Eyring's equation successfully solved.

Carl Eyring's revision of Wallace Sabine's reverberation equation was immediately put to work, and it soon began to appear in articles and texts on acoustics.[194] Bell Labs engineer Walter MacNair spoke for many in 1931 when he noted, "For many years there has been an established science of acoustics which has furnished a basis for the correction of unsatisfactory acoustical conditions in many auditoriums and the proper design of others."[195] When MacNair attempted to apply this established science to the "modern problems" associated with sound motion picture production, however, he discovered—like Eyring—that "the older methods of describing acoustical phenomena were inadequate."[196] Sabine's equation was suddenly perceived to be old and inadequate, as dated as that quaint portrait of the scientist himself from 1906. Acousticians and sound engineers thus turned to Eyring's new equation in order to understand the behavior of sound in the modern world.

When Carl Eyring replaced Wallace Sabine's α_a with $-\ln(1-\alpha_a)$, he signified in a cryptic mathematical code that the material world, the physical world of rooms filled with sound, had fundamentally changed. If only a small cadre of engineers and acousticians were in a position to understand fully the import of the new reverberation equation, many more were able to appreciate the changes that it symbolized. Millions of Americans heard those changes loudly and clearly every week when they went to the movies, where the soundscape of America was celebrated on celluloid. The modern soundscape was also commemorated in more monumental form, in the steel and limestone towers of Rockefeller Center that were then rising in midtown Manhattan.

CHAPTER 7   CONCLUSION: ROCKEFELLER CENTER AND THE END
               OF AN ERA

As the buildings of Rockefeller Center took shape in the early 1930s, it became clear to all that a bold experiment in urban planning was under way. The architects were building a "city within a city,"[1] and an integral component of this self-consciously modern city was its acoustical design. From the soundproofed studios of NBC to the acoustically quieted offices of the RKO Building, state-of-the-art techniques of architectural acoustics were deployed to control and contain the sounds of city life. The RKO Roxy Theatre projected the modern sound track loudly and clearly, and an equally distinct sound signal was heard in the electroacoustically enhanced auditorium of the Radio City Music Hall. Wherever one turned, modern sounds were heard. The creative energy that generated those sounds was celebrated visually, too, in the rich ornamentation that decorated the buildings of the center. As originally conceived, however, this project had been dedicated, not to the electroacoustic excitement of the modern soundscape, but instead to the far more traditional sounds of old-world opera.

The idea that became Rockefeller Center originated in 1926 with the financier and opera patron Otto Kahn, who sought to create a vast urban plaza that would highlight a new house for the Metropolitan Opera of New York. Kahn brought his even wealthier friend, John D. Rockefeller, Jr., in on the project in 1928, when a particularly attractive and expensive piece of midtown property was identified as an ideal location for the new opera house. The parcel of land, which stretched from Forty-eighth Street to Fifty-first Street between Fifth and Sixth Avenues, was owned by Columbia University and leased to an assortment of landlords who were content to collect low rents from the rundown apartment buildings and speakeasies that filled these blocks. Rockefeller agreed to buy up all of the leases and to assume responsibility for future rent payments to Columbia. He stipulated, however, that his contribution would be a business investment, not philanthropy, and he was determined that the new center generate income along with beautiful music.

A series of designs for the opera house and surrounding plaza had already been developed by architects Benjamin Wistar Morris and Joseph Urban by the time Rockefeller became involved. At that time, Rockefeller appointed the building contractors Todd, Robertson & Todd to manage the vast project, and a new emphasis on commercial space was the result. A team of "Associated Architects" was assembled to create what historian Henry-Russell Hitchcock later identified as an "architecture of bureaucracy."[2]

The members of the board of the Metropolitan Opera were unhappy with the increasingly commercial tone of the enterprise, and after the crash of the stock market in 1929, they withdrew from the project. As Rockefeller later recalled, "Thus it came about that in the early part of 1930, with the depression under way and values falling rapidly, I found myself committed to Columbia for a long lease, wholly without the support of the enterprise by which and around which the whole development had been planned."[3]

It wasn't long, however, before Rockefeller would be rescued from his precarious position. By mid-December, talks were under way with the three entertainment subsidiaries of the General Electric Company—the Radio Corporation of America, the National Broadcasting Company, and the Radio-Keith-Orpheum motion picture conglomerate—to become major tenants and a new corporate anchor for the center. Contracts were signed in June 1930, and "Radio City" was soon rising from the rubble of demolition. No longer centered on the live performance of classical opera—the continuation of an aristocratic cultural tradition dating back to the sixteenth century—Radio City now became a celebration of the modern art and science of electroacoustics. The seventy-story RCA Building constituted the new focal point of the complex, towering over "an unprecedented concentration of facilities for the dissemination of sight and sound by radio and by record—through the air, the film, and the disk."[4] (See figure 7.1.)

The architects celebrated the center's new role as the epitome of modern aural culture by decorating their buildings with ornamentation representing all the sounds being created within. Sculptor Lee Lawrie's design for the main entrance to the RCA Building depicts "the genius which interprets to the human race the laws and cycles of the cosmic forces of the universe, and thus rules over all of man's activities."[5] The genius inscribes with his compass the cosmic forces of light and sound, and each force reappears over the doors to the left

7.1
Rockefeller Center, New York (The Associated Architects), facing west, c. 1932. The central tower is the RCA Building and the NBC studios are in a low wing of this building immediately west of the tower. The flat, windowless rear wall of Radio City Music Hall is visible to the right of the tower, and the back of the RKO Roxy Theatre (demolished in 1954) is similarly visible to the tower's left. The RKO Building is the moderately high building immediately west of the Music Hall. Photograph, n.d., Museum of the City of New York, The Wurts Collection.

and right of the central entrance. (See figures 7.2 and 7.3.) A promotional brochure explained, "Although there are other cosmic forces which govern the universe, Mr. Lawrie selected those of Light and Sound because they are an active and vital part of everyday life, and particularly because within contemporary times great discoveries have been made by means of them, and man's technical knowledge of the laws of these two forces has been vastly enlarged."[6]

7.3
Main (East) Entrance to RCA Building with sculpture by Lee Lawrie, "Sound." Photo #8211.20 © Bo Parker photo 1982.

7.2
Main (East) Entrance to RCA Building with sculpture by Lee Lawrie, "Genius, Which Interprets to the Human Race the Laws and Cycles of the Universe, Making the Cycles of Light and Sound." Photo #989. Courtesy Rockefeller Center Archive Center.

7.4

West Side of RCA Building with sculpture by Gaston Lachaise, "Genius Seizing the Light of the Sun" (with motion picture cameras) and "The Conquest of Space" (through radio technology). Photos #593 and 593B. Courtesy Rockefeller Center Archive Center.

At the Sixth Avenue entrance to the RCA Building, a mosaic by Barry Faulkner depicts "Enlightenment" as the radiolike transmission of man's thoughts across space. High above, sculpted stone panels by Gaston Lachaise represent "various aspects of modern civilization," including "Genius seizing the Light of the Sun" (with motion picture cameras) and "The Conquest of Space" (by radio waves).[7] (See figure 7.4.) Leo Friedlander's sculpture at the Fiftieth Street entrance of the building also depicts the transmission and reception of radio signals, and Hildreth Meier's large plaque for the Forty-ninth Street facade of the RKO Roxy Theatre (since demolished) represented "the moving forces in modern civilization," radio and the then-nascent technology of television.[8] (See figure 7.5.)

299 CONCLUSION: ROCKEFELLER CENTER AND THE END OF AN ERA

7.5
North Side of RKO Roxy Theatre with metal and enamel plaque by Hildreth Meier, "The Spirit of Electrical Energy Sending Out Radio and Television." (Building demolished in 1954.) Photo #88-A. Courtesy Rockefeller Center Archive Center.

New technologies of sound control were not only celebrated in the decoration of Radio City, they were also incorporated into the fabric of its buildings. The RKO Building, for example, was equipped with the "Antenaplex System," a centralized antenna network like that in the PSFS tower, which ensured every tenant "efficient reception" of radio transmissions when they plugged their receivers into the special outlets installed in each office.[9] The *New York Times* announced the architects' plans to equip every window in the vast complex with the Maxim-Campbell Silencer and Air Filter, a ventilating unit that admitted fresh air to a room while simultaneously muffling the intake of external noises.[10] Offices were protected from internal noises, too, by means of "scientifically efficient" soundproof partitions that prevented the passage of sound from one room to another. Builder Webster Todd justified this extra expense, noting that "science has definitely established that the absence of disconcerting noises adds to the efficiency of office work."[11] All these techniques for sound control were highlighted in promotional literature for the center and the new complex was compared to the Taj Mahal, "quieting in its serenity," amidst "the swirling life of a great metropolis."[12]

Of course, the technical measures undertaken to control sound in the offices of Radio City paled in comparison to the degree of acoustical control sought and achieved in the NBC studios. The network had just built new studios in 1928, but the industry was rapidly expanding and changing, so as soon as NBC signed on to occupy the as-yet-unbuilt Radio City in 1930, its engineers began to work with the Associated Architects to plan new and better facilities. These studios, which began transmitting programs to listeners in the fall of 1933, were heralded as "a temple to glorify the radio voice;" a "gigantic cathedral of sound" in which even Marconi was impressed.[13]

The NBC studios occupied eleven floors in the west wing of the RCA Building. Stairs led visitors from the tower's main concourse up to the strikingly circular NBC lobby. (See figures 7.6 and 7.7.) Here, glass museum cases displayed and fetishized modern electroacoustical inventions as if they were precious holy relics. A photographic mural by Margaret Bourke-White surrounded the space and enlarged those same devices to monumental proportions.

The studio complex was structurally isolated from the rest of the building to prevent the transmission of noise and vibration. Each individual studio was further isolated to ensure a totally soundproof environment, and a quieted air-conditioning system ventilated the entire windowless complex.[14] The twenty-seven studios ranged in size from small booths to a vast auditorium capable of

7.6
Rockefeller Center, NBC Studios Lobby, showing glass cases displaying technological artifacts and the surrounding photo mural by Margaret Bourke-White, c. 1934. Photograph, n.d., Museum of the City of New York, The Wurts Collection.

7.7
Margaret Bourke-White standing before her photo mural in the lobby of the NBC studios, Rockefeller Center, c. 1933. Bourke-White's mural celebrated the artifacts of electroacoustic technology by enlarging them to monumental proportions. Courtesy Syracuse University Library and Estate of Margaret Bourke-White.

accommodating a large orchestra and an audience of several hundred. (See figure 7.8.) The complex also included audition rooms, performers' lounges, engineering stations, and private clients' booths where the corporate advertisers who sponsored broadcasts could enjoy special access to the programs they paid for. (See figure 7.9.)

Many of the studios had observation areas that were open to the public, and NBC studio tours—capped by an opportunity to sit in on the live performance of one of the network's programs—became a staple item on the agenda of tourists. In most cases, observation galleries were isolated from the studios by large, multipaned soundproof windows. The galleries were wired for sound, so what the members of the audience heard there was not very different from what they heard at home; an electroacoustical reproduction of the live performance that they observed through the glass. When "audience noises" were desired "to

7.8
Rockefeller Center, NBC Studio Audience, c. 1934. This, the largest of NBC's studios, could accommodate a live audience of several hundred. The engineering control booth is located behind the windows above and below the clock on the far wall. Note the loudspeaker horns suspended from the ceiling over the orchestra. Photograph, n.d., Museum of the City of New York, The Wurts Collection.

7.9
Ninth and tenth floor plans of the NBC Studios, Rockefeller Center. The large studio in figure 7.8 is shown here at the right. Note the three sound effects chambers on the tenth floor, at the top center of the drawing. These rooms were used to create artificial reverberation. O. B. Hanson, "Planning the NBC Studios for Radio City," *Proceedings of the Institute of Radio Engineers* 20 (August 1932): 1306, 1307. © 1932 IRE, now IEEE.

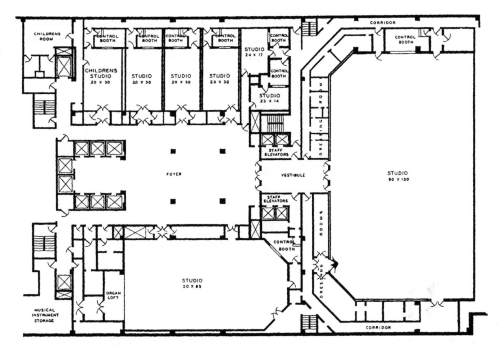

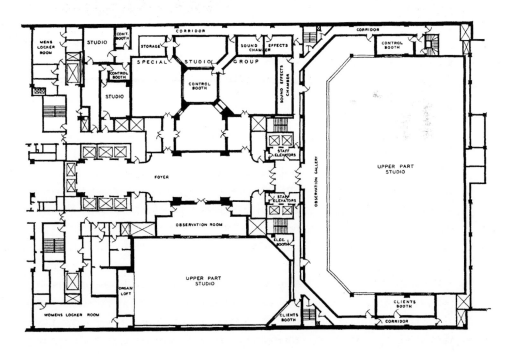

give the production a stamp of authenticity," the glass curtains were raised so that microphones in the studios could pick up the laughter and applause.[15]

Visitors were as interested in the technical side of radio as they were in the celebrities who brought the programs to life, so an observation area was provided for the centralized engineering control room on the sixth floor. From this "nerve centre,"[16] transmission engineers monitored all of the signals arriving from the different studios and adjusted these signals for broadcast. Most of the sound engineering was accomplished in the local control rooms adjacent to each studio, however, and here the engineers exercised an unprecedented degree of control.

The studios were lined with sound-absorbing materials, and were additionally equipped with movable "acoustical units." Depending on their position, these absorbent panels covered or exposed the less-absorptive wall surfaces beneath, thereby changing the reverberation time of the room. Engineers moved the panels by means of electric controls located in their control rooms. They could thus manipulate the architectural acoustics of the studios as easily as they adjusted the electroacoustic signals coming from the microphones placed within them.[17]

But engineering practice dictated that these studios remain acoustically dead even when the panels were set in their most reverberant configuration.[18] Thus, when a significant level of reverberation was required for dramatic effect in a program, the sound engineers employed one of the three echo chambers located on the tenth floor. (See the three rooms labeled "Sound Effects Chambers" in figure 7.9.) NBC engineers had begun to experiment with artificial reverberation at their previous studio complex, and the echo chambers at Radio City were an integral part of the technical equipment of the new studio, allowing the creation of spatial effects that far exceeded the capabilities of the architectural space of the studios.

Within these studios, much of the radio programming that entertained most of America through the 1930s was generated. From the famed symphony broadcasts of Arturo Toscanini to the even more famous mispronunciations of Amos 'n' Andy, NBC programs were broadcast directly to the New York metropolitan region over stations WEAF and WJZ and were transmitted over long-distance telephone lines to distant cities where NBC-affiliates subsequently broadcast them to their own local listeners. The voices of Broadway—including Eddie Cantor and George Burns and Gracie Allen—spoke directly, intimately, and electroacoustically to millions of Americans, from Maine to California. The impact of Radio City thus extended far beyond the bounds of midtown Manhattan, contributing to the nationalization of the modern soundscape.[19]

Radio City Music Hall similarly, if more locally, represented the culmination of the modern soundscape, and the new auditorium was celebrated as "an epitome of the changes that have taken place in American life, manners and taste in the first three fast-moving decades of the twentieth century."[20] When Roxy Rothafel was appointed director of the new musical theater, the nation's most popular showman promised to dazzle audiences with spectacle on an unprecedented scale. He also pledged to maintain the connection between audience and performer that he believed only live theater could offer, a connection increasingly rare in an era whose entertainment was dominated by radio and motion pictures. Roxy was to preside over one of the largest theaters ever built, but he promised that the special character of live performance would not be lost in the vast expanse of the new auditorium. To this end, Roxy had the architects extend the stage beyond the proscenium. A series of rising terraces hugged the side walls of the theater and allowed performers to move out into the space inhabited by the audience. He also called for three shallow balconies rather than one deep one, to create a more open auditorium in which the entire audience could respond as one to the performers on stage. By thus bridging the gap that separated performers from audience, and by uniting the large audience, Roxy hoped to facilitate "a greater degree of intimacy between actors and audience,"[21] an intimacy he claimed was missing from the rival entertainments of motion pictures and radio. (See figure 7.10, also figures 6.2 and 6.3.)

The arches that made up the body of the auditorium were additionally intended to draw the vast audience into the performance by focusing attention on the stage. Roxy claimed to have been inspired to build this form after witnessing a sunset at sea, but the golden contours are similar enough to other projects of the period to suggest the less dramatic influence of auditorium architects and acoustical consultants.[22] The form is reminiscent of the Hollywood Bowl shell, and the Chicago Civic Opera House of 1930 has a similarly stepped series of surfaces constituting the body of its auditorium.[23] Architectural historian Carol Krinsky has pointed more directly to the earlier designs of Joseph Urban for Otto Kahn's new Metropolitan Opera House.[24] Whatever its source, Roxy's auditorium was immediately identified as the quintessential example of a modern auditorium.

Architectural critic Douglas Haskell made explicit the connection between the form of Radio City Music Hall and the modern sound with which it was filled:

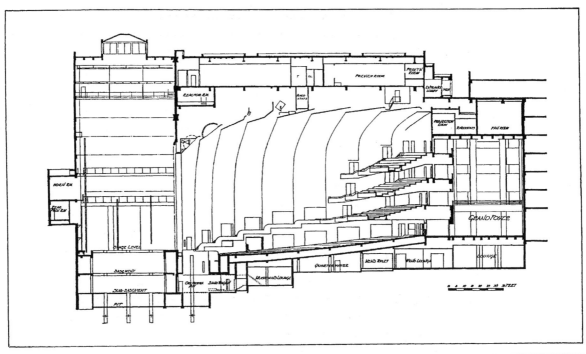

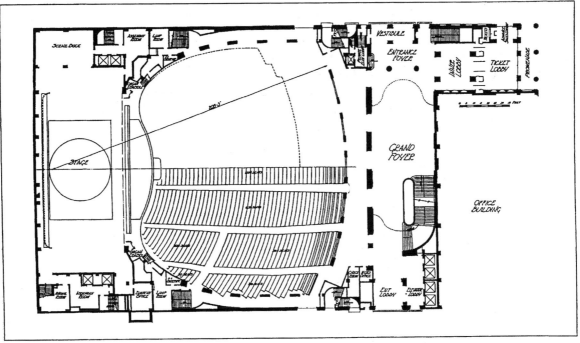

7.10

Section and Plan of Radio City Music Hall, Rockefeller Center, 1932. Roxy Rothafel intended the encompassing arches, the terraced stage extensions, and the open, shallow balconies all to unify the vast audience and to draw them into an intimate relationship with the performance taking place on the stage. *Architectural Forum* 56 (April 1932): 356.

This vault is a delight. Not only the vast space: this nervous energy, this swift radiation. There is something about it that fits. It stands for our thoughts. Picture the Greek, with his serene colonnade topped by the low triangle of his pediment. It is measured and self-contained. Picture the Roman, who commands the round power of the masonry dome. Then the Gothic artist, who thrusts his vaults upwards: his buildings grow like plants. Baroque elaborates on the Roman; twists, turns and moves. It is suited to theaters. But we move in paths of a still greater variety. Our trajectory can be more direct. We have control over forces more abstract and more potent. The investigations of our thinkers are concerned with ethereal radiations and vibrations. It is these that have been manipulated to make possible the whole enterprise of our tremendous industry of sound communication. So it is fitting, almost symbolical, that a great hall of ours, devoted in whatever manner to music, should expand from a focus by waves that follow a great curve.[25]

Radio City Music Hall was not just a symbolic tribute to the tremendous industry of sound communication and control; it was also constructed of the very products of that industry. The curves themselves were constituted of over one thousand tons of Kalite Sound Absorbing Plaster.[26] The rear wall of the auditorium was covered with a thick blanket of sound-absorbing material, and all mechanical systems were structurally isolated in order to eliminate any noise from the enormous air-conditioning system.[27] These elements combined to render silent the vast space of the auditorium. While no original measurements are available, evidence suggests that the reverberation time of the hall was probably less than 1.5 seconds (at 500 cps) unoccupied, and closer to 1.0 seconds with a full house—extremely low values for such a large theater.[28]

This silenced space was then filled with an electroacoustic reproduction of the sounds that were generated on stage. As the *New York Times* reported, "Everything has been done to utilize the latest work of the technicians in sound as well as in vision, so that all the music and all the words of the entertainment can be heard by all of the huge audience. Fifty 'ribbon' microphones are on the stage, each with an amplifier beneath the stage that can be regulated at will."[29] Chief sound engineer Harry Hiller mixed the signals from the various microphones to achieve an appropriate level and balance, and this signal was then broadcast to the audience by RCA loudspeakers that were hidden behind several of the radiant grilles embedded in the acoustical plaster ceiling.[30]

Collier's magazine described the sound engineer's work to its readers:

Sitting at a desk solidly covered with push-buttons and little switches . . . [is] Mr. Hiller. Besides buttons and switches he sees meters, indicators and curious little gadgets which flicker, glow, click, hiss, croon and pop. With that board, he controls the sound. . . . All that he has to do is to see to it that, whether you are in the first row or the fifty-first, you get the same number of sound vibrations and that words and music are distributed impartially to all the audience.[31]

In spite of Roxy's efforts to emphasize the unique aspects of live performance, the sound of Radio City Music Hall was ultimately no different from that heard by listeners to NBC radio broadcasts or sound motion pictures. Captured by microphones, electrically engineered, manipulated, and modified, the sound was finally projected out of loudspeakers into a highly absorbent space.

This kind of sound, with its energy, focus, and direct trajectory, was common enough by 1932 to be unremarkable. Prior to opening night, Roxy had calmly observed, "I think we have made sufficient progress in the science of acoustics to eliminate all possibility of error in reverberation and absorption," and architect Henry Hofmeister similarly expressed his confidence in the "acoustic experts" to create an auditorium "as nearly perfect as possible for sound transmission."[32] Roxy predicted that the amplification system "will be so perfect that you won't be able to tell that the sound is being amplified,"[33] and his prediction proved accurate, as reviews of opening night had surprisingly little to say about the acoustics of the hall. Critics accepted the sound that issued from the concealed loudspeakers without criticism, and no one seemed to care if electroacoustic devices mediated the intimate relationship between performer and audience that had been Roxy's principle goal.[34] Indeed, the sound system, by delivering "close-up" sound to auditors seated as far as 200 feet from the stage, provided the only intimacy that could be claimed for the enormous room. (See figure 7.11.)

While microphones, amplifiers, and loudspeakers ensured that all members of the audience heard everything as if they were right up on stage themselves, there was no visual equivalent to amplify and transmit the performers' emotions, subtle facial gestures, and personality traits across a room that was vast enough to impress even Helen Keller with its size.[35] Opening night was an acoustical success, but Roxy still suffered at the hands of critics who concluded that the Music Hall was simply too big. "In such an enormous auditorium," drama critic Brooks Atkinson argued, "the individual performer labors at great disadvantage. Even from a seat well forward it is difficult to have much response to the personality of the performers." While he noted that the actors worked "valiantly," and were "aided by the sound amplifiers," Atkinson concluded that Roxy would ultimately have to find a new type of entertainment more appropriate for his "tremendous palace."[36] Douglas Haskell agreed. "The impossible remains impos-

7.11
Radio City Music Hall, view of stage in performance, c. 1935. "This nervous energy, this swift radiation. There is something about it that fits. It stands for our thoughts," noted architectural critic Douglas Haskell. Photograph, n.d., Museum of the City of New York, Theater Collection.

sible," he concluded, "and no power of paradox can quite reconcile huge with intimate; so if the stage is to be used for anything much smaller than massed ballets and big orchestras, either it must be covered with a huge lens, or . . . the art director will have to supply the actors with visible facial expressions by means of three-foot masks."[37]

Haskell's words proved prophetic. Radio City Music Hall failed to sustain the interest that had filled the house on opening night, and $180,000 was lost during just its first few weeks of operation, as mediocre box-office receipts failed to keep pace with the cost of staging the spectacular live show.[38] In the harsh economic climate of 1933, too few people were willing or able to part with the premium cost of admission (from 75¢ to $2.50 per seat), especially when a perfectly good sound movie could be enjoyed elsewhere for a fraction of that price, or an equally good radio program could be heard at home for free. In mid-January, the executives at RKO who were in charge of the Music Hall's operation announced that the program format at the Music Hall would immediately be revised. Radio City Music Hall would now offer motion pictures in conjunction with far less extravagant live shows, and the cost of admission would be reduced to standard first-run motion picture theater rates.[39] With this change, the "huge lens" that Haskell had called for was now provided by a motion picture projector, and the "massed ballet" of the Roxyettes (later the Rockettes) was on its way to becoming the defining feature of a live show that eliminated the expensive star performers and emphasized the technical effects made possible by the elaborate stage machinery.[40]

Roxy's dream of an extravagant live theater simultaneously intimate and grand had proved a chimera. The great man himself was unable to defend that dream to the executives in charge, as he had become seriously ill on opening night and spent most of the early part of 1933 in the hospital. When Roxy returned to work later that spring, it was to an organization far different from that over which he had ruled just a few months ago. He attended meetings at which his opinions were ignored, his salary was reduced, and within a few months he was fired.[41]

Roxy was suddenly seen as a vestige of a culture that no longer existed, and his theater was just as suddenly reconceived in the minds of critics and at least some of the public. No longer "an expression of today," Radio City Music Hall was now the "most expensive white elephant in the world," a "monument to the follies of a past age," "thoroughly characteristic of the pre–1929 age of elephantiasis and vulgarity."[42] A letter to the editors of the *New York Times* declared the theater "a symptom of a decadent civilization" in which "the machines multiply and make a god of power," but "the imagination of men becomes duller and duller."[43] Walter Lippmann characterized Radio City as "a monument to a culture in which material power and technical skill have been divorced from human values and the control of reason," an exemplification of "the complete dissociation of means from ends."[44]

Radio City Music Hall embodied the triumph of technical expertise, including acoustical expertise, but this triumph was bittersweet in a world where the machines of production had failed to sustain the material prosperity and faith in progress that had been taken for granted just a few years before. This complicated reaction expanded beyond the walls of the Music Hall itself, reverberating among the rising gray canyons of Rockefeller Center. Critics began to question not just Roxy's misguided efforts, but the larger project of which his theater was a part, and all that it represented.

As early as 1931, an aging Ralph Adams Cram had spoken out bitterly against what he perceived as the philistine nature of the electroacoustically oriented culture to which the complex was now dedicated. "Here then," Cram wrote, "is Radio City. . . . From these belatedly truncated towers will go out even to the ends of the earth the nourishing vitamins of the chosen culture of this climacteric age. Amos 'n' Andy, Mr. Wrigley's Musical Hour, the intimate and revealing details of the latest *crime passionnel* and, in the hours that cannot be profitably disposed of to the exponents of super-salesmanship, such varied propaganda as may covet the high privilege of being 'on the air.'" To Cram, modern culture was no culture at all, and he decried the pervasiveness of electroacoustic media, for now, great newspapers "fold up and die," and "none bothers to go a mile to concert or opera, for, lo, it is in the home for the turning of a screw."[45]

In early 1932, as the rising steel towers of the center only accentuated the downward spiral of the economy, Frederick Lewis Allen reflected on the noises that accompanied the construction. "The sound falls familiarly upon the ear," he observed, "yet already there is a strangeness about it—as if it were an echo from a time gone by."

> [A]ll about you the air is shaken by that sound—compounded of the muffled throb of the air-compressor, the hard staccato of the drill, and the metallic clatter of the riveter.
>
> How it pulls at the memory! Only two or three years ago a New Yorker could hardly escape it. Wherever he went, uptown or downtown, the cacophony of prosperity assailed his ears. But now, in the dark days of business depression, he hears it seldom; . . . [and as] the clamor of the riveter meets him again, he may perhaps be pardoned if he stops to wonder whether he is listening to a portent of the days when hope for the economic future of America shall again return, or simply to a belated echo of the strident nineteen twenties.[46]

In the wake of Radio City Music Hall's opening, the answer to Allen's question was clear. Lewis Mumford characterized the center as a "melancholy pile," and Douglas Haskell similarly compared it to "some giant burial place," whose mood was "gray, unreal and baleful."[47] Haskell found the ornamental sculpture over the entrances silly, "'styled,' one might say, 'in the modern manner,' of a sort that in an up-to-date cemetery might be found on the tomb of a respectable yet progressive family."[48]

The only sign of life in the funereal center had been when the Marxist muralist Diego Rivera was painting his own vivid contribution to the lobby of the RCA Building. Rivera, according to Haskell, "had lodged a huge and highly skilful poster calling down doom—in the name of science, industry and the people—on all the forces that had created the wall on which he worked."[49] Once the face of Vladimir Lenin appeared on that wall, however, Rivera's own work was doomed, and Mr. Rockefeller quickly had it covered, then destroyed.[50] What was left, Haskell concluded, was a paradox with an epitaph: "Every major force at work defeats every other one—just as the architects, for example, canceled one another—as there arises the gigantic opus of congestion in the midst of inactivity, confusion and emptiness. A sort of death by stalemate—a dying not by inches but by the collision and mutual counteraction of tons and miles and millions. All put together entirely according to the rules of business 'economics,' 1920–1930."[51]

Many others besides Douglas Haskell perceived that same collision and death. Prometheus, who stole fire from the gods of ancient Greece in order to bring the civilizing forces of science and technology to man, was given a position of honor at Rockefeller Center; a sculpture of the great Titan by Paul Manship was placed in the heart of the forum in front of the RCA Building. But observers soon noted that, in his placement and his posture, Prometheus appeared to be plummeting from the heights of the great tower. "Leaping Looie," as he came to be known, appeared frozen in time a split-second before his final crushing impact.[52] The fall of Prometheus symbolized the fall from grace of technics; it represented man's folly in thinking that, through his technologies, he could truly control the awesome forces of nature. This message proved far more meaningful than that intended by the artist to the millions of Americans who observed the sculpture during the 1930s.[53]

As many historians of modern culture have noted, Virginia Woolf asserted the origins of the modern era with her bold statement that "on or about December 1910, human character changed."[54] It may be equally audacious but useful to assert that, on or about December 1932, human character changed again, signifying an end to the era whose origins Woolf had identified. When the stock market crashed in October 1929, no one knew the long-term implications of that event. But by December 1932, it was fully evident that a new era in American history was under way and observers were acutely conscious of how dramatically everything had changed. Radio City Music Hall, and Rockefeller Center more generally, prompted many to articulate their thoughts about this change, and the buildings came to symbolize both the epitome of the old culture and the uncertain arrival of something new.

By focusing more narrowly on the particular aspect of modern culture to which this book has been dedicated, the beginning and ending points of modern aural culture can be asserted more specifically, and perhaps more rigorously. On 15 October 1900, with opening night at Symphony Hall, the era of modern acoustics began. The application of Wallace Sabine's reverberation equation to the design of this hall, and the cultural imperative that called for the control it was perceived to provide, rendered it the first auditorium of the modern acoustical era. Sabine's equation provided a key that opened the door to three decades of development in architectural acoustics, and all of these developments—in new materials, electrical inventions, engineering techniques, and scientific theories—resulted in greater degrees of control over sound. On 27 December 1932, with opening night at Radio City Music Hall, this era of modern acoustics came to an end. Here, the culminating technologies of architectural acoustics and electroacoustics combined to transcend completely the physical space of the architecture.

But this demonstration of technical mastery is not what drew the era to a close. Indeed, while a perception of mastery prevailed at the time, acousticians would soon realize they still had much to learn about the behavior of sound in rooms. What brought the era of modern acoustics to a close was a larger cultural shift, a new questioning of the unbounded technological enthusiasm that had sustained the past three decades of acoustical development. The Music Hall's opening precipitated a recognition that those fast-moving decades had now come to a halt. American life had entered a new and uncertain era, and the nation now faced problems that made the long-sought and hard-won mastery of those age-old mysteries of the acoustic seem like mere child's play.

Faith in technology would eventually be restored, but it would never really be quite the same. When engineers were no longer perceived to have all the answers; when their work ceased to inspire artists, writers, and musicians; when the machines they designed no longer challenged people to transform the age-old ways in which they perceived their world, the Machine Age was truly over and the modern soundscape would begin to transform itself again into something new.

Coda

Symphony Hall's opening occurred at the culmination of a culture that had invested science with unprecedented authority. The controlled modern sound that resulted from the application of acoustical science to architecture was something new, and the initially ambivalent critical reception of that sound reflected the unsettling nature of its newness. Radio City Music Hall constituted a culmination of modern control over sound, but this control was now exercised amid a sea of cultural change, and the ambivalent reception that greeted this hall attests to the discomfiture of cultural, not acoustical, change. Still, as with Symphony Hall, the criticisms initially leveled against Radio City gradually fell away as the strange became familiar and as the dislocations of change were replaced by something more stable. Architects and critics soon found much to praise in Rockefeller Center, and even those who had disparaged the complex earlier now began to regard it with respect. "Now that Rockefeller Center has been with us for nearly a decade," Douglas Haskell wrote in 1938, "it seems surprising that its forms should have been so savagely attacked by critics," and in 1941 critic Sigfried Giedion heralded the complex as a "great urban development."[1]

As the critical reputation of Rockefeller Center was restored, so, too, was America's faith in technology, and the country was soon counting on the massive machinations of the New Deal, particularly rural electrification and the construction of hydroelectric dams, to rebuild the economy and to reconstruct national prosperity. By the time prosperity returned after the Second World War, architectural technologies that had been innovative in 1933 had become commonplace, and the total environmental control that had characterized modern buildings like Radio City Music Hall and the PSFS Building became a new paradigm for postwar construction.

William Jordy has described the suspended acoustical plaster ceiling of Radio City Music Hall as a "complex technological apparatus" of lighting, air conditioning, and sound amplification that "demonstrates as spectacularly as any other single work of the period the vastly extended range of interior controls which modern technology gives to the architect."[2] This kind of technological tour de force would become standard practice in the 1950s and 1960s, as American auditoriums, office buildings, schools, and residences were transformed by the widespread adoption of the modern style and its embrace of technologies of acoustical and environmental control. Indeed, by the 1970s some were compelled to react against the dominance of modern architecture and all that it stood for.

Forty years after the elegant etching of acoustical tiles across the ceilings of the PSFS Building, Reyner Banham referred to the "tyranny of the tile format" as he described the relentless march of the gridded and standardized products of modern architecture that now threatened "to become almost absolute."[3] But Banham ultimately only criticized modern architects for not being modern enough, for not designing buildings that lived up to the full promise of environmental technologies. Others were more fundamentally critical of the entire technological enterprise, and Banham himself was taken to task for "insufficient concern for the well-tempered environment as it must eventually affect man's rapport with nature," for conflating "the optimum use of technique" with "the epitome of progress."[4]

While Americans had experienced a brief moment of disillusionment with the promise of modern technology during the early 1930s, the countercultural spirit of the 1970s sustained a far more significant critique. The fundamental equation of technological development with progress was fundamentally shaken by a generation that lived in the shadow of nuclear weapons. A heightened concern for the environment similarly affected the ways that people perceived the role of technology in their lives, and the ecological peril of pesticides like DDT was not the only danger that threatened.[5]

Environmentally conscious critics began to take issue with the energy-intensive technologies that were required to render habitable the hermetically sealed glass boxes of modern life. These buildings were not only bad for the earth and its ecosystem, they also proved more locally harmful to their own inhabitants. Asbestos, a fireproof miracle-material in the 1920s, turned out to be a silent killer, and air-conditioning systems were shown to be equally capable of killing by harboring lethal bacteria like *Legionella,* which caused an epidemic of deadly disease in Philadelphia in 1976. Buildings that had constituted the epitome of healthful practice in the 1920s were now diagnosed as inherently and incurably sick. Once perceived to protect workers from the perils of their environment, those same buildings now incubated illness within. Once dedicated to eliciting healthy and efficient performance from their occupants, they now sapped workers of their vitality and engendered only chronic fatigue.[6] The modern building was reevaluated and transformed from a solution to a problem, and the modern sound heard in those buildings was similarly reevaluated, and redefined as problematic, in the years after the Second World War.

The spaces of Guastavino construction, for example, were no longer considered as acoustically satisfactory as when they had first been built. By the 1960s, the congregations of a number of Akoustolith- and Rumford-lined churches and chapels had become dissatisfied with the sound of their sanctuaries. They wanted to increase the reverberation, to fill their space with the sound of space that the architects had originally banished. Riverside Church and St. Thomas's Church in New York, and the university chapels at Duke and Princeton were all reexamined by acoustical consultants who determined how best to redesign their sound. By painting over the vaults and thereby sealing off their porous surfaces, acousticians were able to reduce the absorptivity of the tiles and increase significantly the reverberation in these spaces.[7]

While Guastavino constructions are currently enjoying a renaissance of critical and popular interest, and while painstaking preservation projects have restored many of these buildings to their full visual glory, there appears so far to be little commitment to restoring and preserving the original acoustical environments created by the Rumford- and Akoustolith-lined vaults and domes. The sound, simply put, has lost its meaning and value.[8]

Standards for reverberation times in concert halls also changed in the years after the war. In 1959, when the New York Philharmonic Society articulated their acoustical criteria for a new hall, they emphasized that "the acoustics of the Hall should approximate as closely as possible those of the Boston Symphony Hall, when filled, but in no event should the reverberation time be shorter. We find the reverberation time of the London Festival Hall too short." "In our opinion," the report continued, "the acoustics of Kleinhans Hall in Buffalo and the Caracas Aula Magna are disappointing, whether due to the fan shape of the hall or the shortness of reverberation time, we are not prepared to say."[9]

The Royal Festival Hall, Kleinhans Music Hall, and the Aula Magna are all examples of fan-shaped, low-reverberation halls that had followed the precedent of the modern American auditoriums of the 1920s. From the time of its opening in 1940, the Kleinhans Hall suffered criticism for being too dead, and the Royal Festival Hall was similarly faulted when it opened ten years later.[10] By this time, criteria for optimum reverberation times had moved up from the values that had defined good sound several decades before. Additionally, acousticians had begun to identify factors other than reverberation that contributed to the acoustical success of an auditorium.

Leo Beranek and his colleagues at the acoustical consulting firm of Bolt, Beranek and Newman undertook a major survey of over 50 concert halls and opera houses from around the world. By 1961, they had collected physical data and subjective opinions on the acoustics of all of these structures.[11] Their study highlighted the importance of a parameter that Beranek labeled the "initial-time-delay gap," the interval of time that a listener experiences between the arrival of direct sound from a source, and the first reflections of sound received off the surfaces of the room. Later studies indicated that lateral reflections, sounds reflected off the side walls of a room, not only contributed significantly to creating a satisfactory initial-time-delay gap, but also played an important role in and of themselves, creating "the subjective impression of being enveloped by the sound, . . . the difference between feeling inside the sound and feeling on the outside observing it, as through a window."[12]

The absorbent and fan-shaped auditoriums of the 1920s and early 1930s not surprisingly failed to meet this new criterion, since they had been designed to achieve just the opposite effect. Then, listeners had sought to achieve an objective, detached mode of listening, "observing" the sound from outside, as if through the window of a monitoring booth. In a culture where engineers were heroes, how better to listen than like a sound engineer?

In 1929, Harold Arnold had summarized the modern acousticians' accomplishments with the proclamation: "Now with one broad sweep the barriers of time and space are gone."[13] Historians of modern culture have agreed with Arnold's conclusion and have emphasized the fundamentally modern character of the transmutation of time and space, the "separation of time and space from the central, premodern preeminence of place," that had characterized earlier eras.[14] The modern sound that Arnold and his colleagues created not only embodied the dislocations of time and space that the new acoustical technologies effected, it also constituted a way of dealing with those dislocations. As Wolfgang Schivelbusch has shown, the earlier disruptions of railroad travel had led travelers to find new ways to experience and to understand their now-rapid passage through space and time. By focusing exclusively on the panorama in the far distance, or by turning away to the virtual landscape captured in the pages of a novel, travelers established means by which to cope with the disorienting and distracting blur of the world passing by just outside their window.[15]

When modern acoustical technologies similarly severed the connections between sound and space, the result was equally exciting and distressing. Fifty years after it was first accomplished, R. Murray Schafer dubbed this splitting "schizophonia," to emphasize its pathological nature.[16] The unsettling possibilities of such a contextless sound were dealt with in the early twentieth century through the definitive identification of one best sound, the modern sound. The infinite possibilities of electroacoustic means were applied to just one end: the construction of a sound that was uniformly clear and direct, controlled and nonreverberant. Rick Altman has described this sound, as it was heard in relation to early sound cinema, as an aural anchor for the constantly changing visual aspect, "providing a satisfying and comfortable base from which the eyes can go flitting about, voyeuristically, satisfying our visual desires without compromising our unity and fixity."[17] That same kind of sound additionally provided an aural anchor in the world outside of the cinema, by establishing a standard that was heard clearly and distinctly above the din and confusion that modern technology had wrought.

This one best sound, the modern sound, was needed at the time it was created; it was a means to deal with change. Today, however, we seem to accept change more easily, and we appear to be more adept at dealing with the endless possibilities that acoustical technology presents. We embrace these possibilities and are unwilling to accept the idea that there exists just one best sound. While our acoustical criteria with respect to reverberation have moved away from the spaceless ideal of the modern sound, we have not simply returned to the aural values of an earlier era. Symphony Hall remains a highly valued place for listening to the nineteenth-century symphonic music to which it was originally dedicated, but we now require more.

Many of the most notable concert halls built in the late twentieth century are designed to be acoustically reconfigurable. They do not embody one best sound, but can instead be physically manipulated to create any one of a range of different acoustical environments. Acoustician Russell Johnson, for example, promotes a type of hall incorporating large moveable arrays of sound-absorbing fabric and sound-reflecting canopies, as well as adjustable walls that modify the size, shape, and therefore the sound, of the room. With these architectural features, a room can be configured prior to each perfomance to achieve a sound best suited to the particular type of music being performed, and reverberation times can be manipulated without affecting the clarity of the sound in the hall.[18]

The technologies deployed in such halls constitute architectural means by which to manipulate the acoustical character of a room. Similar results might be accomplished less expensively with electroacoustic technologies, and in the realm of popular music, such interventions are pervasive to the point that it is only news when a musician performs "Unplugged."[19] But audiences for so-called serious music have remained reluctant to accept such overt forms of technological intervention.

In the 1960s, an electronic system of "assisted resonance" was installed in the Royal Festival Hall in London to compensate for its lack of architectural reverberation. This was not a sound-amplification system, but rather one of tuned microphones and loudspeakers designed to add only reverberation and resonance to the live music in the hall. It was a descendant of the echo chambers and miking techniques that sound engineers had innovated in the 1920s to add spatial effects to otherwise nonreverberant radio broadcasts and motion picture sound tracks.[20] While the system was generally perceived to have improved the sound of the hall, the acousticians in charge maintained a policy of keeping "discussion about the system to a minimum, because of the passions likely to be aroused in some breasts by the thought of loudspeakers in the RFH."[21] Today, the use of such systems is far more pervasive, and sometimes just as secretive.

The openly acknowledged installation in 1999 of an electronic sound enhancement system in the New York State Theater, home of the New York City Opera, caused many critics and some fans to register dismay as they clearly did not share the view of the opera company's chairman, who characterized the system benignly as "electronic architecture."[22] A principal manufacturer of these systems admits that many of its customers ask not to be publicly identified, because the use of their systems in performance venues is not always openly acknowledged.[23] But in fact these systems, as employed in such venues, are not really as acoustically radical as many perceive them to be. While they create the effect of space through electronic rather than architectural means, they are nonetheless deployed to construct one single sound, a new acoustical signature for a space that, in this end if not in the means to achieve it, remains distinctly premodern. Outside of the concert hall, the same kind of technology has been employed in far more innovative ways, and imaginative users have taken advantage of its flexibility to create constantly changing soundscapes that could otherwise never be possible, even with configurable architecture.

In the 1950s, the architect Le Corbusier and the composer Edgard Varèse collaborated with sound engineers from the Philips Company to create electronically an acoustic space that was constantly changing. The result was the Philips Pavilion at the Brussels World's Fair of 1958. Three hundred loudspeakers were distributed throughout the structure and the goal was that listeners within would experience "the illusion that various sound-sources were in motion around them, rising and falling, coming together and moving apart again, and moreover the space in which this took place was to seem at one instant to be narrow and 'dry,' and at another to seem like a cathedral."[24]

The infinite and instantaneous changeability of electroacoustic constructions of space, like that achieved in Brussels, was put to scientific use in 1970, when acousticians at Bolt, Beranek and Newman sought to study the effects of different acoustical parameters on listeners' experiences. To execute this investigation, Thomas Horrall built an auditorium simulator—an array of twelve loudspeakers that were distributed in a nonreverberant room and fed twelve different signals that had been processed through circuitry, tape-loops, and a reverberation chamber—to create the effect of listening in different kinds of rooms. A control panel allowed the listener to switch the sound immediately from one "space" to another.[25] With the development of digital signal processing that has occurred since 1970, far more powerful simulators can now be constructed, and one such device has already appeared on the market.

The Wenger V-Room is a modular, soundproof studio that has been wired with the same electroacoustic technology that has been employed in auditoriums like the New York State Theater to modify the acoustics of their architecture. The V-Room, unlike these auditorium installations, takes full advantage of the flexibility provided by the technology, and by doing so, "Wenger's V-Room® Technology Opens the Door to Virtual Acoustics."[26]

> The Wenger V-Room lets you switch the acoustics of your music space as easily as changing the channels on a television. We call it variable, active acoustics.
>
> V-Room brings you literally dozens of acoustical simulations—each perfect, distinct, and so real that you will not believe your ears. Push a button and you're transported to center stage. Push another and you're standing in a gothic cathedral, a baroque concert hall or an amphitheater.[27]

With the V-Room, we have abandoned the idea of one best sound, and we have renounced the soundscape of modernity.

Andreas Huyssen has characterized the postmodern aesthetic that developed in the 1970s as "an ever wider dispersal and dissemination of artistic practices all working out of the ruins of the modernist edifice, raiding it for ideas, plundering its vocabulary and supplementing it with randomly chosen images and motifs from pre-modern and non-modern cultures as well as from contemporary mass culture."[28] The field of architecture was one of the first to legitimize this new aesthetic, as architects like Robert Venturi began to reject the technological purity of the ubiquitous glass towers and to replace their reductive simplicity with a self-conscious new complexity that was explicitly inspired by the examples of the past.[29]

The V-Room similarly constitutes a postmodern soundscape, treating the past—be it Gothic, baroque, or modern—like an endlessly stimulating old album of phonograph recordings from which we are privileged to pick and choose. Acoustical technology in the modern era had been dedicated to eliminating the effect of space and replacing it with one best sound, the modern sound. Postmodern acoustical technologies, in contrast, summon forth the sound of space so easily and in so many varieties, we hardly know what to listen to first.

The V-Room is currently marketed as an acoustically flexible practice space for music education, but it has also been employed as a rehearsal room in professional venues, and one television network has used the technology to match the sounds of studio announcers to the varying acoustics of live feeds that come from different sporting events.[30] Whether or not this technology will migrate more substantially out into the wider world remains to be seen.

Today's concert halls are poised rather tentatively at the threshold of the postmodern soundscape, still providing anchors in a world filled with change. Some are willing to admit a multitude of sound-spaces, but nonetheless cling conservatively to the idea that those sounds must be based in the architectural materiality of walls, ceilings, and floors. Others embrace the possibilities of digitally constructed sound-space, but they resist the temptation to play with those possibilities. Instead, they set the electronic controls in one "best" position and subsequently pretend that those controls don't exist. What happens next is a story for a future historian to tell.

A history of the soundscape of early twentieth-century America does not provide the means to predict what the soundscape of the twenty-first century will sound like. Nonetheless, it demonstrates clearly that the power with which we are currently endowed is a direct result of the technological accomplishments of that earlier era. The modern belief in one best sound is no longer unquestioned, and the modern sound is now but one of many to explore. By understanding more fully why that particular sound was so compelling to the people who constructed it, however, we may be better equipped to make wise choices from the many options we have inherited from the past.

Notes

Chapter One

1. R. Murray Schafer, *The Soundscape: Our Sonic Environment and the Tuning of the World* (Rochester, Vt.: Destiny Books, 1994), definition on 274–275. This edition is a largely unrevised version of material originally written in the 1960s and 1970s. See also Schafer, *The New Soundscape* (Scarborough, Ont., and New York: Berandol Music and Associated Music Publishers, 1969); and *The Book of Noise* (Wellington, N.Z.: Price Milburn, 1970). Equally stimulating in more theoretical ways is Jacques Attali, *Noise: The Political Economy of Music*, trans. Brian Massumi (1977; Minneapolis: University of Minnesota Press, 1985).

2. Alain Corbin, *Village Bells: Sound and Meaning in the 19th-Century French Countryside*, trans. Martin Thom (1994; New York: Columbia University Press, 1998), ix.

3. See Barry Truax, *Acoustic Communication* (Norwood, N.J.: Ablex, 1984), for a similarly contextualized study of the contemporary soundscape.

4. Dayton Clarence Miller, *Anecdotal History of the Science of Sound* (New York: Macmillan, 1935); Frederick Vinton Hunt, *Origins in Acoustics: The Science of Sound from Antiquity to the Age of Newton* (1978; Woodbury, N.Y.: Acoustical Society of America, 1992), and *Electroacoustics: The Analysis of Transduction, and Its Historical Background* (1954; n.p.: Acoustical Society of America, 1982); Robert Bruce Lindsay, "Historical Introduction" to J. W. S. Rayleigh, *The Theory of Sound*, 2d ed. (1894; New York: Dover, 1945), v–xlii; John W. Kopec, *The Sabines at Riverbank: Their Role in the Science of Architectural Acoustics* (Woodbury, N.Y.: Acoustical Society of America, 1997); Robert T. Beyer, *Sounds of Our Times: Two Hundred Years of Acoustics* (New York: Springer-Verlag, 1999); and the numerous historical articles that have appeared in the *Journal of the Acoustical Society of America* since its founding in 1929. An equally compelling and beautifully illustrated historical account by the architect Michael Forsyth is *Buildings for Music: The Architect, the Musician, and the Listener from the Seventeenth Century to the Present Day* (Cambridge, Mass.: The MIT Press, 1985).

5. Penelope Gouk, *Music, Science and Natural Magic in Seventeenth-Century England* (New Haven: Yale University Press, 1999); Thomas L. Hankins and Robert J. Silverman,

Instruments and the Imagination (Princeton: Princeton University Press, 1995); Timothy Lenoir, "Helmholtz and the Materialities of Communication," *Osiris*, 2d ser., 9 (1994): 184–207; Robert Jacob Silverman, "Instrumentation, Representation, and Perception in Modern Science: Imitating Human Function in the Nineteenth Century," Ph.D. Dissertation, University of Washington, 1992; H. F. Cohen, *Quantifying Music: The Science of Music at the First Stage of the Scientific Revolution, 1580–1650* (Dordrecht: D. Reidel, 1984); Sigalia Dostrovsky, "Early Vibration Theory: Physics and Music in the Seventeenth Century," *Archive for History of Exact Sciences* 14 (December 1975): 169–218. Although not situated within the traditions of the history of science, a thoughtful survey of the inscriptive practices of eighteenth- and nineteenth-century investigators of sound is found in James Lastra, *Sound Technology and the American Cinema: Perception, Representation, Modernity* (New York: Columbia University, 2000).

6. My interest in the material practice of science has been stimulated particularly by the works of Peter Galison, Robert Kohler, and Angela Creager. See Peter Galison, *Image and Logic: A Material Culture of Microphysics* (Chicago: University of Chicago Press, 1997); Robert E. Kohler, *Lords of the Fly: Drosophila Genetics and the Experimental Life* (Chicago: University of Chicago Press, 1994); and Angela N. H. Creager, *The Life of a Virus: Tobacco Mosaic Virus as an Experimental Model, 1930–1965* (Chicago: University of Chicago Press, 2002). I began to explore this approach in my article, "Dead Rooms and Live Wires: Harvard, Hollywood, and the Deconstruction of Architectural Acoustics, 1900–1930," *Isis* 88 (December 1997): 597–626.

7. Hugh G. J. Aitken, *Syntony and Spark: The Origins of Radio* (1976; Princeton: Princeton University Press, 1985), and *The Continuous Wave: Technology and American Radio, 1900–1932* (Princeton: Princeton University Press, 1985); Susan J. Douglas, *Inventing American Broadcasting, 1899–1922* (Baltimore: Johns Hopkins University Press, 1987), and *Listening In: Radio and the American Imagination* (New York: Times Books, 1999). See also the essays collected in Hans-Joachim Braun, ed., *"I Sing the Body Electric": Music and Technology in the 20th Century* (Hofheim: Wolke Verlag, 2000); David Morton, *Off the Record: The Technology and Culture of Sound Recording in America* (New Brunswick: Rutgers University Press, 2000); Alexander Boyden Magoun, "Shaping the Sound of Music: the Evolution of the Phonograph Record, 1877–1950," Ph.D. Dissertation, University of Maryland, College Park, 2000; James P. Kraft, *Stage to Studio: Musicians and the Sound Revolution, 1890–1950* (Baltimore: Johns Hopkins University Press, 1996); Andre Millard, *America on Record: A History of Recorded Sound* (Cambridge: Cambridge University Press, 1995); Joseph O'Connell, "The Fine-Tuning of a Golden Ear: High-End Audio and the Evolutionary Model of Technology," *Technology and Culture* 33 (January 1992): 1–37; and Susan Schmidt Horning, "Chasing Sound: The Culture and Technology of Recording Studios in America, 1877–1977," Ph.D. Dissertation, Case Western Reserve University, 2002.

8. I began to develop this emphasis in "Machines, Music, and the Quest for Fidelity: Marketing the Edison Phonograph in America, 1877–1925," *Musical Quarterly* 79 (spring 1995): 131–171. My ability to expand upon this aspect of my work in the current study has been directly inspired by Douglas, *Listening In*.

9. For a survey of this literature, see Jeffrey K. Stine and Joel A. Tarr, "At the Intersection of Histories: Technology and the Environment," *Technology and Culture* 39 (October 1998): 601–640; and Stine and Tarr, "Technology and the Environment: The Historians' Challenge," *Environmental History Review* 18 (spring 1994): 1–7, introduction to a special issue dedicated to "Technology, Pollution and the Environment." See also Martin V. Melosi, "The Place of the City in Environmental History," *Environmental History Review* 17 (1993): 1–23, and Melosi, ed., *Pollution and Reform in American Cities, 1870–1930* (Austin: University of Texas Press, 1980).

10. See, most notably, the works of Raymond W. Smilor: "Toward an Environmental Perspective: The Anti-Noise Campaign, 1893–1932," pp. 135–151 in Melosi, ed., *Pollution and Reform*; "Personal Boundaries in the Urban Environment: The Legal Attack on Noise: 1865–1930," *Environmental Review* 3 (spring 1979): 24–36; and "Confronting the Industrial Environment: The Noise Problem in America, 1893–1932," Ph.D. Dissertation, University of Texas at Austin, 1978. More in tune with my own approach is Warren Bareiss, "Noise Abatement in Philadelphia, 1907–1966: The Production of a Soundscape," M.A. Thesis, University of Pennsylvania, 1990. See also Hillel Schwartz, "Beyond Tone and Decibel: The History of Noise," *Chronicle of Higher Education* (9 January 1998): B8.

11. One vocal exception to the general silence on the topic of acoustics in architectural history is Reyner Banham, *The Architecture of the Well-tempered Environment*, 2d ed. (Chicago: University of Chicago Press, 1984). My own understanding of the emergence of modern architecture was initially presented in "Listening to/for Modernity: Architectural Acoustics and the Development of Modern Spaces in America," pp. 253–280 in Peter Galison and Emily Thompson, eds., *The Architecture of Science* (Cambridge, Mass.: The MIT Press, 1999). See also Forsyth's specialized study, *Buildings for Music*; Cecil D. Elliott, *Technics and Architecture: The Development of Materials and Systems for Buildings* (Cambridge, Mass.: The MIT Press, 1992); and James Marston Fitch, *American Building 2: The Environmental Forces That Shape It,* 2d ed. (New York: Schocken Books, 1971) for historical works that have considered the acoustic elements of architectural design.

12. See particularly the essays collected in Rick Altman, ed., *Sound Theory/Sound Practice* (New York: Routledge, 1992), and in Elisabeth Weis and John Belton, eds., *Film Sound: Theory and Practice* (New York: Columbia University Press, 1985). See also Lastra, *Sound Technology and the American Cinema*; Donald Crafton, *The Talkies: American Cinema's Transition to Sound, 1926–1931* (New York: Charles Scribner's Sons, 1997); Scott Eyman, *The Speed of Sound: Hollywood and the Talkie Revolution, 1926–1930* (New York: Simon and Schuster, 1997); and Michel Chion, *Audio-Vision: Sound on Screen*, trans. Claudia Gorbman (New York: Columbia University Press, 1994).

13. Leigh Eric Schmidt, *Hearing Things: Religion, Illusion, and the American Enlightenment* (Cambridge, Mass.: Harvard University Press, 2000), and "From Demon Possession to Magic Show: Ventriloquism, Religion, and the Enlightenment," *Church History* 67

(June 1998): 274–304; Mark M. Smith, "Listening to the Heard Worlds of Antebellum America," *Journal of the Historical Society* 1 (spring 2000): 65–99, and "Time, Sound, and the Virginia Slave," pp. 29–60 in John Saillant, ed., *Afro-Virginian History and Culture* (New York: Garland, 1999).

14. Bruce R. Smith, *The Acoustic World of Early Modern England: Attending to the O-Factor* (Chicago: University of Chicago Press, 1999); James H. Johnson, *Listening in Paris: A Cultural History* (Berkeley: University of California Press, 1995); Corbin, *Village Bells*. See also Corbin, *Time, Desire and Horror: Toward a History of the Senses,* trans. Jean Birrell (Cambridge: Polity Press, 1995); and Constance Classen, *Worlds of Sense: Exploring the Senses in History and Across Cultures* (London: Routledge, 1993).

15. This perspective is enhanced by comparing these works to ethnomusicological studies of aural culture in non-Western societies. Most notable here is Steven Feld, *Sound and Sentiment: Birds, Weeping, Poetics, and Song in Kaluli Expression,* 2d ed. (Philadelphia: University of Pennsylvania Press, 1990). See also Feld, "Sound Structure as Social Structure," *Ethnomusicology* 28 (September 1984): 383–409. For ethnographic studies of contemporary Western soundscapes, see Jonathan Sterne, "Sounds Like the Mall of America: Programmed Music and the Architectonics of Commercial Space," *Ethnomusicology* 41 (winter 1997): 22–50; and Thomas Porcello, "The Ethics of Digital Audio-Sampling: Engineers' Discourse," *Popular Music* 10 (1991): 69–84.

16. Schmidt, *Hearing Voices,* 28. For explorations of modernity and visual culture, see the essays collected in David Michael Levin, ed., *Modernity and the Hegemony of Vision* (Berkeley: University of California Press, 1993); and in Hal Foster, ed., *Vision and Visuality* (Seattle: Bay Press, 1988); Jonathan Crary, *Techniques of the Observer: On Vision and Modernity in the Nineteenth Century* (Cambridge, Mass.: The MIT Press, 1993); and Barbara Maria Stafford, *Artful Science: Enlightenment, Entertainment, and the Eclipse of Visual Education* (Cambridge, Mass.: The MIT Press, 1994). Crary's *Suspensions of Perception: Attention, Spectacle, and Modern Culture* (Cambridge, Mass.: The MIT Press, 1999) recognizes the role of aural as well as visual culture in modernity, as does Steven Connor's essay, "The Modern Auditory I," pp. 203–223 in Roy Porter, ed., *Rewriting the Self: Histories from the Renaissance to the Present* (London: Routledge, 1997).

17. Douglas Kahn, *Noise Water Meat: A History of Sound in the Arts* (Cambridge, Mass.: The MIT Press, 1999), 4, 9. See also Kahn, "Sound Awake," *Australian Review of Books* (July 2000): 21–22.

18. Lastra, *Sound Technology and American Cinema,* 4.

19. See, for example, Stephen Kern, *The Culture of Time and Space, 1880–1918* (Cambridge: Harvard University Press, 1983), quote on 119. See also Terry Smith, *Making the Modern: Industry, Art, and Design in America* (Chicago: University of Chicago Press, 1993); Anson Rabinbach, *The Human Motor: Energy, Fatigue, and the Origins of Modernity* (Berkeley: University of California Press, 1990); Miles Orvell, *The Real Thing: Imitation and Authenticity in American Culture, 1880–1940* (Chapel Hill: University of

North Carolina Press, 1989); and Cecelia Tichi, *Shifting Gears: Technology, Literature, Culture in Modernist America* (Chapel Hill: University of North Carolina Press, 1987).

20. "All fixed, fast-frozen relations, with their train of ancient and venerable prejudices and opinions, are swept away, all new-formed ones become antiquated before they can ossify. All that is solid melts into air, all that is holy is profaned, and men at last are forced to face. . . . The real conditions of their lives and their relations with their fellow men." Quoted in Marshall Berman, *All That Is Solid Melts into Air: The Experience of Modernity* (1982; New York: Penguin, 1988), 21.

21. Here my own work is particularly inspired by the material histories of Wolfgang Schivelbusch, whose studies of the technologies of travel and light are both moving and illuminating. Schivelbusch, *The Railway Journey: The Industrialization of Time and Space in the 19th Century* (1977; Berkeley: University of California Press, 1986); and *Disenchanted Night: The Industrialization of Light in the Nineteenth Century* (1983; Berkeley: University of California Press, 1988).

22. Wallace Clement Sabine Correspondence Files, Rudolph Markgraf to Professor G. W. Stewart (6 September 1911); copy sent to Wallace Sabine (8 September 1911). Courtesy of Riverbank Acoustical Laboratories, Illinois Institute of Technology Research Institute, Geneva, Ill. The Harvard University Archives hold photocopies of Sabine's correspondence, the originals of which are held at the Riverbank Acoustical Laboratories.

Chapter Two

1. H. E. K. (Henry E. Krehbiel), "The New Symphony Hall in Boston," *New-York Daily Tribune* (16 October 1900), Clipping Book, Pres 56, vol. 16, p. 21, Boston Symphony Orchestra Archives [BSO Archives].

2. "Boston's New Music-Hall," *New York Evening Post* (16 October 1900), Clipping Book, Pres 56, vol. 16, p. 21, BSO Archives.

3. Henry Lee Higginson to Charles McKim (27 November 1892), Papers of McKim, Mead & White, Folder M-10: Boston Music Hall, New-York Historical Society [MMW Papers]. Emphasis in original. This letter, along with almost all others between Higginson, McKim, and Sabine cited in this chapter, has recently been reproduced in the centenary celebration volume by Richard Poate Stebbins, *The Making of Symphony Hall, Boston* (Boston: Boston Symphony Orchestra, 2000). Stebbins additionally presents his own historical account of the design and construction of the hall. See also *Symphony Hall: The First 100 Years* (Boston: Boston Symphony Orchestra, 2000); and H. Earle Johnson, *Symphony Hall, Boston* (Boston: Little, Brown, 1950).

4. McKim to Higginson (10 July 1893), Papers of Charles Follen McKim, General Correspondence: 1893, Library of Congress, Manuscript Division. I have consulted the microfilm edition of these papers.

5. "New Music Hall Design," *Boston Evening Transcript* (15 January 1894): 1. Photographs of this model are reproduced in Stebbins, *Making of Symphony Hall,* 34–37.

6. Higginson to McKim (23 April 1894), Folder M-10, MMW Papers.

7. Higginson to McKim (27 October 1898), Folder M-10, MMW Papers.

8. This form also followed Boston's old Music Hall, which was considered an acoustical success.

9. Hans Richter is quoted in Charles Moore, *The Life and Times of Charles Follen McKim* (Boston: Houghton Mifflin Co., 1929), 103.

10. Higginson to McKim (23 April 1894), Folder M-10, MMW papers, emphasis in original. Charles Cross was a professor at the Massachusetts Institute of Technology identified by Higginson to McKim in 1892 as "*The* authority on sound here." Higginson to McKim (27 November 1892), Folder M-10, MMW Papers, emphasis in original. For a brief mention of Cross's work at MIT, see Christophe Lécuyer, "MIT, Progressive Reform, and 'Industrial Service,' 1890–1920," *Historical Studies in the Physical and Biological Sciences* 26 (1995): 35–88. Higginson also cited a second scientist, whose name is difficult to discern, in the hand-written letter.

11. Higginson quoted the telegram in his subsequent letter to McKim (26 January 1899), Folder M-10, MMW Papers.

12. Higginson to McKim (26 January 1899), Folder M-10, MMW Papers.

13. McKim to Higginson (27 February 1899), Ser. XII, Box 6A, Folder: 1899-Music, Henry Lee Higginson Collection, Historical Collections, Baker Library, Harvard Business School [Higginson Papers].

14. Quoted in M. A. DeWolfe Howe, *The Boston Symphony Orchestra: 1881–1931* (Boston: Houghton Mifflin, 1931), 113.

15. For surveys of the early history of acoustics, see Frederick Vinton Hunt, *Origins in Acoustics: The Science of Sound from Antiquity to the Age of Newton* (1978; Woodbury, N.Y.: Acoustical Society of America, 1992); and Dayton Clarence Miller, *Anecdotal History of the Science of Sound* (New York: MacMillan, 1935). Vitruvian ideas on theater design are found in his *Ten Books on Architecture*, trans. Morris Hicky Morgan (1914; New York: Dover, 1960), 137–153. See also Robert G. Arns and Bret E. Crawford, "Resonant Cavities in the History of Architectural Acoustics," *Technology and Culture* 36 (January 1995): 104–135.

16. For more on music and acoustics in the development of modern science, see Penelope Gouk, *Music, Science and Natural Magic in Seventeenth-Century England* (New Haven: Yale University Press, 1999), "Performance Practice: Music, Medicine and Natural Philosophy in Interregnum Oxford," *British Journal for the History of Science* 29 (1996): 257–288, "Acoustics in the Early Royal Society, 1660–1680," *Notes and Records of the Royal Society of London* 36 (February 1982): 155–175, and "The Role of Acoustics

and Music Theory in the Scientific Work of Robert Hooke," *Annals of Science* 37 (1980): 573–605; H. F. Cohen, *Quantifying Music: The Science of Music at the First Stage of the Scientific Revolution, 1580–1650* (Dordrecht: D. Reidel, 1984); and Sigalia Dostrovsky, "Early Vibration Theory: Physics and Music in the Seventeenth Century," *Archive for History of Exact Sciences* 14 (December 1975): 169–218.

17. Gouk, "Acoustics in the Early Royal Society." For a fascinating account of how the rationalizing power of science transformed ventriloquial voices from the sacred voice of God (or of the devil) into the mundane deceptions of parlor magicians, see Leigh Eric Schmidt, *Hearing Things: Religion, Illusion, and the American Enlightenment* (Cambridge, Mass.: Harvard University Press, 2000), and "From Demon Possession to Magic Show: Ventriloquism, Religion, and the Enlightenment," *Church History* 67 (June 1998): 274–304. See also Thomas L. Hankins and Robert J. Silverman, *Instruments and the Imagination* (Princeton: Princeton University Press, 1995) for more on talking automata.

18. Alberto Pérez-Gómez, *Architecture and the Crisis of Modern Science* (Cambridge, Mass.: The MIT Press, 1983), 4. See also Pérez-Gómez, "Architecture *as* Science: Analogy or Disjunction?" pp. 337–351, and Antoine Picon, "Architecture, Science, and Technology," pp. 309–335, in Peter Galison and Emily Thompson, eds., *The Architecture of Science* (Cambridge, Mass.: The MIT Press, 1999); and Spiro Kostof, ed., *The Architect: Chapters in the History of the Profession* (New York: Oxford University Press, 1977).

19. The revised and expanded 1894 edition of Rayleigh's treatise is usually cited today. See John William Strutt, Baron Rayleigh, *The Theory of Sound,* 2d ed. (1894; New York: Dover, 1945). For more on Rayleigh's work and the history of acoustics more generally, see the introduction to this edition by Robert Bruce Lindsay. The mathematical analysis of sound in the seventeenth and eighteenth centuries is surveyed in Clifford Truesdell, ed., "Editor's Introduction" to *Leonhardi Euleri Opera Omnia* (Lipsiae: Teubneri, 1911+), ser. 2, 13: xix–lxxii.

20. Miller, *Anecdotal History*, 41–42. See also David Cahan, "From Dust Figures to the Kinetic Theory of Gases: August Kundt and the Changing Nature of Experimental Physics in the 1860s and 1870s," *Annals of Science* 47 (1990): 151–172; and Bernard S. Finn, "Laplace and the Speed of Sound," *Isis* 55 (March 1964): 7–19.

21. Miller, *Anecdotal History*, 85–92.

22. For more on these devices and approaches, see Hankins and Silverman, *Instruments and the Imagination*; James Lastra, *Sound Technology and the American Cinema* (New York: Columbia University Press, 2000), 16–60; and Robert Brain, "Standards and Semiotics," pp. 249–284, 414–425 in Timothy Lenoir, ed., *Inscribing Science: Scientific Texts and the Materiality of Communication* (Stanford: Stanford University Press, 1998).

23. Rayleigh, *Theory of Sound*, vol. 2, 69–72, 328–333, quote from table of contents, x. He concluded, "In large spaces bounded by non-porous walls, roof and floor, and with few windows, a prolonged resonance seems inevitable. The mitigating influence of thick carpets in such cases is well known. The application of similar material to the walls and to the roof appears to offer the best chance of further improvement" (vol. 2, p. 333).

24. Pierre Patte, *Essai sur l'architecture théâtrale. Ou de l'ordonnance la plus avantageuse à une salle de spectacles, relativement aux principes de l'optique et de l'acoustique* (Paris: Chez Moutard, 1782), 154.

25. Jean George Noverre, in his 1781 treatise *Observations sur la Construction d'une Salle d'Opéra*, referred to "La commodité du publique qui paye, et le droit qu'il a de voir et d'entendre." *Oevres de M. Noverre, Tome III* (St. Petersburg: Jean Charles Schnoor, 1804), 5. The theaters of Elizabethan England constitute earlier examples of theaters both large and commercial. But, as Shakespeare himself demonstrated when he proclaimed that "All the world's a stage," a premodern cosmology permeated the meaning of the sounds heard within these structures in ways no longer evident in the theaters of the eighteenth century. For more on the soundscape of Elizabethan drama, see Bruce R. Smith, *The Acoustic World of Early Modern England: Attending to the O-Factor* (Chicago: University of Chicago Press, 1999).

26. Figures from Michael Forsyth, *Buildings for Music: The Architect, the Musician, and the Listener from the Seventeenth Century to the Present Day* (Cambridge, Mass.: The MIT Press, 1985), 334–335.

27. Benjamin Wyatt, *Observations on the Principles of a Design for a Theatre* (London: Lowndes and Hobbs, 1811), 4.

28. Wyatt, *Observations*, 7.

29. Count Francesco Algarotti, *An Essay on the Opera* (London: Davis and Reymers, 1767), 98. This English translation is from the original Italian edition of 1762, which was also translated into German in 1769 and French in 1773. Algarotti's critique may have been directed at the Galli-Bibiena family of theater designers, who were famous for their bell-shaped theaters. See Forsyth, *Buildings for Music*, 80–94. For more on Algarotti, see the biographical sketch by Richard Northcott that accompanies the 1917 reprint of his essay by the Press Printers, London.

30. His acoustical goal was "fair valoir la voix," or "faire valoir la force de retour de la voix." Patte, *Essai sur l'Architecture*, 207 and 23.

31. Patte, *Essai sur l'Architecture*, 157 and 207.

32. George Saunders, *A Treatise on Theatres* (London: I. & J. Taylor, 1790), x.

33. Saunders, *Treatise on Theatres*, 20–22; Algarotti, *Essay on the Opera*, 92–95; Patte, *Essai sur l'Architecture*, 207.

34. This aesthetic, a reaction against the Baroque, also influenced the authors' proscription of ornament, which, like absorbent materials, was believed to weaken the sound by causing "irregular" reflections. For more on neoclassicism and its implications for the design not only of theaters but also of lecture halls and legislative assemblies, see Emil Kaufmann, *Architecture in the Age of Reason* (1955; New York: Dover, 1968), esp. 168 and 206; and Forsyth, *Buildings for Music*, 108–116.

35. Forsyth, *Buildings for Music*, 115. See also Rand Carter, "The Drury Lane Theatres of Henry Holland and Benjamin Dean Wyatt," *Journal of the Society of Architectural Historians* [*JSAH*] 26 (October 1967): 200–216.

36. "Remarks on the best form of a room for hearing and speaking," Benjamin Latrobe to Thomas Parker, c. 1803, pp. 400–408 in John C. Van Horne and Lee W. Formwalt, eds., *The Correspondence and Miscellaneous Papers of Benjamin Henry Latrobe, Vol. 1: 1784–1804* (New Haven: Yale University Press, 1984), quotes on 404.

37. *True American and Commercial Advertiser* (31 October 1807), quoted in John C. Van Horne, ed., *The Correspondence and Miscellaneous Papers of Benjamin Henry Latrobe, Vol. 2: 1805–1810* (New Haven: Yale University Press, 1986) [*Latrobe Papers 2*], 498.

38. Latrobe to William Duane (29 February 1808), *Latrobe Papers 2*, 528.

39. Benjamin Latrobe, "Note," *Edinburgh Encyclopedia* (Philadelphia: Joseph and Edward Parker, 1832) 1: 120–124, on 122. Latrobe's article was written in 1811 for this American edition of the encyclopedia as an addendum to William Campbell's original article on "Acoustics," which did not address architectural problems. See also John C. Van Horne, ed., *The Correspondence and Miscellaneous Papers of Benjamin Henry Latrobe, Vol. 3: 1811–1820* (New Haven: Yale University Press, 1988) [*Latrobe Papers 3*], 208–221.

40. Latrobe, "Report of the Surveyor of the Publick Buildings of the United States at Washington" (23 March 1808), *Latrobe Papers 2*, 565–577, on 569.

41. The source for much of this material is the Curator's Office of the Architect of the Capitol. The Art and Reference File "Old House-Acoustics" contains numerous relevant items, as do the files "House Chamber-Statuary Hall-Alterations," and "House Chamber-Statuary Hall-Seating." Many of these records, including Congressional documents, are reprinted in *Documentary History of the Construction and Development of the United States Capitol Building and Grounds* (Washington: Government Printing Office, 1904). See also George C. Hazelton, Jr., *The National Capitol: Its Architecture, Art and History* (New York: J.F. Taylor, 1902), 227–228 and 275–287.

42. This proposal was initially made in 1832 by the architect Robert Mills, who inherited the challenge of improving the sound in the Hall of the House of Representatives, along with his ideas on how to do so, from his mentor Benjamin Latrobe. See Robert Mills, "Hall of the House of Representatives, U.S." (14 January 1830), 21st Congress, 1st Session, Rep. 83; and "Alteration of Hall House of Representatives" (30 June 1830), 22d Congress, 1st Session, Rep. 495. See also George W. Williams, "Robert Mills' 'Contemplated Addition to St. Michael's Church, Charleston,' and 'Doctrine of Sounds,'" *JSAH* 12 (March 1953): Special Documentary Supplement, 23–31.

43. Jefferson Davis to Montgomery C. Meigs (4 April 1853), reproduced in Serial 691, *Senate Executive Documents,* no. 1, 33d Congress, 1st Session, part 2, pp. 69–70, on 69. Thanks to Kathleen Dorman of the Joseph Henry Papers, Smithsonian Institution, for

providing me with a copy of this report. The architect of the Capitol extensions was Thomas U. Walter.

44. Jefferson Davis to Joseph Henry and Alexander Dallas Bache (20 May 1853), reproduced in Serial 691, p. 85.

45. Joseph Henry and Alexander Dallas Bache to Jefferson Davis (24 June 1853), reproduced in Serial 691, pp. 85–86, on 86. Before approving the plans, Henry and Bache accompanied Meigs on an extended tour of America's acoustically notable buildings in order to identify factors that led to good and bad sound. For Meigs's report on this trip, see Serial 691, pp. 70–71.

46. Meigs had reported that the acoustical survey undertaken by himself, Henry, and Bache included "comparisons between rooms of different forms and sizes," and the collection of "sketches, showing the general form and dimensions" of these rooms. Serial 691, pp. 70, 71.

47. William Shand, "Observations on the Adaptation of Public Buildings to the Propagation of Sound," *Journal of the Franklin Institute* 39 (January 1845): 1–9, on 3. A writer for the *Builder* noted equally critically that the "attractive fact of two persons placed considerably apart, but in the two foci of an ellipse, hearing each other comfortably (almost confidentially), seems to be the turning point at which all go off in the sole pursuit of form in designing an apartment." "A Few Gropings in Practical Acoustics," *Builder* 8 (31 August 1850): 411–412, on 412.

48. A. W. Webster, *On the Principles of Sound; Their Application to the New Houses of Parliament, and Assimilation with the Mechanism of the Ear* (London: The Author, 1840), 30.

49. Shand, for example, argued that the "sonority" of a material was dependent on its density. Another account argued that the intensity of sound in a material was proportional to the square of the velocity of sound in that material, "hence iron is about eighteen times as noisy as lead." Shand, "Observations," 5; "Our Office Table," *Building News* 43 (6 October 1882): 433.

50. Joseph Henry, "On Acoustics Applied to Public Buildings," *Smithsonian Institution Report* (1856): 221–234, on 228. This article also appeared in the *Proceedings of the American Association for the Advancement of Science* [*PAAAS*] 10 (1856): 119–135.

51. Henry, "Acoustics Applied to Buildings," 229. He measured the increase with a thermocouple attached to a sensitive galvanometer.

52. Henry, "Acoustics Applied to Buildings," 233–234. His contributions were based on the results of other experiments that he had carried out. Like Saunders, Henry measured the extent of the voice, and he also determined the distance at which a distinct echo is first perceived off a reflective surface. He called this distance the limit of perceptibility, and he used this dimension—about thirty feet—as a controlling factor in the size and shape of the lecture hall. Robert Shankland credits the limit of perceptibility as an origi-

nal discovery, but numerous earlier authors on acoustics had presented similar measures, in units of space or time, of this phenomenon. See, for example, Archibald Campbell, "Acoustics," *Edinburgh Encyclopedia* (Philadelphia: Joseph and Edward Parker, 1832), 1:104–124, on 117. See also Robert Shankland, "Architectural Acoustics in America to 1930," *JASA* 61 (February 1977): 250–254.

53. Henry, "Acoustics Applied to Buildings," 222. Albert E. Moyer indicates that the Henry family lived in the Smithsonian castle in *Joseph Henry: The Rise of an American Scientist* (Washington: Smithsonian Institution Press, 1997), 274.

54. Histories of historicist architecture include: Talbot Hamlin, *Greek Revival Architecture in America* (1944; New York: Dover, 1964); William H. Pierson, Jr., *Technology and the Picturesque, The Corporate and Early Gothic Styles* (New York: Oxford University Press, 1978); Walter C. Kidney, *The Architecture of Choice: Eclecticism in America, 1880–1930* (New York: George Braziller, 1974); and Richard W. Longstreth, "Academic Eclecticism in American Architecture," *Winterthur Portfolio* 17 (1982): 55–82.

55. For a study of all three projects, see James V. Kavanaugh, "Three American Opera Houses: The Boston Theatre, The New York Academy of Music, The Philadelphia Academy of Music," M.A. Thesis, University of Delaware, 1967. See also John Francis Marion, *Within These Walls: A History of the Academy of Music in Philadelphia* (Philadelphia: Academy of Music, 1984).

56. N. Le Brun and G. Runge, *History and Description of the Opera House or American Academy of Music in Philadelphia* (Philadelphia: G. Andre, 1857), 15–16.

57. Higginson to McKim (26 January 1899), folder M-10, MMW Papers.

58. For more on Adler, see Joan Weil Saltzstein, "The Autobiography and Letters of Dankmar Adler," *Inland Architect* 27 (September–October 1983): 16–27, and "Dankmar Adler: The Man, The Architect, The Author," *Wisconsin Architect* (July–August 1967): 15–19; and Rochelle S. Elstein, "The Architecture of Dankmar Adler," *JSAH* 26 (December 1967): 242–249. Information can also be gleaned from biographies of his more famous partner, esp. Robert Twombley, *Louis Sullivan: His Life and Work* (Chicago: University of Chicago Press, 1986); and Hugh Morrison, *Louis Sullivan: Prophet of Modern Architecture* (New York: Peter Smith, 1952). For Adler & Sullivan's theater projects, see Twombley; Jeffrey Karl Ochsner and Dennis Alan Andersen, "Adler and Sullivan's Seattle Opera House Project," *JSAH* 48 (September 1989): 223–231; and Yvonne Shafer, "The First Chicago Grand Opera Festival: Adler and Sullivan Before the Auditorium," *Theatre Design and Technology* 13 (fall 1977): 9–13, 38.

59. Morrison, *Louis Sullivan*, 289. Frank Lloyd Wright, who worked as a draftsman for Adler & Sullivan during their heyday, claimed that "the old Chief" consulted on fifty-seven different theaters while Wright was employed by the firm. Frank Lloyd Wright, "Acoustics in Building," Recorded lecture to students at Taliesin Fellowship, Scottsdale, Ariz., 1952 (Audio-Forum Sound Seminars, #11021).

60. Twombley indicates that, in 1890, Tuthill made sure that the credit "Adler and Sullivan" was restored to a published drawing of Carnegie Hall that had omitted their names, but he later failed to acknowledge Adler in an article that he wrote on acoustical design. See the pamphlet by William Burnet Tuthill, *Practical Acoustics: A Study of the Diagrammatic Preparation of a Hall of Audience*. Written in 1928, this text was not published until 1946, by his son, Burnet C. Tuthill. Frank Lloyd Wright recalled seeing the plans for Carnegie Hall in Adler & Sullivan's office. Wright, "Acoustics in Building." See Twombley, *Louis Sullivan*, 249–251 for further discussion of Adler's contribution to Carnegie Hall. Adler is not mentioned in Richard Schickel and Michael Walsh, *Carnegie Hall: The First One Hundred Years* (New York: Harry Abrams, 1987); nor in Theodore O. Cron and Burt Goldblatt, *Portrait of Carnegie Hall* (New York: Macmillan, 1966); nor in Ethel Peyser, *Carnegie Hall: The House That Music Built* (New York: Robert M. McBride, 1936), 73–80.

61. Dankmar Adler (Rachel Baron, ed.), "The Theater," *Prairie School Review* 2 (1965): 21–27, on 23 and 22. This article is drawn from manuscript notes left by Adler upon his death in 1900. Baron suggests that he may have been preparing an article for an encyclopedia.

62. Twombley, *Louis Sullivan*, 192. For more on the Auditorium, see David Newton, "Chicago's Auditorium Building," *Historic Illinois* 10 (April 1988): 2–5, 14; David Garrard Lowe, "Monument of an Age," *American Craft* 48 (June 1988): 40–47, 104; Tim Samuelson and Jim Scott, "Auditorium Album," *Inland Architect* 33 (September–October 1989): 64–71; and Joseph M. Siry, "Chicago's Auditorium Building: Opera or Anarchism," *JSAH* 57 (June 1998): 128–159.

63. Patti is quoted in Twombley, *Louis Sullivan*, 177. Ronald L. Davis also indicates that musicians considered the Auditorium "an acoustical dream." *Opera in Chicago* (New York: Appleton-Century, 1966), 45.

64. Montgomery Schuyler, "Great American Architects—Architecture in Chicago," *Architectural Record*, special series, 4 (December 1895): 2–48, on 23.

65. Dankmar Adler, "Theatres," *American Architect and Building News* [*AABN*] 22 (29 October 1887): 206–208, on 207.

66. Sullivan referred here to Adler's work on the Grand Opera House of Chicago (1880). Louis Sullivan, "Development of Construction," *Economist* (Chicago) 55 (24 June 1916): 1252.

67. Dankmar Adler, "Theater-Building for American Cities," *Engineering Magazine* 7 (August 1894): 717–730 and 7 (September 1894): 815–829, on 717.

68. Adler, "Theater-Building for Cities," 829.

69. Adler, "Theater-Building for Cities," 829.

70. Latrobe, "Note," 210.

71. See, for example, Charles Garnier, *Le Théâtre* (Paris: Hachette, 1871), 213.

72. Wallace Clement Sabine Correspondence Files, Rudolph Markgraf to Professor G. W. Stewart (6 September 1911); copy sent to Wallace Sabine (8 September 1911). Courtesy of Riverbank Acoustical Laboratories.

73. William Dana Orcutt, *Wallace Clement Sabine: A Study in Achievement* (Norwood, Mass.: Plimpton Press, 1933), quote on 7. For biographical sketches, see Edwin H. Hall, "Wallace Clement Ware Sabine, 1868–1919," *Biographical Memoirs of the National Academy of Science* 11 (1926): 1–19; and Leo L. Beranek, "Wallace Clement Sabine and Acoustics," *Physics Today* (February 1985): 44–51.

74. Sabine's father Hylas practiced law and politics, but he regularly interrupted his career with less productive forays into farming, invention, and speculation. He also "found peculiar pleasure in associating with local artists, frequenting their studios and encouraging them in their work." Orcutt, *Wallace Sabine*, 12.

75. John Trowbridge and W. C. Sabine, "Wave-Lengths of Metallic Spectra in the Ultra Violet," *PAAAS* 23 (1888): 288–298; "Selective Absorption of Metals for Ultra Violet Light," *PAAAS* 23 (1888): 299–300; "On the Use of Steam in Spectrum Analysis," *American Journal of Science,* 3d ser., 37 (February 1889): 114–116; "Electrical Oscillations in Air," *PAAAS* 25 (1890): 109–123.

76. Trowbridge and Sabine, "Electrical Oscillations in Air," 122–123. For more on Hertz's discovery of electromagnetic waves, and on the analogical resonances between acoustical and electrical studies in the nineteenth century, see Jed Z. Buchwald, *The Creation of Scientific Effects: Heinrich Hertz and Electric Waves* (Chicago: University of Chicago Press, 1994); and Timothy Lenoir, "Helmholtz and the Materialities of Communication," *Osiris,* 2d ser., 9 (1994): 184–207.

77. He did include some basic experiments on sound in his undergraduate lab manual, *A Student's Manual of a Laboratory Course in Physical Measurements* (Boston: Ginn and Co., 1893; with subsequent printings in 1896 and 1898).

78. Harvard University, Presidents Office, Records of President Charles W. Eliot, 1849–1926, Wallace Clement Sabine to Eliot (25 July 1896), Box 120, Folder 329: Wallace C. Sabine [UAI5.150]. Courtesy of the Harvard University Archives [Eliot Papers].

79. W. C. Sabine, "Architectural Acoustics," *American Architect and Building News* 62 (26 November 1898): 71–73, on 71.

80. Sabine to Eliot (25 July 1896), Box 120, Folder 329, Eliot Papers.

81. Sabine, "Architectural Acoustics," 72; and Papers of Wallace Clement Sabine, Research Notebooks, Data on Acoustical Research, 1899–1919, Research Notebook #7, p. 57 [HUG 1761.25]. Courtesy of the Harvard University Archives. See also Sabine, *Student's Manual*, for his lessons in accuracy and precision.

82. Sabine to Eliot (3 November 1897), Box 120, Folder 329, Eliot Papers.

83. Eliot to Sabine (3 November 1897), quoted in Orcutt, *Wallace Sabine*, 125.

84. The tentative and qualitative nature of Sabine's understanding of reverberation at this time is documented in "Architectural Acoustics," the text of a talk he presented to the American Institute of Architects in 1898.

85. Orcutt, *Wallace Sabine*, 133.

86. See the "calibration curve" in Sabine, "Architectural Acoustics," 72.

87. Sabine to Eliot (30 October 1898), Box 120, Folder 329, Eliot Papers.

88. Sabine to Eliot (30 October 1898), Box 120, Folder 329, Eliot Papers.

89. The following outline of this analysis is drawn from Wallace Sabine's own very complete account, "Reverberation," *American Architect and Building News* 68: (7 April 1900): 3–5; (21 April 1900): 19–22; (5 May 1900): 35–37; (12 May 1900): 43–45; (26 May 1900): 59–61; (9 June 1900): 75–76; and (16 June 1900): 83–84. This article also appeared serially in the *Engineering Record* for 1900. Subsequent page citations refer to its reproduction in Wallace Clement Sabine, *Collected Papers on Acoustics* [*CPA*] (Cambridge, Mass.: Harvard University Press, 1922). These page numbers also refer to subsequent editions in 1964 (Dover) and 1998 (Peninsula).

90. Sabine, "Reverberation," *CPA*, 24.

91. Sabine, "Reverberation," *CPA*, 56–60.

92. Sabine, "Reverberation," *CPA*, 41.

93. Sabine, "Reverberation," *CPA*, 34–42.

94. Higginson to McKim (26 January 1899), Folder M-10, MMW Papers.

95. McKim to Higginson (17 March 1899), Ser. XII, Box 6A, Folder: 1899-Music, Higginson Papers.

96. Sabine, "Reverberation," *CPA*, 60–68.

97. Sabine, "Reverberation," *CPA*, 61. In fact, the interior of Symphony Hall does look strikingly like the old Music Hall, but the basic visual parameters of design were set before Sabine became involved, and his statement indicates that maintaining this visual similarity was not an element in the acoustical design of the hall. For a photo of the interior of the old Music Hall, see *Symphony Hall: The First 100 Years*, 30.

98. See, for example, Sabine to McKim (12 November 1899), Folder M-10, MMW Papers; McKim to Higginson (8 November 1899), Sabine to Higginson (13 November 1899), and Sabine to Higginson (8 March 1900), Ser. XII, Boxes 6A and 6B, Folders: 1899-Music, and 1900-Music, Higginson Papers.

99. Sabine to Higginson (8 March 1900), Ser. XII, Box 6B, Folder: 1900-Music, Higginson Papers.

100. The following survey of musical culture in America draws from: Lawrence W. Levine, *Highbrow/Lowbrow: The Emergence of Cultural Hierarchy in America* (Cambridge, Mass.: Harvard University Press, 1988); John F. Kasson, *Rudeness and Civility: Manners in Nineteenth-Century Urban America* (New York: Hill and Wang, 1990); Joseph A. Mussulman, *Music in the Cultured Generation: A Social History of Music in America, 1870–1900* (Evanston: Northwestern University Press, 1971); Philip Hart, *Orpheus in the New World: The Symphony Orchestra as an American Cultural Institution* (New York: W.W. Norton and Co., 1973); H. Wiley Hitchcock, *Music in the United States: A Historical Introduction* (Englewood Cliffs, N.J.: Prentice Hall, 1969); and John H. Mueller, *The American Symphony Orchestra: A Social History of Musical Taste* (Bloomington: Indiana University Press, 1951).

101. Sidney George Fisher, *A Philadelphia Perspective: The Diary of Sidney George Fisher Covering the Years 1834–1871,* ed. Nicholas B. Wainwright (Philadelphia: Historical Society of Pennsylvania, 1967), 148.

102. Louis Moreau Gottschalk (Indianapolis, 15 December 1862), *Notes of a Pianist* (1964; New York: DaCapo Press, 1979), 98.

103. "Charter and Prospectus of the Opera House, or American Academy of Music" (Philadelphia: Crissy and Markley, 1852), 8. William Parker Foulke Papers, American Philosophical Society, Philadelphia. Manuscripts Collection, B-F826 #1, American Academy of Music-Philadelphia.

104. "Charter and Prospectus," 3–4, 5.

105. Algarotti, *Essay on the Opera*, 110.

106. Richard Sennett, *The Fall of Public Man: On the Social Psychology of Capitalism* (New York: Alfred A. Knopf, 1977); James H. Johnson, *Listening in Paris: A Cultural History* (Berkeley: University of California Press, 1995).

107. Algarotti, *Essay on the Opera*, 55.

108. The following survey is indebted to Levine, *Highbrow/Lowbrow* for its rich account of the rise of this new culture of listening. See also Schmidt, *Hearing Voices*, for the complicated acoustical trajectories of science, religion, and romanticism that additionally contributed to this new culture.

109. Levine, *Highbrow/Lowbrow*, 120.

110. William Sydney Thayer, "The New Music Hall," *Dwight's Journal of Music* 2 (27 November 1852): 60. *Dwight's Journal* also carried a series of articles on acoustics that had been prepared by the physician Jabez Baxter Upham, head of the "Committee on Acoustics" for the new auditorium. Upham's series ran from October 1852 through January 1853 in *Dwight's Journal*, and was also published as "A Consideration of Some of the Phenomena and Laws of Sound, and Their Application in the Construction of Buildings Designed for Musical Effect," *American Journal of Science and Art* 65 (1853): 215–226, 348–363; and 66 (1853): 21–33.

111. *Boston Herald* (31 August 1924), quoted in Johnson, *Symphony Hall*, 6.

112. Gottschalk, *Notes of a Pianist,* 251, 77, 187. Of the fantasia offered by the shoemaker flutist, Gottschalk reported that "He did it to his own satisfaction and that of the audience."

113. Levine, *Highbrow/Lowbrow,* 109.

114. Mueller, *American Symphony Orchestra*, 354.

115. "Music Crowned in Its New Home," *New York Herald* (6 May 1891), Clipping File-Carnegie Hall, New York Public Library, Music Division.

116. "Music Crowned in Its New Home."

117. Johnson, *Symphony Hall*, 28.

118. Quoted in Rose Fay Thomas, *Memoirs of Theodore Thomas* (New York: Moffat Yard and Co., 1911), 50.

119. "Symphony Hall's Inaugural," *Boston Evening Transcript* (16 October 1900): 8.

120. For more on the social history of the piano, see Judith Tick, "Passed Away Is the Piano Girl: Changes in American Musical Life, 1870–1900," pp. 325–348 in Jane Bowers and Judith Tick, eds., *Women Making Music: The Western Art Tradition* (Urbana: University of Illinois Press, 1986); Cyril Ehrlich, *The Piano: A History* (London: J.M. Dent and Sons, 1976); and Craig H. Roell, *The Piano in America, 1890–1940* (Chapel Hill: University of North Carolina Press, 1989).

121. *New York Morning Telegraph* (12 June 1906), quoted in Neil Harris, "John Philip Sousa and the Culture of Reassurance," pp. 11–40 in Jon Newsom, ed., *Perspectives on John Philip Sousa* (Washington, D.C.: Library of Congress, 1983), on 39 (note 102); John Philip Sousa, "The Menace of Mechanical Music," *Appleton's Magazine* 8 (September 1906): 278–284, on 281.

122. "The Decline of the Amateur," *Atlantic Monthly* 73 (June 1894): 859–860, on 859. See also Levine, *Highbrow/Lowbrow,* 139–140.

123. Arthur Loesser, *Men, Women and Pianos: A Social History* (New York: Simon and Schuster, 1954), 291. See also Ehrlich, *Piano,* 93.

124. Roell, *Piano in America,* 37–59.

125. Jane Addams, *Twenty Years at Hull-House* (1910; New York: Penguin, 1981), 65.

126. Edward Bellamy, *Looking Backward* (1888; New York: Penguin American Library, 1982), 97.

127. Bellamy, *Looking Backward,* 99. For actual experiments in telephonic entertainment, see Carolyn Marvin, *When Old Technologies Were New: Thinking About Electric Communication in the Late Nineteenth Century* (New York: Oxford University Press, 1988), 209–216.

128. Quoted in Bliss Perry, *Life and Letters of Henry Lee Higginson* (Boston: Atlantic Monthly Press, 1921), 291.

129. W. F. A. (William Foster Apthorp), "Opening of Symphony Hall," *Boston Transcript* (16 October 1900), Clipping Book, Pres 56, vol. 16, p. 18, BSO Archives.

130. "Mr. Higginson's Work," *Boston Herald* (17 October 1900), Clipping Book, Pres 56, vol. 16, p. 18; H. T. F., "Boston's New Music-Hall," *New York Evening Post* (16 October 1900), Clipping Book, Pres 56, vol. 16, p. 21, BSO Archives.

131. H. E. K., "The New Symphony Hall in Boston," *New-York Daily Tribune* (16 October 1900), Clipping Book, Pres 56, vol. 16, p. 20, BSO Archives.

132. Krehbiel had written a letter to Sabine prior to opening night, inquiring about a verification of his calculations. Sabine's response is described in Orcutt, *Wallace Sabine*, 144–145.

133. Higginson to Sabine (22 October 1900), quoted in Orcutt, *Wallace Sabine*, 147.

134. Philip Hale, "First Symphony," *Boston Sunday Journal* (21 October 1900), Clipping Book, Pres 56, vol. 16, p. 24, BSO Archives.

135. "A Complete Success," *Boston Sunday Herald* (21 October 1900), Clipping Book, Pres 56, vol. 16, p. 23, BSO Archives.

136. "Symphony Opens," *Boston Post* (21 October 1900).

137. W. F. A., "Music and Drama: Symphony Hall: Boston Symphony Orchestra," *Boston Evening Transcript* (22 October 1900): 10. Apthorp specified that any conclusions he might draw would relate only to how the hall sounded from his own seat in the first balcony, and not to any other locations in the hall.

138. Apthorp, "Boston Symphony Orchestra" (22 October 1900).

139. Apthorp, "Music and Drama: Symphony Hall: The Handel and Haydn Society" (22 October 1900): 10.

140. "Music in Boston," *Musical Courier* (24 October 1900): 21–23, on 21.

141. W. F. A., "Music and Drama: Symphony Hall: Boston Symphony Orchestra," *Boston Evening Transcript* (4 March 1901).

142. Warren Davenport, "Mr. Gericke Goes Backward," *Boston Sunday Herald* (24 March 1901), Clipping Book, Pres 56, vol. 16, p. 90, BSO Archives.

143. Philip Hale, "Looking Backward," *Boston Sunday Journal* (5 May 1901): 11, 15, on 11.

144. Edmund D. Spear to Higginson (26 May 1902), Ser. XII, Box 7A, Folder: 1902-Music, Higginson Papers. No response is extant in the Higginson correspondence, nor have I encountered any other mention of plans to remodel Symphony Hall.

145. "Boston Symphony Hall: A Scientific Analysis of Its Acoustics," *Boston Evening Transcript* (31 December 1902).

146. See Johnson, *Symphony Hall*, 12–13; Stebbins, *Making of Symphony Hall*, 95–97; and Mary Elliott to Charles McKim (19 February 1902), Folder M-10, MMW Papers.

147. Wallace Sabine to Charles McKim (1 May 1901), Folder M-10, MMW Papers.

148. Sabine to McKim (1 May 1901), Folder M-10, MMW Papers.

149. Wallace C. Sabine, "Architectural Acoustics," *PAAAS* 42 (June 1906): 49–84, and *CPA*, 69–105, quote on 71. The experiments Sabine described here were carried out in 1902.

150. Sabine, "Architectural Acoustics," *CPA*, 69.

151. Sabine, "Architectural Acoustics," *CPA*, 77. The optimum reverberation time was significantly lower than that of Symphony Hall because the rooms tested, intended for small ensembles or soloists, were all much smaller than the concert hall.

152. Mary Elliot to Charles McKim (19 February 1902), Folder M-10, MMW Papers.

153. Leo Beranek has suggested that the gradual increase in the size of the orchestra over the course of the first few decades of the century was responsible for improving the sound of the hall. He asserts that the hall "did not gain the reputation it has today until the orchestra's size increased to 105 players." This occurred in the early 1920s. Leo L. Beranek, "Boston Symphony Hall: An Acoustician's Tour," *Journal of the Audio Engineering Society* 36 (November 1988): 918–930, on 923. For data on orchestra size, see Howe, *Boston Symphony*, 244.

154. The problem with the Auditorium for Thomas was not its acoustical quality, but its capacity. With so many seats available, patrons knew one could always be had, and they resisted purchasing season subscriptions.

155. Thomas, *Memoirs of Theodore Thomas*, 536.

156. Gericke had led the orchestra earlier in its career, from 1884 to 1889. He was replaced by Arthur Nikisch (1889–1893), then Emil Paur (1893–1898), before he returned to replace Paur in 1898. Howe, *Boston Symphony*, 230.

157. See, for example, Apthorp's review in the *Boston Evening Transcript* (22 October 1900); and "Music in Boston," *Musical Courier* (24 October 1900): 21. Both accounts, while highly critical of the sound of the hall, heap praise on the orchestra.

158. Hall, "Wallace Sabine," 8.

CHAPTER THREE

1. Quoted in Don H. Gearheart, "Dr. Miller and Case Host to Acoustical Society," *Case Alumnus* (December 1931): 12.

2. James Loudon, "A Century of Progress in Acoustics," *Science,* n.s., 14 (27 December 1901): 987–995, on 987.

3. W. H. Eccles, "The New Acoustics," *Proceedings of the Physical Society* 41 (15 June 1929): 231–239, on 231, 232. See also N. W. McLachlan, *The New Acoustics: A Survey of Modern Development in Acoustical Engineering* (London: Oxford University Press, 1936).

4. "When Radio Answered a Call to Hollywood," *New York Times* (10 August 1930): sect. 9, p. 12.

5. "Acoustical Engineering as a Career," *Careers,* Research no. 38 (Chicago: Institute for Research, 1931), quotes from pages 9 and 14 in the unpaginated brochure. The phrase "Youth's Inevitable Question: What Shall I Be?" appears as a page heading throughout the pamphlet.

6. "Acoustical Engineering as a Career," 9.

7. "Acoustical Engineering as a Career," 15.

8. "Acoustical Engineering as a Career," 15. Although the pamphlet employed the masculine pronoun throughout, it did indicate, under the heading "Opportunities for Women," that "While no statistics are available to indicate that women have entered upon the profession of acoustical engineering, there appear to be no valid objections to their engaging in this work" (13). This statement was, however, removed from a later edition of the pamphlet that first appeared sometime between 1935 and 1940. I have identified only one woman among the approximately 450 charter members of the Acoustical Society of America, Professor Katherine Frehafer, of Goucher College.

9. Dayton Clarence Miller, *Anecdotal History of the Science of Sound* (New York: Macmillan, 1935), all quotes from 98. This book is an expanded version of Miller's 1932 lecture. For a citation of the original talk, see "Program of the Eighth Meeting of the Acoustical Society of America," *Journal of the Acoustical Society of America* [*JASA*] 4 (January 1933): 171–177, on 174.

10. Wallace C. Sabine, "Architectural Acoustics," *Proceedings of the American Academy of Arts and Sciences* [*PAAAS*] 42 (June 1906): 49–84 and *Collected Papers on Acoustics* [*CPA*] (Cambridge, Mass.: Harvard University Press, 1922), 69–105, quote on 103.

11. Wallace C. Sabine, "Sense of Loudness," *Contributions from the Jefferson Physical Laboratory* 8 (1910) n.p. and *CPA,* 129–130.

12. Wallace Clement Sabine Correspondence Files, Sabine to Frank Bigelow Cook, Jr. (14 February 1916). Courtesy of Riverbank Acoustical Laboratories [Sabine Correspondence].

13. Wallace C. Sabine, "Theatre Acoustics," *American Architect* 104 (31 December 1913): 257–279 and *CPA,* 163–197, quote on 180. For a detailed explanation of the technique, see Arthur L. Foley and Wilmer H. Souder, "A New Method of Photographing Sound Waves," *Physical Review* 35 (November 1912): 373–386.

14. Wallace C. Sabine, "Architectural Acoustics: The Correction of Acoustical Difficulties," *Architectural Quarterly of Harvard University* (March 1912): 3–23 and *CPA*, 131–161, quote on 152–153.

15. Sabine had telephone receivers in his laboratory by 1908, if not earlier, but this is the earliest evidence of his having put them to use. In 1908, one of Sabine's colleagues at Harvard indicated that he had borrowed several telephone receivers from Sabine to carry out his own research. See George W. Pierce, "A Simple Method of Measuring the Intensity of Sound," *PAAAS* 43 (February 1908): 377–395, on 381 (note 2).

16. The contour map appeared in Sabine, "Correction of Acoustical Difficulties," 152, and Wallace C. Sabine, "Architectural Acoustics," *Journal of the Franklin Institute* 179 (January 1915): 1–20 and *CPA*, 219–236, on 233. Sabine's experimental setup is described in Paul E. Sabine, *Acoustics and Architecture* (New York: McGraw-Hill, 1932), 41–46. Paul Sabine indicated that his account was derived from Wallace's unpublished research notes.

17. Sabine did briefly outline the means by which he made his map in a lecture to the Société des Architectes in Paris on 20 June 1917. He described the procedure as "un travail ardu, peut-être le plus ardu de tous les problèmes d'acoustique architecturale," and he indicated that his receiver was "un téléphone spécialement construit d'une sensibilité extrême et en même temps très constant." I thank Bruno Suner for providing me with a copy of the text of this lecture. The original is preserved in the Archives of the Institut Français d'Architecture, quote on p. 15. For Sabine's concern about electrically driven tuning forks, see "Notes on Measurements of the Intensity of Sound," *CPA*, 277–279.

18. "Architects corresponded with in regard to buildings" (list), Sabine Correspondence.

19. The correspondence between Sabine and the architects is preserved in the Papers of McKim, Mead & White, Folder M-5: Sabine, Wallace, New-York Historical Society. Sabine described the Rhode Island House of Representatives project in "Correction of Acoustical Difficulties," *CPA*, 137–138.

20. Stanford White to Wallace Sabine (13 November 1903), quoted in William Dana Orcutt, *Wallace Clement Sabine: A Study in Achievement* (Norwood, Mass.: Plimpton Press, 1933), 234.

21. William Mead to Wallace Sabine (14 April 1903), quoted in Leland Roth, *McKim, Mead and White, Architects* (New York: Harper and Row, 1983), 261.

22. Bertram Grosvenor Goodhue to Wallace Sabine (19 February 1916), Sabine Correspondence.

23. Wallace Sabine to Bertram Grosvenor Goodhue (23 August 1913), 1913: Box 2: 35 (S), Bertram Grosvenor Goodhue Collection, Avery Architectural and Fine Arts Library, Columbia University in the City of New York [Goodhue Papers]. For evidence that Goodhue shared these feelings of friendship, see his letters to Sabine, espe-

cially Goodhue to Sabine (21 August 1913), 1913: Box 2: 35 (S), and (13 December 1916), 1916: Box 5: 28 (S), Goodhue Papers.

24. Stevens & Nelson to Wallace Sabine (7 June 1912), Sabine Correspondence.

25. H. Osgood Holland to Wallace Sabine (15 June 1911), Sabine Correspondence.

26. Wallace Sabine to H. Osgood Holland (5 July 1911), Sabine Correspondence.

27. Wallace Sabine to Alfred Altschuler (1 October 1909), Sabine Correspondence.

28. In 1897, responding to such a prompt, Sabine wrote to Eliot, "You are entitled to a more full explanation than I made this morning. The only expense which has served a useful purpose in the Fogg Lecture Room, the purchase of the cloth now there, has been born by the corporation. . . . In the meantime I do not wish to feel compelled to justify the expense to which I shall go. That situation would be unendurable. Moreover the expenditures have and will vary all the way from the purchase of instruments to admission tickets to various halls and travelling expenses. I am quite unable to separate the necessary from that which has turned out needless and fruitless or personal. I hope that this will free me from the appearance of being unreasonable." Harvard University, President's Office, Records of President Charles W. Eliot, 1849–1926, Wallace Clement Sabine to Eliot (3 November 1897), Folder 329: Wallace C. Sabine [UAI5.150]. Courtesy of the Harvard University Archives [Eliot Papers]. Eliot responded that this explanation was "very far from being satisfactory," and he made clear that Sabine should immediately submit all receipts relating to the investigation, which Sabine eventually did. See Eliot to Sabine (12 November 1897), quoted in Edwin H. Hall, "Wallace Clement Ware Sabine, 1868–1919," *Biographical Memoirs of the National Academy of Science* 11 (1926): 1–19, on 5. See also Sabine to Eliot (17 February 1898) and (27 October 1898), Box 120, Folder 329, Eliot Papers.

29. Wallace Sabine to R. Clipston Sturgis (17 November 1910), quoted in Orcutt, *Wallace Sabine*, 222.

30. Wallace Sabine to Dean Grosvenor (28 March 1912); see also Sabine to Mr. Bacon (5 October 1911), and Sabine to Ralph Adams Cram (29 May 1913), Sabine Correspondence.

31. Quoted in Orcutt, *Wallace Sabine*, 224.

32. Wallace C. Sabine, "Architectural Acoustics: Building Material and Musical Pitch," *Brickbuilder* 23 (January 1914): 1–6 and *CPA*, 199–217, quote on 200–201.

33. Sabine, "Building Material and Musical Pitch," 216.

34. Wallace Sabine to W. L. Krider, United States Gypsum Co. (3 April 1915), Sabine Correspondence.

35. Wallace Sabine to S. G. Webb, Gypsum Industries Association (26 May 1916), Sabine Correspondence. Sabine's correspondence with both lime and gypsum manufac-

turers indicates the competitive nature of the plaster industry. Each hoped to enlist Sabine to substantiate a claim of acoustical superiority for its own type of plaster.

36. Sabine's work with Guastavino is discussed more fully in chapter 5. For now it is sufficient to note that this fee was considerable; Sabine's university salary in 1909 was $4,500. It seems likely that the royalty arrangement was even more lucrative than the initial payment, since it was a direct function of the vast surface areas of the large vaults and domes that would be lined with the new tile. The archives of the Guastavino Company lack any direct evidence of royalty payments made to Sabine over the years, but extant records of tile orders suggest that the royalty for a single project could easily have been hundreds of dollars, and there were dozens of such projects during Sabine's lifetime. One miscellaneous scrap of paper filed with financial figures for the National Academy of Sciences Building in Washington, D.C., indicates an unspecified "Royalty" of $450 that might have been paid to Sabine. See Series II: Projects, Project File 2.55: National Academy of Sciences, Guastavino/Collins Collection, Avery Architectural and Fine Arts Library, Columbia University in the City of New York [GCC]. For the original terms of the agreement, see the signed carbon copy of the contract (20 April 1911), Series I: Administrative and Technical Records, Box 14, Folder 2: Sabine, W. C.-misc; and the proposed modification to that agreement in Wallace C. Sabine to R. Guastavino (27 October 1915), Series I, Box 14, Folder 12: Sabine, W. C.-Jefferson Physical Lab, GCC. Sabine's salary is indicated in Leo L. Beranek, "The Notebooks of Wallace C. Sabine," *JASA* 61 (March 1977): 629–639, on 632.

37. Wallace Sabine to Ralph Adams Cram (18 December 1912), Sabine Correspondence.

38. Sabine, "Correction of Acoustical Difficulties," 132. Sabine was working with Johns-Manville on the development of a sound-absorbing plaster.

39. Wallace Sabine to Albert Kahn (27 June 1911), Sabine Correspondence.

40. Sabine to Kahn (27 June 1911), Sabine Correspondence. Although strongly derived from Sabine's own writings, Tallant's original article apparently failed to cite or otherwise recognize this fact. As published, the article briefly acknowledges Sabine's contribution in an introductory footnote. See Hugh Tallant, "Hints on Architectural Acoustics," *Brickbuilder* 19 (May 1910): 111–116, with subsequent monthly installments through December 1910.

41. Albert Kahn to Wallace Sabine (26 April 1913); see also Sabine to Kahn (22 April 1913), Sabine Correspondence. For more on the Hill Auditorium, see Hugh Tallant, "Acoustic Design in the Hill Memorial Auditorium, University of Michigan," *Brickbuilder* 22 (August 1913): 169–173; and J. T. N. Hoyt, "The Acoustics of the Hill Memorial Hall," *American Architect* 104 (6 August 1913): 49–53.

42. The Mazer affair is discussed more fully in chapter 5.

43. Sabine's mention of the "Toeppler-Boys-Foley" method of sound photography in his 1913 article on theater acoustics is the closest he came to citing any contemporary

work in any of his published papers. Even here, he did not include an explicit reference to Foley and Souder's paper. See Sabine, "Theatre Acoustics," 180.

44. Wallace Sabine to Robert Hope-Jones (27 September 1912), Sabine Correspondence.

45. Wallace Sabine to W. L. Stevens (12 June 1912), Sabine Correspondence; Orcutt, *Wallace Sabine*, 243. Orcutt indicates neither the date nor the context of Sabine's latter statement. Harvard physicist Edwin Hall noted that, at the time of Sabine's death, the man best acquainted with his work was John Connors, the janitor of the Jefferson Physical Lab. Hall, "Wallace Clement Ware Sabine," 4.

46. For more on Sabine's administrative career at Harvard, see Orcutt, *Wallace Sabine*, 194–219. The Harvard-MIT merger was dissolved when the courts ruled that it constituted a misappropriation of the funds that industrialist Gordon McKay had bequeathed to Harvard in support of applied science. For more on this merger, see Hector James Hughes, "Engineering and Other Applied Sciences in the Harvard Engineering School and Its Predecessors, 1847–1929," pp. 413–442 in Samuel Eliot Morison, *The Development of Harvard University Since the Inauguration of President Eliot, 1869–1929* (Cambridge, Mass.: Harvard University Press, 1930). See also Christophe Lécuyer, "MIT, Progressive Reform, and 'Industrial Service,' 1890–1920," *Historical Studies in the Physical and Biological Sciences* 26 (1995): 35–88.

47. The Sorbonne lecture series ran from February to May 1917. The audience, which ranged from twenty to fifty attendees, consisted primarily of professors, architects, and students from the École des Beaux Arts. Orcutt, *Wallace Sabine*, 290–292 and Hall, "Wallace Clement Ware Sabine," 15. Sabine also spoke before the Society of Architects. See Bruno Suner and Jacques-Franck Degioanni, "Architectural Acoustics in France Circa World War I and Today . . . Deja Vu?" *Proceedings of the Wallace Clement Sabine Centennial Symposium* (Woodbury, N.Y.: Acoustical Society of America, 1994), 37–40.

48. For more on Sabine's European war work, see Orcutt, *Wallace Sabine*, 290–306.

49. Quoted in Orcutt, *Wallace Sabine*, 320.

50. For Sabine's American war work and his death, see Orcutt, *Wallace Sabine*, 307–338.

51. For more on Fabyan and his Riverbank estate, see John W. Kopec, *The Sabines at Riverbank: Their Role in the Science of Architectural Acoustics* (Woodbury, N.Y.: Acoustical Society of America, 1997), 19–42.

52. Fred Kranz, who worked at the Riverbank estate in the 1920s, recalled that, "In personal appearance, Mrs. Gallup strongly resembled Whistler's mother." Fred W. Kranz, "Early History of Riverbank Acoustical Laboratories," *JASA* 49 (February 1971): 381–384, on 382.

53. John Kopec indicates that Sabine actually came to Riverbank in 1913 to inspect the device, and that he visited the estate on numerous other occasions over the next few years. Kopec, *Sabines at Riverbank*, 7.

54. See, for example, William B. Ittner to Wallace Sabine (28 November 1911); Dillon, McLellan & Beadel to Wallace Sabine (14 May 1915); and Wallace Sabine to Frank M. Smith, United Electric Light and Power Company (12 February 1915), Sabine Correspondence.

55. Sabine, "Building Material and Musical Pitch," 216.

56. Daniel V. Casey, "Muffling Office Noises," *System: The Magazine of Business* 25 (1913): 246–251, on 248. This article was excerpted as "To Hush Office Noise," in *Literary Digest* 48 (28 March 1914): 696–697.

57. See the letters between Wallace Sabine and Fred J. Miller, Remington Typewriter Company (December 1915–January 1916), Sabine Correspondence, File 382, for more on this consultation. See also Leo L. Beranek and John W. Kopec, "Wallace C. Sabine, Acoustical Consultant," *JASA* 69 (January 1981): 1–16.

58. Kopec suggests that it is unlikely that Sabine would have moved permanently to Illinois. *Sabines at Riverbank*, 61.

59. Kopec, *Sabines at Riverbank*, 13. For biographical details of Paul Sabine, see his obituary, "Paul Earls Sabine (1879–1958)," *JASA* 31 (April 1959): 536; and Kopec, *Sabines at Riverbank*.

60. Paul E. Sabine, "The Wallace Clement Sabine Laboratory of Acoustics, Geneva, Ill.," *American Architect* 116 (30 July 1919): 133–138. Paul Sabine supervised the Riverbank laboratory until 1947, at which time its control was turned over to the Armour Institute of Technology Research Foundation. The AITRF later became the Illinois Institute of Technology Research Institute, and the IITRI continues to operate the facility today.

61. In 1903, for example, Edward Nichols and William Franklin included a summary of Sabine's work as a new chapter on architectural acoustics in their popular textbook on physics. E. L. Nichols and W. S. Franklin, *Elements of Physics* (New York: Macmillan, 1903), 226–237. Sabine's work is also cited in Edwin H. Hall and Joseph Y. Bergen, *Text-Book of Physics*, 3d rev. ed. (New York: Henry Holt, 1903), 408. For evidence of the increasing number of research articles on architectural acoustics in the early twentieth century, see F. R. Watson, "Bibliography of Acoustics of Buildings, *JASA* 3 (July 1931): 14–43. Whereas Watson listed only seventeen citations for the entire nineteenth century, he was able to cite twenty-three articles for the decade 1900–1909. For 1910–1919 he cited 52 articles, and the number increased to 309 for 1920–1929.

62. This presentation is reported in Dayton C. Miller, "American Association for the Advancement of Science. Section B, Physics," *Science,* n.s., 17 (30 January 1903): 170–180, on 174. See also W. S. Franklin, "Derivation of Equation of Decaying Sound in a Room and Definition of Open Window Equivalent of Absorbing Power," *Physical Review* 16 (June 1903): 372–374. Another derivation was provided by G. Jäger, "Zur Theorie des Nachhalls," *Sitzungsberichte der Kaiserliche Akademie der Wissenschaften in*

Wien, Bd. 120, Abt. IIA (1911): 613–634. This article was translated by Floyd Watson as "Acoustics of Auditoriums," *American Architect* 108 (8 December 1915): 369–374.

63. G. W. Stewart, "Architectural Acoustics: Some Experiments in the Sibley Auditorium," *Sibley Journal of Engineering* 17 (May 1903): 295–313. See also Stewart, "Architectural Acoustics: Some Experiments in Reverberation," *Physical Review* 16 (June 1903): 379–380.

64. Stewart, "Experiments in Sibley Auditorium," 305. Even with this basis of comparison established, however, the intensities of both pipes were still only expressed relative to the minimum audible intensity for that frequency, and not in absolute physical units.

65. F. R. Watson to R. Guastavino Company (17 May 1915), Series I, Box 14, Folder 12: Sabine, W. C.-Jefferson Physical Lab, GCC.

66. Watson's progress during these six years is documented in his correspondence with university president Edmund James and university architect James M. White. These letters are preserved in "F. R. Watson Auditorium Acoustics File," Edmund J. James Faculty Correspondence, 1904–1915, R. S. 2/5/6, Box 30, University of Illinois at Urbana-Champaign Archives [James Correspondence]. For more on Watson's long life and career, see his obituary by Daniel W. Martin, "Floyd Rowe Watson, 1872–1972," *JASA* 55 (June 1974): 1362–1363.

67. Abstracts of Watson's American Physical Society papers include: "An Apparatus for Measuring Sound," *Physical Review* 30 (January 1910): 128; "Echoes in an Auditorium," *Physical Review* 32 (February 1911): 231; "The Effects of Gas Currents on Sound," *Physical Review,* 2d ser., 1 (January 1913): 76–77; "The Use of Sounding Boards in an Auditorium," *Physical Review,* 2d ser., 1 (March 1913): 241; and "Transmission of Sound through Fabrics," *Physical Review,* 2d ser., 5 (April 1915): 342. For representative articles by Watson, see "Architectural Acoustics: How Sound Interference in Buildings May Be Cured," *Scientific American Supplement* 68 (18 December 1909): 391; "Inefficiency of Wires as a Means of Curing Defective Acoustics of Auditoriums," *Science,* n.s., 35 (24 May 1912): 833–834; "Air Currents and Their Relation to the Acoustics of Auditoriums," *Engineering Record* 67 (8 March 1913): 265–272; and "The Use of Sounding-Boards in an Auditorium," *Brickbuilder* 22 (June 1913): 139–141.

68. By 1912, Watson had also begun teaching a course on acoustics at the university. Floyd Watson to President E. J. James (9 November 1912), James Correspondence.

69. F. R. Watson to James M. White (29 April 1911), James Correspondence; F. R. Watson, "Acoustics of Auditoriums: An Investigation of the Acoustical Properties of the Auditorium at the University of Illinois," *University of Illinois Engineering Experiment Station Bulletin* no. 73 (1914): 1–32.

70. F. R. Watson to James M. White (25 February 1910), James Correspondence.

71. Loudon, "Century of Progress in Acoustics," 994.

72. George W. Pierce, "A Simple Method of Measuring the Intensity of Sound," *PAAAS* 43 (February 1908): 377–395.

73. For biographical information on Pierce, see his obituary by Frederick A. Saunders and Frederick V. Hunt, "George Washington Pierce," *Biographical Memoirs of the National Academy of Sciences* 33 (1959): 350–380. For more on the concept of syntony, see Hugh G. J. Aitken, *Syntony and Spark: The Origins of Radio* (1976; Princeton: Princeton University Press, 1985).

74. Pierce, "Simple Method of Measuring the Intensity of Sound," 394.

75. Floyd Watson, "An Apparatus for Measuring Sound," *Physical Review* 30 (January 1910): 128. See also Watson, "An Apparatus for Measuring Sound," *Physical Review* 30 (April 1910): 471–473.

76. W. M. Boehm, "A Method of Measuring Intensity of Sound," *Physical Review* 31 (October 1910): 329–331, quotes on 330 and 331.

77. F. L. Tufts, "The Transmission of Sound through Solid Walls," *American Journal of Science,* 4th ser., 13 (June 1902): 449–454. See also Tufts, "The Transmission of Sound through Porous Materials," *American Journal of Science,* 4th ser., 11 (May 1901): 357–364, an earlier study that employed an optically based method of measurement.

78. C. S. McGinnis and M. R. Harkins, "The Transmission of Sound through Porous and Non-Porous Materials," *Physical Review* 33 (August 1911): 128–136.

79. Hawley O. Taylor, "A Direct Method of Finding the Value of Materials as Sound Absorbers," *Physical Review,* 2d ser., 2 (October 1913): 270–287, on 277.

80. F. R. Watson to J. M. White (15 May 1909), James Correspondence.

81. Lord Rayleigh, "On an Instrument Capable of Measuring the Intensity of Aerial Vibrations," *London, Edinburgh, and Dublin Philosophical Magazine,* 5th ser., 14 (September 1882): 186–187.

82. Invented in 1898, Webster's phonometer was revised and improved by the physicist over the next twenty years. For the earliest description of the instrument, see A. G. Webster, "A Portable Apparatus for the Measurement of Sound," *Science,* n.s., 17 (30 January 1903): 175. This abstract of a talk given to the American Association for the Advancement of Science notes that the device was an "improved form of the instrument shown at the Boston meeting, 1898." See also Webster, "A Complete Apparatus for Absolute Acoustical Measurements," *Proceedings of the National Academy of Science* [*PNAS*] 5 (May 1919): 173–179; and Harry B. Miller, "Acoustical Measurements and Instrumentation," *JASA* 61 (February 1977): 274–282.

83. Miller's Lowell Institute lectures were published as *The Science of Musical Sounds* (New York: Macmillan, 1916). The widespread appeal of phonodeik images is documented in Clipping File 19IM2 of the Dayton Clarence Miller Papers, Case Western Reserve University Archives, Cleveland, Ohio. The undated *Boston Evening Transcript* article is found here.

84. For a depiction of World War I soldiers using listening devices to detect enemy aircraft, see the sound motion picture *Hell's Angels* (1930; Caddo/United Artists, dir. Howard Hughes and James Whale, MCA Universal Home Video #80638).

85. For a brief treatment of sound-ranging, see Daniel J. Kevles, *The Physicists* (1978; New York: Vintage, 1979), 127–131.

86. For more on trench warfare, see John Ellis, *Eye-Deep in Hell: Trench Warfare in World War I* (1976; Baltimore: Johns Hopkins University Press, 1989).

87. Erich Maria Remarque, trans. A. W. Wheen, *All Quiet on the Western Front* (1929; New York: Fawcett Crest, 1958), 129–130. See also the sound motion picture *All Quiet on the Western Front* (1930; Universal, dir. Lewis Milestone, MCA Universal Home Video #55018).

88. For more on submarine detection, see Kevles, *Physicists*, 117–126, staff figures on 126. See also Willem Hackmann, "Sonar Research and Naval Warfare 1914–1954: A Case Study of a Twentieth-Century Establishment Science," *Historical Studies in the Physical and Biological Sciences* 16 (1986): 83–110; Hackmann, *Seek and Strike: Sonar, Anti-Submarine Warfare and the Royal Navy, 1914–1954* (London: H. M. S. O., 1984); and Marvin Lasky, "Review of Undersea Acoustics to 1950," *JASA* 61 (February 1977): 283–297.

89. Paper counts cited here were determined from the program abstracts published in the *Physical Review* each year. The connection between the war and the New Acoustics was explicitly cited in McLachlan, *New Acoustics*, 16. See also Gearheart, "Dr. Miller and Case Host Acoustical Society," 12.

90. For more on the formation and activities of the National Research Council during the First World War, see Kevles, *Physicists*, 112–116; and A. Hunter Dupree, *Science in the Federal Government* (1957; Baltimore: Johns Hopkins University Press, 1986), 302–325.

91. "Certain Problems in Acoustics," *Bulletin of the National Research Council* 4 (November 1922), no. 23. Watson's obituary indicated only that he was involved in "military acoustical research" during the war. Miller's obituary cites confidential reports that he submitted to the Committee on Location and Detection of the Submarine Defense Association, as well as reports on "scientific instruments for war uses" for the National Research Council. He also published papers, shortly after the war, on the behavior of sound from large guns. In 1919, George Stewart published a paper on the location of aircraft by sound. See Martin, "Floyd Rowe Watson," 1362; Harvey Fletcher, "Biographical Memoir of Dayton Clarence Miller, 1866–1941," *Biographical Memoirs of the National Academy of Science* 23 (1945): 61–74, on 71; and G. W. Stewart, "Location of Aircraft by Sound," *Physical Review,* 2d ser., 14 (August 1919): 166–167.

92. "Certain Problems in Acoustics," 20.

93. "Certain Problems in Acoustics," 16–19, on 19.

94. Frederick V. Hunt, *Electroacoustics: The Analysis of Transduction, and Its Historical Background* (1954; n.p.: Acoustical Society of America, 1982), 169.

95. For more on the rise of industrial research in the electroacoustical industries, see Leonard S. Reich, *The Making of American Industrial Research: Science and Business at GE and Bell, 1876–1926* (Cambridge: Cambridge University Press, 1985); Neil H. Wasserman, *From Invention to Innovation: Long-Distance Telephone Transmission at the Turn of the Century* (Baltimore: Johns Hopkins University Press, 1985); Stephen B. Adams and Orville R. Butler, *Manufacturing the Future: A History of Western Electric* (Cambridge: Cambridge University Press, 1999); and George Wise, *Willis R. Whitney, General Electric, and the Origins of U.S. Industrial Research* (New York: Columbia University Press, 1985). For a more general account, see Thomas P. Hughes, *American Genesis: A Century of Invention and Technological Enthusiasm, 1870–1970* (New York: Penguin, 1989).

96. For more on the development of microphones, see H. A. Frederick, "The Development of the Microphone," *JASA* 3 (Supplement to July 1931): 1–25; W. C. Jones, "Condenser and Carbon Microphones—Their Construction and Use," *Journal of the Society of Motion Picture Engineers* 16 (January 1931): 3–22; F. S. Goucher, "The Carbon Microphone: An Account of Some Researches Bearing on Its Action," *Bell System Technical Journal* 13 (April 1934): 163–194; and Hunt, *Electroacoustics*, 33–44.

97. General accounts of the early history of telephonic technology and the telephone industry include Steven Lubar, *InfoCulture: The Smithsonian Book of Information Age Inventions* (Boston: Houghton Mifflin, 1993); John Brooks, *Telephone: The First Hundred Years* (New York: Harper & Row, 1976); Robert V. Bruce, *Bell: Alexander Graham Bell and the Conquest of Solitude* (Boston: Little, Brown, 1973); Reich, *American Industrial Research*; and Wasserman, *Invention to Innovation*. More technically oriented accounts include M. D. Fagen, ed., *A History of Engineering and Science in the Bell System: The Early Years (1875–1925)* (n.p.: Bell Telephone Laboratories, 1975); and Hunt, *Electroacoustics*.

98. Reich, *American Industrial Research*, 151–184.

99. Hugh G. J. Aitken, *The Continuous Wave: Technology and American Radio, 1900–1932* (Princeton: Princeton University Press, 1985), 217.

100. Testimony of Harold D. Arnold in *De Forest Radio Company v. General Electric Company*, Supreme Court, October Term, 1930. Quoted in Aitken, *Continuous Wave*, 244.

101. A principal member of the new research group would later recall, "if we could accurately describe every part of the system from the voice through the telephone instruments to and including the ear, we could engineer the parts at our disposal with greater intelligence." Harvey Fletcher, *Speech and Hearing* (New York: Van Nostrand, 1929), v. Fletcher's book constituted the apotheosis of the first fifteen years of that research effort. See also Fagen, ed., *Bell System: Early Years*, 926–958.

102. Fagen, ed., *Bell System: Early Years*, 930; Hunt, *Electroacoustics*, 40–41; H. D. Arnold and I. B. Crandall, "The Thermophone as a Precision Source of Sound," *Physical Review,* 2d ser., 10 (July 1917): 22–38.

103. Quoted in Miller, "Acoustical Measurements and Instrumentation," 276. See also Pierce, "Simple Method of Measuring the Intensity of Sound," 394.

104. E. C. Wente, "A Condenser Transmitter as a Uniformly Sensitive Instrument for the Absolute Measurement of Sound Intensity," *Physical Review,* 2d ser., 10 (July 1917): 39–63.

105. W. West, *Acoustical Engineering* (London: Sir Isaac Pitman & Sons, 1932), 169; "Edward Christopher Wente," *National Cyclopedia of American Biography* (New York: James T. White, 1975) 56: 307–309, on 307.

106. Paul E. Sabine, "The Beginnings of Architectural Acoustics," *JASA* 7 (April 1936): 242–248, on 243.

107. Hunt, *Electroacoustics*, 66. Hunt lists numerous articles and patents that used this analogy to explain acoustical devices and phenomena.

108. Eccles, "New Acoustics," 233.

109. A. E. Kennelly and G. W. Pierce, "The Impedance of Telephone Receivers as Affected by the Motion of Their Diaphragms," *PAAAS* 48 (September 1912): 113–151.

110. A. G. Webster, "Acoustical Impedance, and the Theory of Horns and of the Phonograph," *PNAS* 5 (July 1919): 275–282. This paper was the delayed publication of a talk given in 1914 at a meeting of the American Physical Society. See *Physical Review,* 2d ser., 5 (February 1915): 177, for a listing of this talk.

111. J. P. Maxfield and H. C. Harrison, "Methods of High Quality Recording and Reproducing of Music and Speech Based on Telephone Research," *Transactions of the American Institute of Electrical Engineers* [*TAIEE*] 45 (February 1926): 334–348, on 338. The article also appeared in the *Bell System Technical Journal* 5 (July 1926): 493–523.

112. Miller, "Acoustical Measurement and Instrumentation," 278. Harry Miller characterized the Maxfield and Harrison article as "a textbook which taught both synthesis and analysis of mechanical circuits; and indeed popularized the very concept of the mechanical circuit" (277).

113. Webster had investigated the new instruments, but he was skeptical of Wente's claim that his device responded with equal sensitivity to all sounds. "Who knows," Webster argued, "whether the amplification of that tube is the same for all these frequencies." Arthur Gordon Webster, "The Absolute Measurements of the Intensity of Sound," *TAIEE* 38 (part 1): 701–723, quotes on 721 and 722. See also Webster, "Absolute Measurements of Sound," *Science* 58 (31 August 1923): 149–152, for the text of a talk delivered in 1921, in which the physicist was still promoting the use of his phonometer.

114. Harry B. Miller, ed., *Acoustical Measurements: Methods and Instrumentation* (Stroudsburg, Penn.: Hutchinson Ross, 1982), 86.

115. Biographical information on Knudsen is found in Isadore Rudnick, "Vern Oliver Knudsen, 1893–1974," *JASA* 56 (August 1974): 712–715; Robert Lindsay, "Vern Oliver Knudsen," *Dictionary of Scientific Biography* (New York: Charles Scribner's Sons, 1990) 17 [Supp. II]: 489–490; and Vern Knudsen, *Teacher, Researcher, and Administrator: Vern O. Knudsen*, transcript of oral history conducted 1966–1969 by James V. Mink. Collection 300/101. Department of Special Collections, Young Research Library, University of California, Los Angeles. For Fletcher, see "Harvey Fletcher," transcript of an oral interview by Vern Knudsen with W. J. King (15 May 1964), Niels Bohr Library, Center for History of Physics, American Institute of Physics, College Park, Md. [Fletcher Interview]; Jeff Holland and Ed Butterworth, "Above All, a Graduate of BYU," *BYU Today* (September 1981): 33, 36; and Maureen Meyer Fletcher, *The Caroling of Atoms: The Life's Work of Dr. Harvey Fletcher*, M.F.A. Thesis, Brigham Young University, 1996.

116. Knudsen, *Teacher, Researcher*, 81. Knudsen additionally worked during the war on an effort to speed the transmission of transatlantic telegraph messages.

117. Knudsen, *Teacher, Researcher*, 127.

118. Knudsen, *Teacher, Researcher*, 176.

119. Quoted in Rudnick, "Vern Oliver Knudsen," 712. For an account of this research, see Knudsen, "The Sensibility of the Ear to Small Differences of Intensity and Frequency," *Physical Review*, 2d ser., 21 (January 1923): 84–102; and *Teacher, Researcher*, 170–176.

120. Quoted in Knudsen, *Teacher, Researcher*, 199. Knudsen ultimately played an important role in turning the junior college into a "real university." He became the first Dean of UCLA's Graduate School in 1934, serving until 1956, when he was appointed vice chancellor of the university. He also served as chancellor for one year in 1959, just before he retired.

121. Knudsen, *Teacher, Researcher*, 214.

122. Wallace Clement Sabine, *Collected Papers on Acoustics* (Cambridge, Mass.: Harvard University Press, 1922). Another useful new resource was Floyd Watson's *Acoustics of Buildings* (New York: John Wiley and Sons, 1923), the first general textbook on the topic to appear in the twentieth century.

123. Knudsen, *Teacher, Researcher*, 200–226; 407.

124. Knudsen, *Teacher, Researcher*, 423; 609.

125. The technique as applied to the evaluation of auditoriums is described in Knudsen, "Interfering Effect of Tones and Noise upon Speech Reception," *Physical Review*, 2d ser., 26 (July 1925): 133–138. It will be described more fully in chapter 6.

126. Congressional appropriations for acoustical research at the National Bureau of Standards commenced in 1920; an acoustical division was established in 1922, and a special laboratory for testing acoustical materials was in operation by 1925. Rexmond C. Cochrane, *Measures for Progress: A History of the National Bureau of Standards* (Washington: U.S. Dept. of Commerce, 1966), 263 and Appendix G; and Paul E. Sabine, "Architectural Acoustics: Its Past and Its Possibilities," *JASA* 11 (July 1939): 21–28, on 25. For an account of early work at the NBS, as well as a description of its facilities, see E. A. Eckhardt and V. L. Chrisler, "Transmission and Absorption of Sound by Some Building Materials," *Scientific Papers of the Bureau of Standards* no. 526 (28 April 1926).

127. Knudsen, *Teacher, Researcher*, 325–327. In 1929, when UCLA relocated to its new Westwood campus, Knudsen moved into an acoustical laboratory that he considered to be one of the finest in the nation.

128. Knudsen, *Teacher, Researcher*, 266–267.

129. Knudsen, *Teacher, Researcher*, 385; 664.

130. Knudsen, *Teacher, Researcher*, 663–664.

131. Knudsen, *Teacher, Researcher*, 658.

132. Vern O. Knudsen, *Modern Acoustics and Culture* (Berkeley: University of California Press, 1937).

133. Membership Directory, St. Botolph Club, Boston, 1916. Sabine was also a member of the Century Association in New York. Orcutt, *Wallace Sabine*, 89.

134. Knudsen, *Teacher, Researcher*, 316.

135. Knudsen, *Teacher, Researcher*, 315. Waterfall subsequently served as the Secretary of that society for the next forty years. For his own brief account of the society's formation, see Wallace Waterfall, "History of Acoustical Society of America," *JASA* 1 (October 1929): 5–8. See also R. H. Bolt, "Wallace Waterfall, 1900–1974," *JASA* 56 (December 1974): 1932–1933.

136. All quotes from Knudsen, *Teacher, Researcher*, 313.

137. Fletcher Interview, 44.

138. Miller's claim was based on data he collected while recreating the famous "ether drift" experiment that Albert Michelson and Edward Morley had originally performed at Case Western in 1887. Loyd S. Swenson, Jr., "Dayton Clarence Miller," *American National Biography* (New York: Oxford University Press, 1999) 15: 473–474.

139. Quoted in Robert T. Beyer, *Sounds of Our Times: Two Hundred Years of Acoustics* (New York: AIP/Springer-Verlag, 1999), 223 (note 14).

140. Knudsen, *Teacher, Researcher*, 312–335; Waterfall, "History of Acoustical Society."

141. Knudsen, *Teacher, Researcher*, 333. Membership growth is indicated in Harvey Fletcher, "The Acoustical Society of America: Its Aims and Trends," *JASA* 11 (July

1939): 13–14, on 14. For a list of the charter members, see Waterfall, "History of Acoustical Society," 7–8.

142. "Membership List," *JASA* 2 (April 1931): 401–418, on 401.

143. "Members of the Society," *JASA* 1 (October 1929): 13–23. My analysis of the list of 457 charter members indicates:

    95 (21%) from the telephone industry;

    90 (20%) building materials industry;

    70 (15%) motion picture industry;

    57 (12%) academic affiliations;

    42 (9%) corps. of unknown classification;

    35 (8%) no affiliation indicated;

    35 (8%) radio and phonograph industry;

    17 (4%) musical instrument industry;

    9 (2%) miscellaneous institutions; and

    7 (1%) governmental agencies.

144. "Abstracts of Papers," *JASA* 1 (October 1929): 27–28.

145. Knudsen, *Teacher, Researcher*, 1046–1058.

146. Orcutt, *Wallace Sabine*, vii.

147. Beranek, "Notebooks of Wallace Sabine," 637. Orcutt was a prolific writer whose subjects ranged from Benvenuto Cellini to Mary Baker Eddy. "William Dana Orcutt," *National Cyclopedia* (New York: James White, 1955) 40: 510–511.

148. Historians of science have long grappled with the difficulty of employing concepts like "pure" and "applied" to categorize and understand the science of the past. In the discussion that follows, I use these words as they were used by the people that I describe, which is to say loosely and imprecisely. The terms nonetheless indicate a fundamental distinction that was perceived to be crucial at the time. For an excellent historical analysis of this language, see Ronald Kline, "Construing 'Technology' as 'Applied Science': Public Rhetoric of Scientists and Engineers in the United States, 1880–1945," *Isis* 86 (June 1995): 194–221.

149. Knudsen, *Teacher, Researcher*, 318–319; 328. A separate society for sound engineers was eventually founded in 1948. See Jerry B. Minter, "The AES Begins Its Seventh Year," *Journal of the Audio Engineering Society* 2 (January 1954): 1–2. The need for a distinction between acoustical physicists and engineers had been identified as early as 1911. See Ernest Merritt, "A Plea for Acoustic Engineering," *Sibley Journal of Engineering* 31 (1916): 101–102.

150. His collaborator in this work was Hans Kneser, a German scholar visiting UCLA on a Rockefeller Fellowship. For Knudsen's own account, see Knudsen, *Teacher, Researcher*, 565–600, and 927. Rudnick also highlights this story in his obituary, "Vern Knudsen," 712–713. The prize is mentioned in Lindsay, "Vern Knudsen," 490. See also Vern O. Knudsen, "Absorption of Sound in Air, in Oxygen, and in Nitrogen—Effects of Humidity and Temperature," *JASA* 5 (October 1933): 112–121; and Hans O. Kneser, "Interpretation of the Anomalous Sound-Absorption in Air and Oxygen in Terms of Molecular Collisions," *JASA* 5 (October 1933): 122–126. In a similar vein, generations of acousticians have transmitted the story of the crucial, albeit unrecognized, role that Harvey Fletcher apparently played as a graduate student in designing and performing the famous oil-drop experiment credited to Robert Millikan. This experiment, one of the most significant in the history of modern physics, measured the electrical charge of the electron. For Fletcher's own account, see Fletcher Interview, 13–23.

151. Paul Sabine, "Beginnings of Architectural Acoustics," 242.

152. Hall, "Wallace Clement Ware Sabine," 9.

153. Orcutt, *Wallace Sabine*, 225; 221; 224; 243.

154. "Acoustical Engineering as a Career," 7.

155. "Acoustical Engineering as a Career," 9.

156. Orcutt, *Wallace Sabine* (quoting Adelbert Ames, Jr.), 322.

157. Hall, "Wallace Clement Ware Sabine," 17.

158. The modern photograph of Sabine portrayed in figure 3.16 was taken at the same time as the 1918 portrait reproduced in Orcutt, *Wallace Sabine*, facing 330. The modern Sabine would not reappear until 1954, in an account of the twenty-fifth anniversary of the Acoustical Society of America. At this time, the acousticians were apparently secure enough in their professional standing to poke fun at their founders as well as themselves. The entire account is lighthearted in tone, and Sabine's portrait was captioned "Wallace Clement Sabine (not Adlai Stevenson)," calling attention to the mature physicist's rather striking resemblance to the popular politician. "The Twenty-Fifth Anniversary Celebration," *JASA* 26 (September 1954): 874–905, on 887.

159. Sinclair Lewis, *Arrowsmith* (1925; New York: Penguin, 1980). My reading of *Arrowsmith*, and the connections I draw between Lewis's fictional scientist and the members of the Acoustical Society of America, are indebted to Charles E. Rosenberg, "Martin Arrowsmith: The Scientist as Hero," pp. 123–131 in Rosenberg, *No Other Gods: On Science and American Social Thought* (Baltimore: Johns Hopkins University Press, 1976).

160. C. B. Palmer, "Caves and Cubicles for Modern Black Magic," *Boston Evening Transcript* (30 March 1932). In 1926, the University sought to dismantle the room and in response, the Guastavino and Johns-Manville companies each donated $2,000 toward the cost of constructing a new facility in order that Sabine's shrine might be preserved.

See H. E. Manville to A. Lawrence Lowell (10 June 1926); C. M. Swan to Mr. Blodgett (11 June 1926); R. Guastavino Co. to A. Lawrence Lowell (14 June 1926); and Theodore Lyman to R. Guastavino Co. (15 October 1926), Series I, Box 14, Folder 10: Swan, C. M.-Jefferson Physical Lab, GCC.

Chapter Four

1. F. Scott Fitzgerald, "My Lost City" (written July 1932), 23–33 in *The Crack-Up* (New York: New Directions, 1945), on 31.

2. William L. Chenery, "The Noise of Civilization," *New York Times Magazine* (1 February 1920): 13. The article does not provide the visitor's name.

3. Earley Vernon Wilcox, "To Heal the Blows of Sound," *Harvard Graduates' Magazine* 33 (June 1925): 584–590, on 584.

4. The graffiti is described in Mrs. Isaac L. Rice, "An Effort to Suppress Noise," *Forum* 37 (April-June 1906): 552–570, on 552.

5. Quoted in Jean Gimpel, *The Medieval Machine: The Industrial Revolution of the Middle Ages* (1976; New York: Penguin, 1977), on 84.

6. Jenny Uglow emphasizes the sonorous quality of Hogarth's London scenes in *Hogarth: A Life and a World* (London: Faber and Faber, 1997), 300–301. For studies of the London soundscape during an earlier period, see Bruce R. Smith, *The Acoustic World of Early Modern England: Attending to the O-Factor* (Chicago: University of Chicago Press, 1999); and Eric Wilson, "Plagues, Fairs, and Street Cries: Sounding Out Society and Space in Early Modern England," *Modern Language Studies* 25 (summer 1995): 1–42.

7. Goethe's complaint is cited in Philip G. Hubert, Jr., "For the Suppression of City Noises," *North American Review* 159 (November 1894): 633–635, on 635. See also Arthur Schopenhauer, "On Noise," pp. 216–221 in *The Pessimist's Handbook*, ed. Hazel E. Barnes, trans. T. Bailey Saunders (Lincoln: University of Nebraska Press, 1964). For an account of an early twentieth-century German complainant, see Lawrence Baron, "Noise and Degeneration: Theodor Lessing's Crusade for Quiet," *Journal of Contemporary History* 17 (January 1982): 165–178.

8. Carlyle's battle with noise and the construction of his soundproof room are described in James Anthony Froude, *Thomas Carlyle: A History of His Life in London: 1834–1881* (London: Longmans, Green and Co., 1897), vol. II, 146–147 and 166–167. See also John M. Picker, "The Soundproof Study: Victorian Professionals, Work Space, and Urban Noise," *Victorian Studies* 42 (spring 1999/2000): 427–453.

9. Carlyle's complaint is quoted in Rice, "Effort to Suppress Noise," 552. For an expansive account of how the sounds of the world have changed with industrialization, see R. Murray Schafer, *The Soundscape: Our Sonic Environment and the Tuning of the World* (Rochester, Vt.: Destiny Books, 1994).

10. J. H. Girdner, M.D., "The Plague of City Noises," *North American Review* 163 (September 1896): 296–303, on 300. Girdner did mention, but chose not to emphasize, the sounds of streetcars and elevated trains.

11. "Noise," *Saturday Review of Literature* 2 (24 October 1925): 1.

12. See the category "Vocals, etc." in "Tabulation of Noise Complaints," in Edward F. Brown et al., eds., *City Noise* (New York: Department of Health, 1930), 27. See also Lewis H. Brown, "Attacking City Noises by Science and Law," *American City* 44 (February 1931): 97–101, quotes on 97.

13. In 1893, E. L. Godkin had editorialized that "the reduction of city noise is now one of the most important elements in all city reforms." "Noise," *Nation* 56 (15 June 1893): 433–434, on 433. See also George Ethelbert Walsh, "When Science Banishes City Noises," *Harper's Weekly* 51 (27 July 1907): 1098; and Hollis Godfrey, "The City's Noise," *Atlantic Monthly* 104 (November 1909): 601–610, on 606.

14. Imogen Oakley, "Public Health Versus the Noise Nuisance," *National Municipal Review* 4 (April 1915): 230–237, on 234; "Is It a Symptom?" *New York Times* [*NYT*] (14 September 1929): 18.

15. "Noise" (*Saturday Review of Literature*), 1.

16. Leo Marx, *The Machine in the Garden: Technology and the Pastoral Ideal in America* (London: Oxford University Press, 1964), 3–33 and 242–255; quote on 23. See also John F. Kasson, *Civilizing the Machine: Technology and Republican Values in America, 1776–1900* (New York: Penguin, 1976); and Henry David Thoreau, "Sounds," 79–90 in *Walden* (1854; New York: Penguin, 1960).

17. R. Murray Schafer emphasizes how the very acoustic nature of a hum, which he characterizes as a "flat line" sound, a monotonous, continuous drone, was distinctly new to the industrial era. Schafer, *New Soundscape*, 78–80.

18. "Noise," *NYT* (1 August 1878): 4.

19. "The Elimination of Harmful Noise," *National Safety News* 15 (April 1927): 46, 58, quote on 58. The passage describes the situation in the nineteenth and early twentieth centuries. The article went on to indicate how, in 1927, this approach to noise nuisance lawsuits no longer guaranteed success. See also Raymond Smilor, "Personal Boundaries in the Urban Environment: The Legal Attack on Noise: 1865–1930," *Environmental Review* 3 (spring 1979): 24–36.

20. Chenery, "Noise of Civilization," 13. See also "Of the Right to Make Some Noise," *NYT* (5 March 1909): 8; and "An Earnest Plea for Less Noise," *NYT* (20 September 1913): 10.

21. "Noise" (*Nation*), 433. See also Hubert, "Suppression of City Noises," 634; and Godfrey, "City's Noise," 601.

22. Mrs. Isaac Rice, "Our Most Abused Sense—The Sense of Hearing," *Forum* 38 (April–June 1907): 559–572, on 559 and 560.

23. Julia Barnett studied at the Women's Medical College of the New York Infirmary, then interned for one year before marrying Isaac Rice in 1885 and subsequently dedicating herself to the raising of their six children. Raymond Smilor, "Confronting the Industrial Environment: The Noise Problem in America, 1893–1932" (Ph.D. diss., University of Texas at Austin, 1978), 56.

24. For her own account of the campaign, see Rice, "Effort to Suppress Noise." See also "Oppose Whistling Nuisance," *NYT* (26 January 1906): 1; "Mrs. Rice's Labor for Quiet," *NYT* (16 April 1906): 2; and "Whistling to Be Restricted," *NYT* (8 December 1906): 10. Smilor notes that the Bennet Bill was difficult to enforce, and the sounds of the tugboat whistles were by no means eliminated with this measure. "Confronting the Industrial Environment," 57–64.

25. "Mrs. Rice Now Attacks All Needless Noise," *NYT* (9 December 1906): 12. The name of her organization would later be changed to the Society for the Prevention of Unnecessary Noise.

26. "The Campaign Against Noise," *NYT* (23 December 1906): part 3, p. 4.

27. "Mrs. Rice Attacks Needless Noise," 12.

28. "Campaign Against Noise," 4. Mark Twain also lent his support.

29. Walter B. Platt, "Certain Injurious Influences of City Life and Their Removal," *Journal of Social Science* 24 (April 1888): 24–30, on 27.

30. William Dean Howells, "Editor's Easy Chair," *Harper's Monthly Magazine* 113 (November 1906): 957–960, on 959; "Unnecessary Noises," *NYT* (19 October 1908): 8.

31. For more on smoke abatement, see David Stradling, *Smokestacks and Progressives: Environmentalists, Engineers and Air Quality in America, 1881–1951* (Baltimore: Johns Hopkins University Press, 1999); R. Dale Grinder, "The Battle for Clean Air: The Smoke Problem in Post–Civil War America," pp. 83–103 in Martin V. Melosi, ed., *Pollution and Reform in American Cities, 1870–1930* (Austin: University of Texas Press, 1980); and Joel A. Tarr, *The Search for the Ultimate Sink: Urban Pollution in Historical Perspective* (Akron: University of Akron Press, 1996). Additional sources are cited in Jeffrey K. Stine and Joel A. Tarr, "At the Intersection of Histories: Technology and the Environment," *Technology and Culture* 39 (October 1998): 601–640.

32. Raymond Smilor, "Toward an Environmental Perspective: The Anti-Noise Campaign, 1893–1932," pp. 135–151 in Melosi, ed., *Pollution and Reform*. Smilor's conclusion, that noise abatement ultimately failed because it lacked "a socially and culturally developed environmental attitude," similarly suggests concerns more contemporary than historical. "Confronting the Industrial Environment," 147. Schafer's writings from this

era include *The New Soundscape* (Scarborough, Ont., and New York: Berandol Music and Associated Music Publishers, 1969); *The Book of Noise* (Wellington, N.Z.: Price Milburn, 1970); and *Soundscape*, which was originally published in 1977 as *The Tuning of the World*.

33. Mrs. Rice connected the two but only to focus upon their wasteful aspects. Rice, "Our Most Abused Sense," 560. The term *noise pollution* appears in none of the sources that I have examined. The only explicit connection between noise and "polluting" was made by Imogen Oakley when she wrote, "If [boards of health] can prevent [a man] from polluting his neighbor's water-supply with typhoid germs, they can forbid him from congesting his neighbor's air with sounds that breed insanity." Oakley, "Public Health Versus Noise Nuisance," 236.

34. Samuel Haber, *Efficiency and Uplift: Scientific Management in the Progressive Era 1890–1920* (1964; Chicago: University of Chicago Press, 1973), ix. See also Samuel P. Hays, *Conservation and the Gospel of Efficiency: The Progressive Conservation Movement 1890–1920* (Cambridge, Mass.: Harvard University Press, 1959).

35. *Harper's Weekly* (2 November 1912), quoted in Cecelia Tichi, *Shifting Gears: Technology, Literature, Culture in Modernist America* (Chapel Hill: University of North Carolina Press, 1987), 75.

36. For more on Taylor's life and ideas, see Robert Kanigel, *The One Best Way: Frederick Winslow Taylor and the Enigma of Efficiency* (New York: Viking, 1997); Thomas P. Hughes, *American Genesis: A Century of Invention and Technological Enthusiasm, 1870–1970* (New York: Penguin, 1989); and Frederick Winslow Taylor, *The Principles of Scientific Management* (1911; New York: W.W. Norton and Co., 1967).

37. The class implications of twentieth-century noise reform are emphasized in Warren Bareiss, "Noise Abatement in Philadelphia 1907–1966: The Production of a Soundscape" (M.A. thesis, University of Pennsylvania, 1990). A similar emphasis applied to nineteenth-century London is found in Picker, "Soundproof Study." For a theoretical analysis of the political dimensions of noise, see Jacques Attali, *Noise: The Political Economy of Music*, trans. Brian Massumi (1977; Minneapolis: University of Minnesota Press, 1985).

38. "Barkers at Coney Squelched by Police," *NYT* (24 June 1907): 2.

39. John F. Kasson, *Amusing the Million: Coney Island at the Turn of the Century* (New York: Hill and Wang, 1978), 104.

40. "Barkers Squelched by Police," 2.

41. "Opens War on Noises, Asks Bingham's Aid," *NYT* (27 June 1908): 4; "Bingham Sets Out to Insure Quiet," *NYT* (19 July 1908): part 2, p. 4; "Putting Noise Lid Down," *NYT* (21 July 1908): 1; and "Bingham Hears of Noises," *NYT* (24 July 1908): 12.

42. "Ordinance Puts the Lid on Noise," *NYT* (15 July 1909): 7.

43. Quoted in "Ordinance Puts Lid on Noise," 7.

44. "Ole Clo' Men to Bingham," *NYT* (27 July 1908): 14; "Ol' Clo' Shouters Lose," *NYT* (28 July 1908): 12; and "Peddlers Raid Groceries," *NYT* (28 July 1911): 3. See also Smilor, "Confronting the Industrial Environment," 102–103.

45. Daniel Bluestone, "The Pushcart Evil," pp. 287–312 in David Ward and Olivier Zunz, eds., *The Landscape of Modernity: New York City, 1900–1940* (Baltimore: Johns Hopkins University Press, 1992), on 288.

46. Paul Boyer, *Urban Masses and Moral Order in America, 1820–1920* (Cambridge, Mass.: Harvard University Press, 1978), 266. See also M. Christine Boyer, *Dreaming the Rational City: The Myth of American City Planning* (Cambridge, Mass.: The MIT Press, 1983); Keith Revell, "Regulating the Landscape: Real Estate Values, City Planning and the 1916 Zoning Ordinance," pp. 19–45 in Ward and Zunz, eds., *Landscape of Modernity*; Marc A. Weiss, "Density and Intervention: New York's Planning Traditions," pp. 46–75 in *Landscape of Modernity*; Carol Willis, "Zoning and Zeitgeist: The Skyscraper City in the 1920s," *Journal of the Society of Architectural Historians* 45 (March 1986): 47–59; and Robert Weibe, *The Search for Order, 1877–1920* (New York: Hill and Wang, 1967).

47. "Makes Quiet Zones for City Hospitals," *NYT* (24 June 1907): 7; "Pass Quiet Zone Ordinance," *NYT* (26 June 1907): 6.

48. Oakley was not a doctor herself, but she came from a family of physicians and had actively campaigned for smoke abatement in Pittsburgh before moving to Philadelphia. Bareiss, "Noise Abatement in Philadelphia," 24–29 and passim. For Oakley's own account of her work, see "Public Health Versus Noise Nuisance."

49. "For Quiet School Streets," *NYT* (23 November 1911): 3; "Save the Children by Killing Noise," *NYT* (2 April 1914): 11; "Would Silence the School Blocks," *NYT* (7 June 1914): sect. 3, p. 10.

50. Mrs. Isaac Rice, "'Quiet Zones' for Schools," *Forum* 46 (December 1911): 731–742, on 733.

51. William T. Watson, "Baltimore's Anti-Noise Crusade," *National Municipal Review* 3 (July 1914): 585–589, on 587. For a survey of noise ordinances in other cities, see Carl Henry Mote, "The Effort to Control Municipal Noise," *American City* 10 (February 1914): 147–149.

52. "Commission on Building Districts and Restrictions, Final Report" (2 June 1916), 20, New York Municipal Reference Library. For additional mention of the problem of noise, see pp. 10, 13, 17, 29, 96, and figs. 4, 12, 64, and 65.

53. "Nature and Result of Inspections by Sanitary Inspectors," *Annual Report of the Board of Health of the Department of Health, City of New York* (1912): 31, New York Municipal Reference Library.

54. "Summary of Inspection Work," *Annual Report of the Department of Health of the City of New York* (1913): 26, New York Municipal Reference Library.

55. S. W. Wynne, Commissioner of Health, to Dr. Emanuel Gogel (23 June 1930). New York Municipal Archives, Department of Health, Administration/Subject Files, Box: 1930, "Mo–R" (07-025942), Folder: "Noise, January–July 1930."

56. "Mrs. R. T. Wilson Summoned to Court," *NYT* (4 March 1921): 1, 4; "Topics of the Times," *NYT* (5 March 1921): 12; and (for quote) "Mrs. R. T. Wilson Is Victor in Court," *NYT* (5 March 1921): 28. The Wilsons appear to have celebrated their victory in court with another party. See "R. T. Wilsons Give a Musicale," *NYT* (21 March 1921): 13. While Newton's complaint was dismissed, the *Times* editorialized in his favor. An account of the history of this building, at 130 West 57th Street, also lent sympathy to the complainants. The building was a cooperative of artists' and writers' studios, and the very apartment occupied by Mrs. Wilson (whose only art was that of mingling in New York Society) had previously belonged to none other than William Dean Howells, charter member of the Society for the Suppression of Unnecessary Noises. Hassam had always worn rubber-soled shoes, and even used a rubber-footed easel, to minimize any disturbance to his downstairs neighbor, and other tenants were similarly courteous to each other, at least until nonartists like Mrs. Wilson began moving in. "Let's All Be More Considerate" *NYT* (7 March 1921): 10; Helen Bullitt Lowry, "Noise and Your Neighbors," *New York Times Book Review and Magazine* (20 March 1921): 8.

57. "Wins with Violin in Court," *NYT* (15 January 1925): 12.

58. "Late Music a Nuisance," *NYT* (26 July 1924): 2.

59. "Everybody's 'Rights' Are Involved," *NYT* (5 March 1921): 12. Helen Lowry agreed that interest in the case was widespread, and she noted that Hassam and Newton had received numerous telephone calls from strangers in support of their case. "Noise and Your Neighbors," 8.

60. "Mrs. Wilson Is Victor," 28.

61. J. A. Rogers, "Jazz at Home," pp. 216–224 in Alain Locke, ed., *The New Negro* (1925; New York: Touchstone, 1997) on 219, 218. The rich sounds of African-American aural culture in antebellum America are well documented in Lawrence W. Levine, *Black Culture and Black Consciousness: Afro-American Folk Thought from Slavery to Freedom* (New York: Oxford University Press, 1977); Mark M. Smith, "Listening to the Heard Worlds of Antebellum America," *Journal of the Historical Society* 1 (spring 2000): 65–99; and Smith, "Time, Sound, and the Virginia Slave," pp. 29–60 in John Saillant, ed., *Afro-Virginian History and Culture* (New York: Garland, 1999).

62. Ellington is quoted in Nat Shapiro and Nat Hentoff, eds., *Hear Me Talkin' to Ya* (1955; New York: Dover, 1966), 224–225.

63. Lawrence W. Levine, "Jazz and American Culture," *Journal of American Folklore* 102 (January–March 1989): 6–22, on 12. See also Kathy J. Ogren, *The Jazz Revolution: Twenties America and the Meaning of Jazz* (New York: Oxford University Press, 1989); MacDonald Smith Moore, *Yankee Blues: Musical Culture and American Identity* (Bloomington: Indiana University Press, 1985); and Neil Leonard, *Jazz and the White Americans: The Acceptance of a New Art Form* (Chicago: University of Chicago Press, 1962). Social histories of jazz most often note the sociological influence of cities on the music but neglect to consider the role of the urban soundscape in its genesis. See Burton Peretti, *The Creation of Jazz: Music, Race, and Culture in Urban America* (Urbana and Chicago: University of Illinois Press, 1992); Dan Morgenstern, "Jazz as an Urban Music," pp. 133–143 in George McCue, ed., *Music in American Society: 1776–1976* (New Brunswick, N.J.: Transaction Books, 1977); William Howland Kenney, *Chicago Jazz: A Cultural History, 1904–1930* (New York: Oxford University Press, 1993); LeRoi Jones, *Blues People: The Negro Experience in White America and the Music that Developed from It* (New York: William Morrow and Co., 1963); Leroy Ostransky, *Jazz City: The Impact of Our Cities on the Development of Jazz* (Englewood Cliffs, N.J.: Prentice-Hall, 1978); and Charles Nanry, "Jazz and Modernism: Twin-Born Children of the Age of Invention," *Annual Review of Jazz Studies* 1 (1982): 146–154.

64. Actress Laurette Taylor, quoted in Leonard, *Jazz and White Americans*, 38; Daniel Gregory Mason, "The Jazz Invasion," pp. 499–513 in Samuel D. Schmalhausen, ed., *Behold America!* (New York: Farrar and Rinehart, 1931), quoted in Moore, *Yankee Blues*, 106.

65. Anne Shaw Faulkner, "Does Jazz Put the Sin in Syncopation?" *Ladies' Home Journal* 38 (August 1921): 16, 34, quoted in Leonard, *Jazz and White Americans*, 39; Mason, "Jazz Invasion," quoted in Moore, *Yankee Blues*, 106.

66. Ann Douglas, *Terrible Honesty: Mongrel Manhattan in the 1920s* (New York: Farrar, Straus Giroux, 1995) explores this aspect of modern culture in rich detail.

67. Alain Locke, *The Negro and His Music* (Washington, D.C.: Associated Publishers, 1936), quoted in Levine, "Jazz and American Culture," 14; Ogren, *Jazz Revolution*, 7.

68. Nell Irvin Painter highlights the role of race within larger patterns of social and industrial change in *Standing at Armageddon: The United States, 1877–1919* (New York: W.W. Norton and Co., 1987).

69. Langston Hughes, "The Negro Artist and the Racial Mountain," *Nation* 122 (28 June 1926): 692–694, on 694, quoted in David Levering Lewis, *When Harlem Was in Vogue* (1981; New York: Penguin, 1997), on 191. For more on the meaning of music in the Harlem Renaissance, see Samuel A. Floyd, Jr., "Music in the Harlem Renaissance: An Overview," pp. 1–27 in Floyd, ed., *Black Music in the Harlem Renaissance* (New York: Greenwood, 1990); Nathan Irvin Huggins, *Harlem Renaissance* (New York: Oxford University Press, 1971), 197–198; Houston A. Baker, Jr., *Modernism and the Harlem Renaissance* (Chicago: University of Chicago Press, 1987); and

Ronald M. Radano, "Soul Texts and the Blackness of Folk," *Modernism/Modernity* 2 (January 1995): 71–95.

70. Hermann Helmholtz, *On the Sensations of Tone*, 2d English ed. conforming to the 4th German ed. of 1877, trans. Alexander Ellis (1885; New York: Dover, 1954), 7–8. Rayleigh similarly noted that "sounds may be classed as musical and unmusical; the former for convenience may be called *notes* and the latter *noises*." J. W. S. Rayleigh, *The Theory of Sound*, 2d rev. ed. (1894; New York: Dover, 1945), 4. Helmholtz acknowledged that "Noises and musical tones may certainly intermingle in very various degrees, and pass insensibly into one another" but he emphasized that "their extremes are widely separated" and the distinction between these types of sounds remained fundamental. Rayleigh agreed on both points. Helmholtz, 7; Rayleigh, 4. See also Godfrey, "City's Noise," 601; and H. Weaver Mowery, "Harmful Noises and Their Elimination," *Industrial Psychology* 1 (May 1926): 338–340, on 339, for more popular expressions of the same ideas.

71. Henry Cowell, "The Joys of Noise," *New Republic* 59 (31 July 1929): 287–288, on 287.

72. J. Peter Burkholder, liner notes to Deutsche Grammophon Stereo 429–220–2, works of Charles Ives performed by Leonard Bernstein and the New York Philharmonic. See also J. Peter Burkholder, *Charles Ives: The Ideas Behind the Music* (New Haven; Yale University Press, 1985); and Burkholder, ed., *Charles Ives and His World* (Princeton: Princeton University Press, 1996).

73. Moore, *Yankee Blues*, 29.

74. Ives is quoted in Leon Botstein, "Innovation and Nostalgia: Ives, Mahler, and the Origins of Twentieth-Century Modernism," pp. 35–74 in Burkholder, ed., *Ives and His World*, quote on 43.

75. David Harold Cox and Michael Naslas, "The Metropolis in Music," pp. 173–189 in Anthony Sutcliffe, ed., *Metropolis 1890–1940* (London: Mansell, 1984), on 184. See also Barbara Barthelmes, "Music and the City," pp. 97–105; and Hans-Joachim Braun, "'Movin' On': Trains and Planes as a Theme in Music," pp. 106–120 in Braun, ed., *"I Sing the Body Electric": Music and Technology in the 20th Century* (Hofheim: Wolke Verlag, 2000).

76. For general histories of modern music, see Robert Morgan, ed., *Modern Times: From World War I to the Present* (Englewood Cliffs: Prentice Hall, 1994); and Paul Griffiths, *Modern Music: A Concise History* (New York: Thames and Hudson, 1994).

77. Ferruccio Busoni, "Sketch of a New Esthetic of Music," English trans. of 1911 reprinted in *Three Classics in the Aesthetic of Music* (New York: Dover, 1962), 75–102, quotes on 100 and 93.

78. Busoni, "Sketch," 95. His knowledge of Cahill's Dynamophone (it would later be called the Telharmonium) came from Ray Stannard Baker, "New Music for an Old

World," *McClure's Magazine* 27 (July 1906): 291–301. See also Reynold Weidenaar, *Magic Music from the Telharmonium* (Metuchen, N.J., and London: Scarecrow Press, 1995); and Thomas LaMar Rhea, "The Evolution of Electronic Musical Instruments in the United States" (Ph.D. diss., George Peabody College for Teachers, 1972).

79. F. T. Marinetti, "Let's Murder the Moonshine," pp. 53–62 in R. W. Flint, ed., *Let's Murder the Moonshine: Selected Writings of F. T. Marinetti* (Los Angeles: Sun and Moon Classics, 1991), on 54.

80. For a compelling account of the meaning of sound in the avant-garde, from futurists and Dadaists to John Cage and William Burroughs, see Douglas Kahn, *Noise Water Meat: A History of Sound in the Arts* (Cambridge, Mass.: The MIT Press, 1999). See also Kahn and Gregory Whitehead, eds., *Wireless Imagination: Sound, Radio, and the Avant-Garde* (Cambridge, Mass.: The MIT Press, 1992); Karin Bijsterveld, "'A Servile Imitation': Disputes About Machines in Music, 1910–1930," pp. 121–134 in Braun, ed., *I Sing the Body Electric*; and Robert P. Morgan, "'A New Musical Reality': Futurism, Modernism, and 'The Art of Noises,'" *Modernism/Modernity* 1 (1994): 129–151.

81. F. T. Marinetti, "Geometrical and Mechanical Splendor and the Numeric Sensibility," (1914), pp. 154–160 in Umbro Apollonio, ed., *Futurist Manifestos* (New York: Viking, 1973), on 158.

82. Quoted in Luigi Russolo, *The Art of Noises*, trans. Barclay Brown (1916; New York: Pendragon Press, 1986), 26.

83. Flint, intro. to *Let's Murder the Moonshine*, 26.

84. Carlo Carrà, "The Painting of Sounds, Noises and Smells," pp. 111–115 in Apollonio, ed., *Futurist Manifestos*, on 114. Numerous artists in the teens and twenties, including Vasily Kandinsky, developed new theories relating sound to color and shape. See Kahn, *Noise Water Meat*, 104–109.

85. Quoted in Morgan, "New Musical Reality," 137. Barclay Brown attributes this passage to Marinetti in his introduction to Russolo's *Art of Noises*, 2.

86. Russolo celebrated his lack of "acoustical prejudices," and professed himself "bolder than a professional musician, not worried about my apparent incompetence, and convinced that audacity has all rights and all possibilities." Russolo, *Art of Noises*, 30.

87. The original manifesto, "l'Arte dei rumori," appeared as a pamphlet in 1913. It was reproduced as the first chapter of Russolo's extended treatise of 1916, also titled *l'Arte dei Rumori*. All page numbers cited hereafter refer to Barclay Brown's 1986 translation of this 1916 document, *The Art of Noises*.

88. Russolo, *Art of Noises*, 23, 24.

89. Russolo, *Art of Noises*, 25, 26.

90. The following description of Russolo's instruments follows Barclay Brown, who has hypothesized the construction of the instruments via photographic evidence, a few

vague descriptions found in Russolo's correspondence and in the press, and Brown's own efforts to re-create the instruments. Russolo guarded his designs carefully as he hoped to patent and market the devices, and the instruments themselves disappeared or were destroyed sometime during the Second World War. A phonographic recording of the noise-intoners was made in 1921, but according to Brown, their sounds are overwhelmed by the orchestra that accompanies them. Barclay Brown, "The Noise Instruments of Luigi Russolo," *Perspectives of New Music* 20 (1981–1982): 31–48, and intro. to Russolo, *Art of Noises*. See also Bijsterveld, "Servile Imitation," which includes photos of the instruments on 121 and 124.

91. While he provided a template to guide the musicians in producing the whole and half-tones of a tempered scale, Russolo was also enthusiastic about the ease with which continuously varying tones and microtones could be produced.

92. *Pall Mall Gazette* (18 November 1913), quoted in Brown's introduction to *Art of Noises*, 4–5.

93. Russolo, *Art of Noises*, 33.

94. Russolo, *Art of Noises*, 33–34.

95. *London Times* (16 June 1914): 5, quoted in Michael Kirby and Victoria Nes Kirby, *Futurist Performance* (New York: PAJ Publications, 1986), 38.

96. Russolo, *Art of Noises*, 48.

97. Russolo, *Art of Noises*, 36.

98. Russolo scraped by on a veteran's pension, and he continued to build new instruments which were only occasionally demonstrated, typically in conjunction with a conventional orchestra. He also built a polyphonic instrument that combined many of the different mechanisms of his earlier instruments. In Paris, Russolo found regular employment accompanying silent films with his "noise harmonium," but with the arrival of sound film, this venue was lost. By 1929, Russolo was a hungry vagrant who turned to spiritualism and mysticism for sustenance, as the brutal reality of the material world that had proved so inspiring in 1913 now betrayed him. In 1938, he wrote a book titled *Beyond the Material World*, and in 1945 he died. Brown, intro. to Russolo, *Art of Noises*, 9–10.

99. Liner notes to Vanguard SRV–274SD, quoted in Moore, *Yankee Blues*, 125–126.

100. For more on sirens as a signal of modernity, see Kahn, *Noise Water Meat*, 83–91.

101. Fernand Ouellette, *Edgard Varèse*, trans. Derek Coltman (New York: Orion Press, 1968), 77.

102. Paul Rosenfeld, *An Hour with American Music* (1929; Westport, Conn.: Hyperion, 1979), 163.

103. Stokowski is quoted in Louise Varèse, *Varèse: A Looking-Glass Diary, Vol. I: 1883–1928* (New York: W.W. Norton and Co., 1972), 223; Downes and Gilman on 224.

104. Rosenfeld, review of *Intégrales* in *Dial* (May 1925), quoted in Ouellette, *Edgard Varèse*, 85.

105. Louise Varèse, *Varèse*, 245.

106. Linton Martin, "Catcalls Greet Orchestra Work," *Philadelphia Inquirer* (10 April 1926): 10.

107. Gilman, program notes for premiere, quoted in Ouellette, *Edgard Varèse*, 92.

108. Henderson quoted in Louise Varèse, *Varèse*, 255; Gilman quoted in Ouellette, *Edgard Varèse*, 93.

109. Quoted in Ouellette, *Edgard Varèse*, 93.

110. Quoted in Ouellette, *Edgard Varèse*, 46–47.

111. *Christian Science Monitor* (1922), quoted in Chou Wen-Chung, "Open Rather than Bounded," *Perspectives of New Music* 5 (fall–winter 1966): 1–6, on 1.

112. Varèse is quoted from an application for a Guggenheim Fellowship in Wen-Chung, "Open Rather than Bounded," 2.

113. Varèse later collaborated with the architect Le Corbusier and the sound engineers of the Dutch Philips Company to create the Philips Pavilion, an audiovisual-architectural installation at the Brussels World's Fair of 1958. Marc Treib, *Space Calculated in Seconds: The Philips Pavilion, Le Corbusier, Edgard Varèse* (Princeton: Princeton University Press, 1996).

114. The motto of the International Composers' Guild, founded by Varèse in 1921 to support modern music, was "New Ears for New Music and New Music for New Ears." Louise Varèse, *Varèse*, 207.

115. For more on Antheil's life and career, see his autobiography, *Bad Boy of Music* (1945; New York: Da Capo Press, 1981). See also Linda Whitesitt, *The Life and Music of George Antheil, 1900–1959* (Ann Arbor: UMI Research Press, 1983); Glenda Dawn Goss, "George Antheil, Carol Robinson and the Moderns," *American Music* 10 (winter 1992): 468–485; and Noel Riley Fitch, *Sylvia Beach and the Lost Generation: A History of Literary Paris in the Twenties and Thirties* (New York: Norton, 1983).

116. Ezra Pound, "George Antheil," *Criterion: A Quarterly Review* 2 (April 1924), quoted in Whitesitt, *Life and Music of Antheil*, 18. See also Pound, *Antheil and the Treatise on Harmony* (1927; New York: Da Capo, 1968).

117. "U.S. Composer's Weird Symphony Is Given in Paris," *New York Herald-Tribune* (Paris ed., c. January 1924), quoted in Whitesitt, *Life and Music of Antheil*, 20–21.

118. Antheil, *Bad Boy of Music*, 7, 133.

119. Quoted in Charles Amirkhanian's introduction to Antheil, *Bad Boy of Music*, v. Antheil later learned that this riot had been anticipated and encouraged, and was actually filmed as it occurred for use by director George l'Herbier in his (silent) motion picture *l'Inhumaine*.

120. "Program Notes by Maurice Peress," MusicMasters Classics Compact Disc 01612–67094–2, *Ballet Mécanique: George Antheil's Carnegie Hall Concert of 1927 Recreated and Conducted by Maurice Peress*, 10. For an account of this concert by its promoter, see Donald Friede, *The Mechanical Angel: His Adventures and Enterprises in the Glittering 1920's* (New York: Knopf, 1948), 44–61.

121. William Carlos Williams, "George Antheil and the Cantilene Critics: A Note on the First Performance of Antheil's Music in New York City; April 10-1927," *Transition* 13 (summer 1928): 237–240, on 240..

122. This is the composer's original orchestration, as described in "Boos Greet Antheil Ballet of Machines," *New York Herald Tribune* (11 April 1927): 1, 12, on 1. In fact, ten pianos were used at the New York premiere, as the Baldwin Piano Company, which had sponsored the concert, wanted to maximize their exposure. The piece was performed in front of a backdrop depicting a giant sparkplug surrounded by elevated trains, steam shovels, and skyscrapers.

123. "Boos Greet Antheil," 1; Williams, "Antheil and the Cantilene Critics," 239, 238; "Antheil Art Bursts on Startled Ears," *NYT* (11 April 1927): 23.

124. Lawrence Gilman, "Mr. Antheil Presents His Compliments to New York," *New York Herald Tribune* (11 April 1927): 12.

125. Williams, "Antheil and the Cantilene Critics," 238, 240.

126. Rosenfeld, *Hour with American Music*, 160–162.

127. "Opera to Be Tested by 'Electric Ear,'" *NYT* (27 April 1932): 13.

128. "Lily Pons 'Noisier' Than a Street Car," *NYT* (28 April 1932): 23.

129. "Possibilities of Decibels," *NYT* (29 April 1932): 16. The event was also covered, as "Music and Noise," in the *Literary Digest* 113 (21 May 1932): 17.

130. "Noise," *NYT* (1 August 1878): 4. See also Paul Israel, *Edison: A Life of Invention* (New York: John Wiley & Sons, 1998), 489 (note 32).

131. "Mrs. Rice Now Attacks All Needless Noise," *NYT* (9 December 1906): 12.

132. Elmer S. Batterson, "Progress of the Anti-Noise Movement," *National Municipal Review* 6 (May 1917): 372–378, on 373.

133. Floyd W. Parsons, "Devils of Din," *Saturday Evening Post* 203 (8 November 1930): 16–17, 126, 129–130, on 16.

134. The manual additionally noted that "Even with trained observers the measurements will sometimes vary by as much as 30 or even 50%." "1-A Noise Measuring Set and 3-A Noise Shunt: Instructions for Operating," Western Electric Company Instruction Bulletin no. 145, 1924, quotes on 3–4. Bound volume #249-05, "Instruction Bulletins Supplementary (101–176)," AT&T Archives. See also "Instructions for Operating 2-A Noise Analyzer" (1924) in the same volume.

135. As early as 1896, an article in *Science* reported on "an electrical apparatus, an audiometer, that is useful in determining the sensitiveness to minimal sounds, but is not so satisfactory for determining differential sensibility." No further information on this instrument was provided, except to note that the sound it produced was "very artificial, difficult to listen to, and difficult to reproduce." Fletcher acknowledged that the psychologist Carl Seashore had also built an audiometer before he had, and Vern Knudsen was similarly exploring the problem at this time. Joseph Jastrow, "An Apparatus for the Study of Sound Intensities," *Science,* n.s., 3 (10 April 1896): 544–546, on 545; "Harvey Fletcher," transcript of an oral interview by Vern Knudsen with W. J. King (15 May 1964), Neils Bohr Library, Center for History of Physics, American Institute of Physics, College Park, Md., pp. 47–48; Vern Knudsen, *Teacher, Researcher, and Administrator: Vern O. Knudsen,* transcript of oral history conducted 1966–1969 by James V. Mink. Collection 300/101. Department of Special Collections, Young Research Library, University of California, Los Angeles, 459–489. See also Fletcher, *Speech and Hearing* (New York: D. Van Nostrand, 1929), 133–134; Hallowell Davis, "Psychological and Physiological Acoustics: 1920–1942," *JASA* 61 (February 1977): 264–266; and Robert T. Beyer, *Sounds of Our Times: Two Hundred Years of Acoustics* (New York: AIP/Springer-Verlag, 1999).

136. "The Audiometer: An Instrument for Measuring the Acuity and Quality of Hearing" (c. 1923); "Information for the Care and Operation of the 3-A Audiometer" (1924), #55-07-03, AT&T Archives.

137. "In the field of physical examination, the Audiometer should prove extremely valuable in determining, so far as impaired hearing is concerned, the unfitness of applicants for insurance policies, automobile licenses, and for enlistment in the Army and Navy; also in the life protection tests of railroad and steamship companies, and in the health corrective examinations of schools, colleges and gymnasiums." "Audiometer," 2. It was reported in 1927 that "many plants" were testing the hearing of employees with "a new model audiometer," and testing of schoolchildren in New York also began in 1927. "Ears Good?" *Industrial Psychology Monthly* 2 (August 1927): 431; Morris Fishbein, "The Month in Medical Science," *Scientific American* 140 (February 1929): 124.

138. E. E. Free, "How Noisy Is New York?" *Forum* 75 (February 1926): (illus. sect.) xxi–xxiv, quote on xxi.

139. Free, "How Noisy Is New York?" xxi.

140. Free, "How Noisy Is New York?" xxi; xxii.

141. "It is not easy, nowadays," Free admitted, "to find long lines of horse-drawn traffic, but our investigators accomplished it." "How Noisy Is New York?" xxiv. Free's survey was covered in the *New York Times*, which reported that his audiometer had been "perfected" by Harvey Fletcher at Bell Telephone Laboratories. "Noisiest Spot Here 6th Av. at 34th St." *NYT* (16 January 1926): 7.

142. "Cut-outs" were switches allowing drivers to redirect their engine exhaust to bypass the muffler. It was believed that this improved engine performance.

143. "Noise," *Saturday Review of Literature* 2 (24 October 1925): 1. For representative articles tracking the increasing importance of traffic noise, see "The Campaign Against Noise," *NYT* (23 December 1906): sect. III, p. 4; "Cutting Out Mufflers," *NYT* (2 June 1911): 10; "A Muffler on New York," *NYT* (12 October 1920): 14; and "Police to Free City of Nerve-Racking Noises," *NYT* (15 June 1923): 21. See also "A Crusade for Quiet," *Outlook* 102 (28 September 1912): 157–159; Elmer S. Batterson, "Progress of the Anti-Noise Movement," *National Municipal Review* 6 (May 1917): 372–378; and "Noise Nuisance Suggestions from Civic Club," *American City* 30 (April 1924): 429.

144. "Growth of Bigger Ears Laid to City Life Noise," *NYT* (7 March 1926): sect. VIII, p. 8.

145. "Bedlam on Radio Row," *NYT* (25 May 1930): sect. 10, p. 8.

146. See, for example, A. St. L. Eberle, "Curbing Noise from the Sky," *NYT* (8 October 1928): 22; "Flying Loud-Speaker Chased by Air Police; Dr. Reisner Objects to Noisy Sky Advertising," *NYT* (6 April 1931): 1; "Public Nuisances Aloft," *NYT* (7 April 1931): 26; and "Radio over Brooklyn," *NYT* (29 June 1933): 18.

147. The survey distinguished between radio loudspeakers in the home (774 complaints) and loudspeakers installed outside of radio shops (593 complaints). Brown et al., eds., *City Noise,* 27 (see figure 4.11). The frontispiece to *City Noise*, the official report of the Noise Abatement Commission of New York, illustrated both types of nuisance in the brick building located at the center of the image (see figure 4. 2). My own survey of letters to the New York City Department of Health (1926–1934, with most documents from 1930) indicates that 15 percent of all extant complaints of noise referred to radio loudspeakers in the home or shop. See table 4.1 for the full results of this survey.

148. George MacAdam, "Vision of Fair Taxes," *New York Times Book Review and Magazine* (11 June 1922): 3. The quote is from C. S. F., "Boycotts the Loud Speaker," *NYT* (29 September 1929): sect. III, p. 5.

149. "Father Knickerbocker Warns Loud-Speakers to Be Quiet," *NYT* (13 April 1930): sect. X, p. 15.

150. R. G., "Muzzling Player Pianos," *NYT* (20 October 1927): 28; Robert Grimshaw, "Noises of the Night," *NYT* (24 July 1928): 20; "Would Not Stop Radio, It Is Bombed," *NYT* (2 November 1931): 40.

151. "Bill Would Curb Street Radio Din," *NYT* (5 February 1930): 48.

152. "Brass Band's Blare Ends Anti-Noise Hearing," *NYT* (12 April 1930): 21; "First Loud-Speaker Operator Convicted Here under Law against Unnecessary Street Noise," *NYT* (5 June 1930): 2. See also Bareiss, "Noise Abatement in Philadelphia," 49–55.

153. Quoted in "New City Law Bars Street Radio Noise," *NYT* (9 April 1930): 5.

154. "Fined in First Test of Anti-Noise Law," *NYT* (1 May 1930): 34. While Stand's bill enabled police simply to write summonses to anyone caught operating a loudspeaker outdoors without a permit, the health department's new amendment still required the lengthy procedure in which aggrieved citizens had to file complaints with the health department and then bring corroborating witnesses to court to make their case against the offender.

155. "The Demon in Radio," *Literary Digest* 82 (2 August 1924): 26–27, on 26; Bruce Bliven, "How Radio Is Remaking Our World," *Century Magazine* 108 (June 1924): 147–154, on 151.

156. For a comprehensive history of Leon Theremin and his instrument, see Albert Glinsky, *Theremin: Ether Music and Espionage* (Urbana: University of Illinois Press, 2000), quote on 33. See also "Leon Theremin, Musical Inventor, Is Dead at 97," *NYT* (9 November 1993): B10; and the fascinating documentary film *Theremin: An Electronic Odyssey* (1994; Kaga Bay/Channel 4, dir. Steven M. Martin, Orion Home Video #5080).

157. For accounts of early Theremin recitals in America, see "Talk of the Town," *New Yorker* (4 February 1928): 9; "'Ether Wave' Concert," *NYT* (1 February 1928): 31; and Eugene Bonner, "Music and Musicians," *Outlook* 148 (15 February 1928): 268.

158. "Talk of the Town," *New Yorker* (28 September 1929): 18.

159. "Talk of the Town" (1929), 18. In spite of the clamor of attention that had surrounded the Theremin in 1927–1928, sales of the $175.00 RCA instrument were sluggish in the uncertain economic climate of 1929, and less than 500 of the instruments had been sold when RCA discontinued production in 1931. Glinsky, *Theremin*, 101, 138.

160. Leopold Stokowski, "Wonderful Music Promised by Electrical Production," *Santa Barbara News* (9 June 1928), quoted in Glinsky, *Theremin*, 115.

161. Olin Downes, "Glazounoff Draws a Rising Tribute," *NYT* (4 December 1929): 36.

162. "Chicago Noise Now Registered," *NYT* (12 December 1926): sect. IX, p. 6. See also E. E. Free, "Noise: The Forum's Second Report on City Noise," *Forum* 79 (March 1928): 382–389, on 384. The Burgess Lab, in Madison, Wisc., was "built up of efficient, technically trained engineers for the study and development of scientific methods relating to physics and chemistry." "C. F. Burgess Laboratories," *Sweet's Architectural Trade Catalogue* (1929): A16–A19, on A16.

163. Free, "Forum's Second Report," 384. Also in 1926, the Engineering Section of the National Safety Council formed a committee to investigate noise. The committee's subsequent report highlighted research at Bell Laboratories. "Report of H. W. Mowery, Chairman, Committee on the Elimination of Harmful Noises," *Transactions of the National Safety Council* (1926), vol. 1, 307–329. See also "The Elimination of Harmful Noise," *National Safety News* 15 (April 1927): 46, 58. For a description of a noise meter, see T. G. Castner et al., "Indicating Meter for Measurement and Analysis of Noise," *Transactions of the American Institute of Electrical Engineers* 50 (September 1931): 1041–1047. See also Frank Massa, "Some Personal Recollections of Early Experiences on the New Frontier of Electroacoustics During the Late 1920s and Early 1930s," *Journal of the Acoustical Society of America [JASA]* 77 (April 1985): 1296–1302; Warren Kundert, "Acoustical Measuring Instruments over the Years," *JASA* 68 (July 1980): 64–69; and James Flexner, ed., *City Noise Volume II* (New York: Department of Health, 1932), which mentions the Acoustimeter, Sonometer, Vibrometer, and the Transmit-O-Phone, as well as the "notable Audiometer Devised by the Bell Telephone Laboratories Inc." Appendix, p. 32, New York Municipal Reference Library.

164. Free, "Forum's Second Report," 382, 383.

165. Free, "Forum's Second Report," 388. See also Mowery, "Report," 309 and 317.

166. Donald Laird, "Noise *Does* Impair Production," *American Machinist* 69 (12 July 1928): 59–60.

167. Donald Laird, "The Measurement of the Effects of Noise on Working Efficiency," *Journal of Industrial Hygiene* 9 (October 1927): 431–434, on 432 and 433. See also Laird, "Noise," *Scientific American* 139 (December 1928): 508–510, "Experiments on the Physiological Cost of Noise," *Journal of the National Institute of Industrial Psychology* 4 (January 1929): 251–258, and "The Effects of Noise: A Summary of Experimental Literature," *JASA* 1 (January 1930): 256–262.

168. E. Lawrence Smith and Donald Laird, "The Loudness of Auditory Stimuli Which Affect Stomach Contractions in Health Human Beings," *JASA* 2 (July 1930): 94–98, on 98.

169. This analogy was presented to listeners in a radio address by E. F. Brown, Director of the Noise Abatement Commission of New York, over station WEAF on 17 December 1929. See *City Noise*, 223–231.

170. In 1928, it was reported that the loss resulting from noise-impaired efficiency of workers in the United States amounted to five million dollars per week. "Noise and Health," *American City* 39 (November 1928): 161, citing a report in the *Journal of the American Medical Association*.

171. William Strunk, Jr., *The Elements of Style* (New York: Harcourt, Brace and Co., 1920), 24.

172. Cecelia Tichi examines the interplay between the engineering culture of efficiency and literary culture in early twentieth-century America in *Shifting Gears*. Dui is quoted on p. 67. See also Haber, *Efficiency and Uplift*, 73.

173. Fashion historians have estimated that, in the years immediately following World War I, the amount of clothing worn by the average woman was reduced by 75 percent. Donald Laird noted in 1929 that women's short skirts contributed to the problem of noise by reducing the amount of material available to absorb sound, and published values of the sound-absorbing power of a woman verify his assertion. In 1900, Wallace Sabine calculated that the absorbing power of the average woman was equivalent to 0.54 square meters of open window. By around 1930, according to measurements made at the Bureau of Standards, she was equivalent to only 0.21 square meters. Jane Dorner, *Fashion in the Twenties and Thirties* (New Rochelle: Arlington House, 1974), Foreword by Prudence Glynn; "Noise Slows Growth, Psychologist Says," *NYT* (4 July 1929): 17. Absorption figures from Wallace Sabine, *Collected Papers on Acoustics* (Cambridge, Mass.: Harvard University Press, 1922), 58; and V. L. Chrisler and W. F. Snyder, "Recent Advances in Sound Absorption Measurements," *JASA* 2 (July 1930): 123–128, on 126. Both figures are for sound of 512 cps; I have converted the latter from English to metric units.

174. (Boston) *Traveler* (circa 1930), cited in *City Noise*, 87.

175. Tichi, *Shifting Gears,* 97–170. See also John M. Jordan, *Machine-Age Ideology: Social Engineering and American Liberalism, 1911–1939* (Chapel Hill: University of North Carolina Press, 1994).

176. Shirley Wynne, "Guarding the Health of Seven Million People," *Annual Report of the Department of Health, City of New York* (1929), 15, New York Municipal Reference Library. Shirley was a man. The need for a Noise Commission had been identified in 1928, by a committee appointed by Mayor Walker to evaluate the present status and future of the city. See "Report of the City Committee on Plan and Survey" (1928), 118–120, New York Municipal Reference Library.

177. Wynne, "Guarding Health," 15. The technical members of the commission included Harvey Fletcher of Bell Labs; Lewis Brown, president of the Johns-Manville Corp.; civil engineer Albin Beyer; Alfred Swayne, vice president of General Motors; and E. B. Dennis, acoustical engineer at Johns-Manville. The commission also included doctors, lawyers, and politicians, including Police Commissioner Edward Mulrooney; neurologist Bernard Sachs; Nobel Prize–winning surgeon Alexis Carrel; and Charles Burlingham, president of the Association of the Bar of the City of New York. *City Noise*, v–vii.

178. "Expert Forecasts Sweeter Auto Horn," *NYT* (30 January 1930): 24.

179. *City Noise*, 28. Archival records for the Department of Health are incomplete for this period, and while it is likely that extant letters represent only a portion of those actually received at the time, it is not evident why these letters, and not others, were

preserved. In many cases, the original letter is missing but the correspondence between the Mayor's Office and the Department of Health indicates its contents. The *Times* reported that the Noise Abatement Commission received fifty complaints per week concerning just the "back-yard yowling" of cats, suggesting that the overall volume of mail received circa 1930 and 1931 was significantly greater than that found in the archives today. "Drive on to Silence Night Life of Cats," *NYT* (12 May 1931): 27.

180. Shirley Wynne, "Saving New York From Its Own Raucous Din," *NYT* (3 August 1930) sect. IX, p. 1; *City Noise,* 29–35.

181. W. West, *Acoustical Engineering* (London: Sir Isaac Pitman and Sons, 1932), 169.

182. W. H. Martin, "Decibel—The Name for the Transmission Unit," *Bell System Technical Journal* 8 (January 1929): 1–2. As Martin's title makes clear, the derivation of the new unit had as much, if not more, to do with electrical signals than with physical sounds, and the origins of the transmission unit actually referred back to the loss suffered by a signal as it traveled through one mile of standard telephone cable. For more on the development of the transmission unit and its transformation into the decibel, see M. D. Fagen, ed., *A History of Engineering and Science in the Bell System: The Early Years (1875–1925)* (n.p.: Bell Telephone Laboratories, 1975), 303–309.

183. Free, "Forum's Second Report," 385.

184. "Drive Opens to End Din of Auto Horns," *NYT* (15 July 1930): 25. For additional accounts that define and explain the new unit, see "Finds Noise Lowers Efficiency by 10%," *NYT* (10 May 1930): 5; "Ordinance Forbids Street Amplifiers," *NYT* (21 May 1930): 32; and "The Poultice of Silence," *NYT* (15 July 1930): 22. See also "Street Noises and Skyscrapers," *American City* 42 (June 1930): 118; "The Pest of Noise," *American City* 43 (August 1930): 171; and Parsons, "Devils of Din," 129.

185. In addition to the charts published in *City Noise*, 131 and 158, see, for example, Brown, "Attacking City Noises by Science and Law," 98; and Roger W. Sherman, "Sound Insulation in Apartments," *Architectural Forum* 53 (September 1930): 373–378, on 377. Sherman's chart is topped by that quintessentially modern maker of noise, the saxophone, registering 95 dB.

186. The text of Fletcher's talk, "How Noise Is Measured and Why," is reproduced in *City Noise*, 239–243. Of course, one might question how meaningful this exercise would have been when different listeners would have had their radios set at different volume levels. Fletcher dealt with this problem by directing listeners to set their radio's volume so that his voice sounded like "average conversational speech to a listener who is about three feet away" (239).

187. "Subway Din Worse inside of the Cars," *NYT* (7 January 1931): 27.

188. "Truck Noise Menace Must End," *NYT* (20 May 1931): 27. See also "Tightening Up on Trucks," *NYT* (21 May 1931): 26.

189. "'Silent' Garbage Can Too Lively in Test," *NYT* (30 June 1931): 27. See also "Science Searches for a Quiet Ashcan," *American City* 45 (August 1931): 109.

190. Engineers from the Columbia Broadcasting Co., for example, compared noise levels in Chicago and New York and discovered that Chicago was 6 decibels louder. The *Times* noted, "Perhaps it hurts a little to find that when noise-making is in question we cannot be first." "Times Square Hums but Chicago Roars," *NYT* (18 November 1931): 17; "American Din," *NYT* (19 November 1931): 22. Donald Laird had also concluded that Chicago was noisier when he measured levels in New York, Chicago, and Boston in 1927. "Finds New York Is Less Noisy Than Chicago," *NYT* (20 November 1927), sect. II, p. 1.

191. "A Bouncing New Ash Can," *NYT* (1 July 1931): 24. For examples of letters citing decibels, see "Noises We Have Always with Us," *NYT* (5 August 1930): 22; and "Tenth Avenue Noises," *NYT* (1 June 1932): 22.

192. The report, *City Noise*, was written for a popular audience and reviewed in numerous magazines when it appeared in 1930. Ten thousand copies were printed, and its general availability today, in libraries across the nation, attests to its wide distribution. Further evidence of the pervasiveness of publicity surrounding the campaign is the satire by Corey Ford, "Silence Please!" *American Magazine* 109 (February 1930): 18–19, 160–161. For print run, see Flexner, ed., *City Noise II*, appendix, p. 20.

193. "Manhattan's Noises," *NYT* (4 January 1933): 18. See also "The Anti-Noise Campaign," *NYT* (14 July 1933): 16; and "Police Warnings Ignored," *NYT* (27 July 1934): 16.

194. Flexner, ed., *City Noise II*. See also Smilor, "Confronting the Industrial Environment," 250–267.

195. "Decibels and Others," *NYT* (13 October 1933): 18.

196. "Tenth Avenue Noises," *NYT* (1 June 1932): 22.

197. As of 1934, the city had still failed to enact the ordinance. "Civic Groups Open New Drive on Noise," *NYT* (19 May 1934): 15.

198. Smilor, "Confronting the Industrial Environment," 251. Smilor highlights the fact that this final report contrasted with the first report not only in tone, but also in physical form. Whereas the 1930 report had been professionally edited, typeset, bound, and widely disseminated, the second report exists only in a rough typescript with a hand-scrawled title page. It was apparently never distributed outside of the Department of Health, and the only copy I have seen is that held in the New York Municipal Reference Library.

199. *City Noise II,* part I, p. 1 The frustrating history of the amendment to introduce the system of noise fines is detailed in part III, pp. 11–16.

200. Henry J. Spooner, "The Progress of Noise Abatement," address delivered at a meeting of the American Society of Industrial Engineers (14–16 October 1931), reprinted in *City Noise II*, appendix, pp. 27–37, on 33.

201. *City Noise II*, part I, p. 1.

CHAPTER FIVE

1. "Absorbex Acoustical Corrective for Sound Control and Noise Reduction" (sales pamphlet, Thermax Corp., 1932), 3. Series I: Administrative and Technical Records, Box 13, Folder 11: Other Acoustical Plasters and Tile Materials, Guastavino/Collins Collection, Avery Architectural and Fine Art Library, Columbia University in the City of New York [GCC].

2. For more on the Woolworth Building and the rise of the commercial skyscraper in New York, see Robert A. Jones, "Mr. Woolworth's Tower: The Skyscraper as Popular Icon," *Journal of Popular Culture* 7 (fall 1973): 408–424; and Gail Fenske and Deryck Holdsworth, "Corporate Identity and the New York Office Building: 1895–1915," pp. 129–159 in David Ward and Olivier Zunz, eds., *The Landscape of Modernity: New York City 1900–1940* (Baltimore: Johns Hopkins University Press, 1992). See also Carol Willis, *Form Follows Finance: Skyscrapers and Skylines in New York and Chicago* (New York: Princeton Architectural Press, 1995); Sarah Bradford Landau and Carl W. Condit, *Rise of the New York Skyscraper, 1865–1913* (New Haven: Yale University Press, 1996).

3. A map indicating the highly localized land values in Manhattan in 1903 can be found in Willis, *Form Follows Finance*, 172. For a case study of the role of economics in tall building design, see Sharon Irish's account of another Cass Gilbert building, "A 'Machine That Makes the Land Pay': The West Street Building in New York," *Technology and Culture* 30 (April 1989): 376–397. Her title quotes Gilbert in 1900.

4. Benjamin F. Betts, "Is This Our Next New Big Business?" *American Architect* 140 (December 1931): 21. Betts's editorial was subsequently editorialized upon in "Commercialized Pianissimo," *New York Times* [*NYT*] (3 December 1931): 28; and "Hush!—the Next New Big Business," *Literary Digest* 112 (9 January 1932): 46.

5. Advertisements for the Johns-Manville Company cited hundreds of installations of its sound-absorbing materials as early as 1923, and by 1929 they claimed to have solved "literally thousands of acoustical problems." In 1933, the Celotex Company boasted of over 6,000 sites in which Acousti-Celotex had been employed. *Architectural Forum* 38 (June 1923): advertising sect., 136; *Sweet's Architectural Trade Catalogues* [*Sweet's*](1929): A15; *Architectural Forum* 59 (December 1933): ad sect., 34. See also the "partial list" of 107 installations in "Kalite Sound Absorbing Plasters" (sales pamphlet, Certain-Teed Products Corp., c. 1934), 12, Series I, Box 13, Folder 11, GCC.

6. See, for example, Stephen Kern, *The Culture of Time and Space, 1880–1918* (Cambridge, Mass.: Harvard University Press, 1983); David Gross, "Space, Time and Modern Culture," *Telos* 50 (winter 1981–82): 59–78; and David Harvey, *The Condition*

of Postmodernity: An Enquiry into the Origins of Cultural Change (Cambridge, Mass.: Blackwell, 1990).

7. "Books on Acoustics," *American Architect and Building News* [*AABN*] 47 (12 January 1895): 23.

8. The competition was under way by August 1894. In November 1895, the selection of Gilbert's entry was announced. See *AABN* 45 (11 August 1894): 55; and *AABN* 50 (9 November 1895): 62.

9. H. C. Kent, "A Distinction in the Acoustic Purposes of Public Buildings," *AABN* 39 (7 January 1893): 9.

10. "Samuel Cabot," *Proceedings of the American Academy of Arts and Sciences* 43 (July 1908): 547–556, quote on 550. See also "Cabot, Samuel," *National Cyclopedia of American Biography* (New York: James T. White, 1933) 23: 247.

11. Cabot was also greatly interested in the controversy surrounding the authorship of the plays of Shakespeare. Like George Fabyan, he "espoused the Baconian theory with great vigor, and defended his position by elaborate and costly investigations." "Samuel Cabot," 555.

12. "Sound Deadening: Cabot's Deadening Quilt" (sales pamphlet, Samuel Cabot Inc., c. 1922), 4. Collection of Richard Longstreth.

13. "Sound Deadening," 4. The heat-insulating properties of Quilt were additionally if curiously praised by Antarctic explorer Robert F. Scott, whose journal referred to the "excellent quilted seaweed insulation" of the huts constructed during his ill-fated journey to the South Pole. These were not the huts where Scott and his party would eventually freeze to death. "Cabot's Quilt" (London sales pamphlet, Samuel Cabot Inc., c. 1922), 7. Thanks to Thomas Jester and Michael Tomlan for providing me with a copy of this pamphlet.

14. The other products were Sackett Plaster Wall Board, Keystone Plaster Blocks, National Fireproofing Company Terra Cotta Blocks, and a lath and plaster partition manufactured by J. Russell & Co. Only Cabot's Quilt (here incorporated into a standard lath and plaster wall) was specifically advertised for use in soundproof construction.

15. C. L. Norton, "Sound-Proof Partitions," *American Architect* 78 (4 October 1902): 5–6, quotes on 5. For continued citation of this report in Cabot's advertising, see, for example, "Samuel Cabot Inc.," *Sweet's* (1919): 1670.

16. Wallace Sabine, "Architectural Acoustics: The Insulation of Sound," *Brickbuilder* 24 (February 1915): 31–36 and *Collected Papers on Acoustics* [*CPA*] (Cambridge, Mass.: Harvard University Press, 1922), 237–254; quote on 239.

17. Henry Higginson to Charles McKim (5 May 1899), Papers of McKim, Mead & White, Folder M-10: Boston Music Hall, New-York Historical Society [MMW Papers]; Papers of Wallace Clement Sabine, Research Notebooks, Data on Acoustical

Research, 1899–1919, Research Notebook #9, "Experiments to determine the transmitting power of wall surfaces for sound" (10 February 1915), [HUG 1761.25]. Courtesy of the Harvard University Archives, Cambridge, Mass..

18. Wallace Sabine to Charles McKim (10 June 1901), Folder M-5: Sabine, Wallace, MMW Papers.

19. *Sweet's* (1906): 154–169, entries for F.W. Bird and Son; C.B. Hewitt and Brothers; H.W. Johns-Manville Co.; Union Fibre Co.; and Barrett Manufacturing Co. This was the first edition of *Sweet's*. Compiled by the staff of the *Architectural Record*, this "Book of Catalogues" was intended to provide the practicing architect with "a really scientific standard catalogue and index of building materials and construction" (vii). The easy-to-use reference volume replaced the rapidly growing assortment of trade catalogs and pamphlets that were increasingly frustrating architects with their "heterogeneous distribution," "diversity in shape and form," and "general unsuitableness of contents and arrangement for the purposes of reference" (v). The 760 pages of *Sweet's* 1906 edition expanded to over 6,000 pages by 1931, and it is important to understand the growth of the acoustical materials industry in the context of this larger expansion of the market for all types of building materials. In 1906, the sound-absorbing felts listed above were classified under the general heading of "Building Papers, Roofing and Materials for Insulating." The following year, a special category for "Sound-Deadening Materials" appeared.

20. The description of Keystone comes from "Johns-Manville Building Materials" (sales pamphlet, Johns-Manville Co., 1920), 65, Avery Library, Columbia University. For a description of the project and photographs of the Metropolitan Lecture Room before and after correction, see Wallace Sabine, "Architectural Acoustics: The Correction of Acoustical Difficulties," *Architectural Quarterly of Harvard University* (March 1912): 3–23 and *CPA*, 131–161, on 138–140. For more on the Johns-Manville Company, see "Manville, Hiram Edward," *National Cyclopedia of American Biography* (New York: James T. White, 1934) D: 183; "Manville, Charles Brayton," and "Manville, Thomas Franklyn," *National Cyclopedia of American Biography* (New York: James T. White, 1935), 24: 20.

21. William Dana Orcutt, *Wallace Clement Sabine: A Study in Achievement* (Norwood, Mass.: Plimpton Press, 1933), 236–244.

22. Sabine's own account of these events is found in a letter to architects Allen & Collins (12 February 1915), reproduced in Orcutt, *Wallace Sabine*, 237–238, quote on 238.

23. Orcutt, *Wallace Sabine*, 240. The National Archives in Suitland, Md., hold a file for Mazer's patent, but the folder—also stamped "Withdrawn"—is empty. Sabine's friends had really only bought time with their influence, giving him the opportunity to file an interference against Mazer's patent application and to defend this claim in court. At the hearing, Sabine was represented by patent lawyer and Harvard Overseer Frederick Fish.

Orcutt indicates that the case was heard in the summer of 1911. Sabine's published papers were submitted in evidence, "the plagiarism was clearly proven, and by the time the case before the examiner was closed, Mazer's claims were shown to be so audaciously fraudulent that the verdict in favor of Sabine had become a foregone conclusion." Orcutt, *Wallace Sabine*, 241–242. In his obituary of Sabine, Edwin Hall indicated that Mazer offered no evidence in his defense. "Wallace Clement Ware Sabine, 1868–1919," *Biographical Memoirs of the National Academy of Science* 11 (1926): 1–19, on 12 (note 12).

24. "Business Noise: Its Cost and Prevention" (sales pamphlet, Johns-Manville Inc., 1920), 9, Series I, Box 13, Folder 11, GCC; and Sabine, "Correction of Acoustical Difficulties," 132. In 1913, *Sweet's* introduced the new category of "Architectural Acoustics," distinct from "Sound Deadening Materials," with Johns-Manville as the sole entry. The quoted description of the Johns-Manville Acoustical Department is from *Sweet's* (1914): 303.

25. Orcutt, *Wallace Sabine*, 243, quoting Swan.

26. In 1912, for example, upon receipt of a typical request for advice on acoustical correction, Sabine responded: "In the past I have done not a little consultation work in architectural acoustics, but I am endeavoring to reserve such time as I can spare for consultation in advance of construction. To this end I have urged a Mr. C. M. Swan, formerly an instructor at the Massachusetts Institute of Technology and later for three years a graduate student and associate with me in these investigations, to undertake the correction of acoustical difficulties in a professional capacity and at my suggestion the Johns-Manville Company have placed Mr. Swan at the head of a department designed for architectural acoustics." Wallace Clement Sabine Correspondence Files, Sabine to W. L. Stevens (12 June 1912). Courtesy of Riverbank Acoustical Laboratories [Sabine Correspondence].

27. Wallace Sabine to Albert Kahn (27 June 1911), Sabine Correspondence.

28. Cram, Goodhue & Ferguson to Wallace Sabine (25 March 1911), Sabine Correspondence.

29. Ralph Adams Cram, (editorial), *Knight-Errant* 1 (April 1892): 1, quoted in Richard Oliver, *Bertram Grosvenor Goodhue* (New York and Cambridge, Mass.: The Architectural History Foundation and The MIT Press, 1983), 27. For more on Cram's life and aesthetic philosophy, see Ralph Adams Cram, *My Life in Architecture* (Boston: Little, Brown and Co., 1936). Douglas Shand-Tucci's *Boston Bohemia 1881–1900: Ralph Adams Cram: Life and Architecture* (Amherst: University of Massachusetts Press, 1995) documents the rich mixture of spirituality, sexuality, and aesthetics that constituted the cultural milieu of Cram's early years in fin-de-siècle Boston. See also Shand Tucci, *Ralph Adams Cram: American Medievalist* (Boston: Boston Public Library, 1975) and Robert Muccigrosso, *American Gothic: The Mind and Art of Ralph Adams Cram* (Washington, D.C.: University Press of America, 1980).

30. T. J. Jackson Lears, *No Place of Grace: Antimodernism and the Transformation of American Culture 1880–1920* (New York: Pantheon Books, 1981), 194; Ralph Adams Cram, *Church Building*, 3d ed. (Boston: Marshall Jones Co., 1924), 298.

31. Ralph Adams Cram, *The Gothic Quest* (New York: Baker and Taylor, 1907), 101; Cram, *Church Building* (Boston: Small, Maynard and Co., 1901), 10. For more on the importance of the sermon and the "salient requirement" that the church function as an auditorium, see Rev. Dr. Hugh Birckhead, "Architectural Requirements of the Episcopal Church," *Architectural Forum* 40 (April 1924): 165–166, quote on 165.

32. When Raphael Guastavino, Sr., died in 1908, his son assumed full control of the company. For more on the Guastavino Company, see George R. Collins, "The Transfer of Thin Masonry Vaulting from Spain to America," *Journal of the Society of Architectural Historians* [*JSAH*] 27 (October 1968): 176–201. A rich assortment of drawings and photographs of Guastavino materials and projects has been collected in Janet Parks and Alan G. Neumann, *The Old World Builds the New: The Guastavino Company and the Technology of the Catalan Vault 1885–1962* (New York: Avery Architectural Library, 1996). For contemporary accounts, see R. Guastavino [Sr.], *Essay on the Theory and History of Cohesive Construction* (Boston: Ticknor and Co., 1893) and Peter B. Wight, "The Works of Raphael Guastavino," *Brickbuilder* 10 (1901): 79–81 (April); 100–102 (May); 184–188 (September); 211–214 (October). The historic preservation of Guastavino construction is the focus of a special issue of the *APT Bulletin*. See especially Richard Pounds, Daniel Raichel, and Martin Weaver, "The Unseen World of Guastavino Acoustical Tile Construction: History, Development, Production," *APT Bulletin* 30 (1999): 33–39. See also Ann Katherine Milkovich, "Guastavino Tile Construction: An Analysis of a Modern Cohesive Construction Technique" (M.S. Thesis, University of Pennsylvania, 1992); and Christopher Gray, "An Architect Who Achieved a Vaulting Success," *NYT* (12 May 1996), sect. 9, p. 7.

33. Orcutt, *Wallace Sabine*, 208.

34. See signed carbon copy of contract between Sabine and Guastavino (20 April 1911), Series I, Box 14, Folder 12: Sabine, W. C.-Jefferson Physical Lab, GCC.

35. Sabine, "Building Material and Musical Pitch," *Brickbuilder* 23 (January 1914): 1–6 and *CPA*, 199–217, on 208. Guastavino is quoted in Orcutt, *Wallace Sabine*, 209.

36. Sabine, "Building Material," 208. This building was also designed by Cram, Goodhue & Ferguson.

37. Wallace Sabine to W. E. Blodgett (24 August 1911), Series I, Box 14, Folder 12, GCC.

38. Sabine to Blodgett (24 August 1911), GCC.

39. Wallace Sabine to Ralph Adams Cram (18 December 1912), Sabine Correspondence. See also "St. Thomas's Church, New York," *Architecture* 29 (January 1914): 4–6, which refers to "tile made according to Professor Sabine's suggestions" (4).

40. No official explanation of the name "Rumford" has been found, but it is likely that Sabine and Guastavino named the new material in honor of Benjamin Thompson, Count Rumford, who was born in 1753 in Woburn, the site of the Guastavino kilns that produced the new tile. Thompson was a Royalist who fled to England in 1776, where he became active in politics, military affairs, science, and technology. In England and later in Bavaria (where he received his title), Thompson promoted the use of the steam engine, studied the properties of heat and friction, and invented various devices for making a good cup of coffee. Sabine and Guastavino probably did not know that Thompson had a lifelong habit of "disdaining those who could not promote his career," nor that he "aggressively sought individuals and positions that would speed his advance." They would have known that, upon his death in 1814, Thompson bequeathed to Harvard University funds for the promotion of "the application of science to the useful arts," and this might have been the reason that they named their tile in his honor. Phillip Drennon Thomas, "Thompson, Benjamin," in John A. Garraty and Mark C. Carnes, eds., *American National Biography* (New York: Oxford University Press, 1999) 21: 538–540, quotes on 538.

41. W. C. Sabine and R. Guastavino, "Wall and Ceiling of Auditoriums and the Like," United States Patent #1,119,543 (1 December 1914), quote from p. 2, lines 33–45.

42. "St. Thomas's Church, New York," *Architecture* 27 (15 January 1913): 5, 7, on 7.

43. H. L. Bottomley, "The Story of St. Thomas' Church," *Architectural Record* 35 (February 1914): 101–131, on 126.

44. "St. Thomas's Church," 7.

45. Whether or not this assignation of credit was accurate is open to dispute. Goodhue certainly disputed it, and disagreement over this issue catalyzed his growing dissatisfaction with the partnership, from which he formally withdrew at the close of 1913. After the dissolution of the firm, however, Wallace Sabine remained on good terms with both Cram and Goodhue, and he continued to work with both men. For more on the breakup of the firm, see Oliver, *Bertram Goodhue*, 120–124.

46. "St. Thomas's Church," (1914): 5–6.

47. "The Symbolism in St. Thomas's," *Pencil Points* 2 (September 1921): 9–12, 38. Little of this ornamentation was in place when the otherwise-finished church was dedicated in 1913. It was gradually installed in subsequent years.

48. "St. Thomas's Church" (1913), 7.

49. Montgomery Schuyler, "The New St. Thomas's Church, Fifth Avenue, New York City," *Brickbuilder* 23 (January 1914): 15–20, on 15.

50. Schuyler, "New St. Thomas's Church," 15.

51. "St. Thomas's Church" (1914), 4. See also Bottomley, "Story of St. Thomas' Church," 126.

52. Wallace Sabine to Ralph Adams Cram (18 December 1912), Sabine Correspondence.

53. Sabine's own values for the absorption coefficients of Rumford are cited in Vern Knudsen, *Architectural Acoustics* (New York: Johns Wiley and Sons, 1932), 205.

54. No documentation of the reverberation time of St. Thomas's at the time of its dedication in 1913 has been located. Measurements made in 1955, however, probably do not vary significantly from what the reverberation time would have been circa 1913. In 1955, the reverberation ranged from 3.0 seconds at 250 cps to 1.1 seconds at 8,000 cps. The reverberation time at 500 cps was 2.6 seconds. Thus, while St. Thomas's Church was by no means acoustically "dead," its reverberation was several seconds below what might have been expected from such a space lined had it been lined throughout with stone. Thanks to Gerald Marshall of Marshall/KMK Acoustics for providing me with a copy of the graph that includes this data: "Reverberation Time Characteristics, St. Thomas Church, New York City," Bolt, Beranek and Newman Inc., Project 181635, figure 7 (30 October 1970).

55. See Robert Muccigrosso, "Ralph Adams Cram and the Modernity of Medievalism," *Studies in Medievalism* 1 (spring 1982): 21–38; and Michael Clark, "Ralph Adams Cram and the Americanization of the Middle Ages," *Journal of American Studies* 23 (August 1989): 195–213. Architectural historians have been more inclined to recognize a nascent modernity in the buildings of Goodhue, but they typically focus on the work that he did after leaving the partnership with Cram. See, for example, Oliver, *Bertram Goodhue*; and William J. R. Curtis, *Modern Architecture Since 1900,* 3d ed. (Upper Saddle River, N.J.: Prentice Hall, 1996), 294.

56. Wallace Sabine, "Building Material," 209.

57. See Kern, *Culture of Time and Space*; Gross, "Space, Time and Modern Culture"; and Harvey, *Condition of Postmodernity*.

58. Cram, *My Life in Architecture*, 263.

59. Bertram Grosvenor Goodhue to Wallace Clement Sabine (22 October 1914), Sabine Correspondence. The project to which Goodhue referred was St. Bartholomew's Church in New York. While Sabine did serve as acoustical consultant on this project, it is not evident that he contributed in any way to the design of its exterior. Christine Smith, *St. Bartholomew's Church in the City of New York* (New York: Oxford University Press, 1988).

60. W. C. Sabine and R. Guastavino, "Sound Absorbing Material for Walls and Ceiling," United States Patent #1,197,956 (12 September 1916). The absorptivity of Akoustolith is cited in a letter from Wallace Sabine to Clifford Swan (10 May 1916), Series I, Box 14, Folder 12, GCC.

61. In 1920, Gottlieb responded to a request by the Guastavino Co. for his opinion of the tile. He pronounced it "eminently satisfactory," and elaborated: "When I drew the plans in 1913, I was told by certain so-called experts that on account of the shape of the

building, which was surmounted by a dome about sixty-two feet in diameter, that it would be impossible to make it acoustically good. I then consulted the late Professor Wallace C. Sabine of Harvard University who suggested to me the use of your tile and he assured me that it would solve my problem. He proved to be entirely correct, for there is no trace of an echo or reverberation from the voice of the preacher who when speaking in an ordinary tone can easily be heard in the last pew which is about one hundred ten feet from the pulpit. Members of the choir have also told me that it is a most satisfactory auditorium for the singing voice." Albert Gottlieb to R. Guastavino Company (20 May 1920), Series I, Box 12, Folder 8: Technical: R.G. Co., Theory etc. Acoustics, GCC. The architectural press also reported on the use of Rumford in the temple, and concluded that "the result is most satisfactory." "B'Nai Jeshurun Temple, Newark, N.J.," *Brickbuilder* 24 (December 1915): 305–306, on 306.

62. Bertram Goodhue to Raphael Guastavino (5 May 1916), Series I, Box 12, Folder 8, GCC.

63. In 1933, Raphael Guastavino indicated that there were "several hundred" installations of Akoustolith. Orcutt, *Wallace Sabine*, 209. Photographs of many Rumford and Akoustolith projects are collected in the project binder, "Guastavino Construction," Series I, Oversize Box 1, Folder 1: Guastavino Construction-Sales Manuals, GCC. Guastavino advertisements including photographs of many of these projects can be found in *Architectural Forum*. See, for example, 38 (June 1923): ad sect., 7; and 38 (July 1923): ad sect., 7. See also "Isaiah Temple, Chicago, Ill.," *American Architect and Architectural Review* 126 (31 December 1924): 623–626 and plates; and Parks and Neumann, *Old World Builds the New*.

64. "R. Guastavino Co.," *Sweet's* (1923): 20.

65. According to George Collins, "the rise in the cost of hand labor, the Great Depression, the development of concrete for shells, brought the enterprise down." Raphael Guastavino sold his interest in the company in 1943, and in 1962, its assets were liquidated. Collins, "Transfer of Vaulting," 200.

66. These products are listed and described in the 1931 edition of *Sweet's*. In addition to the specific sources cited throughout this section, my account is based on a more general survey of advertisements and articles in architectural journals, especially the *Architectural Forum* and *American Architect and Architectural Review*, as well as *Sweet's*. I have also examined trade literature for numerous manufacturers, in particular the collection of pamphlets and brochures included in the Guastavino/Collins Collection. Additional sources on the history of acoustical buildings materials include: Hale Sabine, "Manufacture and Distribution of Acoustical Materials Over the Past 25 Years," *Journal of the Acoustical Society of America* [*JASA*] 26 (September 1954): 657–661; Hale Sabine, "Building Acoustics in America, 1920–1940," *JASA* 61 (February 1977): 255–263; and Anne E. Weber, "The Search for Silence: Acoustical Materials of the 1920's and 30's," paper presented at the 1993 meeting of the Association for Preservation Technology. Thanks to Carl Rosenberg and Anne Weber for providing me with a copy of the text of this talk.

67. "Macoustic Controls Sound" (sales pamphlet, Macoustic Engineering Co., 1928), 4, Series I, Box 13, Folder 11, GCC.

68. "U.S.G. Sound Control Service" (sales pamphlet, United States Gypsum Co., c. 1931), 6, Series I, Box 13, Folder 11, GCC.

69. "Absorbex Acoustical Correction for Sound Control and Noise Reduction" (sales pamphlet, Thermax Corp., 1932), 3, Series I, Box 13, Folder 11, GCC.

70. "Artstone Products Inc.," *Sweet's* (1931): B2480.

71. F. R. Watson, "Acoustics of Motion Picture Theaters," *Transactions of the Society of Motion Picture Engineers* 11 (1927): 641–650, on 641.

72. "Little Stories of the Job," *Gypsumist* 5 (August 1926): 21, Series I, Box 13, Folder 11, GCC. *Gypsumist* was published by the United States Gypsum Corporation. While histories of American advertising focus upon the mass-market advertising that was directed toward consumers, it is clear that the same visual and textual techniques used to sell toothpaste, automobiles, and soap were also deployed in the specialized trade advertising for acoustical materials and other building products. For more on the history of advertising, see Roland Marchand, *Advertising the American Dream: Making Way for Modernity, 1920–1940* (Berkeley: University of California Press, 1985); Jackson Lears, *Fables of Abundance: A Cultural History of Advertising in America* (New York: Basic Books, 1994); Pamela Walker Laird, *Advertising Progress: American Business and the Rise of Consumer Marketing* (Baltimore: Johns Hopkins University Press, 1998); and Susan Strasser, *Satisfaction Guaranteed: The Making of the American Mass Market* (Washington, D.C.: Smithsonian Institution Press, 1989).

73. Hale Sabine recalled the "coefficient war" of the 1920s and early 1930s in his 1977 article, "Building Acoustics in America," 256. For a comprehensive list of products and coefficients circa 1932, see Knudsen, *Architectural Acoustics*, 240–251. Knudsen listed materials with coefficients (at 512 cps) as high as 0.75 (1.5" Acousti-Celotex Triple B); and 0.79 (Sanacoustic Tile).

74. Insulite Acoustile advertisement, *American Architect* 138 (July 1930): 5; U.S.G. Co. Acoustone ad, *American Architect* 138 (November 1930): 20.

75. Insulite Acoustile advertisement, *American Architect* 140 (September 1931): 109. For earlier references to engineering departments, see *Sweet's* (1922): "Mechanically Applied Products Co.," 17; "Junius H. Stone Corp.," 473; and "Union Fibre Co. Inc.," 1134.

76. Daniel V. Casey, "Muffling Office Noises," *System: The Magazine of Business* 25 (1913): 246–251; Hale Sabine, "Building Acoustics in America," 258. Casey's article was excerpted as "To Hush Office Noise" in the *Literary Digest* 48 (28 March 1914): 696–697.

77. "Business Noise: Cost and Prevention," 1, 4.

78. "Business Noise: Cost and Prevention," 10, 6.

79. "Business Noise: Cost and Prevention," 6–7. Later in the decade, the quantitative results of Donald Laird's research on the effect of noise on workers would similarly be cited in articles and advertisements. See, for example, Clifford Swan, "Noise Problems in Banks," *Architectural Forum* 48 (June 1928): 913–916; and "United States Gypsum Co.," *Sweet's* (1931): B2675, where text from one of Laird's articles is quoted verbatim.

80. *Architectural Forum*, in particular, regularly devoted issues to particular building types and consistently included articles that addressed acoustical design as part of those special issues. See, for example, Lorentz Schmidt, "Acoustical Treatment of Classrooms," 37 (August 1922): 111; Charles Butler, "Prevention of Sound Travel in Hospitals," 37 (December 1922): 287–288; R. E. Lee Taylor, "Design and Plan of Small City Apartment Buildings," 43 (September 1925): 121–126; Clifford Swan, "Noise Problems in Banks," 48 (June 1928): 913–916; Swan, "Quiet for Hospitals," 49 (December 1928): 935–937; Swan, "The Reduction of Noise in Hotels," 51 (December 1929): 741–744; Roger W. Sherman, "Sound Insulation in Apartments," 53 (September 1930): 373–378; and Charles Neergaard, "Controlling Hospital Noises," 57 (November 1932): 449–450.

81. *Architectural Forum* 38 (June 1923): ad sect., 67 (Armstrong); ad sect., 137 (Gold Seal).

82. Clifford Swan, "The Reduction of Noise in Banks and Offices," *Architectural Forum* 38 (June 1923): 309–310, on 309. By 1923, it appears that Swan was no longer affiliated with Johns-Manville, but was instead working as an independent consultant.

83. Swan, "Reduction of Noise in Banks," 309.

84. The claim is made in L. Green, Jr., "Soundproofing the New York Life Insurance Company Building," *American Architect* 135 (20 March 1929): 411–412, on 411.

85. "Planning for Employees' Welfare in the Design of the New York Life Insurance Company Building," *American Architect* 135 (20 March 1929): 397–401, on 397.

86. Lawrence F. Abbott, *The Story of NYLIC: A History of the Origin and Development of the New York Life Insurance Company from 1845 to 1929* (New York: "The Company," 1930), 241.

87. For more on the history of this site, see "As Cathedrals Were Built," *Eastern Underwriter* (14 December 1928): part 2, special issue, 17–19. See also Robert A. M. Stern, Gregory Gilmartin, and John Massengale, *New York 1900: Metropolitan Architecture and Urbanism, 1890–1915* (New York: Rizzoli, 1983), 202–205; and Stern, Gilmartin, and Thomas Mellins, *New York 1930: Architecture and Urbanism Between the Two World Wars* (New York: Rizzoli, 1987), 541–544.

88. Gilbert is quoted in Geoffrey Blodgett, "Cass Gilbert, Architect: Conservative at Bay," *Journal of American History* 72 (December 1985): 615–636, on 625.

89. The characterization is that of Stern et al. in *New York 1930*, 542.

90. Paul Starrett, *Changing the Skyline: An Autobiography* (New York: McGraw-Hill, 1938), 264–274, on 268. Carol Willis confirms Starrett's point concerning illumination,

arguing that it was not until the development of fluorescent lighting in the 1950s that interior office spaces could be planned independent of natural sources of light. Willis, *Form Follows Finance*, 24, 132–133 and passim. Starrett also claimed that Gilbert personally lost interest in the project after his plan was set aside and subsequently delegated the assignment to subordinates, but the fact that acoustical planning remained a priority even after Gilbert's alleged disengagement only indicates that there was a widely held belief in the importance of this aspect of the design.

91. Abbott, *Story of NYLIC*, 241. See also "Run Big Building Like Small City," *NYT* (25 May 1930): sect. 12, p. 14.

92. All these measures are described in "Planning for Employees' Welfare," quote on 401. The most famous, and most studied, example of an insurance company dedicated to employee welfare is the Metropolitan Life Insurance Company of New York, whose 1909 home office tower (also located on Madison Square) was state-of-the-art commercial architecture in its own day. For more on Metropolitan Life and its home office building, see Olivier Zunz, *Making America Corporate: 1870–1920* (Chicago: University of Chicago Press, 1990); and Angel Kwolek-Folland, *Engendering Business: Men and Women in the Corporate Office, 1870–1930* (Baltimore: Johns Hopkins University Press, 1994).

93. Examples of advertisements in the *Architectural Forum* highlighting the quiet qualities of plumbing and other equipment include: the "Silent Maximus" toilet of the Eljer Company, 37 (September 1922): ad sect., 142; B.F. Sturtevant Co.'s "Silentvane" ventilating fan, 37 (October 1922): ad sect., 119; the "Madera Silent" toilet of Thomas Maddock's Sons, 37 (October 1922): ad sect., 130; W.S. Tyler's "Noiseless Elevator," 37 (December 1922): ad sect., 12; and Adsco Heating for "Silence and Economy," 37 (December 1922): ad sect., 106.

94. Information on the acoustical design of the building is drawn from Green, "Soundproofing"; "For Health, Safety and Efficiency," *Eastern Underwriter* (14 December 1928): part 2, special issue, 12; and Stratford Corbett, "An Office Building of the New Era," *Scientific American* 141 (December 1929): 484–486, quote on 486.

95. Green, "Soundproofing," 411. While no manufacturer is mentioned in any published sources, it is likely that this material was Johns-Manville's Akoustikos Felt. The archives of the New York Life Insurance Company include two account books recording expenses relating to the construction of the building. While no record exists of a large contract for acoustical material (perhaps this was handled directly by the Starrett Brothers), several entries record smaller payments to Johns-Manville for "Acoustical Treatment" of both tenant space and home office space within the building. See "Expense Journal-Building and Construction 1924–1931" and "Home Office Construction Account Book," entries dated (13 March 1929), (2 July 1929), and (6 September 1929), New York Life Insurance Company Archives, New York.

96. Corbett, "Office Building of New Era," 485 and 486. The degree of acoustical control achieved in the Ladies' Cafeteria was sufficient to allow the Fox Film Company to

shoot a Movietone sound newsreel there in 1929. The photograph in figure 5.17 is a frame from this newsreel footage.

97. Corbett, "Office Building of New Era," 484.

98. Corbett, "Office Building of New Era," 485.

99. "Planning for Employees' Welfare," 398.

100. "Planning for Employees' Welfare," 399.

101. "There can be no doubt that here we have a new conception of the business workshop, remarkable in its external beauty and remarkable in its application of scientific principles to modern business planning." Corbett, "Office Building of New Era," 486.

102. New York Life president Darwin P. Kingsley compared the company itself to a "Cathedral of Service," noting that "its great arches support an expanding nave which can cover and protect all who come." "As Cathedrals Were Built," *Eastern Underwriter* (14 December 1928): part 2, special issue, 20.

103. "Notes for Talk" (26 April 1920), quoted in Blodgett, "Cass Gilbert," 627.

104. Robert von Ezdorf, "The Design of the Interior of the New Home Office Building of the New York Life Insurance Company," *American Architect* 135 (20 March 1929): 369–372, on 369.

105. Surveys of modern architecture include: Curtis, *Modern Architecture Since 1900*; Kenneth Frampton, *Modern Architecture: A Critical History,* 3d ed. (London: Thames and Hudson, 1992); Frank Whitford, *Bauhaus* (London: Thames and Hudson, 1984); Barbara Miller Lane, *Architecture and Politics in Germany, 1918–1945* (1968; Cambridge, Mass.: Harvard University Press, 1985); Reyner Banham, *A Concrete Atlantis: U.S. Industrial Building and European Modern Architecture* (Cambridge, Mass.: The MIT Press, 1986); K. Michael Hays, *Modernism and the Posthumanist Subject* (Cambridge, Mass.: The MIT Press, 1992); Hilde Heynen, *Architecture and Modernity: A Critique* (Cambridge, Mass.: The MIT Press, 1999); and Sigfried Giedion, *Space, Time and Architecture: The Growth of a New Tradition* (Cambridge, Mass.: Harvard University Press, 1941).

106. Le Corbusier-Saugnier, *Vers une Architecture* (Paris: G. Cres et Cie, 1923). Quoting from the English translation by Frederick Etchells of the 13th French edition, *Towards a New Architecture* (1931; New York: Dover, 1986), 8.

107. Le Corbusier, *Towards a New Architecture*, 42. The passage quoted here also appeared in the first, French edition of 1923, on 29.

108. Walter Gropius, *Idee und Aufbau des Staatlichen Bauhauses Weimar* (1923), quoted in Reyner Banham, *Theory and Design in the First Machine Age,* 2d ed. (Cambridge Mass.: The MIT Press, 1980), 282. For more on the philosophy of science in the Bauhaus, see Peter Galison, "Aufbau/Bauhaus: Logical Positivism and Architectural Modernism,"

Critical Inquiry 16 (summer 1990): 709–752. See also Jeffrey Herf, *Reactionary Modernism: Technology, Culture and Politics in Weimar and the Third Reich* (Cambridge: Cambridge University Press, 1984).

109. Reyner Banham, *The Architecture of the Well-tempered Environment* (Chicago: University of Chicago Press, 1969), 123–124.

110. Quoted in Banham, *Well-tempered Environment*, 154.

111. Gustave Lyon was head of the Pleyel piano manufacturing firm in Paris. He attempted in 1909 to correct the notoriously bad acoustics of the Trocadero auditorium, and in 1927 he worked with the architects Aubertin, Granel & Mathon on the design for the Salle Pleyel concert hall. Paul Calfas, *La Nouvelle Salle de Concert Pleyel à Paris* (Paris: Publications du Journal le Génie Civil, 1927); and Gustave Lyon, *L'Acoustique Architecturale* (Paris: Bibliothèque Technique du Cinéma, 1932).

112. F. M. Osswald, "The Acoustics of the Large Assembly Hall of the League of Nations at Geneva, Switzerland," *American Architect* 134 (20 December 1928): 833–842, on 840. Lyon's Salle Pleyel, which was based on a similarly parabolic form, so effectively channeled sound to the rear of the hall that the musicians on stage could barely hear each other. There were also problems of echo from the hard rear wall. See Michael Forsyth, *Buildings for Music* (Cambridge, Mass.: The MIT Press, 1985), 262–263 and illustrations on 266–267. Le Corbusier would return to the idea of controlling the projection of sound through space much later, and more successfully, in his Philips Pavilion for the 1958 World's Fair at Brussels. Here, electroacoustic technology substituted for hard, reflective surfaces. The Philips Pavilion, a hyperbolic paraboloid structure, used hundreds of small loudspeakers scattered throughout the interior to project an acoustic signal directly at auditors. The signal so projected was Edgard Varèse's "Poème Electronique." Marc Treib, *Space Calculated in Seconds: The Philips Pavilion, Le Corbusier, Edgard Varèse* (Princeton: Princeton University Press, 1996).

113. Osswald, "Acoustics of Large Assembly Hall," 833.

114. Lyon's *L'Acoustique Architecturale* (1932) did include one ad for a French product, "*Héraclite, le matériau d'élite,*" "*pour la correction acoustique.*" In 1931, British acousticians Hope Bagenal and Alexander Wood admitted, "There is no doubt that we rely in England largely upon American results for information about new types of materials," in *Planning for Good Acoustics* (London: Methuen and Co., 1931), vii.

115. Ludwig Mies van der Rohe, "Industrialized Building" (1924), reproduced in Ulrich Conrads, ed., *Programs and Manifestoes on 20th-Century Architecture*, trans. Michael Bullock (Cambridge, Mass.: The MIT Press, 1970), 81–82, on 82.

116. Banham referred specifically here to the technology of air conditioning, but he developed this argument by also considering acoustical technology. Banham, *Well-tempered Environment*, 162.

117. See, for example, "Less Noise . . . Better Hearing" (sales pamphlet, Celotex Co., 1927), Series I, Box 13, Folder 11, GCC.

118. Henry-Russell Hitchcock and Philip Johnson, *The International Style* (1932; New York: W.W. Norton and Co., 1966), 20. Hitchcock and Johnson were the curators of the MoMA exhibition, and this text was an extension of their catalogue for the show. For more on the history and reception of the exhibition, see Richard Guy Wilson, "International Style: The MoMA Exhibition," *Progressive Architecture* 63 (February 1982): 92–105.

119. "A New Shelter for Savings," *Architectural Forum* 57 (December 1932): 483–498, on 488.

120. "New Shelter for Savings," 483.

121. "New Shelter for Savings," 483. For more on the evolution of the design of the PSFS Building, see William H. Jordy, *The Impact of European Modernism in the Mid-Twentieth Century* (New York: Oxford University Press, 1972); and Jordy, "PSFS: Its Development and Its Significance in Modern Architecture," *JSAH* 21 (May 1962): 47–83. See also Robert A. M. Stern, "PSFS: Beaux-Arts Theory and Rational Expressionism," *JSAH* 21 (May 1962): 84–95 and appendix of primary documents on 95–102; and Frederick Gutheim, "The Philadelphia Saving Fund Society Building: A Re-appraisal," *Architectural Record* (October 1949): 88–95, 180, 182. Architectural historians have correctly pointed to the existence in America of significant "protomodern" precedents to the PSFS Building in America, including buildings by Irving Gill, Richard Neutra, and Raymond Hood. The PSFS Building nonetheless attracted unprecedented attention, and was clearly perceived as something distinctly new in American architecture circa 1932.

122. "Philadelphia's Fancy," *Fortune* 6 (December 1932): 65–69, 130–131, on 65. For the record, Wanamaker's department store fell victim to the merger-mania that swept the retail industry in the 1990s, and there has been a Democratic mayor for the past forty years. Independence Hall is, however, still standing.

123. George Howe to James Willcox (26 May 1930), reproduced in *JSAH* 21 (May 1962): 96–97, quote on 96.

124. George Howe to James Willcox (25 July 1930), reproduced in *JSAH* 21 (May 1962): 98–99, quote on 98.

125. "Architectural Analysis of the Proposed Building for the Philadelphia Saving Fund Society at 12th and Market Streets," undated memorandum, c. July 1930, reproduced in *JSAH* 21 (May 1962): 100–101, quote on 100. For more on the lives and careers of Howe and Lescaze, see Robert A. M. Stern, *George Howe: Toward a Modern American Architecture* (New Haven: Yale University Press, 1975); Helen Howe West, *George Howe, Architect, 1886–1955: Recollections of My Beloved Father* (Philadelphia: William Nunn Co., 1973); and Lorraine Welling Lanmon, *William Lescaze, Architect* (Philadelphia: Art Alliance Press, 1987).

126. "Nothing More Modern" (promotional brochure, PSFS, c. 1932), Accession 2061: PSFS Archives, Record Group VIII: PSFS Offices and Buildings, Subgroup 7: The

PFSF Building, Series H: PSFS Building Promotion, Box 91, Hagley Museum and Library, Wilmington, Del. [PSFS Archive]. Howe's earlier suburban branch banks are described and depicted in George Howe, "The Philadelphia Saving Fund Society Branch Offices," *Architectural Forum* 48 (June 1928): 881–885.

127. "Philadelphia's Fancy," 131. William Lescaze had previously decorated Stokowski's New York apartment in a similarly modern style.

128. "Philadelphia's Fancy," 131; *Philadelphia Sunday Dispatch* (10 October 1932); and *Philadelphia Sunday Transcript* (27 December 1931), *Dispatch* and *Transcript* quoted in Stern, *George Howe*, 131.

129. Robert Reiss, "Mostly Personal," *Philadelphia Record* (21 September 1932), RG VIII, Subgroup 7, Series H, Box 90, Folder: Reactions to PSFS Building, PSFS Archive. Four additional stanzas are not reproduced here.

130. The rubber flooring was used "in most of the service portions of the bank, and in some of the public spaces as well where quietness was particularly desirable." "Planning, Engineering, Equipment: The Philadelphia Saving Fund Society Building," *Architectural Forum* 57 (December 1932): 543–550, on 545.

131. "Philadelphia's Fancy," 130. These partitions were designed by Walter Baerman for the E.F. Hauserman Company.

132. A 1932 advertisement stated clearly, "The place for Acousti-Celotex is on the ceiling." *American Architect* 141 (May 1932): 77. Vern Knudsen agreed, noting that "The absorptive treatment of offices is nearly always accomplished by treating the ceiling with a highly absorptive material." Knudsen, *Architectural Acoustics*, 453.

133. See figure 5.11 for a list of the different finishes offered by Johns-Manville in 1931.

134. "Mechanically Applied Products Co.," *Sweet's* (1922): 17. Sabinite was advertised in *Sweet's* (1924–1925): 2132 as a product of the Keasbey and Mattison Co., but the manufacturing rights were apparently transferred to U.S.G. the following year. In 1926, U.S.G. announced Sabinite in its trade newsletter, *The Gypsumist* 5 (August 1926): 12–13. While the plaster was ostensibly named after Paul Sabine, it is clear that U.S.G. was attempting to evoke the authority of his cousin Wallace to create an instant pedigree for its new product. The announcement noted that the new plaster was developed by Dr. Paul Sabine, who "by the way, is a cousin of Professor Wallace C. Sabine and studied under him at Harvard." See also U.S.G. entries in *Sweet's* from 1926–1927 onward, and "U.S.G. Sound Control Service" (sales pamphlet, United States Gypsum Co., c. 1931), Series I, Box 13, Folder 11, GCC.

135. Vern Knudsen listed twenty-three different acoustical plasters in his 1932 text, *Architectural Acoustics*, 198–201. See also "Sprayo-Flake," *Sweet's* (1931): B2513-B2524; and the sales pamphlets "Kalite Acoustical Plaster" (Kalite Co., c. 1931); "Wyodak Acoustical Plaster" (Wyolite Insulating Products, c. 1932); and "Old Newark Acoustical Plaster" (Newark Plaster Co., 1933). For listings of representative installations, see

"U.S.G. Sound Control Service," 32; and "Kalite Sound Absorbing Plasters" (c. 1934), 12. All pamphlets Series I, Box 13, Folder 11, GCC.

136. For example, an account of a Sabinite installation noted that "This treatment requires constant work; at least three plasterers must be kept on the job or the plaster will set before it is floated and it will have to be removed and a new coat applied." John Klenke, "Construction of a Sound-Proof Studio for Making Sound Motion Pictures," *American Architect* 140 (October 1931): 46–47, 66, on 47.

137. After the unsuccessful outcome of his earlier attempt to patent Sabine's technique, Mazer continued to work as an acoustical consultant. He advertised a free booklet, "Defective Acoustics," in *Literary Digest* 48 (7 March 1914): 515, and identified himself as an acoustical engineer based in Pittsburgh. From 1921 to 1925, the Mazer Acoustile Company of Philadelphia advertised in *Sweet's*. Mazer's 1921 ad claimed hundreds of satisfied clients. Quotes from "Mazer Acoustile Co.," *Sweet's* (1921): 26, 27.

138. "Union Fibre Co.," *Sweet's* (1924–1925): 1240; "Boston Acoustical Engineering Co.," *Sweet's* (1927–1928): A15. Anne Weber notes that Silen-Stone was supplied in 22" x 30" sheets, and was intended to be cut into tiles and installed on furring to resemble ashlar or other stone patterns. Acoustex came in precut units that could either be butted together for a monolithic look, or given beveled edges to emphasize each individual unit. Weber, "Search for Silence," 9.

139. "United States Gypsum Company," *Sweet's* (1931): B2671-B2672, absorption coefficients on B2672. See also "U.S.G. Sound Control Service," 8–9. For additional data on these and twenty-six other brands of acoustical tiles, see Knudsen, *Architectural Acoustics*, 202–205.

140. "Dahlberg, Bror," *National Cyclopedia of American Biography* (New York: James T. White, 1930) C: 327–328, quote on 327. See also "Muench, Carl," *National Cyclopedia of American Biography* (Clifton, N.J.: James T. White, 1979) 58: 532.

141. "Celotex Co.," *Sweet's* (1923): 1180.

142. "Less Noise . . . Better Hearing," 38. According to Hale Sabine, acousticians in the 1920s believed that perforations increased the absorptivity of a material simply by increasing its effective surface area. Experiments soon indicated that the increase in absorptivity of perforated materials surpassed that attributable to surface area alone, but not until 1954 would a full explanation of the phenomenon be developed. Sabine, "Manufacture and Distribution of Acoustical Materials," 657. See also Uno Ingard, "Perforated Facing and Sound Absorption," *JASA* 26 (March 1954): 151–154 and "Sound Absorption by Perforated Porous Tiles," *JASA* 26 (May 1954): 289–293.

143. Mazer's patent on perforated acoustical materials (#1,483,365) was filed in 1918, and he appears to have withdrawn from acoustical consulting upon licensing it to Celotex. He subsequently authored articles, lectured at Cornell, and became active in promoting legislation "to encourage and protect young inventors." "Mazer, Jacob," *National Cyclopedia of American Biography* (New York: James T. White, 1961) 43: 425.

144. "Celotex Co.," *Sweet's* (1925–1926): 1356; "Less Noise . . . Better Hearing," 3. All ellipses in original.

145. Hale Sabine, "Manufacture of Acoustical Materials," 661.

146. "Less Noise . . . Better Hearing," 3.

147. Acousti-Celotex advertisement, *Architectural Forum* 59 (December 1933): ad sect., 34.

148. "C.F. Burgess Laboratories, Inc.," *Sweet's* (1929): A16–A19, quote on A17. R. F. Norris, Chief Engineer at the Burgess Labs, patented Sanacoustic Tile in 1929 (U.S. Patent #1,726,500). Hale Sabine, "Manufacture of Acoustical Materials," 658.

149. "J-M Sanacoustic Tile" advertisement in *American Architect* 136 (November 1929): 77; "Johns-Manville," *Sweet's* (1931): B2664-B2668.

150. "Acoustic Roof," *Architectural Forum* 57 (December 1932): ad sect., 30.

151. "Acoustical Corporation of America," *Sweet's* (1931): B2650.

152. "Acoustical Corporation of America," *Sweet's* (1933): B762–764; quotes on B762. It is not evident why the metal trays of Silent-Seal were replaced with the plaster trays of Mutetile. Plaster may have been chosen to supplement the absorbency of the rock wool fill, or perhaps the Norris patent, controlled by Johns-Manville, forced the change.

153. Specifications, Sub-Division XXIV: Acoustical Treatment, pp. 263–269 (2 September 1931); and Completed Contract no. 28: Acoustical Corporation of America, RG VIII, Subgroup 7, Series C: PSFS Building Contracts, PSFS Archive. In his 1962 article, William Jordy indicated that the banking room ceiling tiles were provided by a "small Philadelphia inventor and manufacturer who went bankrupt during the construction period. His bankruptcy forced the Society to take over his plant and act as its own supplier to finish the job." This suggests that the society chose M. C. Rosenblatt's Acoustical Corporation of America over Johns-Manville in order to support an enterprise in which their funds were already invested. Jordy, "PSFS: Development and Significance," 53.

154. The first office building to be air conditioned was the Milam Building in San Antonio, Texas. The systems in both the Milam and PSFS Buildings were designed and installed by the Carrier Corporation. For more on the history of air conditioning, see Gail Cooper, *Air-Conditioning America: Engineers and the Controlled Environment, 1900–1960* (Baltimore: Johns Hopkins University Press, 1998); Banham, *Well-tempered Environment*; Raymond Arsenault, "The End of the Long Hot Summer: The Air Conditioner and Southern Comfort," *Journal of Southern History* 50 (November 1984): 597–628; and Douglas Gomery, *Shared Pleasures: A History of Movie Presentation in the United States* (Madison: University of Wisconsin Press, 1992).

155. A 1932 advertisement for the PSFS Building noted that "there will be no need to open windows unless force of habit makes you do so." "Quiet! Men Concentrating"

(ad), *Philadelphia Public Ledger* (16 May 1932), clipping, RG VIII, Subgroup 7, Series H, Box 90, PSFS Archive. By 1949, the windows were locked by the landlord, and opened only for cleaning. Frederick Gutheim reported that "Tenants have always accepted this without protest." "PSFS Building: A Re-appraisal," 95.

156. In 1930, the inventor Hiram Percy Maxim, already famous for having designed mufflers and silencers for engines and firearms, invented a forced-air window ventilator that similarly prevented the transmission of noise into a room. "Noise Flies out the Window," *NYT* (25 October 1930): 16. See also "Maxim Demonstrates His Room Silencer," *NYT* (2 September 1931): 23; "New York's Night Symphony," *New York Times Magazine* (18 October 1931): 16; and "A Silencer for Street Noises," *American Architect* 119 (2 February 1921): 131. Maxim's device was manufactured by the Campbell Metal Window Corporation.

157. While electrical and ventilating infrastructures were concealed, they remained easily accessible for maintenance. This combination of concealment and accessibility was a strong selling point for manufacturers of suspended tile ceilings. In 1929, for example, a Johns-Manville Sanacoustic Tile advertisement pointed out that "Any tile may instantly be removed to provide access to pipes, wires or the like in the furred space." *American Architect* 136 (November 1929): 77. For a critique of the inconsistency of this practice of technological concealment within the ideological tenets of modern architecture, see Banham, *Well-tempered Environment*.

158. "Four Reasons Why You Should Investigate Twelve South Twelfth" (ad), *Philadelphia Inquirer* (15 June 1932). See also "Quiet! Men Concentrating," clippings, RG VIII, Subgroup 7, Series H, Box 90, PSFS Archive.

159. "Commercialized Pianissimo," *NYT* (3 December 1931): 28.

160. "Commercialized Pianissimo," 28.

161. "Philadelphia's Fancy," 65.

162. "Nothing More Modern" (c. 1932) 13; "Le Corbusier," *Philadelphia Evening Bulletin* (30 August 1965), clipping, RG VIII, Subgroup 7, Series H, Box 90, PSFS Archive.

163. Hitchcock and Johnson, *International Style*, 61.

164. Signals received by a central antenna on the roof were sent to a central station and transmitted to the offices, which were equipped with special outlets into which employees could plug their own equipment. "Planning, Engineering, Equipment," 549; Contract no. 10-E: R.C.A. Victor Co. Inc., RG VIII, Subgroup 7, Series C, PSFS Archives. See also Karr Parker, "Providing for Radio and Amplifying Installations in Large Buildings," *Architectural Forum* 50 (January 1929): 117–120; and "Western Electric Radio Frequency Distribution Systems" (ad), *Architectural Forum* 56 (June 1932): ad sect., 35.

165. "Opens Tonight!" (ad), *NYT* (27 December 1932): 11. See also the broadcast schedule, "Today on the Radio," on page 10.

Chapter Six

1. Edward W. Kellogg, "Some New Aspects of Reverberation," *Journal of the Society of Motion Picture Engineers* [*JSMPE*] 14 (January 1930): 96–107, on 101.

2. William H. Jordy, *The Impact of European Modernism in the Mid-Twentieth Century* (New York: Oxford University Press, 1972), 78. Jordy emphasizes the difference between the modern style embodied in the PSFS building, and the Art Deco style of Radio City Music Hall. Art Deco, he explains, "is historically interesting as a search for design capable of relating to the modern world, not with the technological purity of more earnest expressions of modernity, but with an abandon calculated to stimulate popular fantasy" (79).

3. Jordy, *Impact of European Modernism*, 78.

4. Edward Angly, "Notables See First Show at Radio City," *New York Herald-Tribune* (28 December 1932): 1, 10 on 10. See also "Radio City Premiere Is a Notable Event," *New York Times* [*NYT*] (28 December 1932): 1, 14; and Charles Francisco, *The Radio City Music Hall: An Affectionate History of the World's Greatest Theater* (New York: E.P. Dutton, 1979), 13–22.

5. Douglas Haskell, "Roxy's Advantage over God," *Nation* 136 (4 January 1933): 11.

6. Brooks Atkinson, "The Play: Music Hall's Opening," *NYT* (28 December 1932): 14.

7. "Radio City Premiere Is Notable Event," 14.

8. Atkinson, "Music Hall's Opening," 14.

9. "Roxy's Gang" began broadcasting locally from the Capitol Theater in New York in 1922, and was soon carried by stations across the nation. Susan Smulyan, *Selling Radio: The Commercialization of American Broadcasting, 1920–1934* (Washington: Smithsonian Institution Press, 1994), 55.

10. "Radio City Premiere Is Notable Event," 14.

11. "Roxy Returns Soon to Broadcasting," *NYT* (16 January 1932): 13.

12. "When Radio Answered a Call to Hollywood," *NYT* (10 August 1930): sect. 9, p. 12.

13. Technical histories of the telephone are cited in chapter 3. For more on the social history of the device, see Claude S. Fischer, *America Calling: A Social History of the Telephone to 1940* (Berkeley: University of California Press, 1992); and Carolyn Marvin, *When Old Technologies Were New: Thinking About Electric Communication in the Late Nineteenth Century* (New York: Oxford University Press, 1988).

14. For more on "annihilation," see Stephen Kern, *The Culture of Time and Space, 1880–1918* (Cambridge, Mass.: Harvard University Press, 1983), 214–215; and Marvin, *When Old Technologies Were New*, 191–231.

15. Telephone companies instructed new users to place the receiver directly against the ear, and to hold the transmitter no more than one inch from the mouth. Fischer, *America Calling*, 70, 162 (Photo 12).

16. For an examination of the cultural tensions that this new intimacy evoked, including the problems of telephonic romance and party-line eavesdropping, see Marvin, *When Old Technologies Were New*, 63–108. Today's cell-phone conversants, who loudly discuss private matters in public places with no sense of embarassment, indicate that this misperception of spatial isolation continues.

17. In 1896, Hall's Drug Store in Palo Alto, Calif., advertised a "sound-proof conversation room" for making telephone calls. Fischer, *America Calling*, 170 (Photo 21). See also F. L. Tufts, "The Transmission of Sound through Solid Walls," *American Journal of Science,* 4th ser., 13 (June 1902): 449–454, a study stimulated by the problem of constructing soundproof telephone booths for noisy cities.

18. For more on the history of the phonograph, see Oliver Read and Walter L. Welch, *From Tin Foil to Stereo: Evolution of the Phonograph* (Indianapolis and New York: Howard Sams and Bobbs-Merrill, 1959); Welch and Leah Brodbeck Stenzel Burt, *From Tinfoil to Stereo: The Acoustic Years of the Recording Industry, 1877–1929* (Gainesville: University Press of Florida, 1994); Roland Gelatt, *The Fabulous Phonograph, 1877–1977,* 2d ed. (New York: Macmillan, 1977); Evan Eisenberg, *The Recording Angel: Explorations in Phonography* (New York: McGraw-Hill, 1987); Andre Millard, *America on Record: A History of Recorded Sound* (Cambridge: Cambridge University Press, 1995), and *Edison and the Business of Innovation* (Baltimore: Johns Hopkins University Press, 1990); William Howland Kenney, *Recorded Music in American Life: The Phonograph and Popular Memory, 1890–1945* (New York: Oxford University Press, 1999); and David Morton, *Off the Record: The Technology and Culture of Sound Recording in America* (New Brunswick: Rutgers University Press, 2000).

19. Edison's original phonograph created embossed records on malleable and fragile tin foil. In 1886, Chichester Bell and Charles Sumner Tainter modified Edison's design, creating a device that engraved a more permanent record into a soft but rigid wax cylinder. The flat disc record, whose grooves were created by a complex process of acid-etching, was introduced by Emile Berliner in 1895. For more on these technical developments, see Read and Welch, *Tin Foil to Stereo*; and Welch and Burt, *Tin Foil to Stereo: Acoustic Years.*

20. For more on the history of radio, see Hugh G. J. Aitken, *Syntony and Spark: The Origins of Radio* (1976; Princeton: Princeton University Press, 1985) and *The Continuous Wave: Technology and American Radio, 1900–1932* (Princeton: Princeton University Press, 1985); Susan J. Douglas, *Inventing American Broadcasting, 1899–1922* (Baltimore:

Johns Hopkins University Press, 1987) and *Listening In: Radio and the American Imagination* (New York: Times Books, 1999); Daniel J. Czitrom, *Media and the American Mind: From Morse to McLuhan* (Chapel Hill: University of North Carolina Press, 1982); Tom Lewis, *Empire of the Air: The Men Who Made Radio* (New York: Harper Collins, 1991); and Smulyan, *Selling Radio*.

21. On Christmas Eve 1906, electrical inventor Reginald Fessenden became the first to broadcast speech and music to an incredulous audience of wireless operators. In 1910, Lee de Forest began to broadcast live music to listeners in New York, and other occasional broadcasts of speech and music occurred over the course of the teens. Aitken, *Continuous Wave*, 469–470. For examples of "proto-broadcasting" telephonic systems for the dissemination of music and speech, see Marvin, *When Old Technologies Were New*, 209–231.

22. Read and Welch, *Tin Foil to Stereo*, 177–187. An alternative method of amplification was to use the movement of the reproducing stylus to control a valve regulating a powerful stream of compressed air. "The Auxetophone for Reinforcing Gramophone Sounds," *Scientific American* 92 (13 May 1905): 382.

23. "The Talking-Machine," *Littel's Living Age* 254 (24 August 1907): 486–489, on 488.

24. The Tone Test campaign is examined more fully in Emily Thompson, "Machines, Music, and the Quest for Fidelity: Marketing the Edison Phonograph in America, 1877–1925," *Musical Quarterly* 79 (spring 1995): 131–171. See also Read and Welch, *Tin Foil to Stereo*, 205–217; and Marsha Siefert, "Aesthetics, Technology, and the Capitalization of Culture: How the Talking Machine Became a Musical Instrument," *Science in Context* 8 (summer 1995): 417–449.

25. Neil Harris explores the derisive characterization of "canned music" in "John Philip Sousa and the Culture of Reassurance," pp. 11–40 in Jon Newsom, ed., *Perspectives on John Philip Sousa* (Washington, D.C.: Library of Congress, 1983), on 39 (note 102). See also John Philip Sousa, "The Menace of Mechanical Music," *Appleton's Magazine* 8 (September 1906): 278–284.

26. Phonograph advertisements regularly sold the social benefits of critical listening along with the acoustical benefits of the machines. For example, an ad for Brunswick phonographs warned "Beware False Tones, Mothers," and admonished "there is grave danger of spoiling a child's 'ear' for music by false tones. The instrument must achieve *true* reproductions." *Ladies' Home Journal* (November 1921): 165.

27. See Steven Lubar, *InfoCulture: The Smithsonian Book of Information Age Inventions* (Boston: Houghton Mifflin, 1993), 215–216; Douglas, *Listening In*, 78; and Aitken, *Continuous Wave*, 432–479, for details on these technological developments.

28. Before loudspeakers were available, networks of headsets were sometimes used so that a number of people could listen simultaneously to a single receiver. More enterprising listeners placed a headset in a large jar or bowl, which apparently amplified the

sound sufficiently for a small group to hear. See Lizbeth Cohen, *Making a New Deal: Industrial Workers in Chicago, 1919–1939* (Cambridge: Cambridge University Press, 1990), 133; and Kathy J. Ogren, *The Jazz Revolution: Twenties America and the Meaning of Jazz* (New York: Oxford University Press, 1989), 103, for accounts of this approach.

29. An example of the gooseneck speaker is the Magnavox R-3, which was introduced around 1921. The earliest hornless loudspeakers include the Western Electric 540-AW (c. 1922) and the General Electric Radiola Loudspeaker Model 104A (1926), which was sold with its own 1-watt amplifier. A concise technical account of the development of loudspeakers is found in Frederick V. Hunt, *Electroacoustics: The Analysis of Transduction, and Its Historical Background* (1954; n.p.: American Institute of Physics and Acoustical Society of America, 1982), 70–91. See also Harry F. Olson, "A Review of Twenty-Five Years of Sound Reproduction," *Journal of the Acoustical Society of America* [*JASA*] 26 (September 1954): 637–643; Leo L. Beranek, "Loudspeakers and Microphones," *JASA* 26 (September 1954): 618–629; and John K. Hilliard, "Electroacoustics to 1940," *JASA* 61 (February 1977): 267–273. Primary sources include C. R. Hanna and J. Slepian, "The Function and Design of Horns for Loud Speakers," *Journal of the American Institute of Electrical Engineers* [*JAIEE*] 43 (March 1924): 250–256; Chester W. Rice and Edward W. Kellogg, "Notes on the Development of a New Type of Hornless Loud Speaker," *JAIEE* 44 (September 1925): 982–991; E. C. Wente and A. L. Thuras, "A High Efficiency Receiver for a Horn-Type Loud Speaker of Large Power Capacity," *Bell System Technical Journal* 7 (January 1928): 140–153; and sales bulletins at the AT&T Archives, including "Western Electric Loud Speaking Telephones (no. 540 Type)," n.d., #T750; and "No. 543-W Loud Speaking Telephone," n.d., #T752, in bound volume "Sales Bulletins T125–T790."

30. Read and Welch, *Tin Foil to Stereo*, 239.

31. Orthophonic Victrola advertisement, *Ladies' Home Journal* (July 1926): 1. Victor and Columbia initially offered their electrically recorded records to be played on improved, but still strictly mechanical, phonographs. Brunswick's Panatrope was the first electrical phonograph to reach the consumer market. Read and Welch, *Tin Foil to Stereo*, 268; Hunt, *Electroacoustics*, 67–70; "Historic Firsts: The Orthophonic Phonograph," *Bell Labs Record* 24 (August 1946): 300–301, AT&T Archives.

32. Orthophonic Victrola advertisement, *McCall's Magazine* (September 1927): 3. Emphasis in original.

33. See, for example, *Edison Retail Sales Laboratory* (1915): 29. Primary Printed Collection, Edison Archives, United States Department of the Interior, National Park Service, Edison National Historic Site, West Orange, N.J.

34. "'Theremin-Voxes' Heard in Open Air," *NYT* (28 August 1928): 27.

35. Sheldon Hochheiser, "AT&T and the Development of Sound Motion-Picture Technology," pp. 23–33 in Mary Lea Bandy, ed., *The Dawn of Sound* (New York: Museum of Modern Art, 1989), 24. For descriptions of systems applications, see the col-

lection of sales and technical bulletins in the bound volume "Sales Bulletins T125-T790," AT&T Archives. For an account of an earlier system, see William P. Kennedy, "Announcing Trains by Magnaphone," *Technical World Magazine* 19 (April 1913): 256–257.

36. Ben M. Hall, *The Best Remaining Seats: The Golden Age of the Movie Palace* (1961; New York: Da Capo, 1988), 68; "Public Address One of ERPI's Functions," *Erpigram* 1 (20 June 1929): 4, AT&T Archives.

37. Samuel L. "Roxy" Rothafel, "What the Public Wants in the Picture Theater," *Architectural Forum* 42 (June 1925): 360–364, on 364.

38. "Public Address One of ERPI's Functions," 4. The Capitol, at Broadway and 51st St., was the largest theater in New York when it was built in 1920, with a capacity of over 5,200 auditors. Emil M. Mlinar, "The Capitol Theater, New York, N.Y.," *Architectural Forum* 32 (January 1920): 21–24.

39. "How the Western Makes the Movies Move," *Western Electric News* 12 (August 1923): 10–11, AT&T Archives.

40. Edison announced in 1888 that he was "experimenting upon an instrument which does for the Eye what the phonograph does for the Ear, which is the recording and reproduction of things in motion." Charles Musser, *The Emergence of Cinema: The American Screen to 1907* (New York: Charles Scribner's Sons, 1990), quote on 64. See also George Parsons Lathrop, "Edison's Kinetograph," *Harper's Weekly* 35 (13 June 1891): 446–447; and "Edison's Kinetograph and Cosmical Telephone," *Scientific American* 64 (20 June 1891): 393.

41. The following survey of motion pictures and sound is drawn from: Musser, *Emergence of Cinema*, *Thomas A. Edison and His Kinetographic Motion Pictures* (New Brunswick: Rutgers University Press, 1995), and *Before the Nickelodeon: Edwin S. Porter and the Edison Manufacturing Company* (Berkeley: University of California Press, 1991); Douglas Gomery, *Shared Pleasures: A History of Movie Presentation in the United States* (Madison: University of Wisconsin Press, 1992) and "The Coming of Sound to the American Cinema: A History of the Transformation of an Industry" (Ph.D. diss., University of Wisconsin–Madison, 1975); David Robinson, *From Peep Show to Palace: The Birth of American Film* (New York: Columbia University Press, 1996); Scott Eyman, *The Speed of Sound: Hollywood and the Talkie Revolution: 1926–1930* (New York: Simon and Schuster, 1997); Donald Crafton, *The Talkies: American Cinema's Transition to Sound, 1926–1931* (New York: Charles Scribner's Sons, 1997); and Hochheiser, "AT&T and Sound Motion Pictures."

42. One surviving film, never released to the public, indicates that Edison had been exploring synchronization. The film, circa 1893, shows Dickson playing the violin into a large phonographic recording horn, as two colleagues dance to the music. The accompanying wax cylinder sound recording is preserved at the Thomas A. Edison National Historic Site; the film at the Library of Congress. See Musser, *Emergence of Cinema*, 88; and

Robinson, *Peep Show to Palace*, 51. The film, minus the missing music, can be seen in the documentary series *Hollywood: A Celebration of the American Silent Film* (1980; Thames Television, dir. David Gill and Kevin Brownlow, Thames Video Collection/HBO Home Video), in Episode 13, "End of an Era."

43. The projector used at the Vitascope premiere was developed by Thomas Armat, who subsequently sold the device to Edison. For more on the development of projection, see Musser, *Emergence of Cinema*, 91–132.

44. For a description of some of these enterprises, including the Humanovo, Actologue, and Talk-o-Photo companies, see Musser, *Before the Nickelodeon*, 396–402.

45. The projector (at the rear of the theater) and the phonograph (at the front, near the screen) were electrically linked, but the synchronization between them still required fine-tuning. "The Gaumont Speaking Kinematograph Films," *Nature* 89 (30 May 1912): 333–334; Gomery, "Coming of Sound," 28–30; Eyman, *Speed of Sound*, 26–28; and Crafton, *Talkies*, 55. Brief excerpts of Gaumont Chronophone films, with accompanying sound, are included in the documentary *Cinema Europe: The Other Hollywood* (1995; Photoplay Productions, dir. Kevin Brownlow and David Gill, DLT Entertainment Video), Episode 1, "Where It All Began," and Episode 6, "End of an Era."

46. Eyman, *Speed of Sound*, 27–29; Gomery, "Coming of Sound," 30–32.

47. Eyman, *Speed of Sound*, 33–37; quotes on 34 and 35; Crafton, *Talkies*, 55–57; Gomery, "Coming of Sound," 34–38; Edward Kellogg, "History of Sound Motion Pictures," pp. 174–220 in Raymond Fielding, ed., *A Technological History of Motion Pictures and Television* (Berkeley: University of California Press, 1967), on 174. A film clip with accompanying sound from an Edison Kinetophone film is included in the documentary *Hollywood*, Episode 13, "End of an Era."

48. In 1904, well before de Forest's work, former Edison employee Eugene Lauste began a long but ultimately unsuccessful series of attempts to record sound on film. Wallace Sabine also devised a technique for creating a motion picture image of sound circa 1910 (see chapter 3), but Sabine was interested only in reading data off this image and he did not attempt to reproduce the sound he had recorded. For more on Lauste, de Forest, and Case, see Crafton, *Talkies*, 63–70; Eyman, *Speed of Sound*, 30–32, 41–54; and Gomery, "Coming of Sound," 40–48.

49. Case and his assistant E. I. Sponable allied with motion picture producer William Fox, and the Fox-Case Corporation would soon be famous for newsreels that used Case's sound-on-film system. For more on the Fox-Case alliance, as well as de Forest's perseverance in the industry, see Crafton, *Talkies*, 89–100; Eyman, *Speed of Sound*, passim; Gomery, "Coming of Sound," 175–183; and E. I. Sponable, "Historical Development of Sound Films," *JSMPE* 48 (April 1947): 275–303 and (May 1947): 407–422.

50. Douglas Gomery, "The Coming of Sound: Technological Change in the American Film Industry," pp. 5–24 in Elisabeth Weis and John Belton, eds., *Film Sound: Theory*

and Practice (New York: Columbia University Press, 1985), on 18–19. See also W. R. G. Baker, "Description of the General Electric Company's Broadcasting Station at Schenectady, New York," *Proceedings of the Institute of Radio Engineers* [*PIRE*] 11 (August 1923): 339–374.

51. Quoted in Richard Koszarski, "On the Record: Seeing and Hearing the Vitaphone," pp. 15–21 in Bandy, ed., *Dawn of Sound*, on 17.

52. The restored Vitaphone shorts can be viewed at the UCLA Film and Television Archive, #A7-378-2. See also *Don Juan* (1926; Warner Bros., dir. Alan Crosland, MGM/UA Home Video #M302162).

53. While spoken words had been heard before in Vitaphone films, Richard Koszarski and others have noted that the dramatic construction of *The Jazz Singer*, and the abrupt way it reverted back to silence after its talking segment, precipitated a new dissatisfaction with nontalking films. See Koszarski, "On the Record," 19, and *The Jazz Singer* (1927; Warner Bros., dir. Alan Crosland, MGM/UA Home Video #M302312).

54. Hochheiser, "AT&T and Sound Motion Pictures," 32.

55. "Sound Pictures . . . a product of the Telephone," special insert in *Erpigram* 1 (20 March 1929), reproducing an advertisement that appeared in *Life*, *Time*, *New Yorker*, *Photoplay*, *Saturday Evening Post*, and other popular magazines. See also Hochheiser, "AT&T and Sound Motion Pictures," 30; and Crafton, *Talkies*, 23–61.

56. "Too Much 'Listening,'" *NYT* (25 July 1930): 16. See also "When Radio Answered a Call to Hollywood," *NYT* (10 August 1930): sect. 9, p. 12. As further evidence of the commodification of sound, AT&T's new sound motion picture subsidiary, Electrical Research Products Inc., was designated in corporate shorthand simply as "Products." An advertisement for Insulite Acoustile also claimed that, "Especially since the advent of the 'talkies' people have become 'sound conscious.'" *American Architect* 140 (September 1931): 109. Western Electric publicized and explained the operation of its new talkie technology through the cartoon *Finding His Voice*. Animated by Max Fleischer, this short film was a popular draw in theaters newly wired for sound in 1929 and 1930. See "Trailers Popular," *Erpigram* 2 (15 June 1930): 5.

57. F. R. Watson, "Ideal Auditorium Acoustics," *Journal of the American Institute of Architects* [*JAIA*] 16 (July 1928): 259–267.

58. Paul Sabine, "Acoustics of the Chicago Civic Opera House," *Architectural Forum* 52 (April 1930): 599–604, on 601.

59. Vern O. Knudsen, "Review of Architectural Acoustics during the Past Twenty-Five Years," *JASA* 26 (September 1954): 646–650, on 646. Knudsen was describing the Royal Festival Hall in London (Matthew & Martin, 1951), a post–World War II auditorium that represents the culmination of the type whose origins are the subject of this section. For more on the Royal Festival Hall, see Michael Forsyth, *Buildings for Music: The Architect, the Musician, and the Listener from the Seventeenth Century to the Present Day* (Cambridge, Mass.: MIT Press, 1985), 270–271.

60. Leo L. Beranek, *Music, Acoustics and Architecture* (New York: John Wiley and Sons, 1962), 102.

61. Forsyth, *Buildings for Music*, 289.

62. "Eastman Theatre and School of Music," *American Architect and Architectural Review* [*AAAR*] 123 (28 February 1923): 181–184. The school and its theater were founded by George Eastman of the Eastman Kodak Company, whose wealth was in part generated by the sale of film to the motion picture industry.

63. In 1923, Grauman's Egyptian Theatre in Hollywood (Meyer & Holler, 1922) was characterized as a "more or less standard design" of the "megaphone type." H. Roy Kelley, "Grauman Theatre, Hollywood, Cal." *AAAR* 123 (31 January 1923): 125–127, on 125.

64. Watson, "Ideal Auditorium Acoustics," 264.

65. The author worked as a recording engineer at the Eastman School in the early 1980s, and George Eastman's own seat, in the center of the mezzanine balcony, was reverently pointed out to her at that time. For more on the Eastman Theatre, see F. R. Watson, "Perfect Acoustics of the Eastman Theatre," *Motion Picture News* 27 (1923): 354, 358, "Acoustics of the Eastman Theatre, Rochester, N.Y.," *American Architect* 128 (1 July 1925): 31–34, and *Acoustics of Buildings* (New York: John Wiley and Sons, 1923), 49–51. See also Allen S. Crocker, "The Heating and Ventilating System and Soundproofing for the Eastman Theatre and School of Music," *AAAR* 123 (28 February 1923): 200–202.

66. Knudsen, "Review of Architectural Acoustics," 648.

67. F. R. Watson, "Optimum Conditions for Music in Rooms," *Science* 64 (27 August 1926): 209–210, on 209.

68. Watson, "Ideal Auditorium Acoustics," 265.

69. Watson, "Optimum Conditions," 210.

70. F. R. Watson, *Acoustics of Buildings*, 2d ed. (New York: John Wiley and Sons, 1930), 34.

71. Knudsen, "Review of Architectural Acoustics," 648.

72. Harvey Fletcher and J. C. Steinberg, "Articulation Testing Methods," *JASA* 1 (April 1930 Supplement): 1–97. Bell Labs researchers carrying out articulation tests on sound-recording equipment can be seen (and heard) in the film *The Birthplace of the Sound Motion Picture* (1930; Bell Telephone Laboratories), a short documentary about the Bell Labs facility at 463 West Street in New York City. AT&T Archives, #415-06-62.

73. Vern O. Knudsen, "The Hearing of Speech in Auditoriums," *JASA* 1 (October 1929): 56–82, on 64, 77. See also Knudsen, "Interfering Effect of Tones and Noise upon Speech Reception," *Physical Review* 26 (July 1925): 133–138.

74. "Open Air Acoustics," *Science* 63 (Supplement, 1926): x.

75. F. R. Watson, "Acoustics of Auditoriums," *Science* 67 (30 March 1928): 335–338, on 337.

76. Watson, *Acoustics of Buildings* (1930), 59.

77. Watson, "Ideal Auditorium Acoustics," 259; Knudsen, *Architectural Acoustics*, 501–504.

78. Arthur T. North, "The Orchestra Shell of the Hollywood Bowl," *Architectural Forum* 51 (November 1929): 549–552, on 549.

79. The history of the Hollywood Bowl is well outlined and illustrated in the souvenir book, "Hollywood Bowl" (Los Angeles: Los Angeles Philharmonic Assoc., 2000), edited by Carol Merrill-Mirsky with text by Merrill-Mirsky and Jeannette Bovard. The historical literature on the Hollywood Bowl is otherwise primarily sentimental. See Isabel Morse Jones, *Hollywood Bowl* (New York: G. Schirmer, 1936); n.a., *The Hollywood Bowl* (Los Angeles: Pepper Tree Press, 1939); John Orlando Northcutt, *Magic Valley: The Story of Hollywood Bowl* (Los Angeles: Osherenko, 1967); Grace G. Koopal, *Miracle of Music* (n.p.: Charles E. Toberman, 1972); and Michael Buckland and John Henken, eds., *The Hollywood Bowl: Tales of Summer Nights* (Los Angeles: Balcony Press, 1996). For more on Lloyd Wright's designs, see David Gebhard and Harriette Von Breton, *Lloyd Wright, Architect: 20th Century Architecture in an Organic Exhibition* (1971; Santa Monica: Hennessey and Ingalls, 1998).

80. "Innovations in Building New Bowl Shell," *Hollywood Citizen,* special ed. (29 June 1929), Tear-sheet Volumes A and B, 1929-Bowl, Los Angeles Philharmonic Archives [LAP Archives]. Many thanks to Steven Lacoste for providing me with copies of reviews from these volumes.

81. Knudsen, *Architectural Acoustics*, 505; Herbert Klein, "Bowl Ready—Waits Opening," *Los Angeles Record* (6 July 1929), LAP Archives.

82. Herbert Klein, "The Bowl," *Los Angeles Record* (20 July 1929), LAP Archives. Knudsen himself pointed out that auditors in certain seats, particularly those at the far sides, heard an imbalanced sound owing to the reflective action of the shell, but, he argued, that same action rendered sound "to very good advantage" in the vast majority of seats. Knudsen, *Architectural Acoustics*, 509. See also Knudsen's interview with Isabel Morse Jones, "Hearing in Hollywood," *Los Angeles Times* (21 July 1929), LAP Archives. Jones was the most vehement of those who criticized the new shell, but her views fluctuated widely during its first season, and even she ultimately conceded in 1936 that "The design pleased everyone." Jones, *Hollywood Bowl*, 136. See also her reviews in the *Los Angeles Times* for 1929, LAP Archives; and Koopal, *Miracle of Music*, 137.

83. North, "Orchestra Shell of Hollywood Bowl," 551.

84. The following survey of motion picture theater architecture, music, and audiences draws from: David Naylor, *American Picture Palaces: The Architecture of Fantasy* (New

York: Van Nostrand Reinhold, 1981); Maggie Valentine, *The Show Starts on the Sidewalk: An Architectural History of the Movie Theatre* (New Haven: Yale University Press, 1994); Q. David Bowers, *Nickelodeon Theatres and their Music* (Vestal, N.Y.: Vestal Press, 1986); James P. Kraft, *Stage to Studio: Musicians and the Sound Revolution, 1890–1950* (Baltimore: Johns Hopkins University Press, 1996); Kevin Brownlow, *The Parade's Gone By . . .* (Berkeley: University of California Press, 1968); Gomery, "Coming of Sound" and *Shared Pleasures*; Hall, *Best Remaining Seats*; and Robinson, *Peep Show to Palace*.

85. See, for example, Charles A. Whittemore, "The Moving Picture Theatre," *Brickbuilder* 23 (February 1914 Supplement): 41–45; and his series, "The Motion Picture Theater," *Architectural Forum* 26 (June 1917): 171–176, 27 (July 1917): 13–18, 27 (August 1917): 39–43, 27 (September 1917): 67–72. See also John J. Klaber, "Planning the Motion Picture Theatre," *Architectural Record* 38 (November 1915): 540–554; and Arthur S. Meloy, *Theatres and Motion Picture Houses* (New York: Architects' Supply and Publishing Co., 1916).

86. These song slides were particularly useful in theaters equipped with just one projector, as the songs filled gaps in the program that occurred during the frequent changing of reels.

87. The Marr & Colton instruments offered mood stops labeled "Love (Mother)," "Love (Passion)," "Pathetic," "Riot," and many others. Hall, *Best Remaining Seats*, 192.

88. Architects were advised in 1915 that "seats need not be upholstered, as the performances are usually short" and that the lighting of the hall "should be sufficient so that spectators may be able to enter or leave at any time." Klaber, "Planning the Motion Picture Theatre," 546–547.

89. Complaining about the state of motion picture music in 1915, Vachel Lindsay wrote that the musicians "make tunes that are most squalid and horrible. With fathomless imbecility, hoochey koochey strains are on the air while heroes are dying. The Miserere is in our ears when the lovers are reconciled. Ragtime is imposed upon us while the old mother prays for her lost boy." *The Art of the Moving Picture* (New York: Macmillan, 1915), 191–192.

90. See, for example, Erno Rapée, *Motion Picture Moods for Pianists and Organists: A Rapid-Reference Collection of Selected Pieces, Adapted to Fifty-Two Moods and Situations* (New York: G. Schirmer, 1924).

91. Gomery, "Coming of Sound," 91–93.

92. Robinson, *Peep Show to Palace*, 174; Gomery, "Coming of Sound," 64. For an excerpt from a cue sheet, see Brownlow, *Parade's Gone By*, 338. As Gomery emphasizes, within this context Warner Brothers' goal for their early Vitaphone sound films, to provide standardized, high-quality scores for otherwise silent pictures, was more evolutionary than revolutionary.

93. F. R. Watson, "Acoustics of Motion Picture Theaters," *Transactions of the Society of Motion Picture Engineers* [*TSMPE*] 11 (1927): 641–650, on 641. Watson wrote this paper

just as sound film was beginning to be heard, but he did not focus upon the transformation taking place. He instead described general aspects of traditional auditorium design, and briefly noted that, for the new sound motion picture theaters, "it is only necessary to consider that the speaker or musician is replaced by the instrument reproducing the recorded sound," 647.

94. See, for example, Whittemore, "Moving Picture Theatre," 43–44, who mentioned only the need to isolate the noisy projector from the audience. As theater organs became popular, advice on how to maximize their effect was offered. Theater organ pipes were housed in closed chambers and volume was controlled by a "Venetian swell," adjustable louvers on the front of the box, to modulate the escape of sound. Architects were advised on the placement and soundproofing of these chambers, to ensure maximum volume range and effect. See, for example, Whittemore, "Motion Picture Theater" (part III); and Robert E. Hall, "Organ Installation in Theaters," *Architectural Forum* 42 (June 1925): 401–402.

95. Clifford M. Swan, "Acoustics of Picture Theaters," *Architectural Forum* 51 (November 1929): 545–548, on 545.

96. Swan, "Acoustics of Picture Theaters," 545. See also Edwin E. Newcomb, "Theatre Acoustics," pp. 41–46 in vol. 2 of R. W. Sexton, ed., *American Theatres of Today* (New York: Architectural Book Publishing Co., 1930).

97. For example, the famed atmospheric theaters of John Eberson, which had been lavishly praised for years, suddenly became "an exaggerated case of faulty acoustic design." Eberson himself argued that the theater owner should "re-decorate, re-equip, re-construct, rip out the insides and build up from the four remaining walls" in order to solve the new problem of acoustics. In his 1932 textbook, Vern Knudsen illustrated a less drastic solution, in which the characteristic plaster ceiling of the Eberson-designed Loew's Theater in Akron, Ohio, was now covered with Macoustic Sound-Absorbing Plaster. G. T. Stanton, "Theater Acoustics, Ventilating and Lighting," *Architectural Record* 68 (July 1930): 87–93, on 87; "Monuments," *Film Daily* 47 (4 January 1929): 1; Knudsen, *Architectural Acoustics*, 532.

98. "Enter the Acoustics Expert!" *Cinema Construction* (September 1929): 5. On page 21 of this issue, the journal began a new series, "Acoustics in the Cinema," offering "practical guidance in a science which, hitherto neglected, has suddenly assumed immense importance."

99. A list of new hires, indicating previous affiliations, was printed in the first edition of the company's newsletter. See "Training School Holds Banquet at Term's End," *Erpigram* 1 (20 December 1928): 1, 3. See also "What Is ERPI?" *Western Electric News* 18 (April 1929): 3–5. Individuals from a variety of backgrounds sought to become ERPI sound men. One unsolicited applicant listed his "salient qualifications" as the following: "he was a blacksmith—he knew something about patent law—he wanted to invent something—he owned a copy of Drake Radio Cyclopedia—he owned a book on

motion picture photography that cost him six dollars—he has an eighty dollar suit and a Java Panama Hat." "Fifty Seventh's Varieties," *Erpigram* 1 (15 August 1929): 7.

100. S. K. Wolf, "Installation School Has Proved Valuable Training Ground for Incoming Engineers," *Erpigram* 1 (20 March 1929): 2, 4.

101. T. L. Dowey, "Sound Laboratories Claim Attention of ERPI's Acoustic Expert Gone A-Visiting," *Erpigram* 1 (20 May 1929): 5.

102. Wolf's hires included Dr. R. E. Martin, a physicist from Lehigh University, and John Chambers, a sales engineer from Johns-Manville. "Fifty-Seventh's Varieties," *Erpigram* 1 (20 July 1929): 11; and "Fifth-Seventh's Varieties," *Erpigram* 1 (15 November 1929): 8. See also G. T. Stanton, "Pioneer in Picture House Acoustics," *Erpigram* 1 (15 October 1929): 1, 4; and "Acoustic Tests for 75 Theatres a Week," *Erpigram* 1 (15 December 1929): 3.

103. "Acoustic Tests for 75 Theatres a Week," 3. See also Stanton, "Pioneer in Picture House Acoustics." A blank "Theatre Survey" form is preserved in the Western Electric Collection, W. H. Schlasman Files, Box 340, Loc. 100-09-01, Folder 3, AT&T Archives. For a full description of an ERPI inspection, see Howard B. Santee, "Installation and Adjustment of Western Electric Sound-Projector Systems," *Bell Laboratories Record* 7 (November 1928): 112–116. RCA's approach is represented in Harry B. Braun, "Sound Motion Picture Requirements," *Architectural Forum* 57 (October 1932): 381–385.

104. "Inspectors War on Enemies of True Sound," *Erpigram* 2 (15 January 1930): 4. See also "Dr. ERPI Takes a Bow," *Western Electric News* 19 (December 1930): 3–5. The three ERPI engineers sent to inspect the Loew's Theatre in Canton, Ohio, may have wished their pistols were loaded with bullets instead of caps when they heard a loud roar coming from the screen. "After considerable time spent in trying to trace the noise through the circuit," they discovered six caged lions backstage. "They Were 'Lyin' Horns," *Erpigram* 1 (20 July 1929): 5.

105. Swan, "Acoustics of Picture Theaters," 545. Few reverberation measurements of motion picture theaters appear to have been made in the pretalkie era, and the ERPI theater surveys have not survived to provide this information. Rick Altman has suggested that a reverberation time of three seconds was typical. Altman, "The Material Heterogeneity of Recorded Sound," pp. 15–31 in Altman, ed., *Sound Theory/Sound Practice* (New York: Routledge, 1992), 28. See also F. L. Hopper, "The Measurement of Reverberation Time and Its Application to Acoustic Problems in Sound Pictures," *JASA* 2 (April 1931): 499–505, figure 4 on 503.

106. "A Technical Inspector Speaks for Himself," *Erpigram* 2 (1 March 1930): 7. The standard per-person absorption coefficient in the late 1920s was 4.7 units.

107. Edward W. Kellogg, "Some New Aspects of Reverberation," *JSMPE* 14 (January 1930): 96–107.

108. Kellogg, "New Aspects of Reverberation," 102.

109. S. K. Wolf, "Theater Acoustics for Sound Reproduction," *JSMPE* 14 (February 1930): 151–160, on 159 (figure 6). See also Wolf, "Theatre Acoustics for Reproduced Sound," pp. 73–84 in *Academy Technical Digest: Fundamentals of Sound Recording and Reproduction for Motion Pictures* (Hollywood: Academy of Motion Picture Arts and Sciences, 1929–30).

110. These values have been read off the graph that is figure 7 in Wolf, "Theatre Acoustics for Reproduced Sound," 83. Seating capacity figures come from Wolf, "Reproduction in the Theatre," pp. 165–183 in *Academy Fundamentals of Sound Recording*, on 171.

111. "One of the leading and very best acoustical Talkie Auditoriums" in England was cited as having a reverberation time of just under one second. "The Plaza, Coventry," *Cinema Construction* (January 1930): 19.

112. Santee, "Installation of Western Electric Systems"; Knudsen, *Architectural Acoustics*, 526–533.

113. Gomery, "Coming of Sound," 283–285.

114. Joseph Maxfield, discussion following Paul E. Sabine, "The Acoustics of Sound Recording Rooms," *TSMPE* 12 (1928): 809–822, on 821. See also Wesley C. Miller, "The Illusion of Reality in Sound Pictures," pp. 101–108 in *Academy Fundamentals of Sound Recording*, on 104; and Knudsen, *Architectural Acoustics*, 530.

115. Stanton, "Theatre Acoustics, Ventilating and Lighting," 88.

116. Edward W. Kellogg, "Means for Radiating Large Amounts of Low Frequency Sound," *JASA* 3 (July 1931): 94–110.

117. Read and Welch note that the first successful acoustic recordings of a full-sized symphony orchestra were not made until 1924. *Tin Foil to Stereo*, 257.

118. For accounts and illustrations of acoustic recording sessions, see "The Manufacture of Edison Phonograph Records," *Scientific American* 83 (22 December 1900): 390 and cover illus.; "Making Phonograph Music," *Current Literature* 33 (1902): 169–170; Yvonne de Treville, "Making a Phonograph Record," *Musician* 21 (November 1916): 658; George Hamlin, "The Making of Records," *Musician* 22 (July 1917): 542; and Austin C. Lescarboura, "At the Other End of the Phonograph," *Scientific American* 119 (31 August 1918): 164, 178, and cover illus. See also photographs in Read and Welch, *Tin Foil to Stereo*, 493–494; and Welch and Burt, *Tin Foil to Stereo: Acoustic Years*, figure 24, following p. 102. The black partitions in the Edison studio depicted in figure 6.13 probably provided some absorption. David Morton had noted that the Edison company was known for making relatively dead, or nonreverberant, recordings, and that Victor in particular favored live conditions. Morton, *Off the Record*, 20.

119. See photos in Baker, "Description of General Electric Station," 365; and D. G. Little, "KDKA: The Radio Telephone Broadcasting Station of the Westinghouse

Electric and Manufacturing Company at East Pittsburgh, Pennsylvania," *PIRE* 12 (June 1924): 255–276, on 273 and 275; and Joseph Maxfield, "Electro-Mechanical Sound Recording," *Bell Laboratories Record* 1 (January 1926): 197–202, on 199.

120. Paul Sabine, "Acoustics of Sound Recording Rooms," 813–814.

121. Maxfield, "Electro-Mechanical Sound Recording," 198.

122. Maxfield's studies of binaural hearing at Bell Labs would soon lead to experiments with stereophonic recording and reproduction. See "The Reproduction of Orchestral Music in Auditory Perspective" (Bell Telephone Laboratories, 1933), #154-04-01, AT&T Archives; and Robert E. McGinn, "Stokowski and the Bell Telephone Laboratories: Collaboration in the Development of High-Fidelity Sound Reproduction," *Technology and Culture* 24 (January 1983): 38–75.

123. As noted above, the reverberation of Chicago's WLS studio in 1926 was no more than 0.64 seconds. NBC engineers recommended a reverberation time of 0.70 seconds for a studio of 2,000–3,000 square feet, and 1.10 seconds for a large studio of 100,000 square feet. ERPI engineers recommended about 0.80 seconds for a studio of 10,000 square feet and 1.10 seconds for 100,000 square feet. O. B. Hanson and R. M. Morris, "The Design and Construction of Broadcast Studios," *PIRE* 19 (January 1931): 17–34, on 34 (NBC); G. T. Stanton and F. C. Schmid, "Acoustics of Broadcasting and Recording Studios," *JASA* 4 (July 1932): 44–55, figure 3 (ERPI).

124. Raymond Hood, "The National Broadcasting Studios, New York," *Architectural Record* 64 (July 1928): 1–6, 25–37, quotes on 1.

125. For examples of structural isolators, see *Sweet's Architectural Trade Catalogue* (1931): B2667 (Johns-Manville System of Sound Isolation); and B2676 (U.S.G. System of Sound Insulation).

126. "Engineering Problems of Radio Broadcasting Studio Design," *American Architect* 135 (5 February 1929): 195–203, on 196. The term was also employed by Hanson and Morris in "Design and Construction of Broadcast Studios," 20.

127. Eyman, *Speed of Sound*, 77–78. See also Stanley Watkins, "Madam, Will You Talk?" *Bell Laboratories Record* 24 (August 1946): 289–295, on 292.

128. *The Voice from the Screen* (1926; Bell Telephone Laboratories, dir. Edward B. Craft, Video Yesteryear #877); Eyman, *Speed of Sound*, 86; Watkins, "Madam, Will You Talk?" 293.

129. Frank Woods, "The Sound Motion Picture Situation in Hollywood," *TSMPE* 12 (1928): 625–632, on 628.

130. Vern Knudsen, *Teacher, Researcher, and Administrator: Vern O. Knudsen*, transcript of oral history conducted 1966–1969 by James V. Mink. Collection 300/101. Department of Special Collections, Young Research Library, University of California, Los Angeles. Quote on 652.

131. Knudsen, *Architectural Acoustics*, 574–577; H. C. Humphrey, "Typical Sound Studio Recording Installations," *TSMPE* 13 (1929): 158–172; and A. S. Ringel, "Sound-Proofing and Acoustic Treatment of RKO Stages," *JSMPE* 15 (September 1930): 352–369.

132. "Mr. Maxfield," discussion after Ringel, "Sound-Proofing of RKO Stages," 368. See also J. P. Maxfield, "Acoustic Control of Recording for Talking Motion Pictures," *JSMPE* 14 (January 1930): 85–95, on 88.

133. A soundstage built by General Electric in Schenectady achieved a reverberation time of 0.60 seconds through the extensive use of Sabinite plaster. By installing portable sound-absorbing panels called "gobos," the reverberation time could be lowered to as little as 0.24 seconds. John Klenke, "Construction of a Sound-Proof Studio for Making Sound Motion Pictures," *American Architect* 140 (October 1931): 46–47, 66. Bell Labs built a soundstage for research with a reverberation time of just 0.35 seconds at 500 cps. Carl F. Eyring, "Reverberation Time in 'Dead' Rooms," *JASA* 1 (January 1930): 217–241, on 239 (figure 13). Perhaps more typical of commercial production sites was the soundstage at RKO, with a reverberation time of about 0.65 seconds at 500 cps. Ringel, "Sound-Proofing of RKO Stages," 363 (figure 6).

134. The material was held in place with a cloth covering or with chicken wire. Early practice had employed cotton batting or other vegetable fibers, but the danger posed by such materials was made tragically evident on 10 December 1929, when an incandescent lamp ignited a fire during the filming of a musical at the Pathé Manhattan Studios in New York. Ten people died, and the subsequent report of the New York Board of Fire Underwriters noted that "the production of sound pictures has developed an additional hazard in the use of considerable quantities of draperies, property and sound-deadening material." "Finds Pathe Fire Was Preventable," *NYT* (21 December 1929): 40. See also "10 Die, 18 Hurt in Film Studio Blaze: Panic Costs Lives of Four Chorus Girls as Quick Fire Cuts Off All but One Exit," *NYT* (11 December 1929): 1–2; and "10 Dead, 20 Hurt in Pathe Studio Fire," *Film Daily* 50 (11 December 1929): 1, 9.

135. Humphrey, "Typical Sound Studio Installations," 160.

136. Little, "KDKA," 256. See also Porter H. Evans, "A Comparison of the Engineering Problems in Broadcasting and Audible Pictures," *PIRE* 18 (August 1930): 1316–1337, on 1318.

137. Hanson and Morris, "Design of Broadcast Studios," 30.

138. Humphrey, "Typical Studio Installations," 160.

139. Kellogg, "New Aspects of Reverberation," 105.

140. See, for example, Maxfield's comments following Ringel, "Sound-Proofing of RKO Stages," 368.

141. Miller, "Illusion of Reality," 108.

142. In 1929, the academy initiated a series of educational lectures on sound and the new motion picture technology, and the talks were published as individual technical papers as well as in compilations. See *Academy Technical Digest: Fundamentals of Sound Recording and Reproduction for Motion Pictures* (Hollywood: Academy of Motion Picture Arts and Sciences, 1929–1930); and Lester Cowan, ed. *Recording Sound for Motion Pictures* (New York: McGraw-Hill, 1931). The *Cinematographic Annual* for 1930 and 1931 also included many papers on sound filmmaking, as did, of course, the *Transactions* (1927–1929) and *Journal* (1930+) of the Society of Motion Picture Engineers. The inaugural meeting of the Acoustical Society of America, held in New York City in May 1929, opened with a joint session with the Society of Motion Picture Engineers. "Program," *JASA* 1 (October 1929): 25.

143. My examination of these ideas in the pages that follow is indebted to the thoughtful analyses found in the essays collected in Altman, ed., *Sound Theory/Sound Practice*; and Weis and Belton, eds., *Film Sound*; as well as Michel Chion, *Audio-Vision: Sound on Screen,* trans. Claudia Gorbman (New York: Columbia University Press, 1994); James Lastra, *Sound Technology and American Cinema: Perception, Representation, Modernity* (New York: Columbia University Press, 2000); Crafton, *Talkies*; and Eyman, *Speed of Sound*. By placing the work of early motion picture sound engineers in the context of the larger soundscape in which they were operating, my own account seeks to provide a wider perspective from which to understand the ideas and actions of those engineers.

144. Paul Sabine, "Acoustics of Sound Recording Rooms," 809. See also Watson, "Acoustics of Motion Picture Theaters."

145. *The Lights of New York* (1928; Warner Bros., dir. Bryan Foy), UCLA Film and Television Archive, #F61-L3-11-17. The quote is taken from one of the intertitles of the film.

146. L. T. Robinson, "Some Thoughts about Motion Pictures with Sound," *TSMPE* 12 (1928): 856–866, on 856.

147. See, for example, Frank S. Crowhurst, "Acoustic Linings for Soundproof Motion Picture Stages and Sets," *TSMPE* 12 (1928): 828–835.

148. This term appears throughout the technical literature. See, for example, Miller, "Illusion of Reality," 105.

149. Richard Koszarski notes that a volume change of this type accompanied a cut in the Marion Talley Vitaphone short that preceded *Don Juan* at its premiere in 1926. Critics were "baffled" by this change, and the film—which was poorly received for a variety of reasons—was dropped from the program. Koszarski, "On the Record," 18.

150. Absorption coefficients determined by Wallace Sabine, Floyd Watson, and Clifford Swan were published in Crowhurst, "Acoustic Linings for Soundproof Stages." The quote is taken from J. P. Maxfield and Ralph Townsend, "Acoustic Measurements of Set Materials," Report #4 of the Academy Producers-Technicians Committee

(Hollywood: Academy of Motion Picture Arts and Science, 1 September 1930), 2. See also R. L. Hanson, "One Type of Acoustic Distortion in Sound Picture Sets," *JSMPE* 15 (October 1930): 460–472.

151. See Maxfield and Townsend, "Acoustic Measurements of Set Materials," which reports upon the work of Knudsen and Hopper.

152. "Mr. Kellogg," discussion following Paul Sabine, "Acoustics of Sound Recording Rooms," 821; "Mr. Maxfield," discussion following Ringel, "Sound-Proofing of RKO Stages," 368; "Mr. Coffman," discussion following R. L. Hanson, "Acoustic Distortion," 471.

153. John L. Cass, "The Illusion of Sound and Picture," *JSMPE* 14 (March 1930): 323–326, on 325.

154. Donald Crafton notes that, in *The Jazz Singer*, a recording of Cantor Josef Rosenblatt singing the Kol Nidre was rerecorded onto a second disc as Al Jolson was filmed lip-syncing to the Rosenblatt record. Crafton, *Talkies*, 240. For more on the role of the "mixer man" or "monitor man," see C. A. Tuthill, "The Art of Monitoring," *TSMPE* 13 (1929): 173–177; K. F. Morgan, "Scoring, Synchronizing and Re-Recording Sound Pictures," *TSMPE* 13 (1929): 268–285; Carl Dreher, "Sound Personnel and Organization," pp. 85–96 in *Academy Fundamentals of Sound Recording*; and Dreher, "Recording, Re-Recording and Editing of Sound," *JSMPE* 16 (June 1931): 756–765.

155. For sound-on-film production, the sound and image were recorded onto separate strips of film that were combined only when the master negative was created to generate the final release prints. Crafton, *Talkies*, 239–241; Barry Salt, "Film Style and Technology in the Thirties: Sound," pp. 37–43 in Weis and Belton, eds., *Film Sound*, on 39–42.

156. Silent's innovation was the realization of an idea originally proposed by Kenneth Morgan. According to D. P. Loye and Joseph Maxfield, numerous others were also working on the problem at this time. See Loye and Maxfield, "Sound in the Motion Picture Industry I. Some Historical Recollections," *Sound* 2 (September–October 1963): 14–27, on 18–19.

157. Scott Eyman quotes Warner Brothers sound engineer George Groves's description of this system: "It was like a telephone dialing system. A system of relays and selector switches was used that would release the turntables, and they would start to spin on preset cues. If there were more records to be dubbed than available turntables, a crew of men would stand there with the next records in a rack, select the next record in sequence, take the old one off, and reset the footage counter so that each turntable operated at the right time." Eyman, *Speed of Sound*, 203–204. See also Morgan, "Scoring, Synchronizing and Re-Recording," 270–271.

158. Dreher, "Recording, Re-Recording and Editing of Sound."

159. Joe W. Coffman, "Art and Science in Sound Film Production," *JSMPE* 14 (February 1930): 172–179, on 178–179.

160. Crafton, *Talkies*, 236. Rick Altman also emphasizes this point in "Sound Space," pp. 46–64 in Altman, ed., *Sound Theory/Sound Practice*, on 53.

161. Carl Dreher, "Microphone Concentrators in Picture Production," *JSMPE* 16 (January 1931): 23–30, on 23 and 24.

162. Harry Olson and Irving Wolff, "Sound Concentrator for Microphones," *JASA* 1 (April 1930): 410–417, on 417.

163. Dreher, "Microphone Concentrators in Picture Production," describes the use of the new technology on these films. See *Danger Lights* (1930; RKO Radio Pictures, dir. George B. Seitz, Nostalgia Family Video, #1722); and *Cimarron* (1930; RKO Radio Pictures, dir. Wesley Ruggles, MGM/UA Home Video, #M300400).

164. Crafton, *Talkies*, 238.

165. Harry F. Olson, "The Ribbon Microphone," *JSMPE* 16 (June 1931): 695–708; Olson, "Mass Controlled Electrodynamic Microphones: The Ribbon Microphone," *JASA* 3 (July 1931): 56–68.

166. Silent films from the late 1920s, such as *The Crowd*, had evoked the urban soundscape with dynamic visual montages, and it is likely that theater musicians provided aurally appropriate accompaniment to these scenes. The recorded sound score released with *Sunrise* includes such an accompaniment of nonsynchronized city sounds, including shouting voices. *The Crowd* (1928; Metro-Goldwyn-Mayer, dir. King Vidor, MGM/UA Home Video, #301357); *Sunrise* (1927; Fox, dir. F. W. Murnau, Critics Choice Video, #4021). Examples of early talkies with scenes and sounds of city noises include *Say It with Songs* (1929; Warner Bros., dir. Lloyd Bacon), *Applause* (1929; Paramount Famous Lasky, dir. Rouben Mamoulian), and *Thunderbolt* (1929; Paramount Famous Lasky, dir. Joseph von Sternberg), all at Celeste Bartos Film Study Center, Museum of Modern Art, New York.

167. Morgan, "Scoring, Synchronizing and Re-Recording," 282.

168. Kellogg, "New Aspects of Reverberation," 105. See also "How Echoes Are Produced: NBC Engineers Perfect Artificial Sound Reflection," *Broadcast News* 13 (December 1934): 26–27. Thanks to Susan Schmidt Horning for providing me with this article.

169. Kellogg, "New Aspects of Reverberation," 105.

170. O. B. Hanson, "Microphone Technique in Radio Broadcasting," *JASA* 3 (July 1931): 81–93, on 91. Here, the primary signal was a nonreverberant pick-up obtained from two microphone concentrators mounted on the side walls of the auditorium. The reverberatory signal was not generated in a special chamber, but instead obtained from the theater itself, by means of a standard, omnidirectional microphone also mounted on

a wall. This natural reverberation was, however, mixed in an artificially controlled manner with the primary signal to create what Hanson himself characterized as "artificial reverberation."

171. *The Black Watch* (1929; Fox, dir. John Ford), Celeste Bartos Film Study Center, Museum of Modern Art; *Platinum Blonde* (1931; Columbia, dir. Frank Capra, Columbia Tri-Star Home Video, #86973).

172. Arthur Knight, "The Movies Learn to Talk: Ernst Lubitch, René Clair, and Rouben Mamoulian," pp. 213–220 in Weis and Belton, eds., *Film Sound*, on 219; *Dr. Jekyll and Mr. Hyde* (1932; Paramount Publix, dir. Rouben Mamoulian, MGM/UA Home Video, #M201462). Mamoulian, who had previously been the musical director for motion picture presentations at the Eastman Theatre, was equally creative in his first talking film, *Applause* (1929). See Lucy Fischer, "*Applause*: The Visual and Acoustic Landscape," pp. 232–246 in Weis and Belton, eds., *Film Sound*; and "Rouben Mamoulian, Director" (reminiscences), pp. 85–97 in Evan William Cameron, ed., *Sound and the Cinema: The Coming of Sound to American Film* (Pleasantville, N.Y.: Redgrave, 1980).

173. Altman, "Sound Space," 53.

174. Examples of volume level changes synchronized with image cuts do exist, but tellingly, these films were primarily made before 1931, when the relevant issues were still in play and the close-up standard was not yet fully established. It is difficult to find a sound track after 1931 that varies from this standard. Examples prior to 1931 include the opening scene of *Disraeli,* in which a long shot of a political orator in Hyde Park cuts to increasingly closer shots of the man, with an abrupt volume level increase at each cut; and the aerial scenes of fighting airplanes in *Hell's Angels*, in which the volume level of the engines' drones changes accordingly as the image cuts toward and away from the planes. *Disraeli* (1929; Warner Bros., dir. Alfred E. Green, MGM/UA Home Video, #M202143); *Hell's Angels* (1930; Caddo/United Artists, dir. Howard Hughes and James Whale, MCA Universal Home Video, #80638).

175. Crafton, *Talkies*, 355–380; James Lastra, "Reading, Writing, and Representing Sound," pp. 65–86 in Altman, ed., *Sound Theory/Sound Practice*, on 82.

176. Lastra, "Reading, Writing, Representing Sound," 82; Cass, "Illusion of Sound and Picture," 325.

177. Carl Dreher noted in 1930 that approximately 80 percent of the sound men working in Hollywood came from outside the film industry, primarily from the fields of radio and telephone engineering. Dreher, "Sound Personnel and Organization," 86.

178. James Lastra, drawing heavily on ideas expressed by Joseph Maxfield about the phonographic recording of concert music, identifies a somewhat different acoustical paradigm, one based on a model of "the invisible auditor," for engineers from the electroacoustical industries. He thus tells a different story about how this paradigm was trans-

formed into the standard sound track of the thirties. See Lastra, *Sound Technology and the American Cinema*, and "Standards and Practices: Aesthetic Norm and Technological Innovation in the American Cinema," pp. 200–225 in Janet Staiger, ed., *The Studio System* (New Brunswick: Rutgers University Press, 1995).

179. Altman, "Sound Space," 61.

180. Kellogg, "New Aspects of Reverberation," 101.

181. "Sound Picture Laboratory" (Bell Telephone Laboratories, 1930), 3. AT&T Archives, #154-04-01.

182. T. E. Shea, "A Modern Laboratory for the Study of Sound Picture Problems," *JSMPE* 16 (March 1931): 277–285, on 278. See also "Research Plant Finest of Kind," *Erpigram* 1 (1 October 1929): 1, 8; and "Quiet on the Set," *Bell Telephone Laboratories Reporter* 14 (May/June 1965): 28–31.

183. "Quiet on the Set," 30.

184. Eyring was born in 1889 in Juarez, Mexico. He received his B.A. from Brigham Young University in 1913, and his 1924 Ph.D. was earned under Robert Millikan's supervision at the California Institute of Technology. His sabbatical at Bell Labs lasted from 1929 to 1931, after which he returned to Brigham Young where he taught physics and mathematics until his death in 1951. In addition to numerous technical papers, he also wrote an undergraduate physics textbook and several volumes on Mormon life and thought. Vern O. Knudsen, "Carl F. Eyring," *JASA* 23 (May 1951): 370–371; "Eyring, Carl," *American Men of Science,* 8th ed. (Lancaster, Penn.: Science Press, 1949), 742.

185. These absorption coefficients for sound at 512 cps are taken from Wallace Sabine, "Reverberation," *Collected Papers on Acoustics* [*CPA*] (Cambridge, Mass.: Harvard University Press, 1922), 3–68, on 56.

186. Reverberation times are from Sabine, "Reverberation," *CPA*, 28–31. The reference to hundreds of reflections is found on p. 39.

187. The absorption coefficient of the rock wool material (for 500 cps) and the reverberation time are taken from Carl F. Eyring, "Reverberation Time in 'Dead' Rooms," *JASA* 1 (January 1930): 217–241, on 238 (table 1) and 239 (figure 13). The reverberation time cited is the average of two measurements, one with the drapes spread out, another with them gathered together.

188. Eyring indicated in his article that the German researchers K. Schuster and E. Waetzmann were the first to note that Sabine's formula was implicitly applicable only to live rooms. Paul Sabine pointed out, in 1939, that Eyring's revision was stimulated by a discussion at the May 1929 meeting of the Acoustical Society of America, in which R. F. Norris suggested the modification to Sabine's equation that Eyring subsequently developed. The revised formula is sometimes referred to as the "Eyring-Norris" equation, but the issue of assigning priority is not particularly relevant for my account.

Indeed, the simultaneous discovery by numerous individuals of the limitations of Sabine's equation circa 1929 only highlights the fact that no such limitation had been identified by the many investigators who had employed the equation in the years between 1900 and 1929. See K. Schuster and E. Waetzmann, "Über den Nachhall in geschlossenen Räumen," *Annalen der Physik*, 5th ser., 1 (12 March 1929): 671–695; Paul E. Sabine, "Architectural Acoustics: Its Past and Its Possibilities," *JASA* 11 (July 1939): 21–28, on 26; and Knudsen, *Architectural Acoustics*, appendix II, pp. 603–605 (for Norris's own derivation of the formula).

189. The apparatus employed by Eyring is more fully described in E. C. Wente and E. H. Bedell, "A Chronographic Method of Measuring Reverberation Time," *JASA* 1 (April 1930): 422–427, quote on 422.

190. Eyring, "Reverberation Time in Dead Rooms," 217.

191. Watson, "Ideal Auditorium Acoustics," 260–261. For earlier examples of the use of acoustical images, see Adolphe Ganot, *Elementary Treatise on Physics*, trans. Edmund Atkinson (London: Longmans, Green and Co., 1867); and Robert A. Millikan and John Mills, *Electricity, Sound, and Light* (Boston: Ginn and Co., 1908), 245–257.

192. Eyring, "Reverberation Time in Dead Rooms," 225.

193. Eyring, "Reverberation Time in Dead Rooms," 230.

194. See, for example, Stanton, "Theater Acoustics, Ventilating and Lighting," 88; Vern Knudsen, "Acoustics of Music Rooms," *JASA* 2 (April 1931): 434–467, on 444; and Knudsen, *Architectural Acoustics*, 128–130. Knudsen and Hopper also used the Eyring equation in their study of the sound-absorbing properties of motion pictures set materials. See Maxfield and Townsend, "Acoustic Measurements of Set Materials," 3.

195. W. A. MacNair, "Some Acoustical Problems of Sound Picture Engineering," *PIRE* 19 (September 1931): 1606–1614, on 1606.

196. MacNair, "Acoustical Problems of Sound Picture Engineering," 1607. See also Ringel, "Sound-Proofing of RKO Stages," 361.

Chapter Seven

1. The phrase is from David Loth, *The City within a City: The Romance of Rockefeller Center* (New York: William Morrow and Co., 1966). The following survey of the history of Rockefeller Center also draws upon: Robert A. M. Stern, Gregory Gilmartin, and Thomas Mellins, *New York 1930: Architecture and Urbanism Between the Two World Wars* (New York: Rizzoli, 1987); Alan Balfour, *Rockefeller Center: Architecture as Theater* (New York: McGraw-Hill, 1978); Carol Herselle Krinsky, *Rockefeller Center* (Oxford: Oxford University Press, 1978); William H. Jordy, *The Impact of European Modernism in the Mid-Twentieth Century* (New York: Oxford University Press, 1972); and Winston Weisman, "Who Designed Rockefeller Center?" *Journal of the Society of Architectural Historians* 10 (March 1951): 11–17.

2. Henry-Russell Hitchcock, "The Architecture of Bureaucracy and the Architecture of Genius," *Architectural Review* (January 1947): 2–6; quoted in Jordy, *Impact of European Modernism*, 55. The firms that made up the Associated Architects were: Corbett, Harrison & MacMurray; Hood, Godley & Fouilhoux; and Reinhard & Hofmeister. See also Raymond Hood, "The Design of Rockefeller Center," *Architectural Forum* 56 (January 1932): 1–8; and L. Andrew Reinhard, "Organization for Cooperation," *Architectural Forum* 56 (January 1932): 78–81.

3. Quoted in Stern et al., *New York 1930*, 638.

4. Alfred N. Goldsmith, "An Entertainment City," *Journal of the Society of Motion Picture Engineers* 16 (February 1931): 220–222, on 221. The complex was not officially designated Rockefeller Center until 1932, and even after this new name was coined, its Sixth Avenue side was called Radio City for many years.

5. *Rockefeller Center* (New York: Rockefeller Center, c. 1934), page 24 of the unpaginated brochure.

6. *Rockefeller Center*, 24.

7. *Rockefeller Center*, 41.

8. *Rockefeller Center*, 42, 57. Quote from "Novel Plaque Placed in New R.-K.-O. Theater," *New York Times* [*NYT*] (9 October 1932): 24. For more on the artwork of Radio City, see Eugene Clute, "The Story of Rockefeller Center X and XI: The Allied Arts," *Architectural Forum* 57 (October 1932): 353–358, and 58 (February 1933): 128–132. See also Balfour, *Rockefeller Center*, 137–191.

9. *Rockefeller Center*, 45–46.

10. "Radio City to Filter Air and Shut Out Noise," *NYT* (14 January 1932): 39. See also *Rockefeller Center* (New York: Rockefeller Center, 1932), 34; and "Materials and Structural Elements in the Buildings of Rockefeller Center," *Engineering News-Record* 109 (17 November 1932): 590–592, on 591. While studios, theaters, and selected offices in Radio City were air conditioned, the majority of the space was not.

11. Webster B. Todd, "Testing Men and Materials for Rockefeller City," *Architectural Forum* 56 (February 1932): 199–204, on 203. Todd invited several partition manufacturers to construct sample offices, and he was in the midst of evaluating their acoustical properties when he wrote this account.

12. *Rockefeller Center* (1932), 34–35 ("Noise Abatement and Fresh Air"), and 36–38 ("Soundproofing"), quotes on 38.

13. "The Temple of Sound," *Popular Mechanics* 60 (December 1933): 818–821, on 818; Orrin E. Dunlap, Jr., "Radio City's Electric Arteries Soon to Pulse with Music," *NYT* (29 October 1933): sect. VIII, p. 11. The *Times*, quoting RCA's Chief Engineer, reported that "The one word Marconi used most as we showed him the machinery and circuits was 'Indeed!' with an exclamation point."

14. O. B. Hanson, "The Story of Rockefeller Center IX: The Plan and Construction of the National Broadcasting Company Studios," *Architectural Forum* 57 (August 1932): 153–160, and "Planning the NBC Studios for Radio City," *Proceedings of the Institute of Radio Engineers* 20 (August 1932): 1296–1309. See also *Rockefeller Center Development Specifications, Vol. II: RCA Building, NBC Studios, Radio City, New York* (New York: Dorothy Brand, 1938), Marquand Library of Art and Archaeology, Princeton University.

15. Hanson, "Planning the NBC Studios," 1304; "Radio City: NBC Studios Move into Elaborate New Quarters," *Newsweek* 2 (18 November 1933): 32–33, on 32.

16. Dunlap, "Radio City's Arteries Soon to Pulse," 11.

17. Hanson, "Story of Rockefeller Center IX," 160. For more on variable acoustics, see Ernst Petzold, "Regulating the Acoustics of Large Rooms," *Journal of the Acoustical Society of America* [*JASA*] 3 (October 1931): 288–291.

18. The range of reverberation times achieved in the adjustable studios is not indicated in any of the sources I have located, but an average reverberation time for a typical studio was given as 0.75 seconds. "How Echoes Are Produced: NBC Engineers Perfect Artificial Sound Reflection," *Broadcast News* 13 (December 1934): 26–27, on 26.

19. For more on the meaning of radio and the culture of listening in America during the Depression, see Susan J. Douglas, *Listening In: Radio and the American Imagination* (New York: Times Books, 1999).

20. "Radio City Music Hall Reflects Three-Decade Growth of Theater," *New-York Herald Tribune* (18 December 1932). Museum of the City of New York, Theater Collection, Radio City Music Hall, Programs and Clippings.

21. "Theater Designs for Rockefeller Center," *Architectural Record* 71 (June 1932): 419. See also S. L. (Roxy) Rothafel, "The Architect and the Box Office," *Architectural Forum* 57 (September 1932): 194–196.

22. No particular acoustical consultant for Radio City Music Hall is credited in published accounts describing the theater, but a brochure promoting Rockefeller Center more generally identified Paul Sabine and Clifford Swan as "consultants on acoustics." *Rockefeller Center* (1934), 2. Paul Sabine's obituary also identified him as a consultant for Radio City Music Hall. "Paul Earls Sabine (1879–1958)," *JASA* 31 (April 1959): 536.

23. See Paul E. Sabine, "Acoustics of the Chicago Civic Opera House," *Architectural Forum* 52 (April 1930): 599–604. Paul Sabine compares the Civic Opera House to Adler & Sullivan's Chicago Auditorium, and the telescoping curves of this nineteenth-century auditorium may, in fact, constitute the original model for the form of Radio City. William Jordy notes the similarity between Radio City Music Hall and the Chicago Auditorium without identifying Paul Sabine as a possible link between the two structures. *Impact of European Modernism*, 82–83.

24. Krinsky also notes that the basic form of the Music Hall—telescoping arches with radiating grilles—was modeled in clay and photographed for the architects six days before Roxy set sail on his allegedly inspirational voyage at sea. Krinsky cites Albert Kahn's Hill Auditorium at the University of Michigan (a precedent first identified by Lewis Mumford in 1933) and the Salle Pleyel in Paris as additional possible influences on the form of the Music Hall. Krinsky, *Rockefeller Center*, 177–181; Lewis Mumford, "The Skyline: Two Theaters," *New Yorker* 8 (14 January 1933): 55–56, on 56. See also Balfour, *Rockefeller Center*, and Stern et al., *New York 1930* for more on the design of the hall.

25. Douglas Haskell, "Roxy's Advantage over God," *Nation* 136 (4 January 1933): 11.

26. *Rockefeller Center Specifications, Vol. III: Radio City Music Hall*, "Acoustical Treatment." See also "Kalite Sound Absorbing Plaster" (sales brochure, Certain-Teed Products Inc., c. 1934), 8, Series I, Box 13, Folder 11, Guastavino/Collins Collection, Avery Architectural and Fine Arts Library, Columbia University in the City of New York; "Materials and Structural Elements in the Building of Rockefeller Center," 591; and Todd, "Testing Men and Materials," 201–202.

27. *Rockefeller Center Specifications, Vol. III: Radio City Music Hall*, "Acoustical Treatment," and "Metal Furring, Lathing and Plastering."

28. These estimates are based on reverberation measurements made in the hall just before its renovation in 1998–1999. At this time, acousticians from Jaffe Holden Scarborough Acoustics Inc. measured a reverberation time of 1.50 seconds at 500 cps in the unoccupied auditorium at orchestra level. Since the acoustical plaster had been painted over long ago (thus losing much of its ability to absorb sound), and since the sound-absorbing rear wall had similarly been altered to reduce its absorptivity, it is likely that the original reverberation time was significantly less than this figure. Thanks to Robert M. Lilkendey of Jaffe Holden Acoustics Inc. and Raymond Pepi of Building Conservation Associates Inc. for providing me with this information. For accounts of the restoration, see Julie V. Iovine, "Piece by Piece, a Faded Icon Regains Its Art Deco Glow," *NYT* (6 September 1999): A1; and Herbert Muschamp, "An Appraisal: Art Deco Authenticity," *NYT* (4 October 1999): E1.

29. "Radio City Premiere Is a Notable Event," *NYT* (28 December 1932): 1, 14, on 14.

30. Hiller is listed in the opening night program reproduced in Charles Francisco, *Radio City Music Hall: An Affectionate History of the World's Greatest Theater* (New York: E.P. Dutton, 1979), 18. See also "Sound Equipment for Radio City," *NYT* (17 June 1932): 37; and Leo L. Beranek, "Loudspeakers and Microphones," *JASA* 26 (September 1954): 618–629, on 623, for brief descriptions of the sound system installed in the auditorium. For references to loudspeaker placement behind the grilles, see Henry Hofmeister, "The Story of Rockefeller Center V: The International Music Hall," *Architectural Forum* 56 (April 1932): 355–360, on 357 (photo caption); "World's Largest Theater in Rockefeller Center Will Seat Six Thousand Persons," *Popular Mechanics* 58 (August

1932): 252–253 (drawing); and "Radio City Music Hall, Rockefeller Center," *Architectural Forum* 58 (February 1933): 153–164, on 157.

31. Walter Davenport, "The Wings of Song Get a Lift," *Collier's* 92 (30 December 1933): 15–17, on 17. By the time of the recent restoration of the hall, the original sound equipment was long gone. Since the majority of shows in the hall today—even the famed Christmas show—now depend on sound systems brought in and operated by outside contractors on an event-by-event basis, the renovators of the sound system focused upon proving a flexible infrastructure to support these temporary systems. Thanks to David W. Robb of Jaffe Holden Acoustics Inc. for providing me with this information.

32. Rothafel, "Architect and Box Office," 196; Hofmeister, "International Music Hall," 358.

33. "Roxy Returns Soon to Broadcasting," *NYT* (16 January 1932): 13.

34. The *Chicago Daily Tribune*, for example, simply noted that "The auditorium is so constructed that one can see and hear in the furthest seats from the stage." "Big Music Hall at Rockefeller Center Opened," *Chicago Daily Tribune* (28 December 1932): 1. For reviews that neglected to mention the sound, see John Mason Brown, "The Play: The Inaugural Program of the Radio City Music Hall," *New York Evening Post* (28 December 1932); and Edward Angly, "Notables See First Show at Radio City," *New York Herald-Tribune* (28 December 1932): 1, 10. One reviewer did refer briefly and elusively to "the microphone problem," but it is unclear whether this problem was one of technical operation or aesthetic philosophy. See "B. B." (Bruce Bliven), "Roxy and His Patron," *New Republic* 73 (11 January 1933): 242–243, on 242.

35. Miss Keller toured the hall in February 1933, and came away "impressed by the sense of magnitude which she received." "Size of Music Hall Amazes Miss Keller," *NYT* (2 February 1933): 21. For patrons who were hard of hearing, the Music Hall provided "Acousticon seat-phones," volume-adjustable headsets that plugged into specially equipped seatbacks and received the same signal that was sent to the loudspeakers. "The Radio City Theatres," page 9 of unpaginated souvenir program, c. 1933. Museum of the City of New York, Theater Collection, Radio City Music Hall, Programs and Clippings File.

36. Brooks Atkinson, "The Play: Music Hall's Opening," *NYT* (28 December 1932): 14.

37. Haskell, "Roxy's Advantage over God," 11.

38. Loth, *City within a City*, 85.

39. "Radio Music Hall to Be Movie House," *NYT* (6 January 1933): 23; "Radio Music Hall Policy," *NYT* (10 January 1933): 27. The cost of admission subsequently ranged from thirty-five cents to ninety-nine cents, typical for a first-run cinema. This new policy doomed the Music Hall's neighbor in Radio City, the RKO Roxy Theatre, which

had been planned specifically to offer the kind of program that was now being offered in the Music Hall. The RKO Roxy was demoted to second-run status, lost its own live show, and then lost even its name when Rothafel's former namesake, the Fox-owned Seventh Avenue Roxy Theater, filed suit for the exclusive use of the name "Roxy." The RKO Roxy was renamed the Center Theater, but it never really made a name for itself and was demolished in 1954 to make way for a new office building. "RKO Roxy to Cut Admission Prices," *NYT* (24 May 1933): 24; Stern et al., *New York 1930*, 656; Krinsky, *Rockefeller Center*, 187–195.

40. For more on the rise of the Rockettes and their role in the Music Hall's survival into the twenty-first century, see Loth, *City within a City*, 88; and Francisco, *Radio City Music Hall*.

41. Balfour, *Rockefeller Center*, 95; Francisco, *Radio City Music Hall*, 34–36.

42. "Back Comes the Music Hall," *NYT* (25 December 1932): sect. IX, p. 1; Cy Caldwell, "Experiment on Sixth Avenue," *New Outlook* 161 (February 1933): 43–45, on 43; Bliven, "Roxy and His Patron," 243; "Error and Amends," *NYT* (7 January 1933): 14.

43. James M. Beck, "Humiliation at Roxy's," *NYT* (29 December 1932): 18.

44. Walter Lippmann, "Radio City," *American Architect* 143 (March 1933): 18, reprinted from *New York Herald Tribune*.

45. Ralph Adams Cram, "Radio City—And After," *American Mercury* 23 (July 1931): 291–296, on 295.

46. Frederick Lewis Allen, "Radio City: Cultural Center?" *Harper's Monthly Magazine* 164 (March 1932): 534–545, on 534.

47. Mumford, "Two Theaters," 56; Douglas Haskell, "Architecture: The Rockefeller Necropolis," *Nation* 136 (31 May 1933): 622, 624, on 622. See also Mumford, "The Skyline: Mr. Rockefeller's Center," *New Yorker* 8 (23 December 1933): 29–30.

48. Haskell, "Rockefeller Necropolis," 624.

49. Haskell, "Rockefeller Necropolis," 624.

50. For more on the Rivera mural, see Balfour, *Rockefeller Center*, 181–191; and Krinsky, *Rockefeller Center*, 145–148. Humorist Will Rogers admonished Rivera, "Never try to fool a Rockefeller in oils." Krinsky, 148.

51. Haskell, "Rockefeller Necropolis," 624.

52. The Federal Writers' Project noted the nickname in 1939. Cited in Stern et al., *New York 1930*, 650.

53. For more on attitudes toward technology during the depression, see John M. Jordan, *Machine-Age Ideology: Social Engineering and American Liberalism, 1911–1939* (Chapel Hill: University of North Carolina Press, 1994); Amy Sue Bix, *Inventing Ourselves out of Jobs?*

America's Debate over Technological Unemployment, 1929–1981 (Baltimore: Johns Hopkins University Press, 2000); and Carroll W. Pursell, Jr., "Government and Technology in the Great Depression," *Technology and Culture* 20 (January 1979): 162–174.

54. Virginia Woolf, "Mr. Bennett and Mrs. Brown" (1924), quoted in Stephen Kern, *The Culture of Time and Space, 1880–1918* (Cambridge, Mass.: Harvard University Press, 1983), 183.

Coda

1. Douglas Haskell, "Architecture: The People Ready to Change" (1938), quoted in Robert A. M. Stern, Gregory Gilmartin, and Thomas Mellins, *New York 1930: Architecture and Urbanism between the Two World Wars* (New York: Rizzoli, 1987), 671; Sigfried Giedion, *Space, Time and Architecture: The Growth of a New Tradition* (Cambridge, Mass.: Harvard University Press, 1941), 568–580, on 569. See also Le Corbusier, *When the Cathedrals Were White*, trans. Francis E. Hyslop, Jr., of 1938 French ed. (1947; New York: McGraw-Hill, 1964), esp. 33–35, where he describes a visit to the "machine age temple," the NBC studios (33). The critical reception of the center is tracked more fully in Alan Balfour, *Rockefeller Center: Architecture as Theater* (New York: McGraw-Hill, 1978).

2. William H. Jordy, *The Impact of European Modernism in the Mid-Twentieth Century* (New York: Oxford University Press, 1972), 83.

3. Reyner Banham, *The Architecture of the Well-tempered Environment* (Chicago: University of Chicago Press, 1969), 216.

4. Kenneth Frampton, "On Reyner Banham's 'The Architecture of the Well-Tempered Environment,'" *Oppositions* (winter 1976): 86–89, on 87. See also Frampton, "The Mutual Limits of Architecture and Science," pp. 353–373 in Peter Galison and Emily Thompson, eds., *The Architecture of Science* (Cambridge, Mass.: The MIT Press, 1999), for a survey and critique of a range of postwar attempts by architects to engage with scientific ideas and approaches.

5. Langdon Winner, *Autonomous Technology: Technics-out-of-Control as a Theme in Political Thought* (Cambridge, Mass.: The MIT Press, 1977); Theodore Roszak, *The Making of a Counterculture* (1968; Berkeley: University of California Press, 1995); Rachel Carson, *Silent Spring* (1962; Boston: Houghton Mifflin, 1994); and Thomas P. Hughes, *American Genesis: A Century of Invention and Technological Enthusiasm, 1870–1970* (New York: Penguin, 1989), 443–472.

6. For more on the history of the sick building syndrome, see Claudette Michelle Murphy, "Sick Buildings and Sick Bodies: The Materialization of an Occupational Illness in Late Capitalism," Ph.D. Dissertation, Harvard University, 1998.

7. The reverberation time of the Duke University Chapel, for example, was increased from three seconds (at 500 cps) to almost seven seconds. Robert B. Newman and James

G. Ferguson, Jr., "Gothic Sound for the Neo-Gothic Chapel of Duke University," unpublished technical paper from Bolt Beranek and Newman, Inc., 1979. Thanks to Carl Rosenberg of Acentech, Inc., for providing me with this paper. See also David Lloyd Klepper, "The Acoustics of St. Thomas Church, Fifth Avenue," *Journal of the Audio Engineering Society* 43 (July/August 1995): 599–601; and Klepper, "The Acoustics of St. Thomas Church Fifth Avenue," undated technical paper from Klepper Marshall King Associates. Thanks to Gerald Marshall of Marshall/KMK Acoustics for providing me with these papers.

8. See "Special Issue: Preserving Historic Guastavino Tile Ceilings, Domes and Vaults," *APT Bulletin* 30 (1999), especially Richard Pounds, Daniel Raichel, and Martin Weaver, "The Unseen World of Guastavino Acoustical Tile Construction: History, Development, Production," 33–39.

9. Philharmonic Society to architect Max Abramovitz (20 April 1959), quoted in Leo L. Beranek et al., "Acoustics of Philharmonic Hall, New York, During Its First Season," *Journal of the Acoustical Society of America* 36 (July 1964): 1247–1262, on 1247. Readers eager to learn more about the infamous acoustical failure of Philharmonic Hall and its subsequent transformation into Avery Fisher Hall may be frustrated by my neglect of this story. Having spoken with the acousticians involved in each phase of the project, however, I agree with Hans Fantel's conclusion that the hall's troubles "stemmed not so much from acoustics as from politics." Hans Fantel, "Back to Square One for Avery Fisher Hall," *High Fidelity and Musical America* (October 1976): 70–80, on 71. See also Beranek, "A Tale of Two Halls," *Sound and Video Contractor* 15 (20 January 1997): 10–12, 93; and Paul Mitchinson, "Bouncing off the Walls: Leo Beranek and the Science of Concert Hall Design," *Lingua Franca* 11 (April 2001): 61–66.

10. Leo Beranek, *Concert and Opera Halls: How They Sound* (Woodbury, N.Y.: Acoustical Society of America, 1996), 83–86 and 315–318; and Michael Forsyth, *Buildings for Music: The Architect, the Musician and the Listener from the Seventeenth Century to the Present Day* (Cambridge, Mass.: The MIT Press, 1985), 263 and 270–271.

11. The results were published by Beranek as *Music, Acoustics and Architecture* (New York: John Wiley and Sons, 1962).

12. Leo L. Beranek, "Concert Hall Acoustics—1992," *JASA* 92 (July 1992): 1–39, on 8. For a description of the numerous other parameters now taken into account in designing and evaluating an auditorium, see Beranek, "Concert Hall Acoustics—1992," and *Concert and Opera Halls*; and James Glanz, "Art + Physics = Beautiful Music," *New York Times* [*NYT*] (18 April 2000): F1.

13. Harold Arnold, Introduction to Harvey Fletcher, *Speech and Hearing* (New York: Van Nostrand, 1929), xi.

14. Roger Friedland and Dierdre Boden, "NowHere: An Introduction to Space, Time and Modernity," pp. 1–60 in Friedland and Boden, eds., *NowHere: Space, Time and Modernity* (Berkeley: University of California Press, 1994), quote on 3.

15. Wolfgang Schivelbusch, *The Railway Journey: The Industrialization of Time and Space in the 19th Century* (1977; Berkeley: University of California Press, 1986).

16. R. Murray Schafer, *The New Soundscape* (Scarborough, Ont., and New York: Berandol and Associated Music Publishers, 1969), 43–47. Schafer focuses on the splitting of sounds from their original sources brought about by recording and electroacoustic transmission, but the splitting of sound from the space of production is also a part of the process he describes. See also Barry Truax, *Acoustic Communication* (Norwood, N.J.: Ablex Publishing, 1984), 120–122.

17. Rick Altman, "Sound Space," pp. 46–64 in Altman, ed., *Sound Theory/Sound Practice* (New York: Routledge, 1992), on 62.

18. Johnson described this work in his Vern O. Knudsen Lecture to the Acoustical Society of America, "Simultaneous Articulation and Reverberation: Some Related Architectural and Acoustics Aspects of a Series of Rooms, 1954–1990," San Diego, Calif. (28 November 1990). I thank Russell Johnson for sharing with me the notes of his unpublished address. See also Mark Alden Branch, "The Artful Science of Acoustical Design," *Yale Alumni News* (October 1990): 44–49. The most recent concert hall that Johnson and his firm, Artec Consultants, Inc., have consulted on is the Kimmel Center for the Performing Arts in Philadelphia, with architect Raphael Viñoly.

19. In fact, the so-called "Unplugged" series of concerts regularly broadcast on MTV is completely dependent on electroacoustic technologies. While musicians strum acoustic, rather than electric, guitars, they perform into microphones whose signals are broadcast via loudspeakers to the audience in the studio as well as at home, and the musicians themselves listen to monitor speakers on stage. For more on the growth of electroacoustic recording techniques in the popular music industry, see Susan Schmidt Horning, "From Polka to Punk: Growth of an Independent Recording Studio, 1934–1977," pp. 148–159 in Hans-Joachim Braun, ed., *"I Sing the Body Electric": Music and Technology in the 20th Century* (Hofheim: Wolke Verlag, 2000); and Edward R. Kealy, "From Craft to Art: The Case of Sound Mixers and Popular Music," *Sociology of Work and Occupations* 6 (February 1979): 3–29.

20. For an account of the development of techniques for creating artificial reverberation circa 1945, see M. Rettinger, "Reverberation Chambers for Rerecording," *Journal of the Society of Motion Picture Engineers* 45 (November 1945): 350–357.

21. P. H. Parkin and K. Morgan, "'Assisted Resonance' in the Royal Festival Hall, London: 1965–1969," *JASA* 48 (May 1970): 1025–1035, on 1033.

22. Martin J. Oppenheimer, quoted in Anthony Tommasini, "Meddling With Opera's Sacred Human Voice," *NYT* (3 August 1999): A1.

23. Anthony Tommasini, "Enhancing Sound in a Hush-Hush Way," *NYT* (18 August 1999): E1. The manufacturer cited is Lexicon, which builds LARES (Lexicon Acoustic Reinforcement and Enhancement System), a system designed by the acoustical consult-

ing firm of Jaffe Holden Acoustics Inc. The New York State Theater installed a Dutch-built system called ACS (Acoustic Control System), and the Vivian Beaumont Theater at Lincoln Theater employs SIAP (System for Improved Acoustic Enhancement), another Dutch product. For more on reactions to the New York State Theater's system, see Tommasini, "Defending the Operatic Voice from Technology's Wiles," *NYT* (3 November 1999): E1; and Dinitia Smith, "Audience Reaction Is More Ho-Hum than Outrage," *NYT* (3 November 1999): E1.

24. For a detailed account of the design, construction, and reception of the pavilion, see Marc Treib, *Space Calculated in Seconds: The Philips Pavilion, Le Corbusier, Edgard Varèse* (Princeton: Princeton University Press, 1996), quoting Willem Tak, the acoustical consultant for the project, on 197. Anna Sophie Christiansen has begun to explore the roughly simultaneous collaboration between Philips and the musician Hermann Scherchen, who was interested in emulating electronically in his studio the acoustical properties of the spaces for which particular types of music were originally written. I thank her for sharing the text of a talk she presented at the History of Science meeting in Vancouver, November 2000, "Hermann Scherchen's *Gravesano Project*: Cultural Globalization through Scientific Verification of Western Art Music."

25. Thomas R. Horrall, "Auditorium Acoustics Simulator: Form and Uses," Audio Engineering Society Preprint no. 761 (J–5), Presented at the 39th Convention, 12–15 October 1970. Many thanks to Thomas Horrall of Acentech Inc. for providing me with this paper.

26. www.wengercorp.com/innovation/html (21 December 1999).

27. www.wengercorp.com/vroom/html (21 December 1999). For reviews of earlier commercial simulators, see Robert Long and Harold A. Rogers, "Four Ways to Put Yourself in the Concert Hall: Ambience-Simulation Devices Can Transform Home Listening," *High Fidelity and Musical America* 26 (October 1976): 63–68.

28. Andreas Huyssen, *After the Great Divide: Modernism, Mass Culture, Postmodernism* (Bloomington: Indiana University Press, 1986), 196. See also David Harvey, *The Condition of Postmodernity* (Cambridge, Mass.: Blackwell, 1990).

29. Robert Venturi, *Complexity and Contradiction in Architecture* (New York: Museum of Modern Art, 1966). For an irreverent popularization of the rejection of modern architecture, see Tom Wolfe, *From Bauhaus to Our House* (New York: Washington Square Press, 1981).

30. "V-Room: Wenger Variable Acoustic Environment: Testimonials—Installations—Articles—Features & Benefits" (sales pamphlet, Wenger Corp., April 1998), 4.

Bibliography

Archival Resources

American Institute of Physics. Center for History of Physics. Niels Bohr Library. College Park, Md.

Architect of the Capitol. Curator's Office. Capitol Building. Washington, D.C.

AT&T Archives. Warren, N.J.

Boston Symphony Orchestra Archives. Symphony Hall. Boston, Mass.

Edison, Thomas A. Archives. National Park Service. Edison National Historic Site. West Orange, N.J.

Eliot, Charles W. Records of President's Office. Harvard University, 1849–1926. Harvard University Archives. Cambridge, Mass.

Foulke, William Parker. Papers. American Philosophical Society. Philadelphia, Penn.

Goodhue, Bertram Grosvenor. Papers. Avery Architectural and Fine Arts Library. Columbia University. New York, N.Y.

Guastavino/Collins Collection. Avery Architectural and Fine Arts Library. Columbia University. New York, N.Y.

Henry, Joseph. Papers. Smithsonian Institution. Washington, D.C.

Higginson, Henry Lee. Historical Collections. Baker Library. Harvard Business School. Boston, Mass.

James, Edmund M. Faculty Correspondence, 1904–1915. University of Illinois at Urbana-Champaign Archives.

Los Angeles Philharmonic Archives. Los Angeles, Calif.

McKim, Mead & White Papers. New-York Historical Society. New York, N.Y.

Miller, Dayton Clarence. Papers. Case Western Reserve University. Cleveland, Ohio.

Municipal Archives of the City of New York, N.Y.

Museum of Modern Art. Celeste Bartos Film Study Center. New York, N.Y.

Museum of the City of New York. Theater Collection. New York, N.Y.

National Archives and Records Administration, United States of America. Washington, D.C.

New York Life Insurance Company Archives. New York, N.Y.

New York Municipal Reference Library. New York, N.Y.

Philadelphia Saving Fund Society Archives. Hagley Museum and Library. Wilmington, Del.

Rockefeller Center Archive Center. New York, N.Y.

Sabine, Wallace Clement. Correspondence, 1899–1919. Harvard University Archives. Cambridge, Mass.; Riverbank Acoustical Laboratories. Illinois Institute of Technology Research Institute. Geneva, Ill.

Sabine, Wallace Clement. Papers. Research Notebooks. Data on Acoustical Research, 1899–1919. Harvard University Archives. Cambridge, Mass.

University of California at Los Angeles. Film and Television Archive. Los Angeles, Calif.

Journal Abbreviations

AAAR	American Architect and Architectural Review
AABN	American Architect and Building News
BSTJ	Bell System Technical Journal
JAES	Journal of the Audio Engineering Society
JAIEE	Journal of the American Institute of Electrical Engineers
JASA	Journal of the Acoustical Society of America
JSAH	Journal of the Society of Architectural Historians
JSMPE	Journal of the Society of Motion Picture Engineers
PAAAS	Proceedings of the American Academy of Arts and Sciences
PIRE	Proceedings of the Institute of Radio Engineers
PNAS	Proceedings of the National Academy of Sciences
TAIEE	Transactions of the American Institute of Electrical Engineers
TSMPE	Transactions of the Society of Motion Picture Engineers

Abbott, Lawrence F. *The Story of NYLIC: A History of the Origin and Development of the New York Life Insurance Company from 1845 to 1929.* New York: "The Company," 1930.

Academy Technical Digest: Fundamentals of Sound Recording and Reproduction for Motion Pictures. Hollywood: Academy of Motion Picture Arts and Sciences, 1929–1930.

"Acoustical Engineering as a Career." *Careers*, Research no. 38. Chicago: Institute for Research, 1931.

"Acoustic Roof." *Architectural Forum* 57 (December 1932): 30 (ad section).

Adams, Stephen B., and Orville R. Butler. *Manufacturing the Future: A History of Western Electric.* Cambridge: Cambridge University Press, 1999.

Addams, Jane. *Twenty Years at Hull-House.* 1910; New York: Penguin, 1981.

Adler, Dankmar. "The Chicago Auditorium." *Architectural Record* 1 (April–June 1892): 415–434.

Adler, Dankmar. "The Theater." Edited by Rachel Baron. *Prairie School Review* 2 (1965): 21–27.

Adler, Dankmar. "Theater-Building for American Cities." *Engineering Magazine* 7 (August 1894): 717–730 and 7 (September 1894): 815–829.

Adler, Dankmar. "Theatres." *AABN* 22 (29 October 1887): 206–208.

Aitken, Hugh G. J. *The Continuous Wave: Technology and American Radio, 1900–1932.* Princeton: Princeton University Press, 1985.

Aitken, Hugh G. J. *Syntony and Spark: The Origins of Radio.* 1976; Princeton: Princeton University Press, 1985.

Algarotti, Count Francesco. *An Essay on the Opera.* London: David and Reymers, 1767, and London: Press Printers, 1917.

Allen, Frederick Lewis. "Radio City: Cultural Center?" *Harper's Monthly Magazine* 164 (March 1932): 534–545.

All Quiet on the Western Front. 1930, Universal Pictures, dir. Lewis Milestone, MCA Universal Home Video #55018.

Altman, Rick. "The Material Heterogeneity of Recorded Sound." In *Sound Theory/Sound Practice*, edited by Rick Altman, 15–31.

Altman, Rick. "Sound Space." In *Sound Theory/Sound Practice*, edited by Rick Altman, 46–64.

Altman, Rick, ed. *Sound Theory/Sound Practice.* New York: Routledge, 1992.

Antheil, George. *Bad Boy of Music.* 1945; New York: Da Capo Press, 1981.

"Anti-Noise Campaign Planned." *American City* 34 (April 1926): 396.

Apollonio, Umbro, ed. *Futurist Manifestos*. Trans. Robert Brain et al. New York: Viking, 1973.

Applause. 1929, Paramount Famous Lasky, dir. Rouben Mamoulian, New York Film Annex Video #1-55881-036-6.

Arnold, H. D., and I. B. Crandall. "The Thermophone as a Precision Source of Sound." *Physical Review,* 2d ser., 10 (July 1917): 22–38.

Arns, Robert G., and Bret E. Crawford. "Resonant Cavities in the History of Architectural Acoustics." *Technology and Culture* 36 (January 1995): 104–135.

Arsenault, Raymond. "The End of the Long Hot Summer: The Air Conditioner and Southern Comfort." *Journal of Southern History* 50 (November 1984): 597–628.

"As Cathedrals Were Built." *Eastern Underwriter* (14 December 1928), Part 2, Special Issue.

Attali, Jacques. *Noise: The Political Economy of Music*. Translated by Brian Massumi. 1977; Minneapolis: University of Minnesota Press, 1985.

"The Auxetophone for Reinforcing Gramophone Sounds." *Scientific American* 92 (13 May 1905): 382.

Bagenal, Hope, and Alexander Wood. *Planning for Good Acoustics*. London: Methuen and Co., 1931.

Baker, Houston A., Jr. *Modernism and the Harlem Renaissance*. Chicago: University of Chicago Press, 1987.

Baker, Ray Stannard. "New Music for an Old World." *McClure's Magazine* 27 (July 1906): 291–301.

Baker, W. R. G. "Description of the General Electric Company's Broadcasting Station at Schenectady, New York." *PIRE* 11 (August 1923): 339–374.

Balfour, Alan. *Rockefeller Center: Architecture as Theater*. New York: McGraw-Hill, 1978.

Bandy, Mary Lea, ed. *The Dawn of Sound*. New York: Museum of Modern Art, 1989.

Banham, Reyner. *The Architecture of the Well-tempered Environment*. Chicago: University of Chicago Press, 1969, and 2d ed., 1984.

Banham, Reyner. *A Concrete Atlantis: U.S. Industrial Building and European Modern Architecture*. Cambridge, Mass.: The MIT Press, 1986.

Banham, Reyner. *Theory and Design in the First Machine Age*. 2d ed. Cambridge, Mass.: The MIT Press, 1980.

Bareiss, Warren. "Noise Abatement in Philadelphia, 1907–1966: The Production of a Soundscape." M.A. Thesis, University of Pennsylvania, 1990.

Baron, Lawrence. "Noise and Degeneration: Theodor Lessing's Crusade for Quiet." *Journal of Contemporary History* 17 (January 1982): 165–178.

Barthelmes, Barbara. "Music and the City." In *I Sing the Body Electric,* edited by Hans-Joachim Braun, 97–105.

Batterson, Elmer S. "Progress of the Anti-Noise Movement." *National Municipal Review* 6 (May 1917): 372–378.

Bellamy, Edward. *Looking Backward.* 1888; New York: Penguin American Library, 1982.

Beranek, Leo L. "Boston Symphony Hall: An Acoustician's Tour." *JAES* 36 (November 1988): 918–930.

Beranek, Leo L. *Concert and Opera Halls: How They Sound.* Woodbury, N.Y.: Acoustical Society of America, 1996.

Beranek, Leo L. "Concert Hall Acoustics—1992." *JASA* 92 (July 1992): 1–39.

Beranek, Leo L. "Loudspeakers and Microphones." *JASA* 26 (September 1954): 618–629.

Beranek, Leo L. *Music, Acoustics and Architecture.* New York: John Wiley and Sons, 1962.

Beranek, Leo L. "The Notebooks of Wallace C. Sabine." *JASA* 61 (March 1977): 629–639.

Beranek, Leo L. "A Tale of Two Halls." *Sound and Video Contractor* 15 (20 January 1997): 10–12, 93.

Beranek, Leo L. "Wallace Clement Sabine and Acoustics." *Physics Today* (February 1985): 44–51.

Beranek, Leo L., et al. "Acoustics of Philharmonic Hall, New York, During Its First Season." *JASA* 36 (July 1964): 1247–1262.

Beranek, Leo L., and John W. Kopec, "Wallace C. Sabine, Acoustical Consultant." *JASA* 69 (January 1981): 1–16.

Berman, Marshall. *All That Is Solid Melts into Air: The Experience of Modernity.* 1982; New York: Penguin, 1988.

Betts, Benjamin F. "Is This Our Next New Big Business?" *American Architect* 140 (December 1931): 21.

Beyer, Robert T. *Sounds of Our Times: Two Hundred Years of Acoustics.* New York: AIP/Springer-Verlag, 1999.

Bijsterveld, Karen. "'A Servile Imitation': Disputes About Machines in Music, 1910–1930." In *I Sing the Body Electric,* edited by Hans-Joachim Braun, 121–134.

Birckhead, Rev. Dr. Hugh. "Architectural Requirements of the Episcopal Church." *Architectural Forum* 40 (April 1924): 165–166.

The Birthplace of the Sound Motion Picture. 1930; Bell Telephone Laboratories. Video transfer of motion picture, AT&T Archives #415-06-02.

Bix, Amy Sue. *Inventing Ourselves out of Jobs? America's Debate over Technological Unemployment, 1929–1981.* Baltimore: Johns Hopkins University Press, 2000.

The Black Watch. 1929; Fox Film Corp., dir John Ford, 16mm film, Museum of Modern Art.

Bliven, Bruce. "How Radio Is Remaking Our World." *Century Magazine* 108 (June 1924): 147–154.

Bliven, Bruce. "Roxy and His Patron," *New Republic* 73 (11 January 1933): 242–243.

Blodgett, Geoffrey. "Cass Gilbert, Architect: Conservative at Bay." *Journal of American History* 72 (December 1985): 615–636.

Bluestone, Daniel. "The Pushcart Evil." In *The Landscape of Modernity,* edited by David Ward and Olivier Zunz, 287–312.

"B'Nai Jeshurun Temple, Newark, N.J." *Brickbuilder* 24 (December 1915): 305–306.

Boehm, W. M. "A Method of Measuring Intensity of Sound." *Physical Review* 31 (October 1910): 329–331.

Bolt, R. H. "Wallace Waterfall, 1900–1974." *JASA* 56 (December 1974): 1932–1933.

Bonner, Eugene. "Music and Musicians." *Outlook* 148 (15 February 1928): 268.

"Books on Acoustics." *AABN* 47 (12 January 1895): 23.

Born, Georgina. *Rationalizing Culture: IRCAM, Boulez, and the Institutionalization of the Musical Avant-Garde.* Berkeley: University of California Press, 1995.

Botstein, Leon. "Innovation and Nostalgia: Ives, Mahler, and the Origins of Twentieth-Century Modernism." In *Charles Ives and His World,* edited by J. Peter Burkholder, 35–74.

Bottomley, H. L. "The Story of St. Thomas' Church." *Architectural Record* 35 (February 1914): 101–131.

Bowers, Q. David. *Nickelodeon Theatres and Their Music.* Vestal, N.Y.: Vestal Press, 1986.

Boyer, M. Christine. *Dreaming the Rational City: The Myth of American City Planning.* Cambridge, Mass.: The MIT Press, 1983.

Boyer, Paul. *Urban Masses and Moral Order in America, 1820–1920.* Cambridge, Mass.: Harvard University Press, 1978.

Brain, Robert. "Standards and Semiotics." In *Inscribing Science: Scientific Texts and the Materiality of Communication,* edited by Timothy Lenoir, 249–284, 414–425. Stanford: Stanford University Press, 1998.

Branch, Mark Alden. "The Artful Science of Acoustical Design." *Yale Alumni News* (October 1990): 44–49.

Braun, Hans-Joachim. "'Movin' On': Trains and Planes as a Theme in Music." In *I Sing the Body Electric,* edited by Hans-Joachim Braun, 106–120.

Braun, Hans-Joachim, ed. *"I Sing the Body Electric": Music and Technology in the 20th Century*. Hofheim: Wolke Verlag, 2000.

Braun, Harry B. "Sound Motion Picture Requirements." *Architectural Forum* 57 (October 1932): 381–385.

Brooks, John. *Telephone: The First Hundred Years*. New York: Harper & Row, 1976.

Brown, Barclay. "The Noise Instruments of Luigi Russolo." *Perspectives of New Music* 20 (1981–1982): 31–48.

Brown, Edward F., et al., eds. *City Noise*. New York: Department of Health, 1930.

Brown, Lewis H. "Attacking City Noises by Science and Law." *American City* 44 (February 1931): 97–101.

Brownlow, Kevin. *The Parade's Gone By. . . .* Berkeley: University of California Press, 1968.

Bruce, Robert V. *Bell: Alexander Graham Bell and the Conquest of Solitude*. Boston: Little, Brown and Co., 1973.

Buchwald, Jed Z. *The Creation of Scientific Effects: Heinrich Hertz and Electric Waves*. Chicago: University of Chicago Press, 1994.

Buckland, Michael, and John Henken, eds. *The Hollywood Bowl: Tales of Summer Nights*. Los Angeles: Balcony Press, 1996.

Burkholder, J. Peter. *Charles Ives: The Ideas behind the Music*. New Haven: Yale University Press, 1985.

Burkholder, J. Peter, ed. *Charles Ives and His World*. Princeton: Princeton University Press, 1996.

Burnett, Charles, Michael Fend, and Penelope Gouk, eds. *The Second Sense: Studies in Hearing and Musical Judgement from Antiquity to the Seventeenth Century*. London: The Warburg Institute, 1991.

Burrows, H. W. "Sound in Its Relation to Buildings." *AABN* 48 (18 May 1895): 65–70.

Busoni, Ferruccio. "Sketch of a New Esthetic of Music." English trans. of 1911 reprinted in *Three Classics in the Aesthetic of Music*. New York: Dover, 1962.

Butler, Charles. "Prevention of Sound Travel in Hospitals." *Architectural Forum* 37 (December 1922): 287–288.

Cahan, David. "From Dust Figures to the Kinetic Theory of Gases: August Kundt and the Changing Nature of Experimental Physics in the 1860s and 1870s." *Annals of Science* 47 (1990): 151–172.

Caldwell, Cy. "Experiment on Sixth Avenue." *New Outlook* 161 (February 1933): 43–45.

Calfas, Paul. *La Nouvelle Salle de Concert Pleyel à Paris*. Paris: Publications du Journal le Génie Civil, 1927.

Campbell, Archibald. "Acoustics." *Edinburgh Encyclopedia* 1: 104–124. Philadelphia: Joseph and Edward Parker, 1832.

Card, James. *Seductive Cinema: The Art of Silent Film*. 1994; Minneapolis: University of Minnesota Press, 1999.

Carson, Rachel. *Silent Spring*. 1962; Boston: Houghton Mifflin, 1994.

Carter, Rand. "The Drury Lane Theatres of Henry Holland and Benjamin Dean Wyatt." *JSAH* 26 (October 1967): 200–216.

Casey, Daniel V. "Muffling Office Noises." *System: The Magazine of Business* 25 (1913): 246–251.

Cass, John L. "The Illusion of Sound and Picture." *JSMPE* 14 (March 1930): 323–326.

Castner, T. G. et al. "Indicating Meter for Measurement and Analysis of Noise." *TAIEE* 50 (September 1931): 1041–1047.

"Certain Problems in Acoustics." *Bulletin of the National Research Council* 4 (November 1922), no. 23.

Chenery, William L. "The Noise of Civilization." *New York Times Magazine* (1 February 1920): 13.

Chion, Michel. *Audio-Vision: Sound on Screen*, trans. Claudia Gorbman. New York: Columbia University Press, 1994.

Chrisler, V. L., and W. F. Snyder. "Recent Advances in Sound Absorption Measurements." *JASA* 2 (July 1930): 123–128.

Cimarron. 1930; RKO Radio Pictures, dir. Wesley Ruggles. MGM/UA Home Video #300400.

Cinema Europe: The Other Hollywood. 1995; Photoplay Productions, dir. Kevin Brownlow and David Gill. Photoplay/DLT Entertainment Video.

"The City's Noise." *Atlantic Monthly* 104 (November 1909): 601–610.

Clark, Michael. "Ralph Adams Cram and the Americanization of the Middle Ages." *Journal of American Studies* 23 (August 1989): 195–213.

Classen, Constance. *Worlds of Sense: Exploring the Senses in History and Across Cultures*. London: Routledge, 1993.

Clute, Eugene. "The Story of Rockefeller Center: X. and XI. The Allied Arts." *Architectural Forum* 57 (October 1932): 353–358 and 58 (February 1933): 128–132.

Cochrane, Rexmond C. *Measures for Progress: A History of the National Bureau of Standards*. Washington: U.S. Dept. of Commerce, 1966.

Coffman, Joe W. "Art and Science in Sound Film Production." *JSMPE* 14 (February 1930): 172–179.

Cohen, H. F. *Quantifying Music: The Science of Music at the First Stage of the Scientific Revolution, 1580–1650*. Dordrecht: D. Reidel, 1984.

Cohen, Lizabeth. *Making a New Deal: Industrial Workers in Chicago, 1919–1939*. Cambridge: Cambridge University Press, 1990.

Collins, George R. "The Transfer of Thin Masonry Vaulting from Spain to America." *JSAH* 27 (October 1968): 176–201.

"Commercialized Pianissimo." *New York Times* (3 December 1931): 28.

Connor, Steven. "The Modern Auditory I." In *Rewriting the Self: Histories from the Renaissance to the Present*, edited by Roy Porter, 203–223. London: Routledge, 1997.

Conrads, Ulrich, ed. *Programs and Manifestoes on 20th-Century Architecture*, trans. Michael Bullock. Cambridge, Mass.: The MIT Press, 1970.

Cooper, Gail. *Air-Conditioning America: Engineers and the Controlled Environment, 1900–1960*. Baltimore: Johns Hopkins University Press, 1998.

Corbett, Stratford. "An Office Building of the New Era." *Scientific American* 141 (December 1929): 484–486.

Corbin, Alain. *The Foul and the Fragrant: Odor and the French Social Imagination*. Cambridge, Mass.: Harvard University Press, 1986.

Corbin, Alain. *Time, Desire and Horror: Toward a History of the Senses*. Translated by Jean Birrell. Cambridge: Polity Press, 1995.

Corbin, Alain. *Village Bells: Sound and Meaning in the 19th-Century French Countryside*. Translated by Martin Thom. New York: Columbia University Press, 1998.

Cowan, Lester, ed. *Recording Sound for Motion Pictures*. New York: McGraw-Hill, 1931.

Cowell, Henry. "The Joys of Noise." *New Republic* 59 (31 July 1929): 287–288.

Cox, David Harold, and Michael Naslas. "The Metropolis in Music." In *Metropolis 1890–1940*, edited by Anthony Sutcliffe, 173–189.

Crafton, Donald. *The Talkies: American Cinema's Transition to Sound, 1926–1931*. New York: Charles Scribner's Sons, 1997.

Cram, Ralph Adams. *Church Building*. Boston: Small, Maynard and Co., 1901; and 3d ed. Boston: Marshall Jones Co., 1924.

Cram, Ralph Adams. *The Gothic Quest*. New York: Baker and Taylor, 1907.

Cram, Ralph Adams. *My Life in Architecture*. Boston: Little, Brown and Co., 1936.

Cram, Ralph Adams. "Radio City—and After." *American Mercury* 23 (July 1931): 291–296.

Crary, Jonathan. *Suspensions of Perception: Attention, Spectacle and Modern Culture*. Cambridge, Mass.: The MIT Press, 1999.

Crary, Jonathan. *Techniques of the Observer: On Vision and Modernity in the Nineteenth Century*. Cambridge, Mass.: The MIT Press, 1990.

Creager, Angela N. H. *The Life of a Virus: Tobacco Mosaic Virus as an Experimental Model, 1930–1965*. Chicago: University of Chicago Press, 2002.

Crocker, Allen S. "The Heating and Ventilating System and Soundproofing for the Eastman Theatre and School of Music." *AAAR* 123 (28 February 1923): 200–202.

Cron, Theodore O., and Burt Goldblatt. *Portrait of Carnegie Hall*. New York: Macmillan, 1966.

The Crowd. 1928; Metro-Goldwyn-Mayer, dir. King Vidor. MGM/UA Home Video #M301357.

Crowhurst, Frank S. "Acoustic Linings for Soundproof Motion Picture Stages and Sets." *TSMPE* 12 (1928): 828–835.

"A Crusade for Quiet." *Outlook* 102 (28 September 1912): 157–159.

Curtis, William J. R. *Modern Architecture Since 1900*. 3d ed. Upper Saddle River, N.J.: Prentice Hall, 1996.

Czitrom, Daniel. *Media and the American Mind: From Morse to McLuhan*. Chapel Hill: University of North Carolina Press, 1982.

Danger Lights. 1930; RKO Radio Pictures, dir. George B. Seitz. Nostalgia Home Video #1722.

Davenport, Walter. "The Wings of Song Get a Lift." *Collier's* 92 (30 December 1933): 15–17.

Davis, Hallowell. "Psychological and Physiological Acoustics: 1920–1942." *JASA* 61 (February 1977): 264–266.

Davis, Ronald L. *Opera in Chicago*. New York: Appleton-Century, 1966.

"The Decline of the Amateur." *Atlantic Monthly* 73 (June 1894): 859–860.

"The Demon in Radio." *Literary Digest* 82 (2 August 1924): 26–27.

"The Discussion on Sound in Its Relation to Buildings." *AABN* 48 (25 May 1895): 78–81.

Disraeli. 1929; Warner Bros., dir. Alfred E. Green. MGM/UA Home Video #M202143.

Dr. Jekyll and Mr. Hyde. 1932; Paramount Publix, dir. Rouben Mamoulian. MGM/UA Home Video #M201642.

Documentary History of the Construction and Development of the United States Capitol Building and Grounds. Washington: Government Printing Office, 1904.

Don Juan. 1926; Warner Bros., dir Alan Crosland. MGM/UA Home Video #M302162.

Dorner, Jane. *Fashion in the Twenties and Thirties*. New Rochelle: Arlington House, 1974.

Dostrovsky, Sigalia. "Early Vibration Theory: Physics and Music in the Seventeenth Century." *Archive for History of Exact Sciences* 14 (December 1975): 169–218.

Douglas, Ann. *Terrible Honesty: Mongrel Manhattan in the 1920s*. New York: Farrar Straus Giroux, 1995.

Douglas, Susan J. *Inventing American Broadcasting, 1899–1922*. Baltimore: Johns Hopkins University Press, 1987.

Douglas, Susan J. *Listening In: Radio and the American Imagination*. New York: Times Books, 1999.

Dreher, Carl. "Microphone Concentrators in Picture Production." *JSMPE* 16 (January 1931): 23–30.

Dreher, Carl. "Recording, Re-Recording and Editing of Sound." *JSMPE* 16 (June 1931): 756–765.

Dreher, Carl. "Sound Personnel and Organization." In *Academy Fundamentals of Sound Recording*, 85–96.

Dupree, A. Hunter. *Science in the Federal Government*. 1957; Baltimore: Johns Hopkins University Press, 1986.

"Ears Good?" *Industrial Psychology Monthly* 2 (August 1927): 431.

"Eastman Theatre and School of Music." *AAAR* 123 (28 February 1923): 181–184.

Eberson, John. "Monuments." *Film Daily* 47 (4 January 1929): 1.

Eccles, W. H. "The New Acoustics." *Proceedings of the Physical Society* 41 (15 June 1929): 231–239.

Eckhardt, E. A., and V. L. Chrisler. "Transmission and Absorption of Sound by Some Building Materials." *Scientific Papers of the Bureau of Standards,* no. 526 (28 April 1926).

"Edison's Kinetograph and Cosmical Telephone." *Scientific American* 64 (20 June 1891): 393.

Ehrlich, Cyril. *The Piano: A History*. London: J.M. Dent and Sons, 1976.

Eisenberg, Evan. *The Recording Angel: Explorations in Phonography*. New York: McGraw-Hill, 1987.

"The Elimination of Harmful Noise." *National Safety News* 15 (April 1927): 46, 58.

Elliott, Cecil D. *Technics and Architecture: The Development of Materials and Systems for Buildings*. Cambridge, Mass.: The MIT Press, 1992.

Ellis, John. *Eye-Deep in Hell: Trench Warfare in World War I*. 1976; Baltimore: Johns Hopkins University Press, 1989.

Elstein, Rochelle S. "The Architecture of Dankmar Adler." *JSAH* 26 (December 1967): 242–249.

"Engineering Problems of Radio Broadcasting Studio Design." *American Architect* 135 (5 February 1929): 195–203.

"Enter the Acoustics Expert!" *Cinema Construction* (September 1929): 5

Evans, Porter H. "A Comparison of the Engineering Problems in Broadcasting and Audible Pictures." *PIRE* 18 (August 1930): 1316–1337.

Eyman, Scott. *The Speed of Sound: Hollywood and the Talkie Revolution, 1926–1930*. New York: Simon and Schuster, 1997.

Eyring, Carl F. "Reverberation Time in 'Dead' Rooms." *JASA* 1 (January 1930): 217–241.

von Ezdorf, Robert. "The Design of the Interior of the New Home Office Building of the New York Life Insurance Company." *American Architect* 135 (20 March 1929): 369–372.

Fagen, M. D., ed. *A History of Engineering and Science in the Bell System: The Early Years (1875–1925)*. N.p.: Bell Telephone Laboratories, 1975.

Fantel, Hans. "Back to Square One for Avery Fisher Hall." *High Fidelity and Musical America* (October 1976): 70–80.

Feld, Steven. *Sound and Sentiment: Birds, Weeping, Poetics, and Song in Kaluli Expression*. 2d ed. Philadelphia: University of Pennsylvania Press, 1990.

Feld, Steven. "Sound Structure as Social Structure." *Ethnomusicology* 28 (September 1984): 383–409.

Fenske, Gail, and Deryck Holdsworth. "Corporate Identity and the New York Office Building: 1895–1915." In *The Landscape of Modernity*, edited by David Ward and Olivier Zunz, 129–159.

"A Few Gropings in Practical Acoustics." *Builder* 8 (31 August 1850): 411–412.

Fielding, Raymond, ed. *A Technological History of Motion Pictures and Television*. Berkeley: University of California Press, 1967.

Finding His Voice. 1929; Western Electric Co., dir. F. Lyle Gordon and Max Fleischer. Video transfer of motion picture, AT&T Archives #374-02-51.

Finn, Bernard S. "Laplace and the Speed of Sound." *Isis* 55 (March 1964): 7–19.

Fischer, Claude. *America Calling: A Social History of the Telephone to 1940*. Berkeley: University of California Press, 1992.

Fischer, Lucy. "*Applause*: The Visual and Acoustic Landscape." In *Film Sound*, edited by Elisabeth Weis and John Belton, 232–246.

Fishbein, Morris. "The Month in Medical Science." *Scientific American* 140 (February 1929): 124.

Fisher, Sidney George. *A Philadelphia Perspective: The Diary of Sidney George Fisher Covering the Years 1834–1871*. Edited by Nicholas B. Wainwright. Philadelphia: Historical Society of Pennsylvania, 1967.

Fitch, James Marston. *American Building 2: The Environmental Forces that Shape It*. 2d ed. New York: Schocken Books, 1971.

Fitch, Noel Riley. *Sylvia Beach and the Lost Generation: A History of Literary Paris in the Twenties and Thirties*. New York: Norton, 1983.

Fitzgerald, F. Scott. *The Crack-Up*. New York: New Directions, 1945.

Fletcher, Harvey. "The Acoustical Society of America. Its Aims and Trends." *JASA* 11 (July 1939): 13–14.

Fletcher, Harvey. "Biographical Memoir of Dayton Clarence Miller, 1866–1941." *Biographical Memoirs of the National Academy of Science* 23 (1945): 61–74.

Fletcher, Harvey. "Harvey Fletcher." Transcript of an oral interview by Vern Knudsen with W. J. King (15 May 1964), Neils Bohr Library, Center for History of Physics, American Institute of Physics, College Park, Md.

Fletcher, Harvey. *Speech and Hearing*. New York: Van Nostrand, 1929.

Fletcher, Harvey, and J. C. Steinberg. "Articulation Testing Methods." *JASA* 1 (April 1930 Supplement): 1–97.

Fletcher, Maureen Meyer. "The Caroling of Atoms: The Life's Work of Dr. Harvey Fletcher." M.F.A. Thesis, Brigham Young University, 1996.

Flexner, James, ed. *City Noise Volume II*. New York: Department of Health, 1932.

Flint, R. W., ed. *Let's Murder the Moonshine: Selected Writings of F.T. Marinetti*. Los Angeles: Sun and Moon Classics, 1991.

Floyd, Samuel A., Jr. "Music in the Harlem Renaissance: An Overview." In *Black Music in the Harlem Renaissance*, edited by Samuel A. Floyd, Jr., 1–27. New York: Greenwood, 1990.

Foley, Arthur L., and Wilmer H. Souder. "A New Method of Photographing Sound Waves." *Physical Review* 35 (November 1912): 373–386.

Ford, Corey. "Silence Please!" *American Magazine* 109 (February 1930): 18–19, 160–161.

"For Health, Safety and Efficiency." *Eastern Underwriter* (14 December 1928): Part 2, Special Issue, p. 12.

Forsyth, Michael. *Buildings for Music: The Architect, the Musician, and the Listener from the Seventeenth Century to the Present Day*. Cambridge, Mass.: The MIT Press, 1985.

Foster, Hal, ed. *Vision and Visuality*. Seattle: Bay Press, 1988.

Frampton, Kenneth. *Modern Architecture: A Critical History*. 3d ed. London: Thames and Hudson, 1992.

Frampton, Kenneth. "The Mutual Limits of Architecture and Science." In *The Architecture of Science*, edited by Peter Galison and Emily Thompson, 353–373.

Frampton, Kenneth. "On Reyner Banham's 'The Architecture of the Well-Tempered Environment.'" *Oppositions* (winter 1976): 86–89.

Francisco, Charles. *The Radio City Music Hall: An Affectionate History of the World's Greatest Theater*. New York: E.P. Dutton, 1979.

Franklin, W. S. "Derivation of Equation of Decaying Sound in a Room and Definition of Open Window Equivalent of Absorbing Power." *Physical Review* 16 (June 1903): 372–374.

Frederick, H. A. "The Development of the Microphone." *JASA* 3 (Supp. to July 1931): 1–25.

Free, E. E. "How Noisy Is New York?" *Forum* 75 (February 1926): (Illust. sect.) xxi–xxiv.

Free, E. E. "Noise: The Forum's Second Report on City Noise. *Forum* 79 (March 1928): 382–389.

Friede, Donald. *The Mechanical Angel: His Adventures and Enterprises in the Glittering 1920's*. New York: Knopf, 1948.

Friedland, Roger, and Dierdre Boden. "NowHere: An Introduction to Space, Time and Modernity." In *NowHere: Space, Time and Modernity*, edited by Roger Friedland and Dierdre Boden, 1–60. Berkeley: University of California Press, 1994.

Froude, James Anthony. *Thomas Carlyle: A History of His Life in London: 1834–1881*. London: Longmans, Green and Co., 1897.

Gaisberg, Fred. *The Music Goes Round*. New York: Macmillan, 1943.

Galison, Peter. "Aufbau/Bauhaus: Logical Positivism and Architectural Modernism." *Critical Inquiry* 16 (summer 1990): 709–752.

Galison, Peter. *Image and Logic: A Material Culture of Microphysics*. Chicago: University of Chicago Press, 1997.

Galison, Peter, and Emily Thompson, eds. *The Architecture of Science*. Cambridge, Mass.: The MIT Press, 1999.

Garnier, Charles. *Le Théâtre*. Paris: Hachette, 1871.

"The Gaumont Speaking Kinematograph Films." *Nature* 89 (30 May 1912): 333–334.

Gearheart, Don H. "Dr. Miller and Case Host to Acoustical Society." *Case Alumnus* (December 1931): 12.

Gebhard, David, and Harriette Von Breton. *Lloyd Wright, Architect: 20th Century Architecture in an Organic Exhibition*. 1971; Santa Barbara: Hennessey and Ingalls, 1998.

Geduld, Harry M. *The Birth of the Talkies: From Edison to Jolson*. Bloomington: Indiana University Press, 1975.

Gelatt, Roland. *The Fabulous Phonograph: 1877–1977*. New York: Macmillan, 1977.

Giedion, Sigfried. *Space, Time and Architecture: The Growth of a New Tradition*. Cambridge, Mass.: Harvard University Press, 1941.

Gimpel, Jean. *The Medieval Machine: The Industrial Revolution of the Middle Ages*. 1976; New York: Penguin, 1977.

Girdner, J. H. "The Plague of City Noises." *North American Review* 163 (September 1896): 296–303.

Gitelman, Lisa. *Scripts, Grooves, and Writing Machines: Representing Technology in the Edison Era*. Stanford: Stanford University Press, 1999.

Glinsky, Albert. *Theremin: Ether Music and Espionage*. Urbana: University of Illinois Press, 2000.

Godfrey, Hollis. "The City's Noise." *Atlantic Monthly* 104 (November 1909): 601–610.

Goldsmith, Alfred N. "An Entertainment City." *JSMPE* 16 (February 1931): 220–222.

Gomery, Douglas. "The Coming of Sound to the American Cinema: A History of the Transformation of an Industry." Ph.D. Dissertation, University of Wisconsin–Madison, 1975.

Gomery, Douglas. "The Coming of Sound: Technological Change in the American Film Industry." In *Film Sound*, edited by Elisabeth Weis and John Belton, 5–24.

Gomery, Douglas. *Shared Pleasures: A History of Movie Presentation in the United States.* Madison: University of Wisconsin Press, 1992.

Gordon, Mel. "Songs from the Museum of the Future: Russian Sound Creation (1910–1930)." In *Wireless Imagination*, edited by Douglas Kahn and Gregory Whitehead, 197–243.

Goss, Glenda Dawn. "George Antheil, Carol Robinson and the Moderns." *American Music* 10 (winter 1992): 468–485.

Gottschalk, Louis Moreau. *Notes of a Pianist.* 1964; New York: DaCapo Press, 1979.

Goucher, F. S. "The Carbon Microphone: An Account of Some Researches Bearing on Its Action." *BSTJ* 13 (April 1934): 163–194.

Gouk, Penelope. "Acoustics in the Early Royal Society, 1660–1680." *Notes and Records of the Royal Society of London* 36 (February 1982): 155–175.

Gouk, Penelope. *Music, Science and Natural Magic in Seventeenth-Century England.* New Haven: Yale University Press, 1999.

Gouk, Penelope. "Performance Practice: Music, Medicine and Natural Philosophy in Interregnum Oxford." *British Journal for the History of Science* 29 (1996): 257–288.

Gouk, Penelope. "The Role of Acoustics and Music Theory in the Scientific Work of Robert Hooke." *Annals of Science* 37 (1980): 573–605.

Green, L., Jr. "Soundproofing the New York Life Insurance Company Building." *American Architect* 135 (20 March 1929): 411–412.

Griffiths, Paul. *Modern Music: A Concise History.* New York: Thames and Hudson, 1994.

Grinder, R. Dale. "The Battle for Clean Air: The Smoke Problem in Post–Civil War America." In *Pollution and Reform*, edited by Martin Melosi, 83–103.

Gross, David. "Space, Time and Modern Culture." *Telos* 50 (winter 1981–82): 59–78.

Guastavino, Raphael (Sr.). *Essay on the Theory and History of Cohesive Construction.* Boston: Ticknor and Co., 1893.

Gutheim, Frederick. "The Philadelphia Saving Fund Society Building: A Re-appraisal." *Architectural Record* (October 1949): 88–95, 180, 182.

Haber, Samuel. *Efficiency and Uplift: Scientific Management in the Progressive Era 1890–1920.* 1964; Chicago: University of Chicago Press, 1973.

Hackmann, Willem. *Seek and Strike: Sonar, Anti-Submarine Warfare and the Royal Navy, 1914–1954.* London: H.M.S.O., 1984.

Hackmann, Willem. "Sonar Research and Naval Warfare 1914–1954: A Case Study of a Twentieth-Century Establishment Science." *Historical Studies in the Physical and Biological Sciences* 16 (1986): 83–110.

Hall, Ben M. *The Best Remaining Seats: The Golden Age of the Movie Palace.* 1961; New York: Da Capo, 1988.

Hall, Edwin H. "Wallace Clement Ware Sabine, 1868–1919." *Biographical Memoirs of the National Academy of Science* 11 (1926): 1–19.

Hall, Edwin H., and Joseph Y. Bergen. *Text-book of Physics.* 3d rev. ed. New York: Henry Holt, 1903.

Hall, Robert E. "Organ Installation in Theaters." *Architectural Forum* 42 (June 1925): 401–402.

Hamlin, George. "The Making of Records." *Musician* 22 (July 1917): 542.

Hamlin, Talbot. *Greek Revival Architecture in America.* 1944; New York: Dover, 1964.

Hankins, Thomas L., and Robert J. Silverman. *Instruments and the Imagination.* Princeton: Princeton University Press, 1995.

Hanna, C. R., and J. Slepian. "The Function and Design of Horns for Loud Speakers." *JAIEE* 43 (March 1924): 250–256.

Hanson, O. B. "Microphone Technique in Radio Broadcasting." *JASA* 3 (July 1931): 81–93.

Hanson, O. B. "Planning the NBC Studios for Radio City." *PIRE* 20 (August 1932): 1296–1309.

Hanson, O. B. "The Story of Rockefeller Center: IX. The Plan and Construction of the National Broadcasting Company Studios." *Architectural Forum* 57 (August 1932): 153–160.

Hanson, O. B., and R. M. Morris. "The Design and Construction of Broadcast Studios." *PIRE* 19 (January 1931): 17–34.

Hanson, R. L. "One Type of Acoustic Distortion in Sound Picture Sets." *JSMPE* 15 (October 1930): 460–472.

Harris, Neil. "John Philip Sousa and the Culture of Reassurance." In *Perspectives on John Philip Sousa*, edited by Jon Newsom, 11–40. Washington, D.C.: Library of Congress, 1983.

Hart, Philip. *Orpheus in the New World: The Symphony Orchestra as an American Cultural Institution.* New York: W.W. Norton and Co., 1973.

Harvey, David. *The Condition of Postmodernity: An Enquiry into the Origins of Cultural Change.* Cambridge, Mass.: Blackwell, 1990.

Haskell, Douglas. "Architecture: The Rockefeller Necropolis." *Nation* 136 (31 May 1933): 622, 624.

Haskell, Douglas. "Roxy's Advantage over God." *Nation* 136 (4 January 1933): 11.

Hays, K. Michael. *Modernism and the Posthumanist Subject*. Cambridge, Mass.: The MIT Press, 1992.

Hays, Samuel P. *Conservation and the Gospel of Efficiency: The Progressive Conservation Movement 1890–1920*. Cambridge, Mass.: Harvard University Press, 1959.

Hazelton, George C., Jr. *The National Capitol: Its Architecture, Art and History*. New York: J.F. Taylor, 1902.

Hell's Angels. 1930; Caddo/United Artists, dir. Howard Hughes and James Whale, MCA Universal Home Video #80638.

Helmholtz, Hermann. *On the Sensations of Tone*. 2d English ed., trans. Alexander Ellis. 1885; New York: Dover, 1954.

Henry, Joseph. "On Acoustics Applied to Public Buildings." *Smithsonian Institution Report* (1856): 221–234. Also in *PAAAS* 10 (1856): 119–135.

Herf, Jeffrey. *Reactionary Modernism: Technology, Culture and Politics in Weimar and the Third Reich*. Cambridge: Cambridge University Press, 1984.

Herschel, J. W. F. "Sound." In *Encyclopedia Metropolitana*, edited by Edward Smedley, 2d div., vol. 2, pp. 747–824. London: Baldwin and Craddock, 1830.

Heynen, Hilde. *Architecture and Modernity: A Critique*. Cambridge, Mass.: The MIT Press, 1999.

Hilliard, John K. "Electroacoustics to 1940." *JASA* 61 (February 1977): 267–273.

Hitchcock, H. Wiley. *Music in the United States: A Historical Introduction*. New Jersey: Prentice Hall, 1969.

Hitchcock, Henry-Russell. *Architecture: Nineteenth and Twentieth Centuries*. New York: Penguin, 1977.

Hitchcock, Henry-Russell, and Philip Johnson. *The International Style*. 1932; New York: W.W. Norton and Co., 1966.

Hochheiser, Sheldon. "AT&T and the Development of Sound Motion-Picture Technology." In *The Dawn of Sound*, edited by Mary Lea Bandy, 23–33.

Hochheiser, Sheldon. "What Makes the Picture Talk: AT&T and the Development of Sound Motion Picture Technology." *IEEE Transactions on Education* 35 (November 1992): 278–285.

Hofmeister, Henry. "The Story of Rockefeller Center: V. The International Music Hall." *Architectural Forum* 56 (April 1932): 355–360.

Holland, Jeff, and Ed Butterworth. "Above All, a Graduate of BYU." *BYU Today* (September 1981): 33, 36.

Hollywood: A Celebration of the American Silent Film. 1980; Thames Television, dir. David Gill and Kevin Brownlow, Thames Video Collection/HBO Video.

The Hollywood Bowl. Los Angeles: Pepper Tree Press, 1939.

Hood, Raymond. "The Design of Rockefeller Center." *Architectural Forum* 56 (January 1932): 1–8.

Hood, Raymond. "The National Broadcasting Studios, New York." *Architectural Record* 64 (July 1928): 1–6, 25–37.

Hopper, F. L. "The Measurement of Reverberation Time and Its Application to Acoustic Problems in Sound Pictures." *JASA* 2 (April 1931): 499–505.

Horning, Susan Schmidt. "Chasing Sound: The Culture and Technology of Recording Studios in America, 1877–1977." Ph.D. Dissertation, Case Western Reserve University, 2002.

Horning, Susan Schmidt. "From Polka to Punk: Growth of an Independent Recording Studio, 1934–1977." In *I Sing the Body Electric,* edited by Hans-Joachim Braun, 148–159.

"How Echoes Are Produced: NBC Engineers Perfect Artificial Sound Reflection." *Broadcast News* 13 (December 1934): 26–27.

"How the Western Makes the Movies Move." *Western Electric News* 12 (August 1923): 10–11.

Howe, George. "The Philadelphia Saving Fund Society Branch Offices." *Architectural Forum* 48 (June 1928): 881–885.

Howe, M. A. DeWolfe. *The Boston Symphony Orchestra: 1881–1931.* Boston: Houghton Mifflin, 1931.

Howells, William Dean. "Editor's Easy Chair." *Harper's Monthly Magazine* 113 (November 1906): 957–960.

Hoyt, J. T. N. "Acoustics of the Hill Memorial Hall." *American Architect* 104 (6 August 1913): 49–53.

Hubert, Philip G., Jr. "For the Suppression of City Noises." *North American Review* 159 (November 1894): 633–635.

Huggins, Nathan Irvin. *Harlem Renaissance.* New York: Oxford University Press, 1971.

Hughes, Hector James. "Engineering and Other Applied Sciences in the Harvard Engineering School and Its Predecessors, 1847–1929." In *The Development of Harvard University Since the Inauguration of President Eliot, 1869–1929,* edited by Samuel Eliot Morison, 413–442. Cambridge, Mass.: Harvard University Press, 1930.

Hughes, Thomas P. *American Genesis: A Century of Invention and Technological Enthusiasm, 1870–1970.* New York: Penguin, 1989.

Humphrey, H. C. "Typical Sound Studio Recording Installations." *TSMPE* 13 (1929): 158–172.

Hunt, Frederick Vinton. *Electroacoustics: The Analysis of Transduction, and Its Historical Background.* 1954; n.p.: Acoustical Society of America, 1982.

Hunt, Frederick Vinton. *Origins in Acoustics: The Science of Sound from Antiquity to the Age of Newton.* 1978; Woodbury, N.Y.: Acoustical Society of America, 1992.

"Hush!—the Next New Big Business." *Literary Digest* 112 (9 January 1932): 46.

Huyssen, Andreas. *After the Great Divide: Modernism, Mass Culture, Postmodernism.* Bloomington: Indiana University Press, 1986.

Ingard, Uno. "Perforated Facing and Sound Absorption." *JASA* 26 (March 1954): 151–154.

Ingard, Uno. "Sound Absorption by Perforated Porous Tiles." *JASA* 26 (May 1954): 289–293.

Inman, W. S. *Report of the Committee of the House of Commons, on Ventilation, Warming and Transmission of Sound.* London: John Weale, 1836.

Irish, Sharon. "A 'Machine That Makes the Land Pay': The West Street Building in New York." *Technology and Culture* 30 (April 1989): 376–397.

"Isaiah Temple, Chicago, Ill." *AAAR* 126 (31 December 1924): 623–626.

Israel, Paul. *Edison: A Life of Invention.* New York: John Wiley & Sons, 1998.

Jäger, G. "Zur Theorie des Nachhalls." *Sitzungsberichte der Kaiserliche Akademie der Wissenschaften in Wien,* Bd. 120, Abt. IIA (1911): 612–634. Translated by F. R. Watson as "Acoustics of Auditoriums." *American Architect* 108 (8 December 1915): 369–374.

Jastrow, Joseph. "An Apparatus for the Study of Sound Intensities." *Science,* n.s., 3 (10 April 1896): 544–546.

The Jazz Singer. 1927; Warner Bros., dir. Alan Crosland. MGM/UA Home Video #M302312.

Johnson, H. Earle. *Symphony Hall, Boston.* Boston: Little, Brown and Co., 1950.

Johnson, James H. *Listening in Paris: A Cultural History.* Berkeley: University of California Press, 1995.

Jones, Isabel Morse. *Hollywood Bowl.* New York: G. Schirmer, 1936.

Jones, LeRoi. *Blues People: The Negro Experience in White America and the Music That Developed from It.* New York: William Morrow and Co., 1963.

Jones, Robert A. "Mr. Woolworth's Tower: The Skyscraper as Popular Icon." *Journal of Popular Culture* 7 (fall 1973): 408–424.

Jones, W. C. "Condenser and Carbon Microphones—Their Construction and Use." *JSMPE* 16 (January 1931): 3–22.

Jordan, John M. *Machine-Age Ideology: Social Engineering and American Liberalism, 1911–1939.* Chapel Hill: University of North Carolina Press, 1994.

Jordy, William H. *The Impact of European Modernism in the Mid-Twentieth Century.* New York: Oxford University Press, 1972.

Jordy, William H. "PSFS: Its Development and Its Significance in Modern Architecture." *JSAH* 21 (May 1962): 47–83.

Kahn, Douglas. *Noise Water Meat: A History of Sound in the Arts.* Cambridge, Mass.: The MIT Press, 1999.

Kahn, Douglas. "Sound Awake." *Australian Review of Books* (July 2000): 21–22.

Kahn, Douglas, and Gregory Whitehead, eds. *Wireless Imagination: Sound, Radio, and the Avant-Garde.* Cambridge, Mass.: The MIT Press, 1992.

Kanigel, Robert. *The One Best Way: Frederick Winslow Taylor and the Enigma of Efficiency.* New York: Viking, 1997.

Kasson, John F. *Amusing the Million: Coney Island at the Turn of the Century.* New York: Hill and Wang, 1978.

Kasson, John F. *Civilizing the Machine: Technology and Republican Values in America, 1776–1900.* New York: Penguin, 1976.

Kasson, John F. *Rudeness and Civility: Manners in Nineteenth-Century Urban America.* New York: Hill and Wang, 1990.

Kauffmann, Emil. *Architecture in the Age of Reason.* 1955; New York: Dover, 1968.

Kavanaugh, James V. "Three American Opera Houses: The Boston Theatre, The New York Academy of Music, The Philadelphia Academy of Music." M.A. Thesis, University of Delaware, 1967.

Kealy, Edward R. "From Craft to Art: The Case of Sound Mixers and Popular Music." *Sociology of Work and Occupations* 6 (February 1979): 3–29.

Kelley, H. Roy. "Grauman Theatre, Hollywood, Cal." *AAAR* 123 (31 January 1923): 125.

Kellogg, Edward W. "History of Sound Motion Pictures." In *Technological History of Motion Pictures,* edited by Raymond Fielding, 174–220.

Kellogg, Edward W. "Means for Radiating Large Amounts of Low Frequency Sound." *JASA* 3 (July 1931): 94–110.

Kellogg, Edward W. "Some New Aspects of Reverberation." *JSMPE* 14 (January 1930): 96–107.

Kelly, Eugene Henri. *Architectural Acoustics or the Science of Sound Application Required in the Construction of Audience Rooms.* Buffalo: Bensler & Wesley, 1898.

Kennelly, A. E., and G. W. Pierce. "The Impedance of Telephone Receivers as Affected by the Motion of Their Diaphragms." *PAAAS* 48 (September 1912): 113–151.

Kenney, William Howland. *Chicago Jazz: A Cultural History, 1904–1930*. New York: Oxford University Press, 1993.

Kenney, William Howland. *Recorded Music in American Life: The Phonograph and Popular Memory, 1890–1945*. New York: Oxford University Press, 1999.

Kent, H. C. "A Distinction in the Acoustic Purposes of Public Buildings." *AABN* 39 (7 January 1893): 9.

Kern, Stephen. *The Culture of Time and Space, 1880–1918*. Cambridge, Mass.: Harvard University Press, 1983.

Kevles, Daniel J. *The Physicists*. 1978; New York: Vintage, 1979.

Kidney, Walter. *The Architecture of Choice: Eclecticism in America, 1880–1930*. New York: George Braziller, 1974.

Kirby, Michael, and Victoria Nes Kirby. *Futurist Performance*. New York: PAJ Publications, 1986.

Klaber, John J. "Planning the Motion Picture Theatre." *Architectural Record* 38 (November 1915): 540–554.

Klenke, John. "Construction of a Sound-Proof Studio for Making Sound Motion Pictures." *American Architect* 140 (October 1931): 46–47, 66.

Klepper, David Lloyd. "The Acoustics of St. Thomas Church, Fifth Avenue." *JAES* 43 (July/August 1995): 599–601.

Kline, Ronald. "Construing 'Technology' as 'Applied Science': Public Rhetoric of Scientists and Engineers in the United States, 1880–1945." *Isis* 86 (June 1995): 194–221.

Kneser, Hans O. "Interpretation of the Anomalous Sound-Absorption in Air and Oxygen in Terms of Molecular Collisions." *JASA* 5 (October 1933): 122–126.

Knight, Arthur. "The Movies Learn to Talk: Ernst Lubitch, René Clair, and Rouben Mamoulian." In *Film Sound*, edited by Elisabeth Weis and John Belton, 213–220.

Knudsen, Vern O. "Absorption of Sound in Air, in Oxygen, and in Nitrogen—Effects of Humidity and Temperature." *JASA* 5 (October 1933): 112–121.

Knudsen, Vern O. "Acoustics of Music Rooms." *JASA* 2 (April 1931): 434–467.

Knudsen, Vern O. *Architectural Acoustics*. New York: John Wiley and Sons, 1932.

Knudsen, Vern O. "Carl F. Eyring." *JASA* 23 (May 1951): 370–371.

Knudsen, Vern O. "The Hearing of Speech in Auditoriums." *JASA* 1 (October 1929): 56–82.

Knudsen, Vern O. "Interfering Effect of Tones and Noise upon Speech Reception." *Physical Review,* 2d ser., 26 (July 1925): 133–138.

Knudsen, Vern O. *Modern Acoustics and Culture.* Berkeley: University of California Press, 1937.

Knudsen, Vern O. "Review of Architectural Acoustics during the Past Twenty-Five Years." *JASA* 26 (September 1954): 646–650.

Knudsen, Vern O. "The Sensibility of the Ear to Small Differences of Intensity and Frequency." *Physical Review,* 2d ser., 21 (January 1923): 84–102.

Knudsen, Vern O. *Teacher, Researcher, and Administrator: Vern O. Knudsen.* Transcript of oral history conducted 1966–1969 by James V. Mink. Collection 300/101. Department of Special Collections, Young Research Library, University of California, Los Angeles.

Knudsen, Vern O., and Cyril M. Harris. *Acoustic Designing in Architecture.* 1950; n.p.: Acoustical Society of America, 1978.

Kohler, Robert E. *Lords of the Fly: Drosophila Genetics and the Experimental Life.* Chicago: University of Chicago Press, 1994.

König, Rudolph. *Quelques Expériences d'Acoustique.* Paris: A. Lahure, 1882.

Koopal, Grace G. *Miracle of Music.* N.p.: Charles E. Toberman, 1972.

Kopec, John W. *The Sabines at Riverbank: Their Role in the Science of Architectural Acoustics.* Woodbury, N.Y.: Acoustical Society of America, 1997.

Kostof, Spiro, ed. *The Architect: Chapters in the History of the Profession.* New York: Oxford University Press, 1977.

Koszarski, Richard. "On the Record: Seeing and Hearing the Vitaphone." In *Dawn of Sound,* edited by Mary Lea Bandy, 15–21.

Kraft, James P. *Stage to Studio: Musicians and the Sound Revolution, 1890–1950.* Baltimore: Johns Hopkins University Press, 1996.

Kranz, Fred W. "Early History of Riverbank Acoustical Laboratories." *JASA* 49 (February 1971): 381–384.

Krinsky, Carol Herselle. *Rockefeller Center.* Oxford: Oxford University Press, 1978.

Kundert, Warren. "Acoustical Measuring Instruments over the Years." *JASA* 68 (July 1980): 64–69.

Kwolek-Folland, Angel. *Engendering Business: Men and Women in the Corporate Office, 1870–1930.* Baltimore: Johns Hopkins University Press, 1994.

Lachèz, Théodore. *Acoustique et Optique des Salles de Réunions Publiques.* Paris, Chez l'Auteur, 1848. See also 2d expanded ed., 1879.

Laird, Donald. "The Effects of Noise: A Summary of Experimental Literature." *JASA* 1 (January 1930): 256–262.

Laird, Donald. "Experiments on the Physiological Cost of Noise." *Journal of the National Institute of Industrial Psychology* 4 (January 1929): 251–258.

Laird, Donald. "The Measurement of the Effects of Noise on Working Efficiency." *Journal of Industrial Hygiene* 9 (October 1927): 431–434.

Laird, Donald. "Noise." *Scientific American* 139 (December 1928): 508–510.

Laird, Donald. "Noise *Does* Impair Production." *American Machinist* 69 (12 July 1928): 59–60.

Laird, Pamela Walker. *Advertising Progress: American Business and the Rise of Consumer Marketing*. Baltimore: Johns Hopkins University Press, 1998.

Lance, Adolphe. *Rapport au Nom de la Commission Chargée d'Examiner l'Ouvrage de M. Th. Lachèz*. Paris: Société Centrale des Architectes, 1849.

Landau, Sarah Bradford, and Carl W. Condit. *Rise of the New York Skyscraper, 1865–1913*. New Haven: Yale University Press, 1996.

Lane, Barbara Miller. *Architecture and Politics in Germany, 1918–1945*. 1968; Cambridge, Mass.: Harvard University Press, 1985.

Lanmon, Lorraine Welling. *William Lescaze, Architect*. Philadelphia: Art Alliance Press, 1987.

Lasky, Marvin. "Review of Undersea Acoustics to 1950." *JASA* 61 (February 1977): 283–297.

Lastra, James. "Reading, Writing, and Representing Sound." In *Sound Theory/Sound Practice*, edited by Rick Altman, 65–86.

Lastra, James. *Sound Technology and the American Cinema: Perception, Representation, Modernity*. New York: Columbia University Press, 2000.

Lastra, James. "Standards and Practices: Aesthetic Norm and Technological Innovation in the American Cinema." In *The Studio System*, edited by Janet Staiger, 200–225. New Brunswick: Rutgers University Press, 1995.

Lastra, James. "Standards and Practices: Technology and Representation in the American Cinema." Ph.D. Dissertation, University of Iowa, 1992.

Lathrop, George Parsons. "Edison's Kinetograph." *Harper's Weekly* 35 (13 June 1891): 446–447.

Latrobe, Benjamin. "Note." In *Edinburgh Encyclopedia* 1: 120–124. Philadelphia: Joseph and Edward Parker, 1832.

Lears, T. J. Jackson. *Fables of Abundance: A Cultural History of Advertising in America*. New York: Basic Books, 1994.

Lears, T. J. Jackson. *No Place of Grace: Antimodernism and the Transformation of American Culture 1880–1920.* New York: Pantheon Books, 1981.

Le Brun, N., and G. Runge. *History and Description of the Opera House or American Academy of Music in Philadelphia.* Philadelphia: G. Andre, 1857.

Le Corbusier. *Towards a New Architecture.* Translation of 13th French edition. 1931; New York: Dover, 1986.

Le Corbusier. *Vers une Architecture.* Paris: G. Cres et Cie, 1923.

Le Corbusier. *When the Cathedrals Were White.* Translation by Francis E. Hyslop, Jr. of 1938 French edition. 1947; New York: McGraw-Hill, 1964.

Lécuyer, Christophe. "MIT, Progressive Reform, and 'Industrial Service,' 1890–1920." *Historical Studies in the Physical and Biological Sciences* 26 (1995): 35–88.

Lenoir, Timothy. "Helmholtz and the Materialities of Communication." *Osiris,* 2d ser., 9 (1994): 184–207.

Leonard, Neil. *Jazz and the White Americans: The Acceptance of a New Art Form.* Chicago: University of Chicago Press, 1962.

Lescarboura, Austin C. "At the Other End of the Phonograph." *Scientific American* 119 (31 August 1918): 164, 178.

Levin, David Michael, ed. *Modernity and the Hegemony of Vision.* Berkeley: University of California Press, 1993.

Levine, Lawrence W. *Black Culture and Black Consciousness: Afro-American Folk Thought from Slavery to Freedom.* New York: Oxford University Press, 1977.

Levine, Lawrence W. *Highbrow/Lowbrow: The Emergence of Cultural Hierarchy in America.* Cambridge, Mass.: Harvard University Press, 1988.

Levine, Lawrence W. "Jazz and American Culture." *Journal of American Folklore* 102 (January–March 1989): 6–22.

Lewis, David Levering. *When Harlem Was in Vogue.* 1981; New York: Penguin, 1997.

Lewis, Sinclair. *Arrowsmith.* 1925; New York: Penguin, 1980.

Lewis, Tom. *Empire of the Air: The Men Who Made Radio.* New York: Harper Collins, 1991.

The Lights of New York. 1928; Warner Bros., dir. Bryan Foy. Video transfer, UCLA Film and Television Archive #F61-L3-11-17.

Lindsay, Robert Bruce. "Historical Introduction" to J. W. S. Rayleigh, *The Theory of Sound.* 2d ed. 1894; New York: Dover, 1945.

Lindsay, Vachel. *The Art of the Moving Picture.* New York: Macmillan, 1915.

Lippman, Walter. "Radio City." *American Architect* 143 (March 1933): 18.

Little, D. G. "KDKA: The Radio Telephone Broadcasting Station of the Westinghouse Electric and Manufacturing Company at East Pittsburgh, Pennsylvania." *PIRE* 12 (June 1924): 255–276.

Locke, Alain, ed. *The New Negro*. 1925; New York: Touchstone, 1997.

Loesser, Arthur. *Men, Women and Pianos: A Social History*. New York: Simon and Schuster, 1954.

Long, Robert, and Harold A. Rogers. "Four Ways to Put Yourself in the Concert Hall: Ambience-Simulation Devices Can Transform Home Listening." *High Fidelity and Musical America* 26 (October 1976): 63–68.

Longstreth, Richard W. "Academic Eclecticism in American Architecture." *Winterthur Portfolio* 17 (1982): 55–82.

Loth, David. *The City within a City: The Romance of Rockefeller Center*. New York: William Morrow and Co., 1966.

Loudon, James. "A Century of Progress in Acoustics." *Science,* n.s. 14 (27 December 1901): 987–995.

Lowe, David Garrard. "Monument of an Age." *American Craft* 48 (June 1988): 40–47, 104.

Lowry, Helen Bullitt. "Noise and Your Neighbors." *New York Times Book Review and Magazine* (20 March 1921): 8.

Loye, D. P., and Joseph Maxfield. "Sound in the Motion Picture Industry I. Some Historical Recollections." *Sound* 2 (September–October 1963): 14–27.

Lubar, Steven. *InfoCulture: The Smithsonian Book of Information Age Inventions*. Boston: Houghton, Mifflin, 1993.

Lyon, Gustave. *L'Acoustique Architecturale*. Paris: Bibliothèque Technique du Cinéma, 1932.

MacNair, W. A. "Some Acoustical Problems of Sound Picture Engineering." *PIRE* 19 (September 1931): 1606–1614.

Magoun, Alexander Boyden. "Shaping the Sound of Music: The Evolution of the Phonograph Record, 1877–1950." Ph.D. Dissertation, University of Maryland–College Park, 2000.

"Making Phonograph Music." *Current Literature* 33 (1902): 169–170.

Mamoulian, Rouben. "Rouben Mamoulian, Director." In *Sound and the Cinema: The Coming of Sound to American Film,* edited by William Cameron, 85–97. Pleasantville, N.Y.: Redgrave, 1980.

"The Manufacture of Edison Phonograph Records." *Scientific American* 83 (22 December 1900): 390.

Marchand, Roland. *Advertising the American Dream: Making Way for Modernity, 1920–1940*. Berkeley: University of California Press, 1985.

Marchand, Roland. *Creating the Corporate Soul: The Rise of Public Relations and Corporate Imagery in American Big Business*. Berkeley: University of California Press, 1998.

Marion, John Francis. *Within These Walls: A History of the Academy of Music in Philadelphia*. Philadelphia: Academy of Music, 1984.

Martin, Daniel W. "Floyd Rowe Watson, 1872–1972." *JASA* 55 (June 1974): 1362–1363.

Martin, W. H. "Decibel—The Name for the Transmission Unit." *BSTJ* 8 (January 1929): 1–2.

Marvin, Carolyn. *When Old Technologies Were New: Thinking About Electric Communication in the Late Nineteenth Century*. New York: Oxford University Press, 1988.

Marx, Leo. *The Machine in the Garden: Technology and the Pastoral Ideal*. London: Oxford University Press, 1964.

Massa, Frank. "Some Personal Recollections of Early Experiences on the New Frontier of Electroacoustics During the Late 1920s and Early 1930s." *JASA* 77 (April 1985): 1296–1302.

"Materials and Structural Elements in the Buildings of Rockefeller Center." *Engineering News-Record* 109 (17 November 1932): 590–592.

Maxfield, Joseph P. "Acoustic Control of Recording for Talking Motion Pictures." *JSMPE* 14 (January 1930): 85–95.

Maxfield, Joseph P. "Electro-Mechanical Sound Recording." *Bell Laboratories Record* 1 (January 1926): 197–202.

Maxfield, Joseph P., and H. C. Harrison. "Methods of High Quality Recording and Reproducing of Music and Speech Based on Telephone Research." *TAIEE* 45 (February 1926): 334–348. Also in *BSTJ* 5 (July 1926): 493–523.

Maxfield, Joseph P., and Ralph Townsend. "Acoustic Measurements of Set Materials." Report #4 of the Academy Producers-Technicians Committee. Hollywood: Academy of Motion Picture Arts and Science, 1 September 1930.

McGinn, Robert E. "Stokowski and the Bell Telephone Laboratories: Collaboration in the Development of High-Fidelity Sound Reproduction." *Technology and Culture* 24 (January 1983): 38–75.

McGinnis, C. S., and M. R. Harkins. "The Transmission of Sound through Porous and Non-Porous Materials." *Physical Review* 33 (August 1911): 128–136.

McLachlan, N. W. *The New Acoustics: A Survey of Modern Development in Acoustical Engineering*. London: Oxford University Press, 1936.

Melosi, Martin. "The Place of the City in Environmental History." *Environmental History Review* 17 (1993): 1–23.

Melosi, Martin, ed. *Pollution and Reform in American Cities, 1870–1930.* Austin: University of Texas Press, 1980.

Meloy, Arthur S. *Theatres and Motion Picture Houses.* New York: Architects' Supply and Publishing Co., 1916.

Merrill-Mirsky, Carol, and Jeanette Bovard. *Hollywood Bowl.* Los Angeles: Los Angeles Philharmonic Assoc., 2000.

Merritt, Ernest. "A Plea for Acoustic Engineering." *Sibley Journal of Engineering* 31 (1916): 101–102.

Milkovich, Ann Katherine. "Guastavino Tile Construction: An Analysis of a Modern Cohesive Construction Technique." M.S. Thesis, University of Pennsylvania, 1992.

Millard, Andre. *America on Record: A History of Recorded Sound.* Cambridge: Cambridge University Press, 1995.

Millard, Andre. *Edison and the Business of Innovation.* Baltimore: The Johns Hopkins University Press, 1990.

Miller, Dayton Clarence. *Anecdotal History of the Science of Sound.* New York: Macmillan, 1935.

Miller, Dayton Clarence. *The Science of Musical Sounds.* New York: Macmillan Co., 1916.

Miller, Harry B. "Acoustical Measurements and Instrumentation." *JASA* 61 (February 1977): 274–282.

Miller, Harry B., ed. *Acoustical Measurements: Methods and Instrumentation.* Stroudsburg, Penn.: Hutchinson Ross, 1982.

Miller, Wesley C. "The Illusion of Reality in Sound Pictures." In *Academy Fundamentals of Sound Recording*, 101–108.

Minter, Jerry B. "The AES Begins Its Seventh Year." *JAES* 2 (January 1954): 1–2.

Mitchinson, Paul. "Bouncing Off the Walls: Leo Beranek and the Science of Concert Hall Design." *Lingua Franca* 11 (April 2001): 61–66.

Mlinar, Emil M. "The Capitol Theater, New York, N.Y." *Architectural Forum* 32 (January 1920): 21–24.

Moderwell, Hiram K. "Ragtime." *New Republic* 4 (16 October 1915): 284–286.

Moore, Charles. *The Life and Times of Charles Follen McKim.* Boston: Houghton Mifflin Co., 1929.

Moore, MacDonald Smith. *Yankee Blues: Musical Culture and American Identity*. Bloomington: Indiana University Press, 1985.

Morgan, K. F. "Scoring, Synchronizing and Re-Recording Sound Pictures." *TSMPE* 13 (1929): 268–285.

Morgan, Robert. "'A New Musical Reality': Futurism, Modernism, and 'The Art of Noises.'" *Modernism/Modernity* 1 (1994): 129–151.

Morgan, Robert, ed. *Modern Times: From World War I to the Present*. Englewood Cliffs: Prentice Hall, 1994.

Morgenstern, Dan. "Jazz as an Urban Music." In *Music in American Society: 1776–1976*, edited by George McCue, 133–143. New Brunswick, N.J.: Transaction Books, 1977.

Morrison, Hugh. *Louis Sullivan: Prophet of Modern Architecture*. New York: Peter Smith, 1952.

Morton, David. *Off the Record: The Technology and Culture of Sound Recording in America*. New Brunswick: Rutgers University Press, 2000.

Mote, Carl Henry. "The Effort to Control Municipal Noise." *American City* 10 (February 1914): 147–149.

Mowery, H. Weaver. "Harmful Noises and Their Elimination." *Industrial Psychology* 1 (May 1926): 338–340.

Moyer, Albert E. *Joseph Henry: The Rise of an American Scientist*. Washington: Smithsonian Institution Press, 1997.

Muccigrosso, Robert. *American Gothic: The Mind and Art of Ralph Adams Cram*. Washington D.C.: University Press of America, 1980.

Muccigrosso, Robert. "Ralph Adams Cram and the Modernity of Medievalism." *Studies in Medievalism* 1 (spring 1982): 21–38.

Mueller, John H. *The American Symphony Orchestra: A Social History of Musical Taste*. Bloomington: Indiana University Press, 1951.

Mumford, Lewis. "The Skyline: Mr. Rockefeller's Center." *New Yorker* 8 (23 December 1933): 29–30.

Mumford, Lewis. "The Skyline: Two Theaters." *New Yorker* 8 (14 January 1933): 55–56.

Murphy, Claudette Michelle. "Sick Buildings and Sick Bodies: The Materialization of an Occupational Illness in Late Capitalism." Ph.D. Dissertation, Harvard University, 1998.

"Music and Noise." *Literary Digest* 113 (21 May 1932): 17.

Musser, Charles. *Before the Nickelodeon: Edwin S. Porter and the Edison Manufacturing Company*. Berkeley: University of California Press, 1991.

Musser, Charles. *The Emergence of Cinema: The American Screen to 1907*. New York: Charles Scribner's Sons, 1990.

Musser, Charles. *Thomas A. Edison and His Kinetographic Motion Pictures*. New Brunswick: Rutgers University Press, 1995.

Mussulman, Joseph A. *Music in the Cultured Generation: A Social History of Music in America, 1870–1900*. Evanston: Northwestern University Press, 1971.

Nanry, Charles. "Jazz and Modernism: Twin-Born Children of the Age of Invention." *Annual Review of Jazz Studies* 1 (1982): 146–154.

Naylor, David. *American Picture Palaces: The Architecture of Fantasy*. New York: Van Nostrand Reinhold, 1981.

Neergaard, Charles. "Controlling Hospital Noises." *Architectural Forum* 57 (November 1932): 449–450.

Newcomb, Edwin E. "Theatre Acoustics." In *American Theatres of Today*, edited by R. W. Sexton, vol. 2, pp. 41–46. New York: Architectural Book Publishing Co., 1930.

"A New Shelter for Savings." *Architectural Forum* 57 (December 1932): 483–498.

Newton, David. "Chicago's Auditorium Building." *Historic Illinois* 10 (April 1988): 2–5, 14.

Nichols, E. L., and W. S. Franklin. *Elements of Physics*. New York: Macmillan, 1903.

"Noise." *Nation* 56 (15 June 1893): 433–434.

"Noise." *Saturday Review of Literature* 2 (24 October 1925): 1.

"Noise and Health." *American City* 39 (November 1928): 161.

"Noise Nuisance Suggestions from Civic Club." *American City* 30 (April 1924): 429.

North, Arthur T. "The Orchestra Shell of the Hollywood Bowl." *Architectural Forum* 51 (November 1929): 549–552.

Northcutt, John Orlando. *Magic Valley: The Story of Hollywood Bowl*. Los Angeles: Osherenko, 1967.

Norton, C. L. "Sound-Proof Partitions." *American Architect* 78 (4 October 1902): 5–6.

Noverre, Jean George. *Observations sur la Construction d'une Salle d'Opéra* (1787). In *Oevres de M. Noverre, Tome III*. St. Petersburg: Jean Charles Schnoor, 1804.

Oakley, Imogen. "Public Health Versus the Noise Nuisance." *National Municipal Review* 4 (April 1915): 231–237.

Ochsner, Jeffrey Karl, and Dennis Alan Andersen. "Adler and Sullivan's Seattle Opera House Project." *JSAH* 48 (September 1989): 223–231.

O'Connell, Joseph. "The Fine-Tuning of a Golden Ear: High-End Audio and the Evolutionary Model of Technology." *Technology and Culture* 33 (January 1992): 1–37.

Ogren, Kathy J. *The Jazz Revolution: Twenties America and the Meaning of Jazz*. New York: Oxford University Press, 1989.

Oliver, Richard. *Bertram Grosvenor Goodhue*. New York and Cambridge, Mass.: The Architectural History Foundation and The MIT Press, 1983.

Olson, Harry F. "Mass Controlled Electrodynamic Microphones: The Ribbon Microphone." *JASA* 3 (July 1931): 56–68.

Olson, Harry F. "A Review of Twenty-Five Years of Sound Reproduction." *JASA* 26 (September 1954): 637–643.

Olson, Harry F. "The Ribbon Microphone." *JSMPE* 16 (June 1931): 695–708.

Olson, Harry F., and Irving Wolff. "Sound Concentrator for Microphones." *JASA* 1 (April 1930): 410–417.

"Open Air Acoustics." *Science* 63 (Supplement, 1926): x.

Orcutt, William Dana. *Wallace Clement Sabine: A Study in Achievement*. Norwood, Mass.: Plimpton Press, 1933.

Orvell, Miles. *The Real Thing: Imitation and Authenticity in American Culture, 1880–1940*. Chapel Hill: University of North Carolina Press, 1989.

Osswald, F. M. "The Acoustics of the Large Assembly Hall of the League of Nations at Geneva, Switzerland." *American Architect* 134 (20 December 1928): 833–842.

Ostransky, Leroy. *Jazz City: The Impact of Our Cities on the Development of Jazz*. Englewood Cliffs, N.J.: Prentice-Hall, 1978.

Ouellette, Fernand. *Edgard Varèse*. Translated by Derek Coltman. New York: Orion Press, 1968.

"Our Office Table." *Building News* 43 (6 October 1882): 433.

Painter, Nell. *Standing at Armageddon: The United States, 1877–1919*. New York: W.W. Norton and Co., 1987.

Parker, Karr. "Providing for Radio and Amplifying Installations in Large Buildings." *Architectural Forum* 50 (January 1929): 117–120.

Parkin, P. H., and K. Morgan. "'Assisted Resonance' in the Royal Festival Hall, London: 1965–1969." *JASA* 48 (May 1970): 1025–1035.

Parks, Janet, and Alan G. Neumann. *The Old World Builds the New: The Guastavino Company and the Technology of the Catalan Vault 1885–1962*. New York: Avery Architectural Library, 1996.

Parsons, Floyd W. "Devils of Din." *Saturday Evening Post* 203 (8 November 1930): 16–17, 126, 129–130.

Patte, Pierre. *Essai sur l'architecture théâtrale. Ou de l'ordonnance la plus avantageuse à une salle de spectacles, relativement aux principes de l'optique et de l'acoustique*. Paris: Chez Moutard, 1782.

"Paul Earls Sabine (1879–1958)." *JASA* 31 (April 1959): 536.

Peretti, Burton. *The Creation of Jazz: Music, Race, and Culture in Urban America*. Urbana and Chicago: University of Illinois Press, 1992.

Pérez-Gómez, Alberto. *Architecture and the Crisis of Modern Science*. Cambridge, Mass.: The MIT Press, 1983.

Pérez-Gómez, Alberto. "Architecture *as* Science: Analogy or Disjunction?" In *The Architecture of Science*, edited by Peter Galison and Emily Thompson, 337–351.

Perry, Bliss. *Life and Letters of Henry Lee Higginson*. Boston: Atlantic Monthly Press, 1921.

"The Pest of Noise." *American City* 43 (August 1930): 171.

Peterson, Frederick. "Dire Is the Noise." *Saturday Review of Literature* 7 (27 December 1930): 484.

Petzold, Ernst. "Regulating the Acoustics of Large Rooms." *JASA* 3 (October 1931): 288–291.

Pevsner, Nikolaus. *The Sources of Modern Architecture and Design*. London: Thames and Hudson, 1968.

Peyser, Ethel. *Carnegie Hall: The House That Music Built*. New York: Robert M. McBride, 1936.

"Philadelphia's Fancy." *Fortune* 6 (December 1932): 65–69, 130–131.

Picker, John M. "The Soundproof Study: Victorian Professionals, Work Space, and Urban Noise." *Victorian Studies* 42 (spring 1999/2000): 427–453.

Picon, Antoine. "Architecture, Science, and Technology." In *The Architecture of Science*, edited by Galison and Thompson, 309–335.

Pierce, George W. "A Simple Method of Measuring the Intensity of Sound." *PAAAS* 43 (February 1908): 377–395.

Pierson, William H., Jr. *Technology and the Picturesque, The Corporate and Early Gothic Styles*. New York: Oxford University Press, 1978.

"Planning, Engineering, Equipment: The Philadelphia Saving Fund Society Building." *Architectural Forum* 57 (December 1932): 543–550.

"Planning for Employees' Welfare in the Design of the New York Life Insurance Company Building." *American Architect* 135 (20 March 1929): 397–401.

Platinum Blonde. 1931; Columbia Pictures, dir. Frank Capra. Columbia Tristar Home Video #86973.

Platt, Walter B. "Certain Injurious Influences of City Life and Their Removal." *Journal of Social Science* 24 (April 1888): 24–30.

"The Plaza, Coventry." *Cinema Construction* (January 1930): 19.

Pocock, William Wilmer. "The Acoustics of Buildings." *AABN* 22 (31 December 1887): 312–313.

Porcello, Thomas. "The Ethics of Digital Audio-Sampling: Engineers' Discourse." *Popular Music* 10 (1991): 69–84.

Pound, Ezra. *Antheil and the Treatise on Harmony*. 1927; New York: Da Capo, 1968.

Pounds, Richard, Daniel Raichel, and Martin Weaver. "The Unseen World of Guastavino Acoustical Tile Construction: History, Development, Production." *APT Bulletin* 30 (1999): 33–39.

"The PSFS Building." *Perspecta* 25 (1989): 78–141.

Pursell, Carroll W., Jr. "Government and Technology in the Great Depression." *Technology and Culture* 20 (January 1979): 162–174.

"Quiet on the Set." *Bell Telephone Laboratories Reporter* 14 (May/June 1965): 28–31.

Rabinbach, Anson. *The Human Motor: Energy, Fatigue and the Origins of Modernity*. Berkeley: University of California Press, 1990.

Radano, Ronald M. "Soul Texts and the Blackness of Folk." *Modernism/Modernity* 2 (January 1995): 71–95.

"Radio City: NBC Studios Move into Elaborate New Quarters." *Newsweek* 2 (18 November 1933): 32–33.

"Radio City Music Hall, Rockefeller Center." *Architectural Forum* 58 (February 1933): 153–164.

Rapée, Erno. *Motion Picture Moods for Pianists and Organists: A Rapid-Reference Collection of Selected Pieces, Adapted to Fifty-two Moods and Situations*. New York: G. Schirmer, 1924.

Rayleigh, Lord. "On an Instrument Capable of Measuring the Intensity of Aerial Vibrations." *London, Edinburgh and Dublin Philosophical Magazine,* 5th ser., 14 (September 1882): 186–187.

Rayleigh, Lord. *The Theory of Sound.* 2d ed. 1894; New York: Dover, 1945.

Read, Oliver, and Walter L. Welch. *From Tin Foil to Stereo: Evolution of the Phonograph.* Indianapolis and New York: Howard Sams and Bobbs-Merrill, 1959.

Reich, Leonard S. *The Making of American Industrial Research: Science and Business at GE and Bell, 1876–1926.* Cambridge: Cambridge University Press, 1985.

Reinhard, L. Andrew. "Organization for Cooperation." *Architectural Forum* 56 (January 1932): 78–81.

Remarque, Erich Maria. *All Quiet on the Western Front.* Translated by A. W. Wheen. 1929; New York: Fawcett Crest, 1958.

"Report of H. W. Mowery, Chairman, Committee on the Elimination of Harmful Noises." *Transactions of the National Safety Council* (1926), vol. 1: 307–329.

Rettinger, Michael. "Reverberation Chambers for Rerecording." *JSMPE* 45 (November 1945): 350–357.

Revell, Keith. "Regulating the Landscape: Real Estate Values, City Planning and the 1916 Zoning Ordinance." In *The Landscape of Modernity*, edited by David Ward and Olivier Zunz, 19–45.

Rhea, Thomas LaMar. "The Evolution of Electronic Musical Instruments in the United States." Ph.D. Dissertation, George Peabody College for Teachers, 1972.

Rice, Chester W., and Edward W. Kellogg, "Notes on the Development of a New Type of Hornless Loud Speaker." *JAIEE* 44 (September 1925): 982–991.

Rice, Mrs. Isaac L. "An Effort to Suppress Noise." *Forum* 37 (April–June 1906): 552–570.

Rice, Mrs. Isaac L. "Our Most Abused Sense—The Sense of Hearing." *Forum* 38 (April–June 1907): 559–572.

Rice, Mrs. Isaac L. "'Quiet Zones' for Schools." *Forum* 46 (December 1911): 731–742.

Ringel, A. S. "Sound-Proofing and Acoustic Treatment of RKO Stages." *JSMPE* 15 (September 1930): 352–369.

Robinson, David. *From Peep Show to Palace: The Birth of American Film.* New York: Columbia University Press, 1996.

Robinson, L. T. "Some Thoughts about Motion Pictures with Sound." *TSMPE* 12 (1928): 856–866.

Rockefeller Center. New York: Rockefeller Center Inc., 1932.

Rockefeller Center. New York: Rockefeller Center Inc., c. 1934.

Roell, Craig H. *The Piano in America, 1890–1940.* Chapel Hill: University of North Carolina Press, 1989.

Rogers, J. A. "Jazz at Home." In *The New Negro*, edited by Alain Locke, 216–224.

Rose, Tricia. *Black Noise: Rap Music and Black Culture in Contemporary America*. Hanover, N.H.: Wesleyan University Press, 1994.

Rosenberg, Charles E. "Martin Arrowsmith: The Scientist as Hero." In Rosenberg, *No Other Gods: On Science and American Social Thought*, 123–131. Baltimore: Johns Hopkins University Press, 1976.

Rosenfeld, Paul. *An Hour with American Music*. 1929; Westport, Conn.: Hyperion, 1979.

Roszak, Theodore. *The Making of a Counterculture*. 1968; Berkeley: University of California Press, 1995.

Roth, Leland. *McKim, Mead and White, Architects*. New York: Harper and Row, 1983.

Rothafel, S. L. (Roxy). "The Architect and the Box Office." *Architectural Forum* 57 (September 1932): 194–196.

Rothafel, S. L. (Roxy). "What the Public Wants in the Picture Theater." *Architectural Forum* 42 (June 1925): 360–364.

Rudnick, Isadore. "Vern Oliver Knudsen, 1893–1974." *JASA* 56 (August 1974): 712–715.

Russell, John Scott. "Elementary Considerations of Some Principles in the Construction of Buildings Designed to Accommodate Spectators and Auditors." *Edinburgh New Philosophical Journal* 27 (April–October 1839): 131–136.

Russell, John Scott. "The Laws of Sound in the Construction of Buildings." *Builder* 4 (23 May 1846): 248.

Russolo, Luigi. *The Art of Noises*. Translated by Barclay Brown. 1916; New York: Pendragon Press, 1986.

Sabine, Hale J. "Building Acoustics in America, 1920–1940." *JASA* 61 (February 1977): 255–263.

Sabine, Hale J. "Manufacture and Distribution of Acoustical Materials over the Past 25 Years." *JASA* 26 (September 1954): 657–661.

Sabine, Paul E. *Acoustics and Architecture*. New York: McGraw-Hill, 1932.

Sabine, Paul E. "Acoustics of the Chicago Civic Opera House." *Architectural Forum* 52 (April 1930): 599–604.

Sabine, Paul E. "The Acoustics of Sound Recording Rooms." *TSMPE* 12 (1928): 809–822.

Sabine, Paul E. "Architectural Acoustics: Its Past and Its Possibilities." *JASA* 11 (July 1939): 21–28.

Sabine, Paul E. "The Beginnings of Architectural Acoustics." *JASA* 7 (April 1936): 242–248.

Sabine, Paul E. "The Wallace Clement Sabine Laboratory of Acoustics, Geneva, Ill." *American Architect* 116 (30 July 1919): 133–138.

Sabine, Wallace C. "Architectural Acoustics." *AABN* 62 (26 November 1898): 71–73.

Sabine, Wallace C. *Collected Papers on Acoustics*. Cambridge, Mass.: Harvard University Press, 1922. Also New York: Dover, 1968; and Malabar, Fla.: Peninsula Press, 1992.

Sabine, Wallace C. *A Student's Manual of a Laboratory Course in Physical Measurements*. Boston: Ginn and Co., 1893. Also 1896 and 1898.

"St. Thomas's Church, New York." *Architecture* 27 (15 January 1913): 5, 7.

"St. Thomas's Church, New York." *Architecture* 29 (January 1914): 4–6.

Salt, Barry. *Film Style and Technology: History and Analysis*. London: Starword, 1992.

Salt, Barry. "Film Style and Technology in the Thirties: Sound." In *Film Sound*, edited by Elisabeth Weis and John Belton, 37–43.

Saltzstein, Joan Weil. "The Autobiography and Letters of Dankmar Adler." *Inland Architect* 27 (September–October 1983): 16–27.

Saltzstein, Joan Weil. "Dankmar Adler: The Man, The Architect, The Author." *Wisconsin Architect* (July–August 1967): 15–19.

"Samuel Cabot." *PAAAS* 43 (July 1908): 547–556.

Samuelson, Tim, and Jim Scott. "Auditorium Album." *Inland Architect* 33 (September–October 1989): 64–71.

Santee, Howard B. "Installation and Adjustment of Western Electric Sound-Projector Systems." *Bell Laboratories Record* 7 (November 1928): 112–116.

Saunders, Frederick A., and Frederick V. Hunt. "George Washington Pierce." *Biographical Memoirs of the National Academy of Sciences* 33 (1959): 350–380.

Saunders, George. *A Treatise on Theatres*. London: I. and J. Taylor, 1790.

Say It with Songs. 1929; Warner Bros., dir. Lloyd Bacon. Video transfer, Museum of Modern Art.

Schafer, R. Murray. *The Book of Noise*. Wellington, N.Z.: Price Milburn and Co., 1970.

Schafer, R. Murray. *The New Soundscape*. Scarborough, Ont., and New York: Berandol Music Ltd. and Associated Music Publishers, 1969.

Schafer, R. Murray. *The Soundscape: Our Sonic Environment and the Tuning of the World*. Rochester, Vt.: Destiny Books, 1994.

Schickel, Richard, and Michael Walsh. *Carnegie Hall: The First One Hundred Years*. New York: Harry Abrams, 1987.

Schivelbusch, Wolfgang. *Disenchanted Night: The Industrialization of Light in the Nineteenth Century*. 1983; Berkeley: University of California Press, 1988.

Schivelbusch, Wolfgang. *The Railway Journey: The Industrialization of Time and Space in the 19th Century*. 1977; Berkeley: University of California Press, 1986.

Schmidt, Leigh Eric. "From Demon Possession to Magic Show: Ventriloquism, Religion, and the Enlightenment." *Church History* 67 (June 1998): 274–304.

Schmidt, Leigh Eric. *Hearing Things: Religion, Illusion, and the American Enlightenment*. Cambridge, Mass.: Harvard University Press, 2000.

Schmidt, Lorentz. "Acoustical Treatment of Classrooms." *Architectural Forum* 37 (August 1922): 111.

Schopenhauer, Arthur. *The Pessimist's Handbook*. Translated by T. Bailey Saunders. Lincoln: University of Nebraska Press, 1964.

Schuster, K., and E. Waetzmann. "Über den Nachhall in geschlossenen Räumen." *Annalen der Physik*, 5th ser., 1 (12 March 1929): 671–695.

Schuyler, Montgomery. "Great American Architects—Architecture in Chicago." *Architectural Record*, Special Series, 4 (December 1895): 2–48.

Schuyler, Montgomery. "The New St. Thomas's Church, Fifth Avenue, New York City." *Brickbuilder* 23 (January 1914): 15–20.

Schwartz, Hillel. "Beyond Tone and Decibel: The History of Noise." *Chronicle of Higher Education* (January 1998): B8.

"Science Searches for a Quiet Ashcan." *American City* 45 (August 1931): 109.

Sennett, Richard. *The Fall of Public Man: On the Social Psychology of Capitalism*. New York: Alfred A. Knopf, 1977.

Shafer, Yvonne. "The First Chicago Grand Opera Festival: Adler and Sullivan before the Auditorium." *Theatre Design and Technology* 13 (fall 1977): 9–13, 38.

Shand, William. "Observations on the Adaptation of Public Buildings to the Propagation of Sound." *Journal of the Franklin Institute* 39 (January 1845): 1–9.

Shand-Tucci, Douglas. *Boston Bohemia 1881–1900: Ralph Adams Cram: Life and Architecture*. Amherst: University of Massachusetts Press, 1995.

Shand-Tucci, Douglas. *Ralph Adams Cram: American Medievalist*. Boston: Boston Public Library, 1975.

Shankland, Robert. "Architectural Acoustics in America to 1930." *JASA* 61 (February 1977): 250–254.

Shapiro, Nat, and Nat Hentoff, eds. *Hear Me Talkin' to Ya*. 1955; New York: Dover, 1966.

Shea, T. E. "A Modern Laboratory for the Study of Sound Picture Problems." *JSMPE* 16 (March 1931): 277–285.

Sherman, Roger W. "Sound Insulation in Apartments." *Architectural Forum* 53 (September 1930): 373–378.

Siefert, Marsha. "Aesthetics, Technology, and the Capitalization of Culture: How the Talking Machine Became a Musical Instrument." *Science in Context* 8 (summer 1995): 417–449.

"A Silencer for Street Noises." *American Architect* 119 (2 February 1921): 131.

Silverman, Robert Jacob. "Instrumentation, Representation, and Perception in Modern Science: Imitating Human Function in the Nineteenth Century." Ph.D. Dissertation, University of Washington, 1992.

Siry, Joseph M. "Chicago's Auditorium Building: Opera or Anarchism." *JSAH* 57 (June 1998): 128–159.

Smilor, Raymond. "Cacophony at 34th and 6th: The Noise Problem in America, 1900–1930." *American Studies* 18 (1977): 23–38.

Smilor, Raymond. "Confronting the Industrial Environment: The Noise Problem in America, 1893–1932." Ph.D. Dissertation, University of Texas at Austin, 1978.

Smilor, Raymond. "Personal Boundaries in the Urban Environment: The Legal Attack on Noise: 1865–1930." *Environmental Review* 3 (spring 1979): 24–36.

Smilor, Raymond. "Toward an Environmental Perspective: The Anti-Noise Campaign, 1893–1932." In *Pollution and Reform*, edited by Martin Melosi, 135–151.

Smith, Bruce R. *The Acoustic World of Early Modern England: Attending to the O-Factor*. Chicago: University of Chicago Press, 1999.

Smith, Christine. *St. Bartholomew's Church in the City of New York*. New York: Oxford University Press, 1988.

Smith, E. Lawrence, and Donald Laird. "The Loudness of Auditory Stimuli Which Affect Stomach Contractions in Health Human Beings." *JASA* 2 (July 1930): 94–98.

Smith, Mark M. "Listening to the Heard Worlds of Antebellum America." *Journal of the Historical Society* 1 (spring 2000): 65–99.

Smith, Mark M. "Time, Sound, and the Virginia Slave." In *Afro-Virginian History and Culture*, edited by John Saillant, 29–60. New York: Garland, 1999.

Smith, T. Roger. *Acoustics in Relation to Architecture and Building*. London: Crosby, Lockwood and Son, 1895.

Smith, T. Roger. *A Rudimentary Treatise on the Acoustics of Public Buildings*. London: John Weale, 1861.

Smith, Terry. *Making the Modern: Industry, Art, and Design in America*. Chicago: University of Chicago Press, 1993.

Smulyan, Susan. *Selling Radio: The Commercialization of American Broadcasting, 1920–1934*. Washington: Smithsonian Institution Press, 1994.

Sousa, John Philip. "The Menace of Mechanical Music." *Appleton's Magazine* 8 (September 1906): 278–284.

Stanton, G. T. "Theatre Acoustics, Ventilating and Lighting." *Architectural Record* 68 (July 1930): 87–93.

Stanton, G. T., and F. C. Schmid. "Acoustics of Broadcasting and Recording Studios." *JASA* 4 (July 1932): 44–55.

Stapleton, Darwin H., and Edward C. Carter II. "'I have the itch of Botany, of Chemistry, of Mathematics . . . strong upon me': The Science of Benjamin Henry Latrobe." *Proceedings of the American Philosophical Society* 128 (September 1984): 173–192.

Starrett, Paul. *Changing the Skyline: An Autobiography*. New York: McGraw-Hill, 1938.

Stebbins, Richard Poate. *The Making of Symphony Hall, Boston*. Boston: Boston Symphony Orchestra, 2000.

Stern, Robert A. M. *George Howe: Toward a Modern American Architecture*. New Haven: Yale University Press, 1975.

Stern, Robert A. M. "PSFS: Beaux-Arts Theory and Rational Expressionism." *JSAH* 21 (May 1962): 84–95 and appendix, 95–102.

Stern, Robert A. M., Gregory Gilmartin, and John Massengale. *New York 1900: Metropolitan Architecture and Urbanism, 1890–1915*. New York: Rizzoli, 1983.

Stern, Robert A. M., Gregory Gilmartin, and Thomas Mellins. *New York 1930: Architecture and Urbanism Between the Two World Wars*. New York: Rizzoli, 1987.

Sterne, Jonathan. "Sounds Like the Mall of America: Programmed Music and the Architectonics of Commercial Space." *Ethnomusicology* 41 (winter 1997): 22–50.

Stewart, G. W. "Architectural Acoustics: Some Experiments in Reverberation." *Physical Review* 16 (June 1903): 379–380.

Stewart, G. W. "Architectural Acoustics: Some Experiments in the Sibley Auditorium." *Sibley Journal of Engineering* 17 (May 1903): 295–313.

Stewart, G. W. "Location of Aircraft by Sound." *Physical Review* 14 (August 1919): 166–167.

Stine, Jeffrey K., and Joel A. Tarr. "At the Intersection of Histories: Technology and the Environment." *Technology and Culture* 39 (October 1998): 601–640.

Stine, Jeffrey K., and Joel A. Tarr. "Technology and the Environment: The Historians' Challenge." *Environmental History Review* 18 (spring 1994): 1–7.

Stradling, David. *Smokestacks and Progressives: Environmentalists, Engineers and Air Quality in America, 1881–1951.* Baltimore: Johns Hopkins University Press, 1999.

Strasser, Susan. *Satisfaction Guaranteed: The Making of the American Mass Market.* Washington, D.C.: Smithsonian Institution Press, 1989.

"Street Noises and Skyscrapers." *American City* 42 (June 1930): 118.

Strunk, William, Jr. *The Elements of Style.* New York: Harcourt, Brace and Co., 1920.

Sullivan, Louis. "Development of Construction." *Economist* 55 (24 June 1916): 1252.

Suner, Bruno, and Jacques-Franck Degioanni. "Architectural Acoustics in France Circa World War I and Today . . . Deja Vu?" In *Proceedings of the Wallace Clement Sabine Centennial Symposium*, 37–40. Woodbury, N.Y.: Acoustical Society of America, 1994.

Sunrise. 1927; Fox Film Corp., dir. F. W. Murnau. Killiam Collection of Critics' Choice Video #4021.

Sutcliffe, Anthony, ed. *Metropolis 1890–1940.* London: Mansell, 1984.

Swan, Clifford M. "Acoustics of Picture Theaters." *Architectural Forum* 51 (November 1929): 545–548.

Swan, Clifford M. "Noise Problems in Banks." *Architectural Forum* 48 (June 1928): 913–916.

Swan, Clifford M. "Quiet for Hospitals." *Architectural Forum* 49 (December 1928): 935–937.

Swan, Clifford M. "The Reduction of Noise in Banks and Offices." *Architectural Forum* 38 (June 1923): 309–310.

Swan, Clifford M. "The Reduction of Noise in Hotels." *Architectural Forum* 51 (December 1929): 741–744.

"The Symbolism in St. Thomas's." *Pencil Points* 2 (September 1921): 9–12, 38.

Symphony Hall: The First 100 Years. Boston: Boston Symphony Orchestra, 2000.

"The Talking-Machine." *Littel's Living Age* 254 (24 August 1907): 486–489.

Tallant, Hugh. "Acoustic Design in the Hill Memorial Auditorium, University of Michigan." *Brickbuilder* 22 (August 1913): 169–173.

Tallant, Hugh. "Hints on Architectural Acoustics." *Brickbuilder* 19 (May 1910): 111–116; (July 1910): 155–158; (August 1910): 177–180; (September 1910): 199–203; (October 1910): 221–225; (November 1910): 243–247; (December 1910): 265–268.

Tarr, Joel A. *The Search for the Ultimate Sink: Urban Pollution in Historical Perspective*. Akron: University of Akron Press, 1996.

Taylor, Frederick Winslow. *The Principles of Scientific Management*. 1911; New York: W.W. Norton and Co., 1967.

Taylor, Hawley O. "A Direct Method of Finding the Value of Materials as Sound Absorbers." *Physical Review,* 2d ser., 2 (October 1913): 270–287.

Taylor, R. E. Lee. "Design and Plan of Small City Apartment Buildings." *Architectural Forum* 43 (September 1925): 121–126.

"The Temple of Sound." *Popular Mechanics* 60 (December 1933): 818–821.

Thayer, William Sydney. "The New Music Hall." *Dwight's Journal of Music* 2 (27 November 1852): 60.

"Theater Designs for Rockefeller Center." *Architectural Record* 71 (June 1932): 419.

Thomas, Rose Fay. *Memoirs of Theodore Thomas*. New York: Moffat Yard and Co., 1911.

Thompson, Emily. "Dead Rooms and Live Wires: Harvard, Hollywood, and the Deconstruction of Architectural Acoustics, 1900–1930." *Isis* 88 (December 1997): 597–626.

Thompson, Emily. "Listening to/for Modernity: Architectural Acoustics and the Development of Modern Spaces in America." In *The Architecture of Science*, edited by Peter Galison and Emily Thompson, 253–280.

Thompson, Emily. "Machines, Music, and the Quest for Fidelity: Marketing the Edison Phonograph in America, 1877–1925." *Musical Quarterly* 79 (spring 1995): 131–171.

Thoreau, Henry David. *Walden*. 1854; New York: Penguin, 1960.

Thunderbolt. 1929; Paramount Famous Lasky, dir. Josef von Sternberg. Museum of Modern Art.

Tichi, Cecelia. *Shifting Gears: Technology, Literature, Culture in Modernist America*. Chapel Hill: University of North Carolina Press, 1987.

Tick, Judith. "Passed Away Is the Piano Girl: Changes in American Musical Life, 1870–1900." In *Women Making Music: The Western Art Tradition*, edited by Jane Bowers and Judith Tick, 325–348. Urbana: University of Illinois Press, 1986.

Todd, Webster B. "Testing Men and Materials for Rockefeller City." *Architectural Forum* 56 (February 1932): 199–204.

"To Hush Office Noise." *Literary Digest* 48 (28 March 1914): 696–697.

Treib, Marc. *Space Calculated in Seconds: The Philips Pavilion, Le Corbusier, Edgard Varèse*. Princeton: Princeton University Press, 1996.

de Treville, Yvonne. "Making a Phonograph Record." *Musician* 21 (November 1916): 658.

Trowbridge, John, and W. C. Sabine. "Electrical Oscillations in Air." *PAAAS* 25 (1890): 109–123.

Trowbridge, John, and W. C. Sabine. "On the Use of Steam in Spectrum Analysis." *American Journal of Science*, 3d ser., 37 (February 1889): 114–116.

Trowbridge, John, and W. C. Sabine. "Selective Absorption of Metals for Ultra Violet Light." *PAAAS* 23 (1888): 299–300.

Trowbridge, John, and W. C. Sabine. "Wave-Lengths of Metallic Spectra in the Ultra Violet." *PAAAS* 23 (1888): 288–298.

Truax, Barry. *Acoustic Communication*. Norwood, N.J.: Ablex Publishing, 1984.

Truesdell, Clifford. "Editor's Introduction." In *Leonhardi Euleri Opera Omnia*. Ser. 2, vol. 13, xix–lxxii. Lipsiae: Teubneri, 1911+.

Tufts, F. L. "The Transmission of Sound through Porous Materials." *American Journal of Science*, 4th ser., 11 (May 1901): 357–364.

Tufts, F. L. "The Transmission of Sound through Solid Walls." *American Journal of Science,* 4th ser., 13 (June 1902): 449–454.

Tuthill, C. A. "The Art of Monitoring." *TSMPE* 13 (1929): 173–177.

Tuthill, William Burnet. *Practical Acoustics: A Study of the Diagrammatic Preparation of a Hall of Audience*. 1928; New York: Burnet C. Tuthill, 1946.

"The Twenty-Fifth Anniversary Celebration." *JASA* 26 (September 1954): 874–905.

Twombley, Robert. *Louis Sullivan: His Life and Work*. Chicago: University of Chicago Press, 1986.

Uglow, Jenny. *Hogarth: A Life and a World*. London: Faber and Faber, 1997.

Upham, J. B. "A Consideration of Some of the Phenomena and Laws of Sound, and Their Application in the Construction of Buildings Designed Especially for Musical Effect." *American Journal of Science and Art* 65 (1853): 215–226; 348–363; and 66 (1853): 21–33. Also in *Dwight's Journal of Music* from October 1852 through January 1853.

Valentine, Maggie. *The Show Starts on the Sidewalk: An Architectural History of the Movie Theatre*. New Haven: Yale University Press, 1994.

Van Horne, John C., ed. *The Correspondence and Miscellaneous Papers of Benjamin Henry Latrobe, Vol. 2: 1805–1810*. New Haven: Yale University Press, 1986.

Van Horne, John C., ed. *The Correspondence and Miscellaneous Papers of Benjamin Henry Latrobe, Vol. 3: 1811–1820*. New Haven: Yale University Press, 1988.

Van Horne, John C., and Lee W. Formwalt, eds. *The Correspondence and Miscellaneous Papers of Benjamin Henry Latrobe, Vol. 1: 1784–1804.* New Haven: Yale University Press, 1984.

Varèse, Louise. *Varèse: A Looking-Glass Diary. Vol. 1: 1883–1928.* New York: W.W. Norton and Co., 1972.

Venturi, Robert. *Complexity and Contradiction in Architecture.* New York: Museum of Modern Art, 1966.

Vitruvius. *Ten Books on Architecture.* Translated by Morris Hicky Morgan. 1914; New York: Dover, 1960.

The Voice from the Screen. 1926; Bell Telephone Laboratories, dir. Edward B. Craft. Video Yesteryear #877.

Walker, Alexander. *The Shattered Silents: How the Talkies Came to Stay.* New York: William Morrow and Co., 1979.

Walsh, George Ethelbert. "When Science Banishes City Noises." *Harper's Weekly* 51 (27 July 1907): 1098.

Ward, David, and Olivier Zunz. *The Landscape of Modernity: New York City 1900–1940.* Baltimore: Johns Hopkins University Press, 1992.

Wasserman, Neil H. *From Invention to Innovation: Long-Distance Telephone Transmission at the Turn of the Century.* Baltimore: Johns Hopkins University Press, 1985.

Waterfall, Wallace. "History of Acoustical Society of America." *JASA* 1 (October 1929): 5–8.

Watkins, Stanley. "Madam, Will You Talk?" *Bell Laboratories Record* 24 (August 1946): 289–295.

Watson, Floyd R. "Acoustics of Auditoriums." *Science* 67 (30 March 1928): 335–338.

Watson, Floyd R. "Acoustics of Auditoriums: An Investigation of the Acoustical Properties of the Auditorium at the University of Illinois." *University of Illinois Engineering Experiment Station Bulletin* no. 73 (1914): 1–32.

Watson, Floyd R. *Acoustics of Buildings.* New York: John Wiley and Sons, 1923; and 2d rev. ed., 1930.

Watson, Floyd R. "Acoustics of the Eastman Theatre, Rochester, N.Y." *American Architect* 128 (1 July 1925): 31–34.

Watson, Floyd R. "Acoustics of Motion Picture Theaters," *TSMPE* 11 (1927): 641–650.

Watson, Floyd R. "Air Currents and Their Relation to the Acoustics of Auditoriums." *Engineering Record* 67 (8 March 1913): 265–272.

Watson, Floyd R. "An Apparatus for Measuring Sound." *Physical Review* 30 (April 1910): 471–473.

Watson, Floyd R. "Architectural Acoustics: How Sound Interference in Buildings May Be Cured." *Scientific American Supplement* 68 (18 December 1909): 391.

Watson, Floyd R. "Bibliography of Acoustics of Buildings, *JASA* 3 (July 1931): 14–43.

Watson, Floyd R. "Ideal Auditorium Acoustics." *JAIA* 16 (July 1928): 259–267.

Watson, Floyd R. "Inefficiency of Wires as a Means of Curing Defective Acoustics of Auditoriums." *Science* 35 (24 May 1912): 833–834.

Watson, Floyd R. "Optimum Conditions for Music in Rooms." *Science* 64 (27 August 1926): 209–210.

Watson, Floyd R. "Perfect Acoustics of the Eastman Theatre." *Motion Picture News* 27 (1923): 354, 358.

Watson, Floyd R. "The Use of Sounding-Boards in an Auditorium." *Brickbuilder* 22 (June 1913): 139–141.

Watson, William T. "Baltimore's Anti-Noise Crusade." *National Municipal Review* 3 (July 1914): 585–589.

Webster, A. W. *On the Principles of Sound; Their Application to the New Houses of Parliament, and Assimilation with the Mechanism of the Ear.* London: The Author, 1840.

Webster, Arthur Gordon. "The Absolute Measurement of the Intensity of Sound." *TAIEE* 38 (Part 1): 701–723.

Webster, Arthur Gordon. "Absolute Measurement of Sound." *Science* 58 (31 August 1923): 149–152.

Webster, Arthur Gordon. "Acoustical Impedance, and the Theory of Horns and of the Phonograph." *PNAS* 5 (July 1919): 275–282.

Webster, Arthur Gordon. "A Complete Apparatus for Absolute Acoustical Measurements." *PNAS* 5 (May 1919): 173–179.

Weibe, Robert. *The Search for Order, 1877–1920.* New York: Hill and Wang, 1967.

Weidenaar, Reynold. *Magic Music from the Telharmonium.* Metuchen, N.J., and London: Scarecrow Press, 1995.

Weis, Elisabeth, and John Belton, eds. *Film Sound: Theory and Practice.* New York: Columbia University Press, 1985.

Weisman, Winston. "Who Designed Rockefeller Center?" *JSAH* 10 (March 1951): 11–17.

Weiss, Marc A. "Density and Intervention: New York's Planning Traditions." In *Landscape of Modernity*, edited by David Ward and Olivier Zunz, 46–75.

Welch, Walter, and Leah Brodbeck Stenzel Burt. *From Tinfoil to Stereo: The Acoustic Years of the Recording Industry, 1877–1929.* Gainesville: University Press of Florida, 1994.

Wen-Chung, Chou. "Open Rather than Bounded." *Perspectives of New Music* 5 (fall–winter 1966): 1–6.

Wente, Edward C. "A Condenser Transmitter as a Uniformly Sensitive Instrument for the Absolute Measurement of Sound Intensity." *Physical Review,* 2d. ser., 10 (July 1917): 39–63.

Wente, Edward C., and E. H. Bedell. "A Chronographic Method of Measuring Reverberation Time." *JASA* 1 (April 1930): 422–427.

Wente, Edward C., and A. L. Thuras. "A High Efficiency Receiver for a Horn-Type Loud Speaker of Large Power Capacity." *BSTJ* 7 (January 1928): 140–153.

West, Helen Howe. *George Howe, Architect, 1886–1955: Recollections of My Beloved Father.* Philadelphia: William Nunn Co., 1973.

West, W. *Acoustical Engineering.* London: Sir Isaac Pitman & Sons, 1932.

White, John J. *Literary Futurism: Aspects of the First Avant Garde.* Oxford: Clarendon Press, 1990.

Whitesitt, Linda. *The Life and Music of George Antheil, 1900–1959.* Ann Arbor: UMI Research Press, 1983.

Whitford, Frank. *Bauhaus.* London: Thames and Hudson, 1984.

Whittemore, Charles A. "The Motion Picture Theater." Parts I-IV, *Architectural Forum* 26 (June 1917): 171–176; 27 (July 1917): 13–18; 27 (August 1917): 39–43; 27 (September 1917): 67–72.

Whittemore, Charles A. "The Moving Picture Theatre." *Brickbuilder* 23 (February 1914): Supplement, 41–45.

Wight, Peter B. "The Works of Raphael Guastavino." *Brickbuilder* 10 (April 1901): 79–81; (May 1901): 100–102; (September 1901): 184–188; (October): 211–214.

Wilcox, Earley Vernon. "To Heal the Blows of Sound." *Harvard Graduates' Magazine* 33 (June 1925): 584–590.

Williams, George W. "Robert Mills' 'Contemplated Addition to St. Michael's Church, Charleston' and 'Doctrine of Sounds,'" *JSAH* 12 (March 1953): Special Documentary Supplement, 23–31.

Williams, William Carlos. "George Antheil and the Cantilene Critics: A Note on the First Performance of Antheil's Music in New York City, April 10-1927." *Transitions* 13 (summer 1928): 237–240.

Willis, Carol. *Form Follows Finance: Skyscrapers and Skylines in New York and Chicago.* New York: Princeton Architectural Press, 1995.

Willis, Carol. "Zoning and Zeitgeist: The Skyscraper City in the 1920s." *JSAH* 45 (March 1986): 47–59.

Wilson, Eric. "Plagues, Fairs, and Street Cries: Sounding Out Society and Space in Early Modern England." *Modern Language Studies* 25 (summer 1995): 1–42.

Wilson, Richard Guy. "International Style: The MoMA Exhibition." *Progressive Architecture* 63 (February 1982): 92–105.

Winner, Langdon. *Autonomous Technology: Technics-out-of-Control as a Theme in Political Thought.* Cambridge, Mass.: The MIT Press, 1977.

Wise, George. *Willis R. Whitney, General Electric, and the Origins of U.S. Industrial Research.* New York: Columbia University Press, 1985.

Wolf, S. K. "Reproduction in the Theatre." In *Academy Fundamentals of Sound Recording,* 165–183.

Wolf, S. K. "Theater Acoustics for Sound Reproduction." *JSMPE* 14 (February 1930): 151–160.

Wolf, S. K. "Theatre Acoustics for Reproduced Sound." In *Academy Fundamentals of Sound Recording,* 73–84.

Wolfe, Tom. *From Bauhaus to Our House.* New York: Washington Square Press, 1981.

Woods, Frank. "The Sound Motion Picture Situation in Hollywood." *TSMPE* 12 (1928): 625–632.

"World's Largest Theater in Rockefeller Center Will Seat Six Thousand Persons." *Popular Mechanics* 58 (August 1932): 252–253.

Wright, Frank Lloyd. "Acoustics in Building." Recorded lecture to students at Taliesin Fellowship, Scottsdale, Arizona, 1952. Audio-Forum Sound Seminars, #11021.

Wyatt, Benjamin (Dean). *Observations on the Design for the Theatre Royal, Drury Lane.* London: J. Taylor, 1813.

Wyatt, Benjamin (Dean). *Observations on the Principles of a Design for a Theatre.* London: Lowndes and Hobbs, 1811.

Zunz, Olivier. *Making America Corporate: 1870–1920.* Chicago: University of Chicago Press, 1990.

Index

"Abating the Noise Evil" (Soglow), *165*
Absorbex, 169, 191
Absorption coefficients. *See* Coefficient of absorption
Academy of Motion Picture Arts and Sciences, 272–273, 409n142
"Accuracy of Musical Taste in Regard to Architectural Acoustics, The" (Sabine), 55
Acoustical building materials, 169–228. *See also* Acoustical tile; Coefficient of absorption; Felt; Plaster
    Cabot's Quilt, 173, 174–175, 378n13
    efficiency promoted by, 7
    in Europe, 209–210
    frequency and sound absorption, 62–63, *63*
    Henry's experiments on, 27–28, 34, 37
    Knudsen's research on, 101
    measuring sound-absorbing properties of, 84–85
    in modern auditorium, 249, 251
    in motion picture theaters, 262
    in NBC studios, 306
    and new acoustical criteria, 250
    in New York Life Building, 198–207
    perforations in, 218, 220, 392n142
    in Philadelphia Saving Fund Society Building, 210–226
    proliferation of, 190–198
    in Radio City Music Hall, 309
    in Rockefeller Center offices, 301
    Sabine's use of, 78
    in St. Thomas's Church, 172, 180–187
    sound and space dissociated by, 2–3, 171–172, 321
    in soundstages, 269, 409n134
    in *Sweet's Architectural Trade Catalogue*, 218, 379n19, 379n24, 384n66
    at the turn of the century, 173–179
Acoustical Corporation of America, 221, 393n153
Acoustical engineering. *See also* Architectural acoustics; Electroacoustic devices
    and Acoustical Society of America, 108
    Audio Engineering Society, 356n149
    children urged to consider as career, 59–60
    in college curricula, 60
    in noise abatement, 144–157
    women in, 343n8
Acoustical impedance, 97
Acoustical recording, 236, 263–264, *265*
Acoustical Society of America
    corporate support for, 105, 356n143
    first official meeting of, 105–106
    founding of, 5, 60, 62, 104–105
    and motion picture sound, 273, 410n142
    the New Acoustics and, 107
    organizational meeting of, 105, *106*
    physics and engineering balanced in, 108

Acoustical tile
   Acousti-Celotex, 170, 218–220, *219,*
      377n5, 391n132
   Acoustone, 170, 218
   Akoustolith, 170, 188–190, *188,* 217,
      319, 384n63
   Banham on "tyranny of the tile
      format," 318
   Guastavino sound-absorbing tiles,
      73–74, 180–187, 346n36
   Insulite Acoustile, 190, 193, 216, 218
   Mutetile, 221, *222,* 393n152
   Rumford tile, 183, 185, 186–187, 189,
      319, 382n40
   Sanacoustic Tile, 170, 220–221,
      393n148, 394n157
   variety of, 170, 217–221
Acoustic Control System (ACS), 424n23
Acousti-Celotex, 170, 218–220, *219,*
   377n5, 391n132
Acoustics. *See also* Acoustical Society of
   America; Acoustical engineering;
   New Acoustics
   as changing after World War I, 62
   as "has-been branch of physics," 104
   Knudsen on modern era in, 104
   Loudon's address of 1901 on, 59
   new tools in development of modern,
      90–99
   and noise control, 119
   origins of modern, 13–57
   papers at American Physical Society,
      89, 104
   World War I as catalyst for, 87–89
Acoustics of Buildings (Watson), 252, *253,*
   354n122
Acoustone, 170, 218
ACS (Acoustic Control System), 424n23
Addams, Jane, 50
Adler, Dankmar, 29–33, 45, 335n59,
   336n60
Advertising
   of building products, 196–198, 385n72

   of Philadelphia Saving Fund Society
      Building, 224, *225, 226,* 393n155
   of phonographs, 397n26
African American migration to northern
   cities, 131, 132
Air conditioning
   first air-conditioned office building,
      393n154
   noise from, 103
   in Philadelphia Saving Fund Society
      Building, 221–222
   for recording studios, 267
Airplanes, broadcasting from, 150–151
Aitken, Hugh G. J., 8, 92–93
Akoustikos Felt, 221, 387n95
Akoustolith, 170, 188–190, *188,* 217,
   319, 384n63
Algarotti, Francesco, 20, 24, 46, 332n29
Allen, Frederick Lewis, 313
All Quiet on the Western Front
   (Remarque), 88
Altman, Rick, 283, 284, 321, 406n105
Altschuler, Alfred, 72, 189
American Physical Society, 89, 104
American Society of Cinematographers,
   272–273
American Standards Association, 106
American Telephone & Telegraph
   Company (AT&T). *See also* Bell
   Telephone Laboratories; Electrical
   Research Products Incorporated;
   Western Electric
   acoustical engineers employed at, 60
   Acoustical Society of America
      supported by, 105
   becoming a monopoly, 91
   in-house research department of, 91
   public address and sound motion sys-
      tems developed by, 241
   radio as threat to, 91–92
   subway noise measurement by, 162,
      163
   vacuum-tube amplifier of, 94, 95, 245

Amériques (Varèse), 138–140
Ames, Winthrop, 72, 109
Amplifiers
  AT&T research on, 92
  in recording studios, 264
  reproducing horns as, 237
  vacuum-tube amplifiers, 94, 95, 99, 241, 244, 245
Amusement parks, 124
Animals
  complaints about noise from, 115, 127, 160
  horse-drawn traffic noise, 148, 149
  organic sound created by, 116
Antenaplex System, 301
Antheil, George, 141–144
  Ballet pour Instruments Mécanique et Percussion, 141, 142–143, 145, 369n122
  Mechanisms, 142, 369n119
  Symphony for Five Instruments, 141
Applause, 48
Applied science, 356n148
Apthorp, William Foster, 51, 52–53, 54, 56, 57, 341n137
Arcana (Varèse), 140, 141
Architectural acoustics. *See also* Acoustical building materials; Reverberation equation
  desire for control driving, 2
  in eighteenth and nineteenth centuries, 18–33
  growing literature on, 81, 82, 348n61
  modern acoustical science in construction of Boston Symphony Hall, 4, 5, 57
  for motion picture theaters, 256–263
  for recording studios, 263–267
  as reformulated for new aural environment, 234–235
  Sabine as consultant to architects, 69–73
  Sabine on physical and subjective aspects of, 55

  Sabine's *Collected Papers on Acoustics,* 100
  for soundstages, 267–272
  Watson's *Acoustics of Buildings,* 252, 253, 354n122
Architectural Forum (journal), 198, 385n80, 387n93
Architecture. *See also* Architectural acoustics; Modernism; *and architects and buildings by name*
  architects as won over to acoustical engineering, 60
  Art Deco, 229, 394n1
  eclecticism, 28
  historicism, 28
  neo-Gothicism, 169, 180, 185, 207, 208
  neoclassicism, 24, 27, 28, 332n34
  postmodern aesthetic in, 323–324
Armat, Thomas, 400n43
Arnold, Harold, 92, 93, 106, 320
Arrowsmith (Lewis), 112–113, 357n159
Art Deco, 229, 394n1
Articulation testing, 101, 252–254
"Art of Noises, The" (Russolo), 136, 366n87
Asbestos, 176, 318
Ash can, "semi-noiseless," 162
Assisted resonance, 322
Astor, John Jacob, 69
AT&T. *See* American Telephone & Telegraph Company
Atkinson, Brooks, 310
Attali, Jacques, 325n1
Audience
  applause, 48
  listening habit developed by, 247
  psychology affecting judgments of acoustics in, 44, 55
  Symphony Hall gradually accepted by, 57
  transformation of, 47–48
Audio Engineering Society, 356n149

Audiometers, 146–148, *147*, 158, 370n135, 370n137, 371n141
Audion, 92–93, 245
Auditorium Building and Theater (Chicago), 29–33
   auditorium interior, *31*
   Chicago Symphony Orchestra moves from, 56, 342n154
   cross-section of, *30*
   opinions on acoustics of, 30, 32
   as precedent for Radio City Music Hall, 417n23
Auditoriums. *See also* Concert halls; Theaters
   Horall's auditorium simulator, 323
   initial-time-delay gap criteria for, 320
   the modern, 248–256
   new ideal type of, 248
   Watson's "Ideal Auditorium Acoustics," 254
Aula Magna (Caracas, Venezuela), 319
Automobiles, noise created by, 124, 148, 149, *160*
Avant-garde music
   modern noises inspiring, 119, 133–144
   sound's meaning redefined in, 6, 119
Avery Fisher Hall (New York City), 422n9
Awakening of a City (Russolo), 137

Bache, Alexander Dallas, 25, 334n45
Bacon, C. N., 175
Bagasse, 218, 220
Bagenal, Hope, 389n114
Ballet pour Instruments Mécanique et Percussion (Antheil), 141, 142–143, 145, 369n122
Baltimore, anti-noise policeman in, 126
Banham, Reyner, 209, 210, 318, 327n11, 389n116
Banks, acoustical correction in, *197,* 198
Bareiss, Warren, 327n10, 361n37
Barrymore, John, 246

Bauhaus, 208, 209
Beethoven, Ludwig van, 4, 15, 51, 52, 53
Bell, Alexander Graham, 90–91, 93, 107, 108, 158
Bell, Chichester, 396n19
Bellamy, Edward, 50
Bell Telephone Laboratories
   Acoustical Society of America organizational meeting at, 105
   electrical recording developed by, 240
   Noise Abatement Commission of New York City supported by, 158
   noise investigated by, 146
   Sound Picture Laboratory, *285,* 286, *287*
Bennet, William, 121
Bennet Act (1907), 121, 360n24
Beranek, Leo, 248, 320, 342n153
Berliner, Emile, 91, 396n19
Betts, Benjamin, 170, 171, 224
Bingham, Thomas, 124
Birth of a Nation, The (film, dir. Griffith), 242
Black Watch, The (film, dir. Ford), 283
Blodgett, William, 182
Bluestone, Daniel, 125
B'Nai Jeshurun Synagogue (Newark, N.J.), 189, *189,* 383n61
Boccioni, Umberto, 135
Boehm, W. M., 84
Bolt, Beranek and Newman, 320, 323
Boston. *See also* Harvard University; Music Hall; Symphony Hall
   First Church of Christ, Scientist, *179*
   Massachusetts Institute of Technology, 60, 76, 347n46
   New England Conservatory of Music, 62, 174, 175, 261
   St. Botolph's Club, 104
Boston Acoustical Engineering Company, 218
Boston Symphony Orchestra, 48, 342n153

Boston Theatre, 28
Bourke-White, Margaret, 301, *302, 303*
Brown, Barclay, 366n90
Brown, Edward F., *118,* 373n169
Brunswick Phonograph Company, 397n26, 398n31
Buffalo (New York)
   Kleinhans Music Hall, 248, 319
   New York Central Railroad Terminal, 189
Building materials, acoustical. *See* Acoustical building materials
Burgess, C. F., Laboratories, 155, 220, 259, 372n162, 393n148
Burnham, Daniel, 56
Busoni, Ferruccio, 134

Cabot, Samuel, 173–175, 378n11
Cabot's Quilt, 173, 174–175, 378n13
Cahill, Thaddeus, 134, 140
Cameraphone, 244
Campbell, William, 100
Capacitor (condenser), 94
Capra, Frank, 283
Carbon transmitter, 91, 94
Careers (pamphlets), 60, 109–110
Carlyle, Thomas, 116, 117, 358n8
Carnegie Hall (New York City), 29, 48–49, 142, 154, 238, 336n60
Carrà, Carlo, 135
Carrier, Willis, 221
Carrier Corporation, 103, 393n154
Cars, noise created by, 124, 148, 149, *160*
Carty, John J., 91–92
Case, Theodore, 245, 247, 400n49
Cass, John, 277, 284
Cathedral of St. John the Divine (New York City), 70, 72, 181
Cell phones, 396n16
Celotex Company, 218
   Acoustical Society of America supported by, 105
   Acousti-Celotex, 170, 218–220, *219,* 377n5, 391n132

   "Less Noise" slogan of, 210
Central Park in the Dark (Ives), 133–134
Century Theater (New Theater) (New York City), 65
Chaloner Theater (New York City), 275
Chicago. *See also* Auditorium Building and Theater
   Civic Opera House, 248, 417n23
   Isaiah Temple, 189
   noise survey in, 155
   as noisier than New York, 376n190
   noisy radio incident in, 151
   Sabinite at 300 West Adams St. building, *217*
   street vendors' noise ban, 124–125
   WLS radio, 266, 408n123
Chicago Symphony Orchestra, 48, 56
Chladni, Ernst, 19
Chronophone, 243–244, 400n45
Cinema. *See* Motion pictures
Cities. *See also* City noise; *and cities by name*
   African American migration to, 131, 132
   as engines of change, 132
City Beautiful movements, 125–126
City noise
   in avant-garde music, 133–144
   becoming a problem in 1920s, 6, 61, 115–120
   engineering approach to noise abatement, 144–157
   in jazz, 130–132
   in motion pictures, 280–281, 412n166
   progressive movement for noise abatement, 120–130
City Noise (Brown et al.), 376n192
   charts showing noise levels, *163*
   Day cartoon in, *150*
   Fletcher's radio address reproduced in, 375n186
   frontispiece of, *118,* 371n147
   noise abatement questionnaire in, *159*

Civic Club of Philadelphia, 126
Civic Opera House (Chicago), 248, 417n23
Clothing, women reducing amount of, 157, 374n173
Coefficient of absorption
   the "coefficient wars," 193
   at first meeting of Acoustical Society of America, 106
   Johns-Manville acoustics department using, 178
   in reverberation equation, 40–42
   Sabine calculating, 40, 84, 286
   of sound-absorbing materials, 288, 414n187
   for stage sets, 276–277, 410n150
   Stewart calculating cocoa-matting's, 81
Coffman, Joe, 279
Collected Papers on Acoustics (Sabine), 100–101
Columbia Broadcasting Company, 376n190
Columbia Phonograph Company, 145, 398n31
Concert halls. *See also by name*
   becoming solemn places, 48
   Bolt, Beranek and Newman survey of, 320
   Hi-Fi concert halls, 248
   late-twentieth-century halls as configurable, 321–322
   the modern auditorium, 248–256
   reverberation time standards changing for, 319–320
Condenser (capacitor), 94
Condenser transmitter, 90–96, *95*, 353n113
Cone-type loudspeakers, 239, *240*, 398n29
Coney Island, 123–124
Connors, John, 347n45
Conservation of energy, 27, 32

Constant Temperature Room (Jefferson Physical Laboratory, Harvard University), 66–69, 83, 113, 357n160
Coolidge, Calvin, 102
Copeland, Aaron, 142
Corbin, Alain, 1
Cornell University Sibley Auditorium, 81
Corporate research programs, 90
Cotting, Charles, 17
Cowell, Henry, 133
Crafton, Donald, 279, 283, 411n154
Cram, Ralph Adams
   design of St. Thomas's Church, 185
   on neo-Gothic style, 180–181
   modernity in work of, 186, 187, 383n55
   on Rockefeller Center, 313
   and Rumford tile, 186
   and Sabine, 382n45
Cram & Ferguson, 189
Cram, Goodhue & Ferguson
   Cathedral of St. John the Divine, 70, 72, 181
   neo-Gothic churches of, 172
   Sabine and Guastavino introduced by, 180, 182
   Sabine consulting for, 70–71, 72
   St. Thomas's Church, 70, 183–186
   West Point chapel, 182
Crandall, Irving, 93–94, 146
Crowd, The (film, dir. Vidor), 412n166
Cubism, 277, 284
Cut-outs (automobile), 149, 371n142

Dahlberg, Bror, 218, 220
Darlington, Thomas, 122
Davis, Jefferson, 25
Day, Robert, *150*
Decibel, 158, 162, *163*, 164, 375n182
De Forest, Lee, 92–93, 245, 397n21
De Forest Phonofilm Corporation, 245
Depero, Fortunato, 135
Dewey, Melville, 157

Diamond Disc Phonograph, 237–239, 240
Dickson, William K. L., 242, 399n42
Disraeli (film, dir. Green), 413n174
Don Juan (film, dir. Crosland), 246, 273
Douglas, Susan J., 8, 236, 326n8
Downes, Olin, 139, 154
Dreher, Carl, 279, 280, 413n177
Dr. Jekyll and Mr. Hyde (film, dir. Mamoulian), 283
Drury Lane Theatre (London), 24
Duchamp, Marcel, 142
Duke University Chapel, 319, 421n7
Dwight, John Sullivan, 47, 48, 339n110
Dynamophone, 134, 140

Eastman, George, 401n62, 402n65
Eastman Theatre (Rochester, New York), 249–251
  mezzanine balcony of, 249–251, *250, 251,* 402n65
  new acoustical criteria embodied in, 234, 250
  and new ideal type of auditorium, 248, 249
  plan of, *249*
  Watson as acoustical consultant for, 249
Eberson, John, 257–258, 405n97
Eccles, William, 59, 61, 96, 107
Echo chambers, 281–283, *282,* 306, 322
Echoes. *See* Reverberation
Eclecticism, 28
Edison, Thomas
  elevated train noise studied by, 145
  light bulb, 92
  motion pictures, 242, 399n40
  phonograph, 236
  sound motion pictures, 243, 244, 399n42
  telephone mouthpiece, 91
Edison Company
  Diamond Disc Phonograph, 237–239, 240
  recording studio of, *265,* 407n118
  suggesting music to be played with their films, 258
  Vitascope, 243
Efficiency
  acoustical building materials promoting, 7
  modern architecture as celebration of, 172
  New York Life Insurance Company Building as epitome of business, 207
  noise affecting, 118, 122, 155–157, 373n170
  in Philadelphia Saving Fund Society Building, 212, 214
  reverberation as inefficient, 171
  stylistic turn in concept of, 156–157
  turn-of-the-century craze for, 123
Einstein, Albert, 105, 133, 355n138
Electrical recording, 234, 240, 398n31
Electrical Research Products Incorporated (ERPI)
  acoustical survey procedure of, 260
  as "Products," 401n56
  on reverberation time for recording studios, 408n123
  sound picture systems installed by, 259–263
  staff of, 259, 405n99
Electroacoustic devices, 229–293. *See also* Amplifiers; Loudspeakers; Microphones; Telephone
  for configurable concert halls, 322
  Knudsen employing, 102
  listening affected by, 7, 233–248
  musical instruments compared with, 145
  new generation of acoustical scientists coming out of electroacoustic industries, 62
  in noise abatement, 145
  noise created by, 149
  noise measured by, 119

in origins of modern acoustics, 90–99
Sabine using in sound intensity experiments, 66–69, 334n15
sounds reconceived as signals in, 3, 5, 96–97, *98*, 262–263, *263*, 269–271, *272*, 289, 291
World War I and development of, 76–77
Elements of Style (Strunk), 157
Elevated trains, 120, 145, 148
Eliot, Charles W.
and Mazer's patent application, 75, 176
and reimbursement for Sabine's research, 72, 345n28
Sabine as unofficial assistant to, 75
Sabine recommended to Higginson by, 17, 37, 38
and Sabine's Fogg Art Museum work, 34, 37
Elizabethan theaters, 332n25
Ellington, Duke, 131
Elliot, Mary, 56
Elliot, Bowen & Walz, 254
Emerson, Ralph Waldo, 120
Energy, conservation of, 27, 32
"Enraged Musician, The" (Hogarth), *116*
Environmentalism, 318
ERPI. *See* Electrical Research Products Incorporated
Ethnomusicology, 328n15
Eyring, Carl E., 414n184
method of images employed by, *290*, 290–291
method of measuring reverberation, *289,* 289–290
reverberation equation reformulated by, 286–293, 414n188
Eyring-Norris equation, 414n188

Fabyan, George, 77–78, 100
Fabyan, Marshall, 77, 78
Feature films, 257, 258
Feedback, 152

Fellheimer & Wagner, 189
Felt
frequency and sound absorption in, *63*
in Mazer Acoustical Sound Controlling System, 218
in New York Life Insurance Company Building, 202
Sabine on use of, 178
for sound deadening, 175, 216
Fessenden, Reginald, 397n21
First Airphonic Suite (Schillinger), 154
First Baptist Church (Pittsburgh), 182
First Church of Christ, Scientist (Boston), *179*
First Congregational Church (Montclair, N.J.), 189
Fisher, Sidney George, 45
Fitzgerald, F. Scott, 115
Flappers, 157
Fleming, John Ambrose, 92
Fletcher, Harvey
at Acoustical Society of America organizational meeting, 105, *106*
on acoustics at American Physical Society, 104
audiometers designed by, 146–147, 371n141
Eyring as student of, 286
and Knudsen, 99
on microphone research at Western Electric, 94
and Millikan oil-drop experiment, 357n150
radio address on noise by, 162, 375n186
Speech and Hearing, 352n101
Varèse corresponding with, 141
Fogg Art Museum (Harvard University), 34–37, *35,* 345n28
Foley, Arthur, 89
Ford, John, 283
Forsyth, Michael, 248, 325n4
Forum (magazine), 121, 148

Fox, William, 247, 400n49
Fox-Case Corporation, 400n49
Franklin, William S., 81, 348n61
Free, Edward Elway, 148–149, 155, 158, 162
Frequency analyzers, 96, 158
Futurism, 134–138

Gables Beach Club (Santa Monica, California), 104
Galileo, 18
Galli-Bibiena family, 332n29
Gallup, Elizabeth Wells, 77, 347n52
Gaumont, Leon, 243–244
General Electric, 90, 144, 245, 296
General Order 47 (New York City), 124
Gericke, Wilhelm, 52, 55, 56, 342n156
Germain, Sophie, 19
Gewandhaus (Neues) (Leipzig), 16, 29, 42, *43*
Giedion, Sigfried, 317
Gigli, Beniamino, 144
Gilbert, Cass
    acoustics information sought by, 173
    medievalesque skyscrapers of, 172
    Minnesota State Capitol, 173, 181, 200, 377n8
    on modernism, 207
    New York Life Insurance Company Building, 199–200, 207–208, 387n90
    Woolworth Building, 169, 183, 200
Gill, Irving, 390n121
Gilman, Lawrence, 139, 140, 143, 144
Girdner, J. H., 117
Godkin, E. L., 359n13
Goethe, Johann Wolfgang von, 116
Gogel, Emanuel, 128
Goodhue, Bertram Grosvenor
    First Congregational Church of Montclair, N.J., 189
    The Knight Errant, 180
    modernity in work of, 187, 383n55
    National Academy of Sciences Building, 189
    Nebraska State Capitol, 189
    Sabine consulting for, 70–71, 187, 344n23
    St. Bartholomew's Church, 70, 187, 383n59
    St. Thomas's Church, 185, 382n45
    St. Vincent Ferrer Church, 189
Gooseneck loudspeakers, 239, *240,* 398n29
Goossens, Eugene, 142
Gothic style, 169, 180, 185, 207, 208
Gottlieb, Albert, 189, *189,* 383n61
Gottschalk, Louis Moreau, 45–46, 340n112
Graham, Anderson, Probst & White, 248
Grand pianos, 46
Grauman, Sid, 258
Gray, Elisha, 91
Graybar Electric Company, 155
Greek amphitheaters, 254
Griffith, D. W., 242
Gropius, Walter, 208–209
Groves, George, 411n157
Guastavino, Raphael, Jr.
    advertising in *Architectural Forum,* 198
    Akoustolith, 170, 188–190, *188,* 217, 319, 384n63
    B'Nai Jeshurun Synagogue, *189,* 383n61
    in Constant Temperature Room preservation, 357n160
    criticism of spaces of, 319
    "Pittsburgh tile," 182
    restoration of works of, 319
    Rumford tile of, 183, 185, 186–187, 189, 319, 382n40
    Sabine collaborating with, 73–74, 109, 180–187, 346n36
Guastavino, Raphael, Sr., 181, 381n32
Gypsum Industries Association, 73

Haber, Samuel, 123
Hale, Philip, 52

Hall, Edwin, 347n45, 380n23
Hanson, O. B., 412n170
Harding, Warren, 241
Harkins, M. R., 85
Harlem Air Shaft (Ellington), 131
Harlem Renaissance, 131–132
Harlem Symphony (Johnson), 131
Harrison, Henry, 97, *98,* 353n112
Harvard University. *See also* Sanders Theater
   architectural acoustics taught at, 60
   Constant Temperature Room of Jefferson Physical Laboratory, 66–69, 83, 113, 357n160
   Fogg Art Museum, 34–37, *35,* 345n28
   merger with MIT, 76, 347n46
   Sabine as student at, 33–34
   Sabine's administrative duties at, 75–76
   Thompson's bequest to, 382n40
Haskell, Douglas
   on focus and energy of Radio City Music Hall, 231
   on form and sound of Radio City Music Hall, 307, 309
   on Radio City Music Hall as too large, 310–311
   on Rivera murals, 314
   on Rockefeller Center, 313, 317
Hassam, Childe, 128–129, *129,* 130, 363n56, 363n59
Hawthorne, Nathaniel, 120
Hays, Will, 246
Hearing, audiometers for studying, 146–148, *147,* 370n137
Heaviside, Oliver, 96
Heins & LaFarge, 181
Heisenberg, Werner, 133
Held, John, Jr., *129*
Hell's Angels (film, dir. Hughes and Whale), 413n174
Helmholtz, Hermann, 132, 365n70
Henderson, W. J., 140

Henry, Joseph
   consultation on House of Representatives, 25–27, 334n45
   experiments on effects of materials on sound, 27–28, 34, 37
   on limit of perceptibility, 334n52
   Smithsonian lecture hall design of, 27–28, 334n52
Herringbone Rigid Metal Lath, *194*
Hertz, Heinrich, 34, 97
Higginson, Henry Lee
   in Boston Symphony Orchestra's establishment, 48
   on Carnegie Hall, 29
   Greek theater plan rejected by, 15–16, 29
   and Mazer's patent application, 75, 176
   McKim engaged to design Symphony Hall by, 15
   new Symphony Hall planned by, 13
   Sabine consulted on Symphony Hall by, 16–17, 37, 42, 44
   Sabine thanked for his work by, 52
   Spear's letter on remodeling Symphony Hall, 54, 341n144
   as sponsoring musicians more talented than himself, 50–51
High modernism, 10
Hill, Thomas, 152
Hill Auditorium (University of Michigan), 74, 418n24
Hiller, Harry, 309–310
Historicism, 28
Hitchcock, Henry-Russell, 296, 390n118
Hofmeister, Henry, 310
Hogarth, William, 116, *116*
Holland, H. Osgood, 71–72
Hollywood. *See also* Hollywood Bowl
   motion picture studios relocating in, 267
Hollywood Bowl, 254–256
   capacity of, 256
   Knudsen as acoustical consultant for, 254, 256

new acoustical criteria embodied in, 234
orchestra shell for, 254, *255,* 256
as precedent for Radio City Music Hall, 307
sound in, 256, 403n82
Hood, Raymond, 266, 390n121
Hooligan, Pop, 124
Hope-Jones, Robert, 75
Hopper, F. L., 276–277, 415n194
Horall, Thomas, 323
Horse-drawn traffic, noise created by, 148, 149
Hospitals
acoustical correction in, 198
noise affecting patients in, 121–122, 145
quiet zones for, 126
House of Representatives (U.S. Capitol), 24–27, *26*
Howard, John Galen, 15
Howe, George, 214
Howe & Lescaze, 210, 214
Howells, William Dean, 122
Hoxie, Charles, 245
Hughes, Langston, 131–132
Humphrey, H. C., 271
Hunchback of Notre Dame, The (film, dir. Worsley), 242
Huyssen, Andreas, 323
Hyperprism (Varèse), 139

"Ideal Auditorium Acoustics" (Watson), 254
Images, method of, *290,* 290–291, *291*
Impedance, 96–97
Industrialization, 117, 120
Initial-time-delay gap, 320
Insulite Acoustile, 190, 193, 216, 218
International Composers' Guild, 368n114
International Style, 210, 227, 390n118
Isaiah Temple (Chicago), 189
Ives, Charles, 133–134

Jaffe Holden Scarborough Acoustics Inc., 418n28, 424n23
Jazz
modern noises inspiring, 119, 130–132, 364n63
racism in criticism of, 131
sound's meaning redefined in, 6, 119
in Wilson–Hassam–Childe dispute, 130
Jazz Singer, The (film, dir. Crosland), 247, 273, 401n53, 411n154
Jessel, George, 247
Jewett, Frank, 92, 93
Johns-Manville, H. W., Co.
acoustical corrections department formed, 178, *179,* 380n26
acoustical products and services of, 194–198, *195*
Acoustical Society of America supported by, 105
advertisement for acoustical correction in banks, *197,* 198
Akoustikos Felt, 221, 387n95
Banroc wool, 221
coated fabric finishes for acoustical materials, 216
in Constant Temperature Room preservation, 357n160
installations of sound-absorbing materials by, 377n5
Keystone Hair Insulator, 175–176
and Mazer, 176
Noise Abatement Commission of New York supported by, 158
Sabine collaborating with, 74
Sanacoustic Tile distributed by, 220–221, 394n157
Johnson, James H., 10, 47
Johnson, James P., 131
Johnson, Philip, 390n118
Johnson, Russell, 321, 423n18
"Joint Is Jumpin', The" (Waller), 131
Jolson, Al, 247, 273
Jordy, William H., 317–318, 390n121, 393n153, 394n1, 417n23

Journal of the Acoustical Society of America, 106, 325n4
"Joys of Noise, The" (Cowell), 133
Jullien, Louis Antoine, 48

K (hyperbolic constant), 38, 39, 41
Kahn, Albert, 74, 178, 418n24
Kahn, Douglas, 10
Kahn, Otto, 295
Kalite Sound Absorbing Plaster, 193, *193*, 217, 309
Kandinsky, Vasily, 366n84
Kasson, John, 124
KDKA radio (Pittsburgh), 239, *265, 272*
Keller, Helen, 310, 419n35
Kellogg, Edward
    on microphone placement in motion picture studios, 277
    on reverberation chamber, 281–283
    on reverberation in motion picture theaters, 261–262, 271
    on Sabine and reverberation, 284–285
    theater loudspeaker proposed by, 263, *263*
Kelvin, Lord, 99
Kennelly, Arthur, 96–97
Kidd, F. J. & W. A., 248
Kinetophone, 243, 244
Kinetoscope, 242, 256
King, Louis, 89
Kingsley, Darwin P., 388n102
Kipling, Rudyard, 174
Kleinhans Music Hall (Buffalo), 248, 319
Kneser, Hans, 357n150
Knight, Arthur, 283
Knight Errant, The (quarterly), 180
Knudsen, Vern O., 99–107
    on absorption coefficients, 106
    in Acoustical Society of America founding, 104–105, *106*
    on acoustical treatment of ceilings, 391n132
    Architectural Acoustics, 391n135
    articulation testing used by, 101, 252–254
    commercial aspect of expertise of, 102
    education of, 99–100
    Eyring equation used by, 415n194
    in Gables Beach Club, 104
    on Greek and Roman amphitheaters, 254
    high school auditorium consultation of, 101
    as Hollywood Bowl acoustical consultant, 254, 256, 403n82
    on modern era in acoustics, 104
    as motion picture industry consultant, 103, 259, 269, 276–277, 405n97
    on new acoustical criteria, 248, 250, 401n59
    in new generation of acoustical scientists, 62, 99
    on outdoor listening, 254
    pure scientific research of, 108, 357n150
    on reverberation time, 252–254
    Sabine as influence on, 100–101
    sound-absorbing materials research of, 101
    at UCLA, 100, 354n120
    at Western Electric, 99
    Western Electric supporting research of, 102
Koenig, Rudolph, 19, 35
Koszarski, Richard, 401n53, 410n149
Krantz, Fred W., 347n52
Krauss, Joseph, 151
Krehbiel, Henry, 51–52, 341n132
Kreisler, Fritz, 49
Krinsky, Carol, 307, 418n24

Lachaise, Gaston, 299, *299*
Laird, Donald, 155–156, 374n173, 376n190, 385n79
Laloux, Victor, 15
Lamb, Thomas, 257

Lamoureux, Charles, 15
La Música (Russolo), 135
LARES (Lexicon Acoustic Reinforcement and Enhancement System), 423n23
La Scala (Milan), 20, 28
Lastra, James, 10, 283–284, 413n178
Latrobe, Benjamin, 24–25, 33, 333n39
Lauste, Eugene, 400n48
Lawrie, Lee, 296–297, *298*
League of Nations assembly hall, 209
Lears, Jackson, 180
Le Corbusier
  on engineers, 208
  League of Nations assembly hall design, 209
  on NBC studios, 421n1
  Pavillon Suisse, 209
  Philips Pavilion, 322–323, 368n113, 389n112
  on PSFS Building, 227
  sound control as concern of, 209
  white villas of, 209
Legionnaires' disease, 318
Leipzig Gewandhaus (Neues), 16, 29, 42, *43*
Lenin, Vladimir, 154, 314
Lescaze, William, 214, 390n127
Leschatzsky, Isaac, 125
Levine, Lawrence, 47
Lewis, Sinclair, 112–113, 357n159
Lewis, Wyndham, 135
Lexicon Acoustic Reinforcement and Enhancement System (LARES), 423n23
Light bulb, 92
Lights of New York, The (film, dir. Foy), 247, 273, 274
Lime Manufacturers Association, 73
Limit of perceptibility, 334n52
Lindsay, Vachel, 404n89
Lippmann, Walter, 312
Listening. *See also* Audience
  Algarotti on new attitude toward, 46

  audiences developing listening habit, 247
  culture of in turn-of-the-century America, 45–51
  electroacoustical devices affecting, 7, 233–248
  new trends in culture of, 2
  outdoors as best environment for, 254
  to reproductions, 238
Locke, Alain, 131
Lodge, Henry Cabot, 176
Loesserman, Arthur, 129
Loew's Theater (Akron, Ohio), 405n97
Loew's Theater (Canton, Ohio), 406n104
London
  din of eighteenth-century, 116, *116*
  Drury Lane Theatre, 24
  Royal Festival Hall, 319, 322, 401n59
Looking Backward (Bellamy), 50
Loudness
  frequency-dependence of human sense of, 64
  in motion picture sound track, 276, 283, 410n149, 413n174
  in motion picture theaters, 261
Loudon, James, 59, 83
Loudspeakers. *See also* Public address (P.A.) systems
  cone-type loudspeakers, 239, *240,* 398n29
  deployment into the soundscape, 233
  donated to UCLA, 102
  gooseneck loudspeakers, 239, *240,* 398n29
  in motion picture theaters, 262, 263, *263*
  noise created by, 117, 149, 151–152, *160,* 371n147, 372n154
  in Philips Pavilion, 322–323
  in phonographs, 233, 240
  in Radio City Music Hall, 231, 242, 309
  in radios, 239, *240,* 398n29
  at Theremin-Vox concert of 1928, 241

Lumière, Louis and Auguste, 243
Lyon, Gustave, 209, 389n111, 389n112

McGinnis, C. S., 85
Machine Age, The
  aural culture and texture of, 1
  as ending in 1932, 4, 5, 8, 166, 315
  as the Jazz Age, 132
  new sources of noise of, 149
McKay, Gordon, 347n46
McKim, Charles
  Greek theater plan for Symphony Hall, 15–17, *16,* 29
  in Rhode Island State Capitol acoustics project, 69, 175
  Sabine advising on Symphony Hall, 17, 42, 44–45
  and statues in Symphony Hall, 55, 56
McKim, Mead & White. *See also* McKim, Charles
  Boston Public Library, 15, 181
  Boston Symphony Hall commission for, 15
  Guastavino working with, 181
  Mead, William, 69–70, 78, 175–176
  Sabine consulting for, 57, 69–70
  White, Stanford, 69, 199
MacNair, Walter A., 292
Macoustic Plaster, 217, 405n97
Madison Square Garden (New York City), 199
Mamoulian, Rouben, 258, 283, 413n172
Manhattan, value of real estate in, 169–170, 377n3
Manhattan Opera House (New York City), 267, *268*
Mannix, Eddie, 103
Manship, Paul, 314
Marconi, Guglielmo, 301, 416n13
Margrave's Opera House (Bayreuth), 20
Marinetti, F. T., 135, 137
Markgraf, Rudolph, 12, 33
Marr & Colton organs, 257, 404n87

Marx, Karl, 11, 329n20
Marx, Leo, 120
Massachusetts Institute of Technology, 60, 76, 347n46
Matsui, Yasuo, 200
Maxfield, Joseph
  binaural hearing research of, 408n122
  on microphone placement in motion picture studios, 277
  phonograph improvements of, 97, *98,* 353n112
  on reverberation in motion picture soundstages, 269, 271
  on reverberation in recording studios, 266
Maxim, Hiram Percy, 394n156
Maxim-Campbell Silencer and Air Filter, 301
Mayer, Louis B., 269
Mazer, Jacob
  as acoustical consultant, 392n137, 392n143
  Mazer Acoustile Sound Controlling System, 217–218
  patent application based on reverberation equation, 75, 176, *177,* 379n23
  perforated acoustical materials patented by, 220, 392n143
Mead, M. S., 144
Mead, William, 69–70, 78, 175–176
Mechanically Applied Products Co., 216–217
Mechanisms (Antheil), 142, 369n119
Meeting of Automobiles and Airplanes (Russolo), 137
Megaphones, 123–124
Meier, Hildreth, *300*
Meigs, Montgomery, 25, 334n45, 334n46
Mellor, Meigs & Howe, 214
Mendenhall, Thomas Corwin, 33
Mersenne, Marin, 18
Method of images, *290,* 290–291, *291*
Metro-Goldwyn-Mayer, 103, 269, *270*

Metropolitan Life Insurance Company Building (New York City), 387n92
Metropolitan Museum of Art (New York City), 175–176
Metropolitan Opera House (New York City), 49, 144, 267, 295, 296, 307
Meyer & Holler, 257
MGM. *See* Metro-Goldwyn-Mayer
Microphones
  deployment into the soundscape, 233
  monaural, 266
  for motion picture sound tracks, 275, 277, 279–280
  omnidirectional, 264
  in recording studios, 240, 264
  ribbon microphones, 280, 309
  Wente's development of, 94–95
Microtones, 367n91
Mies van der Rohe, Ludwig, 209, 210, *210*
Milam Building (San Antonio, Texas), 393n154
Miller, Dayton Clarence
  at Acoustical Society of America organizational meeting, *106*
  on development of acoustics, 59, 107
  on Einstein's theory, 105, 355n138
  Lowell Institute lectures of, 86–87, 350n83
  on National Research Council Committee on Acoustics, 89–90
  phonodeik of, 86
  on Sabine, 60
  war work of, 351n91
  as "Wizard of Visible Sound," 86, 105
Miller, Harry, 99, 353n112
Miller, Wesley, 272
Millikan, Robert, 99–100, 357n150, 414n184
Mills, Robert, *27,* 333n42
Minnesota State Capitol, 173, 181, 200, 377n8
Mixing, 277–278

Modernism
  arrival of modern architecture in America, 172, 210
  Cram on, 187
  criticism of hermetically sealed glass boxes of, 318–319
  European modernists turning to technology, 208–210
  Goodhue on, 187
  Gilbert on, 207
  high modernism, 10
  International Style, 210, 227, 390n118
  in New York Life Insurance Company Building kitchen, 207
  of Philadelphia Saving Fund Society Building, 210–226
  sound control associated with, 210
Modernity. *See also* Modernism
  aural dimension requiring study, 10
  noise and modern culture, 115–168
  noise and modern music, 130–144
  nonreverberant sound as characteristic of, 3–4, 171–172, 284
  Radio City Music Hall celebrating sound of, 4, 8
  space and time perceptions reformulated in, 187, 320–321
  stock market crash of 1929 and, 314–315
  technology in defining, 4, 11
Monitor booth, 271, *271*
Morgan, Kenneth, *281,* 411n156
Morris, Benjamin Wistar, 296
Motion pictures. *See also* RKO; Sound motion pictures; Warner Brothers
  development of, 242–243
  feature films, 257, 258
  Kinetoscope, 242, 256
  Lumière brothers, 243
  Metro-Goldwyn-Mayer, 103, 269, *270*
  music played with, 257, 258, 404n89
  Mutoscope, 243
  newsreels, 247, 400n49

organs in theaters, 257, 404n94
public address systems used by directors, 242
Vitascope, 243
Motorcycles, 149
Movies. *See* Motion pictures
Muench, Carl, 218, 220
Mumford, Lewis, 313, 418n24
Museum of Modern Art (New York City), 210, 227, 390n118
Music. *See also* Avant-garde music; Concert halls; Jazz; Symphony orchestras
amateur's decline in, 49–50
complaints about noise from, 127–129, *160*
defined as harmonious and orderly, 132–133, 365n70
electroacoustical devices and musical instruments, 145
in motion picture theaters, 257, 258, 404n89
noise and modern, 130–144
pianos, 46, 141, 257
popular, 322, 423n19
sober attitude in domestic, 49
the Theremin-Vox, 152–154, *153,* 241, 372n159
in turn-of-the-century America, 45–51
Musical Courier (newspaper), 53–54
Music Hall (Boston)
as acoustical model for Symphony Hall, 42, *43,* 338n97
acoustics of Symphony Hall compared with, 52
committee on acoustics for, 339n110
demolition of, 13
Dwight on, 47
Mutetile, 221, *222,* 393n152
Mutoscope, 243
"My Lost City" (Fitzgerald), 115

National Academy of Sciences Building (Washington, D.C.), 189

National Broadcasting Company (NBC)
Echo System of, *282,* 283, 412n170
New York studios of, 266
in Rockefeller Center development, 296
National Broadcasting Company (NBC) studios (Rockefeller Center), 301–306
engineering control room of, 306
Le Corbusier on, 421n1
lobby of, 301, *302, 303*
NBC studio audience, *304*
ninth and tenth floor plans of, *305*
in RCA Building, *297,* 301
reverberation times in, 306, 417n18
sound effects chambers, *305,* 306
tours of, 304
National Bureau of Standards, 101, 355n126
National Lime Manufacturers Association, 73
National Research Council Committee on Acoustics, 89–90
NBC. *See* National Broadcasting Company
Nebraska State Capitol, 189
Neoclassicism, 24, 27, 28, 332n34
Neo-Gothicism, 169, 180, 185, 207, 208
Neues Gewandhaus (Leipzig), 16, 29, 42, *43*
Neutra, Richard, 390n121
New Acoustics, 59–113
and Acoustical Society of America founders, 107
commercial aspects of, 62
in creation of modern soundscape, 113
Eccles on, 59, 61, 107
proclamation of, 5
World War I and, 351n89
New England Conservatory of Music, 62, 174, 175, 261
New London (Connecticut) Experimental Station, 89

Newsreels, 247, 400n49
New Theater (Century Theater) (New York City), *65*
Newton, Francis, 128, 130, 363n56, 363n59
Newton, Isaac, 18, 133
New York Academy of Music, 28, 48
New York Central Railroad Terminal (Buffalo), 189
New York City. *See also* New York Life Insurance Company Building; Noise Abatement Commission of New York City; Rockefeller Center; St. Thomas's Church
    Avery Fisher Hall, 422n9
    Bell Laboratories' Sound Picture Laboratory, *285, 286, 287*
    Carnegie Hall, 29, 48–49, 142, 154, 238, 336n60
    Cathedral of St. John the Divine, 70, 72, 181
    Chaloner Theater, *275*
    Chicago noise level compared with, 376n190
    Church of St. Vincent Ferrer, 189
    din of, 115
    elevated train noise in, 120, 145, 148
    first commercial motion picture showing in, 243
    first Vitaphone presentation in, 246
    Free's audiometer study of, 148–149
    General Order No. 47, 124
    Gilbert on noise and congestion of, 200
    Ives's *Central Park in the Dark,* 133–134
    Madison Square Garden, 199
    Manhattan Opera House, 267, *268*
    megaphones banned on Coney Island, 123–124
    Metropolitan Life Insurance Company Building, 387n92
    Metropolitan Museum of Art, 175–176
    Metropolitan Opera House, 49, 144, 267, 295, 296, 307
    Museum of Modern Art, 210, 227, 390n118
    NBC studios, 266
    New (Century) Theater, *65*
    noise complaint procedure in, 127–130
    noise complaint summary in, 158, *159, 160*
    noise complaint survey in, 151, 158, *159*
    noise sources in, 148–152, *150*
    Pathé studio fire, 409n134
    Pennsylvania Station, 181
    quiet zones around hospitals, 126
    Rice's noise abatement movement in, 121–122
    St. Bartholomew's Church, 70, 187, 383n59
    Sanitary Code Section 215a, 151–152, 166
    Society for the Suppression of Unnecessary Noise, 121, 122, 126, 145
    street vendors' noise ban, 124–125
    Varèse's *Amériques* as tribute to, 138–139
    Woolworth Building, 169, 183, 200
    zoning law of 1916, 127, 170
New Yorker (magazine), *165*
New York Life Insurance Company Building (New York City), 198–207
    artificial ventilation in, 202
    cafeterias of, 201
    as city-in-a-building, 200–202
    dish washing room in, *206*
    as epitome of business efficiency, 207
    exterior, *201,* 207
    exterior and interior contrasted, 207–208, 214
    Gilbert in design of, 199–200, 207–208, 387n90
    kitchen in, 205, *206*
    ladies' dining room in, *204*
    men's dining room in, *204*

noise control in, 172–173, 199, 202, 205, 216
offices in, 202, *203*
pneumatic mail-tube system, 202, 205
site of, 199
as state-of-the-art corporate architecture, 199
as world's largest installation of sound-absorbing materials, 198
New York Philharmonic Society, 319, 422n9
New York State Theater (New York City), 322, 424n23
Nichols, Edward, 348n61
Noise, 115–168. *See also* City noise; Noise abatement
  in Boehm's sound detector, 84
  cultural context of, 9
  defined as discordant and disorderly, 132–133, 365n70
  efficiency affected by, 118, 122, 155–157, 373n170
  electroacoustical instruments measuring, 96, 119, 148, 155, *155*, 158, *161*
  health effects of, 118
  and music, 130–144
  in offices, 195–196
  physiological effect of, 156
  psychological effect of, 155
  in recording studios, 266–267
  reverberation as, 3, 171, 172, 234, 284
  Sabine's work on, 77–78
  smoke compared with, 122–123
  on soundstages, 103
  technology as source of, 2, 6, 117, *118*, *150*
  in telephone transmission, 146–147, 235
  women's sound-absorbing power, 374n173
  in World War I, 88
Noise abatement
  electroacoustic devices in, 145
  engineers in, 144–157
  failure of, 6, 157–168, 360n32
  progressive approach to, 6, 120–130
  Society for the Suppression of Unnecessary Noise, 121, 122, 126, 145
  zoning as method of, 125–127
Noise Abatement Commission of New York City. *See also City Noise*
  advisory nature of, 166–167
  appointment of, 157
  charts showing noise levels, 162, *163*
  dissolution of, 166
  final report of, 166, 167
  impact of, 164–167
  members of, 374n177
  noise abatement questionnaire of, 158, *159*
  noise measuring truck of, 158, *161*
  purpose of, 157–158
Noise complaints
  New York procedure for, 127–130
  questionnaire about New Yorkers', 158, *159*
  summary of New Yorkers', 158, *159*, 160
Noise-intoners (*intonarumori*), 136, 366n90
Noise machines, 281, *281*
Noise of the Street Penetrates the House, The (Boccioni), 135
"Noise pollution," 123, 361n33
"Noise units," 146, 158, 162
Norcross, Otto, 17
Norris, R. F., 393n148, 414n188
Noverre, Jean George, 332n25

Oakley, Imogen, 126, 361n33, 362n48
O'Brien, John, 231
Offices
  in New York Life Insurance Company Building, 202, *203*
  noise in, 195–196

School Department office space in PSFS Building, *224*
sound control in Rockefeller Center, 301
typing efficiency affected by noise, 155–156
Ogren, Kathy, 131
1-A Noise Measuring Set, 146–147, 369n134
Open-window unit of absorption, 39–40
Orcutt, William Dana, 107–109, 111, 347n45, 356n147, 380n23
Organic sounds, 116–117
Organs, theater, 257, 404n92
Orthophonic Victrola, 240, 398n31
Oscillograph, 96
Osswald, F. M., 209

Paderewski, Ignacy Jan, 49
"Painting of Sounds, Noises, and Smells, The" (Carrá), 135
Pallophotophone, 245
Panatrope phonograph, 398n31
Parrino, Nunzio, 162
Partial tones, 134
Pathé studio (New York City), 409n134
Patte, Pierre, 19–21, *21,* 24
Patti, Adelina, 30, 47
Pease, Maurice, 126
Pennsylvania Station (New York City), 181
Perceptibility, limit of, 334n52
Pérez-Gómez, Alberto, 19
Philadelphia. *See also* Philadelphia Saving Fund Society Building
  Legionnaires' disease outbreak in, 318
  Mazer as acoustical consultant in, 392n137
  quiet zones around hospitals, 126
  Varèse's work performed in, 139–140
  Wanamaker's, 214, 390n122
Philadelphia Academy of Music, 28, 29, 46

Philadelphia Saving Fund Society Building (PSFS), 210–226
  advertising of, 224, *225, 226,* 393n155
  air conditioning in, 221–222
  efficiency in, 212, 214
  escalators and stairs to main banking room, *213*
  exterior, *212*
  functionally differentiated spaces of, 210
  Howe & Lescaze as designers of, 210, 214
  Le Corbusier on, 227
  main banking room of, *213,* 393n153
  as model for postwar construction, 317
  Mutetile used in, 221, *222,* 393n153
  noise control in, 173, 216
  protomodern precedents of, 390n121
  radio in, *222,* 227–228, 394n164
  reactions to, 215
  School Department office space in, *224*
  Stokowski on, 214–215
  as too quiet for comfort, 227
  total environmental control in, 221–226
Philips Pavilion (Brussels World Fair), 322–323, 368n113, 389n112, 423n24
Phonautograph, 145
Phonodeik, 86, *86, 87*
Phonofilm, 245
Phonograph. *See also* Sound recording
  advertising of, 397n26
  Bell/Tainter and Berliner improvements of, 396n19
  Columbia Phonograph Company, 145, 398n31
  competition between manufacturers, 237
  and decline of amateur music, 49, 50
  Edison Diamond Disc Phonograph, 237–239, 240
  loudspeakers in, 233, 240
  Maxfield and Harrison improvement of, 97, *98*

reproducing horns in, 237
Rice using in anti-noise campaign, 145
sound and space dissociated by, 236
Victor Talking Machine Company, 237, 240, 398n31, 407n118
Victrola, 237, 240
Phonometer, 86, 89, 98, 350n82
Photion, 245
Photophone, 247
Pianos
in Antheil's compositions, 141
grand pianos, 46
in motion picture theaters, 257
Pierce, George, 83–84, 85, 96–97
Pittsburgh
Edison Tone Test in, *238*
First Baptist Church, 182
KDKA radio, 239, *265, 272*
"Plague of City Noises, The" (Girdner), 117
Plaster
absorption coefficients for, 40, 286
acoustical, 170, 216–217
in Eberson theaters, 257, 258, 405n97
Kalite Sound Absorbing Plaster, 193, *193,* 217, 309
Sabine interested in absorbing characteristics of old, 73
Sabinite, 170, 217, *217,* 391n134, 392n136
Simpson Brothers Cal-Acoustic Plastering Company, 101–102
Sprayo-Flake Acoustical Plaster, 170, *191,* 217
United States Gypsum Company advertising of, 191–192, *192*
Plastic Motor-Noise Construction (Depero), 135
Platinum Blonde (film, dir. Capra), 283
Platt, Walter, 122
Pompeii, 115–116
Pons, Lily, 144
Popular music, 322, 423n19

Postmodern aesthetic, 323–324
Pound, Ezra, 141
Pratella, Balilla, 135–136
Princeton University Chapel, 189, 319
PSFS Building. *See* Philadelphia Saving Fund Society Building
Public address (P.A.) systems
in military aircraft, 99
motion picture directors using, 242
in public places, 233, 241
theater directors using, 241–242
Western Electric in development of, 99, 241
Pulitzer, Joseph, 70, 78
Pure science, 356n148

Quiet zones, 125–127

Radio
Antenaplex radio antenna network in RKO Building, 301
AT&T attempting to gain control over, 92
commercial broadcasting, 239
DX listening, 236–237
headsets for listening to, 236, 397n28
improved receivers, 239
live broadcasts, 239
loudspeakers in, 239, *240,* 398n29
noise created by, 149, 151, 152, 371n147
in Philadelphia Saving Fund Society Building, *222,* 227–228, 394n164
Pierce's *Principles of Wireless Telegraphy,* 83
speech and music transmitted over, 236, 397n21
tuning of, 239
as wireless telegraphy, 236
Radio City Music Hall (New York City)
acoustical consultants for, 417n22
Acousticon seat-phones for hearing-impaired patrons, 419n35

acoustics of, 231, 233, 310
Art Deco architecture of, 229, 394n1
auditorium of, 231, *232*
chief sound engineer of, 309–310
connection between audience and performer in, 307
as culmination of modern soundscape, 233, 307, 317
as economic failure, 311–312
lobby of, *230*
as model for postwar construction, 317
modern acoustical era ending with, 4, 8, 315
as motion picture theater, 312, 419n39
opening night at, 229–231
plan of, *308*
precedents for, 307, 417n23, 418n24
reverberation times in, 309, 418n28
"Roxy" Rothafel as director of, 233, 307, 312
section of, *308*
as signaling end of period of change, 5
sound absorbing materials in, 309
sound system in, 231, 242, 309, 419n31
stage of, *232*
stage performance, c. 1935, *311*
suspended acoustical plaster ceiling of, 317–318
as too large, 310–311
Radio Corporation of America (RCA). *See also* RCA Building
formation of, 245
Photophone, 247
in Rockefeller Center development, 296
Theremin bought by, 154, 372n159
Ragtime, 130
Railroads
elevated trains, 120, 145, 148
noise created by, 117, *160*
streetcars, 124, 149
subways, *160,* 162, *163*
travelers coping with disorientation of, 320

Rapp & Rapp, 257
Rappold, Marie, *238*
Ray, Man, 142
Ray, Veronica, 129–130
Rayleigh, Lord, 19, 85, 104, 331n23, 365n70
Rayleigh disc, 85–86, 89
RCA. *See* Radio Corporation of America
RCA Building (Rockefeller Center). *See also* National Broadcasting Company (NBC) studios
Faulkner sculpture on Sixth Avenue entrance of, 299, *299*
as focal point of Rockefeller Center, 296, *297*
Lachaise sculpture on west side of, 299, *299*
Lawrie sculptures for main entrance of, 296–297, *298*
Manship's Prometheus sculpture in front of, 314
Rivera murals in, 314
Recording studios, 263–267
microphones in, 240, 264
monitor booth, 271, *271*
noise in, 266–267
reverberation in, 264, 266
Reinforced concrete construction, 78
Reisenfeld, Hugo, 258
Relativistic physics, 105, 133, 355n138
Remarque, Erich Maria, 88
Remington Typewriter Company, 78
Renwick, James, 28
Reverberation. *See also* Reverberation equation
acoustical building materials minimizing, 7
as acoustic signature of particular place, 3, 171, 186–187
articulation and, 252–254
assisted resonance, 322
Eyring's method of measuring, *289,* 289–290

in Fogg Art Museum lecture room, 34
"good sounds" as nonreverberant, 7, 284
Guastavino attempting to reduce, 181–182
in House of Representatives, 24–27, *26*
as inefficient, 171
as matter of taste to Sabine, 55
in the modern auditorium, 248
modernity as characterized by nonreverberant sound, 3–4, 171–172, 284
in motion picture theaters, 260–262, 406n105
in NBC studios, 306, 417n18
as noise to be eliminated, 3, 171, 172, 234, 284
in Radio City Music Hall, 309, 418n28
in recording studios, 264, 266, 408n123
reverberant chambers, 281–283, *282*, 306, 322
Sabine expanding his work on, 62–81
in St. Thomas's Church, 186, 383n54
in soundstages, 269, 409n133
standards for reverberation time changing, 319–320
Watson on reverberation time, 252, *253*
"Reverberation" (Sabine), 60, 81, 338n89
Reverberation equation, 33–45
Franklin's theoretical derivation of, 81
frequency and, 63
Johns-Manville acoustical corrections department applying, 178
Knudsen applying, 101
Mazer applying for patent based on, 75, 176
revised to accommodate electronically produced sound, 7–8, 235, 285–293, 414n188
Sabine applying to architectural projects, 69–73
Sabine's derivation of, 38–42
Stewart's confirmation of, 81

Rhode Island State Capitol, 69, *70*, 175
Ribbon microphones, 280, 309
Rice, Isaac, 121, 148
Rice, Julia Barnett
class and reform movement of, 123
medical training of, 359n23
noise abatement campaign of, 120–122
on noise and smoke, 361n33
on noise in schools, 126
noise recorded by, 145
Rivera, Diego, 314
Riverbank Acoustical Laboratory (Geneva, Illinois), 77–80
exterior of, *80*
Knudsen visiting, 100
National Research Council Committee on Acoustics meeting at, 89
plan and section of, *79*
Paul Sabine as director of, 80, 348n60
in Sabinite development, 217, 391n134
sound-absorbing materials research at, 101
Wolf visiting, 259
RKO (Radio-Keith-Orpheum)
Antenaplex radio network in RKO Building, 301
in Rockefeller Center development, 296
Roxy Theatre, 283, *300*, 419n39
sound concentrators used by, 279–280, *280*
Roaring Twenties, 61, 115, 120
Rockefeller, John D., 295–296, 314
Rockefeller Center (New York City), 295–315. *See also* Radio City Music Hall; RCA Building
as city within a city, 295
critical reappraisal of, 317
criticisms of, 313
Manship's Prometheus sculpture at, 314
Maxim-Campbell Silencer and Air Filter in, 301
Roxy Theatre, 283, *300*, 419n39
sound control in, 301

Rockettes, 312
Rogers, Joel, 130–131
Roman amphitheaters, 254
Rosenblatt, M. C., 221, 393n153
Rosenfeld, Paul, 139, 140, 143–144
Rothafel, Samuel "Roxy"
  broadcasting of stage shows of, 231, 395n9
  music enhancing films exhibited by, 258
  opening show at Radio City produced by, 231
  public address systems used by, 241–242
  as Radio City Music Hall director, 233, 307, 312
  on sound control, 310
Roxyettes, 231, 312
Roxy Theatre (New York City), 283, 300, 419n39
Royal Festival Hall (London), 319, 322, 401n59
Rumford tile, 183, 185, 186–187, 189, 319, 382n40
Russian Music Lovers' Association, 129–130
Russolo, Luigi, 136–138
  "The Art of Noises," 136, 366n87
  Awakening of a City, 137
  first public performance of noise orchestra of, 137–138
  as lacking acoustical prejudices, 366n86
  La Música, 135
  Meeting of Automobiles and Airplanes, 137
  noise-intoners of, 136, 366n90
  after World War I, 138, 367n98
Ryder, George Hope, 121–122

Saarinen, Eliel, 248
Sabine, Hale J., 385n73, 392n142
Sabine, Hylas, 337n74
Sabine, Jane Kelly, 76
Sabine, Paul E.
  on architectural acoustics for motion picture studios, 273, 276
  on Chicago Auditorium and Radio City, 417n23
  on electroacoustic devices in acoustics research, 96
  on National Research Council Committee on Acoustics, 89
  Radio City Music Hall consulting of, 417n22
  on reverberation time in recording studios, 264, 266
  on revision of reverberation equation, 414n188
  as Riverbank Laboratory director, 80, 348n60
  on Wallace Sabine's practical side, 109
  Sabinite development, 217, 391n134
  sound-absorbing materials research of, 101, 106
Sabine, Wallace Clement
  "The Accuracy of Musical Taste in Regard to Architectural Acoustics," 55
  acoustics shifting away from work of, 61–62
  administrative duties at Harvard, 75–76
  in Akoustolith development, 188–189
  "Architectural Acoustics," 338n84
  on audience psychology in judgments of acoustical quality, 44, 55–56, 342n151
  Careers pamphlet on, 109–110
  Collected Papers on Acoustics, 100–101
  commercial aspect of expertise of, 72–74, 102, 109, 345n28
  as consultant to architects, 69–73
  and Cram, Goodhue & Ferguson, 70–71, 72
  and criticism of acoustics of Symphony Hall, 52, 53–57
  death of, 61, 76
  education of, 33–34

electroacoustic devices used by, 66–69, 334n15
and Fabyan's Riverbank Acoustical Laboratory, 77–78
as failing to train succeeding generation of acoustical researchers, 75
Fogg Art Museum lecture room acoustics work of, 34–37, 345n28
French lectures of, 76, 347n47
on frequency and sound absorption, 62–63, *63*
and Guastavino sound-absorbing tiles, 73–74, 109, 180–197, 346n36
influence of work of, 81–90
and Johns-Manville acoustical corrections department, 178, 380n26
Kellogg on, 285
Knudsen influenced by, 100–101
Lewis's *Arrowsmith* compared with, 113
and lime and gypsum industry, 73, 345n35
McKim, Mead & White reenlisting services of, 57, 69–70
and Mazer patent application, 75, 176, 379n23
Metropolitan Museum of Art consultation, 175–176
Miller on, 60
New England Conservatory of Music consultation, 175, 261
noise as research topic of, 77–78
obituary of, 111
Orcutt's biography of, 107–109, 111
photograph of 1906 of, 111, *111*
photograph of 1918 of, 111–112, *112*, 357n158
publications in architectural journals, 72–73
pure and applied science in, 107–113
recording sound on film, 67, *67*, 400n48
reputation as concern of, 74–75
"Reverberation," 60, 81, 338n89
reverberation equation of, 7, 33–45
in Rumford tile development, 183, 185, 186–187, 189, 382n40
and Sabinite, 391n134
in St. Botolph's Club, 104
sound intensity in Constant Temperature Room mapped by, 66–69, *66, 67, 68,* 344n17
sound waves photographed by, 64, *65*
Swan as only student of, 75
after Symphony Hall, 62–81
Symphony Hall acoustical consultation of, 4, 5, 17–18, 37–39, 42–45
Symphony Hall plaque dedicated to, 13, 57
System interview, 196
on women's sound-absorbing power, 374n173
as working alone, 61, 75, 77, 81
in World War I, 76
Sabinite, 170, 217, *217,* 391n134, 392n136
St. Bartholomew's Church (New York City), 70, 187, 383n59
St. Botolph's Club (Boston), 104
St. Thomas's Church (New York City), 180–187
  acoustical redesign of, 319
  exterior, *184*
  Guastavino ceramic tiles in, 172, 183, 185, 186–187, 189
  interior, *184*
  modernity of, 186–187
  ornamentation of, 185
  reverberation in, 186, 383n54
  Sabine consulting on, 70
St. Vincent Ferrer Church (New York City), 189
Salle Pleyel (Paris), 389n111, 389n112, 418n24
Sanacoustic Tile, 170, 220–221, 393n148, 394n157
Sanders, Martha, 129

Sanders Theatre (Harvard University)
  Fogg Art Museum lecture room contrasted with, 34, 36, 37
  seat cushions of, 36, 38, 39, 40, 56
Sanitary Code Section 215a (New York City), 151–152, 166
Satie, Erik, 142
Saunders, George, 21–23, *22, 23*
Savarin, Inc., 205
Schafer, R. Murray, 1, 123, 321, 325n1, 423n16
Scherchen, Hermann, 424n24
Schillinger, Joseph, 154
Schivelbusch, Wolfgang, 320, 329n21
Schizophonia, 321
Schmidt, Leigh, 10
Schools
  acoustical correction in, 198
  quiet zones around, 126
Schopenhauer, Arthur, 116
Schuster, K., 414n188
Schuyler, Montgomery, 32, 185–186
Scott, Robert F., 378n13
Seashore, Carl, 370n135
Sennett, Richard, 47
Severance Hall (Cleveland), 248
Shand, William, 334n47, 334n49
Shankland, Robert, 334n52
SIAP (System for Improved Acoustic Enhancement), 424n23
Sibley Auditorium (Cornell University), 81
Silent, H. C., 278, 411n156
Silentaire, 222, *223*
Silent-Ceal, 221, 393n152
Simplified spelling, 157
Simpson Brothers Cal-Acoustic Plastering Company, 101–102
Smilor, Raymond W., 123, 167, 327n10, 360n24, 360n32, 376n198
Smith, Bessie, 132
Smith, Bruce, 10
Smith, Mark, 10

Smithsonian Institution (Washington, D.C.), 27–28, 334n52
Smoke, noise compared with, 122–123
Society for the Suppression of Unnecessary Noise, 121, 122, 126, 145
Society of Motion Picture Engineers, 272–273, 410n142
Soglow, Otto, *165*
Sound-absorbing materials. *See* Acoustical building materials
Sound concentrators, 279–280, *280*
Sound meters, 96, *155*, 158
Sound motion pictures. *See also* Soundstages; Sound tracks
  Bell Laboratories' Sound Picture Laboratory, *285, 286, 287*
  Cameraphone, 244
  Chronophone, 243–244, 400n45
  city noise in, 280–281, 412n166
  cultural context of development of, 9
  development of, 243–247
  Don Juan, 246, 273
  early films as filmed stage performances, 273
  first all-talking film, 247, 273
  The Jazz Singer, 247, 273, 401n53, 411n154
  Kinetophone, 243, 244
  Knudsen as consultant for, 103
  The Lights of New York, 247, 273, 274
  Pallophotophone, 245
  Phonofilm, 245
  Photophone, 247
  sound/space relationship affected by, 7
  theaters for, 247, 256–263
  Vitaphone, 102, 241, 245–247
Soundproof materials. *See* Acoustical building materials
Sound ranging systems, 88
Sound recording. *See also* Phonograph; Recording studios
  acoustical recording, 236, 263–264, *265*

electrical recording, 234, 240, 398n31
stereophonic recording, 106, 408n122
Sounds. *See also* Acoustics; Loudness; Noise; Reverberation; Soundscape
decibel, 158, 162, 164, 375n182
differential equations for representing, 96
electroacoustic devices and modern, 229–293
ephemeral quality of, 12
in face-to-face conservation, 235
good sounds, 7, 61, 284
measuring intensity of, 82–87, 96
musicians redefining meaning of, 6, 119
organic sounds, 116–117
Sabine mapping variation in intensity of, 66–69, *66, 67, 68,* 344n17
Sabine photographing, 64, *65*
schizophonia, 321
as signals, 3, 5, 61, 96–97
"sound consciousness" developing, 59, 233
space dissociated from, 2–3, 7, 171, 172, 186–187, 235, 236, 321
technological mediation in, 2
Soundscape
cultural aspects of, 1–2
defined, 1
historical studies of, 10
nationalization of, 306
New Acousticians in creation of modern, 113
noise in the modern, 115–168
physical aspects of, 1
Radio City Music Hall as culmination of modern, 233, 307, 317
V-Room constituting postmodern, 324
Soundstages, 267–272
at Bell Laboratories' Sound Picture Lab, *287*
increasing number in 1928 and after, 247
Knudsen consulting on construction of, 102–103

of Metro-Goldwyn-Mayer, c. 1929, *270*
for *The Voice from the Screen,* 267, *268*
Sound tracks, 272–285
as assembled in post-production, 279
background music, 278
dubbing, 278–279
editing, 278
microphones for, 275, 277, 279–280
mixing, 277–278
noise machines for, 281, *281*
reverberant chambers for, 281–283
sound concentrators for, 279–280
and space, 273–274, 284
volume of, 276, 283, 410n149, 413n174
Sousa, John Philip, 49
Space
modernity reformulating perception of, 187, 320–321
motion picture sound tracks and, 273–274, 284
postmodern acoustical technologies and, 324
reverberation as acoustic signature of particular place, 3, 171, 186–187
sound engineers creating "virtual," 234
sounds dissociated from, 2–3, 7, 171, 172, 186–187, 235, 236, 321
Spear, Edmund, 54, 341n144
Speech and Hearing (Fletcher), 352n101
Spelling, simplified, 157
Sponable, E. I., 400n49
Sprayo-Flake Acoustical Plaster, 170, *191,* 217
Stand, Murray, 151, 166, 372n154
Standardization, 106
Stanton, G. T., 262–263
Starrett, Paul, 200, 387n90
Starrett Brothers' construction company, 199
Steam whistles, 120, 121
Stepanoff, Alexandra, *153*

Stereophonic recording, 106, 408n122
Stevens & Nelson, 71
Stewart, George, 81, 89
Stokowski, Leopold, 139, 141, 154, 214–215, 229, 391n127
Streetcars, 124, 149
Street vendors, 124–125
Strunk, William, 157
Studios. *See* Recording studios; Soundstages
Sturgis, R. Clipston, 72
Submarines, detection of, 88–89
Subways
  complaints of noise of, *159*
  measuring noise of, 162, *163*
Sullivan, "Little Tim," 126
Sullivan, Louis, 29, 30, 32
Sunrise (film, dir. Murnau), 412n166
Suspended ceiling systems, 220–221, 317–318, 394n157
Swan, Clifford M.
  as consultant, 386n82
  in Johns-Manville acoustics department, 178, *179,* 380n26
  on motion picture theater acoustics, 259, 260–261
  on noise reduction for banks, 198
  Radio City Music Hall consulting of, 417n22
  as Sabine's "only student," 75
Sweet's Architectural Trade Catalogue, 218, 379n19, 379n24, 384n66
Symphony for Five Instruments (Antheil), 141
Symphony Hall (Boston), 13–18
  acoustical models for, 42–44, *43*
  as acoustical standard, 319
  audience at opening night at, 49
  Cabot's Quilt considered for, 175
  criticism of acoustics of, 51–57
  exterior, *14*
  first regular season concert at, 52–53
  Greek theater proposal for, 15–16, *16,* 29
  interior, *14*
  modern acoustical science in construction of, 4, 5, 57, 315, 317
  musicians adjusting to, 56
  for nineteenth-century music, 321
  opening night at, 51–52
  orchestra size affecting acoustical quality of, 342n153
  plaque dedicated to Sabine, 13, 57
  Sabine consulting on, 4, 5, 17–18, 37–39, 42–45
  as secular temple to music of the past, 4, 15
  statues in niches of, 55
Symphony orchestras
  Boston Symphony Orchestra, 48, 342n153
  Chicago Symphony Orchestra, 48, 56
  recording of, 264, 407n117
System for Improved Acoustic Enhancement (SIAP), 424n23

Tainter, Charles Sumner, 396n19
Talkies. *See* Sound motion pictures
Tallant, Hugh, 74, 75, 346n40
Taylor, Frederick Winslow, 123
Taylor, Hawley, 85
"Technical Manifesto of Futurist Music" (Pratella), 135–136
Technology. *See also* Acoustical engineering
  countercultural critique of, 318
  European modernists turning to, 208–210
  modernity as defined by, 4, 11
  postwar restoration of faith in, 317
  science contrasted with, 108
  stock market crash affecting faith in, 315
Telephone. *See also* American Telephone & Telegraph Company
  articulation testing, 101, 252
  carbon transmitters in, 91, 94
  cell phones, 396n16

fundamental research required for development of, 93, 352n101
noise studies, 146–147, 235
sound and space reconfigured by, 235
transcontinental telephone service, 92, 93
transducers in, 90–91
Theater organs, 257, 404n92
Theaters. *See also by name*
acoustics as problem for, 20–24
Adler on design of, 29–30
commercialization of, 20
Elizabethan theaters, 332n25
the modern auditorium, 248–256
Patte's design for, 20–21, *21*
royal tradition in design of, 20
Saunders's design for, 21–23, *22, 23*
for sound films, 247, 256–263
Theory of Sound (Rayleigh), 19, 104, 331n23
Theremin, Leon, 152, 154, 241
Theremin-Vox, 152–154, *153,* 241, 372n159
Thermophone, 93, 95
Thin-shelled (timbrel) vaults, 181
Thomas, Theodore, 48, 56, 199, 342n154
Thompson, Benjamin, Count Rumford, 382n40
Thoreau, Henry David, 120
3-A Audiometer, 146–147, *147,* 369n134
Time, modernity reformulating perception of, 187, 320–321
Todd, Webster B., 301, 416n11
Todd, Robertson & Todd, 296
Toeppler-Boys-Foley method, 64, *65,* 346n43
Tone Tests, 237–239, *238*
Toscanini, Arturo, 306
Traffic, noise created by, 148–149
Trains. *See* Railroads
Transcontinental telephone service, 92, 93
Transducers, 90–91

Transite, 254
Trowbridge, John, 34, 37
Trucks, noise created by, 148, 149, *160, 162*
Tufts, F. L., 85
Tuthill, William Burnet, 29, 336n60
Twombley, Robert, 336n60
Typing, 155–156, *156*

Uncertainty Principle, 133
Underwater sound detectors, 88–89
United Research Corporation, 105
United States Gypsum Company
Acoustone, 170, 218
advertising by, 191–192, *192,* 210, *211*
Sabinite, 170, 217, *217,* 391n134
University of California at Los Angeles (UCLA)
architectural acoustics taught at, 60
acoustical laboratory at, 355n127
Knudsen joins faculty at, 100, 354n120
loudspeakers donated to, 102
University of Illinois, 60, 82, 349n66
University of Michigan Hill Auditorium, 74, 418n24
"Unplugged" concert series, 322, 423n19
Upham, Jabez Baxter, 339n110
Urban, Joseph, 296
Urban noise. *See* City noise

Vacuum-tube amplifiers, 94, 95, 99, 241, 244, 245
Vail, Theodore, 91
Varèse, Edgard, 138–141
Amériques, 138–140
Arcana, 140, 141
Hyperprism, 139
and Philips Pavilion, 322–323, 368n113
Rosenfeld on, 143–144
writing for the Theremin, 154
Venturi, Robert, 324
Victor Talking Machine Company, 237, 240, 398n31, 407n118

Victrola, 237, 240
Virtual sound, 7, 281–283, *282*, 322–324
Vitaphone, 102, 241, 246–247, 267, *268*, 273
Vitascope, 243
Vitruvius, 18
Voice from the Screen, The (documentary), 267, *268*
Volume. *See* Loudness
V-Room, 323, 324

Waetzmann, E., 414n188
Waldo, Frank, 54
Walker, Jimmy, 158
Waller, Fats, 131
Walter & Weeks, 248
Wanamaker's (Philadelphia), 214, 390n122
Warner, Sam, 246
Warner Brothers
   Don Juan, 246, 273
   editing disc-recorded sound, 278, 411n157
   The Jazz Singer, 247, 273, 401n53, 411n154
   Knudsen consulting for, 103
   The Lights of New York, 247, 273, 274
   New York studios of, 267
   relocating to Hollywood, 267
   Western Electric Vitaphone adopted by, 102, 246, 404n92
Washington, D.C.
   House of Representatives, 24–27, *26*
   National Academy of Sciences Building, 189
   noise survey in, 155
   Smithsonian Institution, 27–28, 334n52
Waterfall, Wallace, 104, 105, *106*, 355n135
Watson, Floyd R.
   on acoustical materials advertising, 191
   in Acoustical Society of America founding, 104–105, *106*

Acoustics of Buildings, 252, *253*, 354n122
bibliography on architectural acoustics, 82, 348n61
Eastman Theatre consulting by, 249–251
electrical sound detector of, 84
"Ideal Auditorium Acoustics," 254
method of images employed by, 290, *291*
on motion picture theater acoustics, 258, 405n93
on National Research Council Committee on Acoustics, 89–90
on new type of auditorium, 250–252
on outdoor listening, 254
on reverberation time, 252, *253*
University of Illinois auditorium work of, 81–82, 349n66
war work of, 351n91
Watson, Thomas, 93
Webster, Arthur Gordon
   on acoustical impedance, 97
   on electroacoustic instruments, 98–99, 353n113
   on National Research Council Committee on Acoustics, 89–90
   phonometer of, 86, 89, 98, 350n82
   suicide of, 105
Wenger V-Room, 323, 324
Wente, Edward
   on absorption coefficients, 106
   at Acoustical Society of America organizational meeting, *106*
   condenser transmitter of, 90, 94–95, 353n113
Western Electric Company. *See also* Electrical Research Products Incorporated
   American Telephone & Telegraph Company purchasing, 91
   articulation testing developed by, 101
   Knudsen as employee of, 99

  Knudsen's research supported by, 102
  noise investigated by, 146–147
  public address systems developed by,
   99, 241
  science and corporate concerns coexist-
   ing at, 102
  sound meter of 1931, *155*
  sound-on-film recording system, 278
  Vitaphone, 102, 241, 245–247
  Wente as employee of, 90
Westinghouse, 90, 245
West Point chapel, 182
White, Stanford, 69, 199
Willcox, James, 214
Williams, William Carlos, 122, 142, 143,
  144
Willis, Carol, 386n90
Wilson, Mrs. Richard T., 128–129, *129,*
  130, 363n56
WLS radio (Chicago), 266, 408n123
Wolf, S. K., 259–260, 262
Women
  in acoustical engineering, 343n8
  clothing reduced in 1920s, 157,
   374n173
  Ladies' Dining Room in New York
   Life Building, *204*
  sound-absorbing power of, 374n173
Wood, Alexander, 389n114
Woolf, Virginia, 314
Woolworth Building (New York City),
  169, 183, 200
Worsley, Wallace, 242
Wright, Frank Lloyd, 335n59, 336n60
Wright, Lloyd, 254
Wurlitzer organs, 257
Wyatt, Benjamin Dean, 20, 24
Wynne, Shirley, 157, 158, 374n176

Zang-Tumb-Tumb (Marinetti), 135
Zoning, 125–127, 170